# Child Psychiatry

## A DEVELOPMENTAL APPROACH

### Philip Graham

*Professor of Child Psychiatry, Institute of Child Health, London
Honorary Consultant Child Psychiatrist, Hospital for Sick
Children, Great Ormond Street, London*

## SECOND EDITION

OXFORD   NEW YORK   TOKYO
OXFORD UNIVERSITY PRESS

Oxford University Press, Walton Street, Oxford OX2 6DP

Oxford New York Toronto
Delhi Bombay Calcutta Madras Karachi
Kuala Lumpur Singapore Hong Kong Tokyo
Nairobi Dar es Salaam Cape Town
Melbourne Auckland Madrid
and associated companies in
Berlin Ibadan

Oxford is a trade mark of Oxford University Press

Published in the United States by
Oxford University Press Inc., New York

First published 1991
Reprinted 1993, 1994

A catalogue record for this book is
available from the British Library

Library of Congress Cataloging in Publication Data
Graham, P. J. (Philip Jeremy)
Child psychiatry: a developmental approach/Philip Graham.—
2nd edn.
p. cm.
Includes bibliographical references and index.
1. Child psychiatry. 2. Child development. I. Title.
[DNLM: 1. Child Development. 2. Child Psychiatry. 3. Mental
Disorders—in infancy & childhood. WS 350 G741c]
RJ499.C716 1991 618.92'89—dc20 91–345 CIP
ISBN 0 19 262143 2 (Hbk)
ISBN 0 19 262142 4 (Pbk)

Printed in Great Britain by
Bookcraft Ltd, Midsomer Norton, Avon

*To Nori, Anna, Daniel, and David*

# Preface to the second edition

Encouraged by the response to the first edition of this book, I have prepared the second edition with the same intention, namely, to provide a book that is helpful to a wide variety of professionals dealing with disturbed children, but particularly those working in paediatrics, psychiatry, general practice, psychology, and social work, as well as those training to enter these fields. I hope that readers who find themselves expected at times to take a perspective of the subject from an unfamiliar professional vantage-point will find this an enriching as well as a disconcerting experience.

I should once again like to thank my colleagues at the Hospital for Sick Children, Great Ormond Street, and the Institute of Child Health, London, for the opportunities they provide for stimulating discussion. Michael Rutter's work at the Institute of Psychiatry, London, has continued to be a rich source of intellectual inspiration.

I owe particular thanks to those who have read and commented on sections of the book, particularly Arnon Bentovim, Roger Freeman, Roy Howarth, Richard Lansdown, Bryan Lask, Tessa Leverton, Stuart Logan, Jeanne Magagna, Tony McShane, Celia Mostyn, David Skuse, Jeremy Turk, and Dieter Wolke.

Finally, my thanks are due to Jackie Moore, whose secretarial assistance has once again been invaluable.

*London*                                                                                           P.G.
July, 1990

# Preface to the first edition

This book is intended to be of use to all doctors dealing with children and their families, but especially to paediatricians and psychiatrists in training. I hope that a number of others, such as medical students, clinical medical officers, general practitioners, medical and psychiatric social workers, nurses, psychologists, general and child psychiatrists, will also find it helpful. The level of the book is intended to be suitable for those with an elementary as well as a somewhat more advanced knowledge of paediatrics, and I have tried to bear in mind the very different settings in which professionals dealing with children and families may carry out their work.

Throughout the book, with few exceptions, I have referred to both children and professionals as if they were all male. The exceptions consist of children suffering from conditions with a heavy female preponderance (such as anorexia nervosa) and when reference is made to professionals such as nurses who, at least in the children's field, are nearly all women. I realize this is unsatisfactory, but the alternatives seemed even more so.

In the section on physical conditions, I was aware that, especially with less common disorders, many clinicians such as family doctors, clinical medical officers, and child psychiatrists (as well as non-physicians such as psychologists and social workers) might find it helpful to have available an introductory résumé of the physical aspects of the condition before psychosocial aspects were discussed. I have therefore included a certain amount of such information, but it should be emphasized that this book is in no way a substitute for a textbook of paediatrics. I hope it will be seen as complementary to such a textbook, filling out in more detail an aspect of paediatric care which has so far received less attention than the prevalence of psychosocial disorders in childhood suggests it should.

*London*                                                                                    P.G.
June, 1986

# Contents

## 3 Adult-type psychiatric disorders

## 4 Psychosocial aspects of physical disorders: general

## 5 Psychosocial aspects of specific physical conditions

## Contents

# 6 Prevention and treatment

# 7 Services

# 1

# Overview

## 1.1 Introduction

What is paediatric or child psychiatry? More than most branches of medical practice, the field is difficult to define. The central focuses of the subject are the behavioural and emotional disorders of childhood, but many would include those 'psychosomatic' physical symptoms, such as non-organic headache and stomach pains in which stress or other environmental factors appear to play an important causative role. Delays and deviations in development, as well as general and specific learning problems lie in the borderland of child psychiatric practice. Environmental factors are often important in their causation, and they are often, but by no means always, accompanied or followed by significant emotional and/or behavioural disturbance.

Because children's development is so closely bound up with the quality of care given by parents, failures of parenting and the fostering of parental skills are often seen as an integral part of child psychiatric practice. There are indeed some child psychiatrists who prefer to see themselves as family psychiatrists. This can often be a helpful approach, but it is important to remember that, while usually children's problems can best be seen as arising from an interaction between the child, the family, and the wider environment, in a significant number of cases as, for example, in childhood autism, a serious primary disturbance does indeed lie in the child.

Defining the territory of child psychiatry in this way makes it clear that most professional assessment and treatment of child psychiatric disorder is carried out by non-psychiatrists and indeed by professionals not medically qualified. These include family doctors, paediatricians, and clinical medical officers, as well as psychologists, nurses, social workers, and teachers. The situations which call for the particular skills of a child psychiatrist (P. Graham 1984) will vary, but, in general, psychiatrists are most likely to be appropriately involved where a disorder is severely handicapping and persistent. With training, other medical and non-medical professionals who are interested can be effective in assessing and treating problems in this field of work.

Finally, an important characteristic of child psychiatric disorders lies in the degree to which they affect different aspects of the life of the child and the family. The family of a seriously disturbed child will require the services of many different types of professional, drawn especially from the health, education, and social welfare services. An important principle of practice involves the need for all practitioners, no matter to which discipline they belong, to respect the contribution that other disciplines may make, to be aware of the point at which another discipline may need to be involved, and to maintain as close communication with other disciplines as is compatible with professional confidentiality.

### 1.1.1 THEORIES OF DEVELOPMENT AND ITS DISORDERS

The multiplicity of theoretical models available to explain and treat child psychiatric disorders is confusing not just for the beginner, but even sometimes for the experienced practitioner. Behavioural/psychodynamic, individual/family, medical/sociological theories and part theories do indeed abound. Most experienced practitioners recognize that the multiplicity of models means that none is totally satisfactory and that different models can be helpful in different situations. It is not surprising that there is no one theory capable of explaining all child and family behaviour, and one of the skills of the practitioner lies in selecting a way of looking at a child's problems that helps both him and the family to understand it better and, if possible, relieve the distress associated with it. A brief account will be given here of the main theories used to explain development and its disorders.

**Biological theories**

Strictly speaking, biology is the study of living organisms and includes the investigation of environmental as well as physical influences but, by convention, in psychiatry, the use of the term 'biological' usually refers only to physical or physiological influences. Biological theories in this field place particular emphasis on expression of the genotype, and on various aspects of brain function. In most areas of child development, such as mobility and language, the process of maturation can only be understood if the role of genetic factors is taken into account. In understanding behaviour problems, the explanatory power of biological theories depends on the condition in question. In childhood autism, for example, biological factors probably contribute considerably to the development of the disorder, whereas in conduct problems they probably contribute rather little. In conditions such as the hyperkinetic disorder (attention deficit disorder/hyperactivity) their role is intermediate.

**Psychoanalytic theories**

Developmental concepts derived from the work of Sigmund Freud (1856–1939) are referred to as psychoanalytic or, because of the central role of psychic energy, psychodynamic theories. These theories have been modified by analysts such as Freud's daughter Anna, and by Melanie Klein. Other analysts such as Erik Eriksen and John Bowlby have transformed psychoanalytic theory in concepts bearing only a tenuous relationship to Freud's original ideas.

Freud's first formulations describe the way in which early traumatic experiences, especially sexual seduction in childhood, were repressed and then unconsciously transformed (converted) into bodily symptomatology, such as paralysis and fainting fits. Freud later developed the view that sexual abuse had not actually occurred, but had been fantasized, partly out of wish fulfilment. Controversy still exists as to whether Freud actually believed that his patients' memories were only fantasies, or whether he deliberately or unconsciously falsified his findings in order to make his theories more acceptable.

In any event, repression and conversion were early corner-stones of the *functional* part of Freud's theories. These functional psychological mechanisms were later elaborated by both Freud and his daughter to explain the various ways in which anxiety about unacceptable thoughts could be transformed into symptomatology (see also page 247). Such thoughts were unacceptable because they involved sexual or aggressive (perhaps murderous) ideas about loved ones, especially parents. The psychological mechanisms described included projection (transferring unacceptable thoughts into another person), displacement (transferring aggression from the loved person to someone else), reaction formation (experiencing the opposite emotion to that unconsciously felt (e.g. being extremely protective towards someone who is really hated), intellectualization (cutting out emotional feelings), and denial (pretending that the unacceptable thoughts and the reason for them do not exist).

Freud later elaborated a *structural* theory to describe the way in which different components of the mind related to each other. The structure consisted of the id, the source of basic instinctual drives, the ego, which finds acceptable ways for the expression of these drives, and the super-ego (conscience). Initially the child obeys his parents, especially his father, because he is frightened to do otherwise. Later, as a result of the process of identification, the child internalizes the rule-making part of his parents so that his internal conscience plays the part of the absent, punishing parent: the super-ego has been formed.

To this description of Freud's functional and structural concepts must be added his theory of infant development over the first five years. As the child matures, so the different parts of his body (first the mouth, then the anus, and finally the genitalia) represent the primary sources of gratification or pleasure. If for some reason, the child fails to progress from one phase to the next, he may become fixated at the oral, anal, or genital level.

Melanie Klein, whose views have dominated the main British school of analytic child psychotherapy for many years, puts great emphasis on the role of intensely emotionally charged, unconscious fantasies in the first year of life. The child develops these relationships with 'part-objects', split off parts of other people. The notion of the infant forming a relationship with the 'good breast' and the 'bad breast' of his mother is developed in this way.

Erik Eriksen's concepts (Eriksen 1965) place greater emphasis on the achievement of mastery at different stages of development. The infant must first develop 'basic trust' in those around him, then 'autonomy' as a toddler before he begins to exercise 'initiative'. The early school years are characterized by 'industry' and adolescence by the achievement of a solid 'identity'. Failure at any of these stages brings its own characteristic psychological problem, mistrust, shame and doubt, guilt, inferiority and, in adolescence, role confusion rather than a strong identity.

Freud's ideas have permeated not just psychiatric thought, but all aspects of Western cultural life. Although the therapeutic value of psychoanalysis (four or five times a week treatment derived from his theories) has been seriously brought in question, there are few transactions or communications between children, parents and professionals which are not enriched by psychodynamic understanding. Further, professionals who have themselves undertaken a course of psychodynamic psychotherapy for training purposes often feel greater understanding of their own unconscious processes in dealing with children and families.

## Attachment theory

This theory has been developed by John Bowlby (1908–90) and elaborated in a trilogy describing psychological processes related to attachment, separation, and loss (Bowlby 1971, 1975, 1980).

Bowlby describes attachment as a complex two-way process in which the child becomes emotionally linked to members of his or her family, usually the mother, father, and sibs in diminishing order of intensity. Bowlby described attachment, especially to the mother, as an adaptive, biological process serving the needs of the child for protection and nurture. He suggested that it can best be understood as a 'control system' in which proximity to the mother is maintained through a series of signals emitted by both mother and child. The later capacity of the child to develop social relationships is considered to be based on the way with which attachment behaviour is established. Bowlby put particular emphasis on the central importance of the biological and social ties with the mother.

Although the infant's tendency to form an attachment to its mother is genetically determined, the behaviour of those around the child will influence the security of the attachment. According to Mary Ainsworth and her co-workers (Ainsworth *et al.* 1978) there are three main patterns of attachment that can be characterized by observation of the toddler's behaviour with the mother. These observations are based on the reactions to the child's being separated from the mother and left with a stranger, following which his behaviour is observed when he is reunited with his mother.

i. Secure attachment. The care-giver provides a secure base for exploration, and the child is readily comforted if distressed. When reunited with his mother, if he has been distressed, he will immediately seek and maintain contact.

ii. Anxious/resistant attachment. The child is too anxious to leave his mother to explore, and is wary of new situations and people. When reunited after a separation he may be aggressive, cry and refuse to be comforted or show considerable passivity.

iii. Anxious/avoidant attachment. The child explores readily, and indeed is unduly friendly with a stranger. After a period of separation he will ignore or avoid his mother.

Theories of attachment are particularly useful in increasing our understanding of disorders of social relationships (Section 2.1.2), of anxiety states (both generalized and specific) and of situations such as admission to hospital or placement in day care when brief or prolonged separations are in question.

### Piaget's theory of cognitive development

Piaget's work will be discussed particularly briefly, because although it has had a major influence on educational practice and stimulated numerous studies of children's thought processes, it has had relatively little impact on our understanding of child psychiatric disorders.

Jean Piaget (1896–1980) carried out countless observational experiments on children of different ages. He described four major phases of cognitive development with quite sharp discontinuity between each phase.

i. Sensori-motor phase. Below 2 years the infant understands the world directly through perception and action. He learns directly from what he can see, hear, feel, and taste, as well as from the results of his own activity impinging on the external world.

ii. Concrete pre-operational. From 2–6 years the child is now able to make mental representation of objects, and imagine actions related to them, but he is still completely egocentric, unable to imagine the world from the perspective of other people.

iii. Concrete operational. From 7–11 years the child is able to think logically, but only in concrete terms. The perspective of others is now fully appreciated.

iv. Formal operational. From 12 years the child is able to think in abstract terms and develop concepts of, for example, the relativity of man-made rules.

Piaget's ideas have needed considerable modification since he first formulated them. It is clear that his emphasis on laboratory observation and on the results of formal experiment led Piaget to underestimate what children can do in real life. If a child understands the meaning of a term in real-life terms, he is much more likely to perform well. Piaget's ideas have however cast light on the egocentricity of the young child, and on the process of moral development in the child with a conduct disorder. An awareness of Piaget's concepts can alert the child health professional to take into account a child's level of understanding, when, for example, communicating with a child about his chronic illness (Section 4.3).

### Family theories

Although the classification systems in child psychiatry refer exclusively to pathology within the individual child, many clinicians regard the problems they see as better understood by considering the family as the unit in question. Family theories provide ways of conceptualizing the processes that go on within families that may result in one family member, especially a child, being presented as a problem or a patient. Such theories include the following.

1. *Systems theory.* A family system is conceived as a functional unit operating according to its own rules, and with certain special characteristics. Such a concept can be applied to systems of family functioning that would not be applicable to the functioning of individuals. In particular, those who use systems theory often also borrow from the science of cybernetics to explain how a family system regulates itself. Every change in the relationship between two family members is likely to result in changes in other relationships within the family, before the system reaches equilibrium again. Some family systems are dysfunctional and work badly, especially those that are rigid and inflexible and cannot react to changing demands of children.

2. *Communication theory.* For a family to function well, messages between family members need to be transmitted and received unambiguously. Family life consists of a constant series of demands, requests, and injunctions to do or not to do this or that. There is also a flow of actual information not requiring immediate action but perhaps important for the

future. Finally, there is a need for the communication of feelings. Children need to know when their parents feel warm and approving, as well as when they are upset and condemnatory. In dysfunctional families, one or more types of communication are often transmitted poorly or ambiguously. One characteristic type of dysfunction involves passing double messages—for example a parent might laugh at a child's naughtiness as if it were trivial or even a lovable fault, but moments later shout at the child for the same behaviour as though bitterly disapproving of it. The 'double-bind' theory of schizophrenia involves a suggestion that such double messages are important in the causation of schizophrenia. This has now been discredited as a cause of schizophrenia (Leff 1978), but the transmission of ambivalent messages from parents to children (and sometimes vice versa) certainly causes confused feelings in family members, who may then suffer significant symptoms.

3. *Structural theory.* The relationships in a family can be regarded as organized or 'structured' in a characteristic way. Minuchin (1974) describes a variety of ways in which family organization or structure can be dysfunctional. The intergenerational boundaries may be blurred, and parents behave like the brothers and sisters of the children. Conversely children may take care-taking roles in relation to their parents. Emotional distress may be inappropriate. Parents and children may be overinvolved or emotionally 'enmeshed' with each other, or they may be too uninvolved emotionally, disengaged or detached. There may be alliances or coalitions between family members—a mother and son ganging up against a father, for example. Finally, structure may be impaired by detouring of conflicts rather than dealing with them. Characteristically, disputes between the parents are not settled by open and honest disagreement, argument, and negotiation between them. Instead, when for example a mother feels angry with her husband for sexual inadequacy, she may aggressively focus instead on her son's shortcomings which otherwise would have passed unnoticed.

Family theories have been directly applied in therapeutic work (Section 6.2.4). They are particularly useful in situations where the primary pathology almost certainly lies in family interaction rather than in the individual child.

### Behaviour or learning theories

Early learning theories, such as those put forward by John Watson (1878–1958), suggested that all forms of behaviour and emotional expression were *learned* as a result of experience. The role of internal mental processes could thus be ignored. One could understand behaviour simply by studying the stimuli to which the organism was exposed, and noting the responses. Two main mechanisms for learning were described.

1. *Classical conditioning.* In this form of conditioning a stimulus was paired or associated with a response a number of times, until the mere presence of part of the stimulus would elicit the response. Thus Pavlov sounded a bell at the same time as he produced food for his dogs. After several repetitions of this experience, the sound of the bell alone was sufficient to produce salivation. Children who develop fears of places where they have previously been frightened by dogs show a similar phenomenon.

2. *Operant conditioning.* This form of learning occurs as a result of a process whereby a positive outcome following a particular piece of behaviour leads to a repetition of the behaviour so as to reproduce the same desirable outcome. B. F. Skinner (1904–90) demonstrated this phenomenon with pigeons who learned to peck at a button because they had previously been given grain for this behaviour. *Positive reinforcement* (usually a reward) occurs when behaviour is followed by an event that tends to make the behaviour more likely to occur in the future. *Negative reinforcement* occurs when a behaviour is followed by the removal of an unpleasant stimulus. Intermittent positive reinforcement is said to occur when there is only occasional 'reward' following an event; such reinforcement is often more powerful than if it is continuous.

More recent learning theories have stressed the role of other processes on the way behaviour is shaped. Imitation following observation is clearly one such process. Mental processes have now been introduced into learning theory by cognitive behaviourists who stress the way learning can be altered by what a child thinks, or is encouraged to think, about experiences. They have also accepted the importance of the quality of the child's relationship with whoever may be providing the experiences—parent, teacher, or experimenter. Humans and other animals, it has been suggested, may learn more complex behaviour and forms of emotional expression than was originally proposed. Thus Seligman (1975) demonstrated that rats could learn to behave in a helpless way if they were put into painful situations over which they had no control. This theory of 'learned helplessness' has been used to explain the occurrence of depressive feelings in individuals who have learned that nothing they can do alters their fate.

Early behavioural theories were useful in explaining some aspects of child behaviour such as the acquisition of simple habits, and the development of monosymptomatic phobias. Later elaboration of learning theory by the introduction of cognitive concepts and processes such as imitation and observation has widened the possible applications very considerably.

## Child development and life experiences

These theories of child development that have been described all emphasize processes that unfold *within* the child. As the theories have been subjected to critical evaluation, it has become clear that each of them is capable, at best, only of telling part of the story. They have all needed to be modified to take account of children's life experiences in all their variety. The practitioner seeing a child with a problem must ask not just 'How has this child developed?', but also 'What has happened to this child?'. The ways in which experience makes and is made by individual personality is a fascinating one (Goodyer 1990). In the clinical situation, understanding this interaction with the help of one or more of the theories described here is often the key to success.

## 1.1.2 CLASSIFICATION

There is a need for a classification of the psychiatric and psychological disorders seen in practice by family doctors, paediatricians, psychiatrists, and other clinicians for a number of reasons. First, typing a disorder allows one to draw on knowledge available from studies of children with a similar label. This may give a lead to management or the prediction of outcome. Second, a typology of disorders can help to clarify thinking about the nature of the problems with which a child may present. Confusion produced in the mind of someone faced by the multiplicity of difficulties often shown by a child and family may be helped by the application of a classification which allows separation of one problem from another. Classification also assists in record-keeping, and in communication between professionals.

### Available classification systems

There are two main, and rather similar, systems of classification of child psychiatric disorder—the *International classification of diseases* (10th version) (ICD-10) (World Health Organization 1990), and the *Diagnostic and statistical manual* (DSM III-R) (American Psychiatric Association 1987). They are subject to periodic revision. The latter is used particularly in the USA, Canada, and Australasia, and the former in the rest of the world. Both these classifications are multiaxial and have five axes—that is, they require those who use them to make five different judgements about the child (Table 1).

The multiaxial approach has not, in fact, been fully incorporated into the latest version of ICD-10 but, with some modifications, the scheme above can be derived from it.

The use of one of these multiaxial classifications means that different components of a child's problems can readily be recorded separately. Thus a child with autism, with moderately retarded intellectual ability, and epilepsy from a home characterized by marked interpersonal tension, would receive a positive coding on four areas of both the ICD-10 and the DSM III-R classifications.

**Table 1**

| Axis no. | Axis ICD-10 (modified)* | DSM III-R** |
|---|---|---|
| 1 | Behavioural and emotional disorders with onset usually occurring in childhood and adolescence | Clinical syndromes |
| 2 | Developmental disorders | Developmental disorders: personality disorders |
| 3 | Intellectual level | Physical disorders and conditions |
| 4 | Medical condition | Severity of psychosocial stressors |
| 5 | Associated abnormal psychosocial situations | Global assessment of functioning |

\* Rutter (1989*a*)
\*\* American Psychiatric Association (1987)

**Classification of psychiatric syndromes**

In both the classifications mentioned above, the first axis concerns psychiatric disorders, and this is the axis most likely to be employed by physicians. Most disorders can be classified in the following scheme, an abbreviation of the two most relevant sections of ICD-10.

*F90–99. Behavioural and emotional disorders with onset usually occur-
        ring in childhood or adolescence (abbreviated)*
—*F90. Hyperkinetic disorder*
.0      Disorder of activity and attention
.1      Hyperkinetic conduct disorder
—*F91. Conduct disorders*
.0      Conduct disorder confined to the family context
.1      Unsocialized conduct disorder
.2      Socialized conduct disorder
.3      Oppositional defiant disorder
—*F92. Mixed disorders of conduct and emotions*
.0      Depressive conduct disorder
—*F93. Emotional disorders with onset specific to childhood*
.0      Separation anxiety disorder
.1      Phobic disorder of childhood
.2      Social sensitivity disorder
.3      Sibling rivalry disorder

—*F94. Disorders of social functioning with onset specific to childhood or adolescence*
.0    Elective mutism
.1    Reactive attachment disorder of childhood
.2    Disinhibition attachment disorder of childhood
—*F95. Tic disorders*
.0    Transient tic disorder
.1    Chronic motor or vocal tic disorder
.2    Combined vocal and multiple motor tic (Tourette syndrome)
—*F98.* Other emotional and behavioural disorders with onset usually occurring during childhood
.0    Enuresis
.1    Encopresis
.2    Feeding disorder of infancy or childhood
.3    Pica
.4    Stereotyped movement disorder
.5    Stuttering
.6    Cluttering
*F80–89. Developmental disorders (abbreviated)*
—*F80. Specific developmental disorders of speech and language*
.0    Specific speech articulation disorder
.1    Expressive language disorder
.2    Receptive language disorder
.3    Acquired aphasia with epilepsy
—*F81. Specific developmental disorders of scholastic skills*
.0    Specific reading disorder
.1    Specific spelling disorder
.2    Specific disorder of arithmetical skills
.3    Mixed disorder of scholastic skills
—*F82. Specific developmental disorders of motor function*
—*F83. Mixed specific developmental disorders*
—*F84. Pervasive developmental disorders*
.0    Childhood autism
.1    Atypical autism
.2    Rett syndrome
.3    Other childhood disintegrative disorders
.4    Overactive disorder associated with mental retardation and stereotyped movements
.5    Asperger syndrome

Each of these categories has, to some degree, been validated, i.e. there is evidence available that the type of disorder mentioned has a characteristic presentation, aetiology, and course.

**Drawbacks of existing classification schemes**

There are various problems with the existing schemes which make them
less than satisfactory (P. Graham 1982) and indeed discourage some practitioners from using them at all.

1. They necessarily involve a judgement of the nature of the problem
'within the child'. For many practitioners the essential nature of the disturbance is usually seen elsewhere—within the family, or within the interaction between the child and family members, or the child and wider
society.

2. The existing diagnostic categories, although reasonably well validated, provide only a very rough guide to aetiology, treatment, and prognosis. They generally need to be amplified by a formulation, giving a more
extended account of the child's condition and background features before
action can be taken.

3. Many of the situations of psychiatric concern with which family
practitioners and paediatricians deal quite frequently, cannot readily be
categorized within this classification system. Non-accidental injury and
monosymptomatic abdominal pain of psychogenic or uncertain origin are
examples.

4. The positioning of some of the categories is occasionally arbitrary
and might be misleading. For example, the introductory section to
Developmental Disorders in ICD-10 defines such disorders in terms of their
early onset, delay in functions strongly related to biological maturation,
and steady course; yet enuresis, which often fills all these criteria, is
classified as a behavioural or emotional disorder. The problem is that some
symptoms, like enuresis and perhaps hyperactivity, may occur either as
developmental disorders or as signs of emotional disorder, and consequently
their place in the classification is inevitably somewhat arbitrary.

Nevertheless, the framework of these new classification systems in child
psychiatry does represent a real advance on previous attempts, and the
organization of much of this book is based upon them.

1.1.3 PREVALENCE OF PSYCHOLOGICAL PROBLEMS

**Introduction**

Information about the incidence (number of new cases) and prevalence
(rate in the population over a defined period of time) of psychiatric disorders, general and specific learning difficulties, and other psychosocial
problems is available from a variety of sources.

In the UK a number of national surveys have provided relevant information. Three cohorts have been studied. The National Survey of Health and Development (1946 births), the National Child Development Study (1958 births), and the Child Health and Education Survey (1970 births) have all involved the longitudinal study of children born during one week during the year in question.

In addition, a number of local surveys have also been undertaken. These include particularly a study of preschool children living in the north of London (N. Richman *et al.* 1982) and a study of 10- and 11-year-olds living on the Isle of Wight. The Isle of Wight study has been a particularly rich source of data (Rutter 1989*c*). All the children aged 10–11 years living on the island were screened for behavioural deviance and learning problems by the use of parent and teacher questionnaires as well as group tests of intelligence and educational attainment. The children also had a general medical examination. Children who, on the basis of the question- naire scores or group test performance, were at risk either for psychiatric disorder or for specific or general learning difficulties were selected for further investigation. In this second stage of the study, these children, together with a randomly selected control group, were seen individually for a psychiatric interview, and were also tested for IQ and educational attain- ment by psychologists. Their parents were interviewed in detail on their social circumstances, parental mental and physical health, and the chil- dren's behaviour.

Following the Isle of Wight study, a number of related investigations were carried out. The cohort was followed up to 14 years of age to examine the outcome of disorders that had been identified. A comparison study of London children was undertaken to examine differences between children living in urban and rural settings. The methods originally used in the Isle of Wight study have been widely applied elsewhere.

## Prevalence of disorders in the general population

Findings described in this section relate particularly to children living in the UK, but there is good evidence that very similar rates are found in other economically developed countries, especially in Europe (e.g. Verhulst *et al.* 1986), North America (e.g. Offord *et al.* 1989), Australasia (J. Anderson *et al.* 1986), and South-east Asia (Luk *et al.* 1988).

1. *Overall prevalence.* Behaviour and emotional disorders are common in the general population. Criteria for disturbance vary, so that it is difficult to make comparisons, especially between different age groups. Twenty-two per cent of preschool children have been found to have significant be- haviour problems, of which about 7 per cent are thought to be severe enough to require specialist assessment (N. Richman *et al.* 1982). The corresponding figures for middle childhood are in the region of 25 per cent

for those living in big cities and about half this figure for those living in small towns (Rutter *et al.* 1975). About 20 per cent of young adolescents have been found to show significant psychological problems (Leslie 1974; Rutter *et al.* 1976*b*), and the figure is about the same for older adolescent girls. Thus, throughout childhood and adolescence, about one in five children shows a significant psychological problem in any one year.

2. *Prevalence of specific psychiatric disorders.* This is discussed under the various headings in previous chapters. By far the most common types of problem occurring in childhood are emotional disorders, characterized by symptoms of anxiety and depression, and conduct disorders, characterized by aggression and other forms of antisocial behaviour. In middle childhood these constitute over 90 per cent of all psychiatric disorders.

3. *Prevalence of mental retardation* (see p. 128). About 3 per 1000 children show severe mental retardation (IQ less than 50) and about 2.5 per cent show mild retardation (IQ 50–70) (Laxova *et al.* 1977; Rutter *et al.* 1970*a*).

4. *Learning difficulties.* About 10 per cent of 10-year-old children living in inner cities and about half this number in small towns have a reading age two standard deviations (or roughly 2 years) behind their expected reading age, based on their non-verbal ability (Berger *et al.* 1975).

5. *Chronic physical illness.* About 5 per cent of children have chronic physical disorders (Rutter *et al.* 1970*a*). While most of these children do not have psychological or psychiatric disorders, the health care of all of them certainly has psychosocial implications, in order that the impact of the physical condition on their lives and that of their families can be minimized (Breslau 1985).

### Psychosocial problems in ethnic minority groups

Rates of psychiatric disorder in ethnic minority groups have generally been found to be rather similar to those in the indigenous population within the UK. This holds particularly for the Afro-Caribbean (West Indian) population, which has been well studied (Rutter *et al.* 1974; Earls and Richman 1980). The Asian minority group has been less intensively studied, but indications are that rates are somewhat lower than those in indigenous children (Kallarackal and Herbert 1976). In the USA, studies of Puerto Rican children have shown similar rates of disturbance, and similar correlates of disturbance to those in the indigenous population (Bird *et al.* 1989).

Cognitive development and educational attainment in Asian and West Indian children is similar to that in the indigenous population in the UK, once social factors have been taken into account. For example, in the UK, West Indian families suffer poor social circumstances, working in low-paid jobs, and living in cramped accommodation. The children are frequently educationally retarded, but perform about as well as white children living in similar social circumstances. Asian families, although also often living in deprived circumstances, have a higher proportion of more affluent self-employed, working as shop-keepers, in business, or the professions. Educational attainment of this group is higher.

The fact that rates of psychiatric and educational problems are similar to those in the general population should not be taken to mean that the stresses that children in minority groups face are also necessarily similar. They are often very different, and professionals dealing with immigrant families need to be aware of the common problems that occur. In the UK, African, Caribbean, and Asian groups still face racial discrimination and prejudice in matters concerning housing, employment, and police practice. Racist attacks on black families inevitably elicit both anxious and aggressive reactions in their children, and these are sometimes marked. Both groups, but especially the Asian subgroup, are also likely to resort to traditional rather than Western methods of dealing with family crises when they arise. A Muslim father, for example, is more likely to seek help from an Islamic priest than a psychiatrist if his son is in trouble. In both groups teenagers are likely to be torn between respect for the values of their parents, and a wish to conform to the values they see in the wider world in school, the neighbourhood, or on television.

Children of West Indian origin are more likely to be living in one-parent families. Both mothers and fathers often work very long hours to achieve better material conditions, and substitute child-care arrangements are not infrequently unsatisfactory. A relative lack of toys and opportunity for parent–child interaction sometimes makes the early years of the child's life relatively unstimulated. These disadvantages are often shared, it should be stressed, by white children living in similar circumstances.

Asian children are more likely to be living in families where the mother speaks little or no English and has little contact with the outside world. Teenagers, especially girls, often take main responsibility for contact with health services for the younger children. Dietary observances further complicate communication by professionals about nutritional problems. The tradition of arranged marriage may be particularly stressful for teenagers living in a wider society in which romantic love is thought to be the basis for marriage.

These and other cultural factors will be relevant not only to those UK professionals working with African, Caribbean, and Asian groups, but also to those involved with Greek and Turkish-Cypriot families, Orthodox Jews, Chinese immigrant families, and the many other less well-represented minority groups. An awareness of the relevant cultural factors, close working links with members of the minority groups, and, when appropriate, the ready availability of interpreters are factors likely to improve the quality of service provided. The availability of translated printed material concerning health education, child-rearing practices, and local services is invaluable.

### Prevalence of disorders in children living in developing countries

A number of studies examining rates of behavioural and emotional disorders in children living in developing countries have been carried out (P. Graham 1979). The findings suggest that children living in big cities in these countries have very similar rates of problems to those living in economically developed countries. Rates in the order of 10–20 per cent of disturbed children have been commonly found. Types of disorder are also rather similar, although in some South-east Asian countries hysterical and psychosomatic reactions have been recorded at unusually high levels. In these areas, the language in which emotional states are described is often more closely linked to bodily change than is the case in developed countries. A mother may, for example, complain there is something wrong with her child's ears when she means he does not pay attention to what she says. Culture-based psychiatric syndromes such as koro and latah, that occur in adults, are only rarely seen in children.

There is little reliable information on rates of mental retardation and specific learning difficulties in children living in developing countries. This is partly because of problems in applying IQ tests standardized in developed countries, and partly because, not unnaturally, in countries where universal schooling has still not been achieved, the identification of rates of learning difficulty is regarded as a somewhat valueless exercise. It is probable that, because of unsatisfactory antenatal care and delivery conditions at birth, the incidence of severe mental retardation is higher in developing countries, but low survival rates for the handicapped child may mean that the prevalence of mental handicap is less than it otherwise might be.

Background factors leading to emotional and behavioural disorders are probably also rather similar to those operating in developed countries. It was once thought that children living in extended rather than close-knit nuclear families were less likely to show psychiatric problems because of the more relaxed emotional atmosphere they encountered. However, the evidence suggests that the stresses acting on children living in extended families are not particularly lessened, only different in nature. For example, in a polygamous marriage, children of a less-favoured wife may be particularly discriminated against. Further, certain other stresses, such as those arising from uncontrolled urbanization with the development of large, shanty areas on the fringe of big cities, are more frequently experienced in developing countries. Poor social conditions and health care, with the breakdown of traditional patterns of belief, may compound problems experienced by children living in these conditions.

## Sex differences in psychological disorders

There are significant sex differences in prevalence in many child psychological problems. Delays and deviations in development are more common in boys. This is the case for language delay, clumsiness, the hyperkinetic syndrome (attention-deficit disorder), enuresis, encopresis, and childhood autism. Boys are also more frequently affected when a disorder has an aggressive component. Thus boys more commonly show conduct disorder and delinquency. Anorexia nervosa occurs much more commonly in girls. All other conditions, such as emotional disorders and preschool adjustment problems, have a roughly equal sex incidence.

Biological factors are probably mainly responsible for the greater vulnerability of boys to developmental disorders. The reasons are uncertain, but it is possible that cerebral and psychological development are mainly under control of the X chromosome and that, if, for some reason, an X chromosome is defective, the presence of another provides a safeguard. This is certainly the case in the fragile X syndrome (Section 5.3), and may be important in other conditions. The preponderance of males with aggressive disorders is also influenced by biological factors such as greater physical strength and testosterone levels, but to a much lesser degree. Studies of children with intersex conditions demonstrate that the sex the child is assigned (whether the child is brought up as a boy or girl) is more important in development than chromosomal sex, the form of the gonads, or even the appearance of the external genitalia (Money and Erhardt 1972). The slightly raised rate of seriously aggressive behaviour in XYY individuals (Section 5.3) is an unexplained exception to this rule. There is also no evidence of hormonal imbalance or other physiological changes in unusually aggressive children.

In contrast, environmental and social influences do seem of major importance in determining sex differences in aggression. Parents are less inclined to deter signs of aggression in boys than in girls, and physical strength is a major source of self-esteem in boys throughout childhood. From an early age, boys are more likely than girls to respond to frustration by hitting out. Teachers also respond differently to boys and girls showing aggressive behaviour. Men and women generally provide very different models for children of the two sexes to imitate, though in the Western world this pattern has changed significantly in the 1960s and 1970s.

# 1.2 Assessment

## 1.2.1 GENERAL PRINCIPLES

Although the specialized skills of a child and adolescent psychiatrist are necessary for complex problems, most appraisals of behaviour and emotional disorders are carried out by non-psychiatrists—family practitioners, nurses, psychologists, social workers and other professional groups. This section aims to provide guidance to all professionals carrying out such assessments.

The aims of a psychiatric assessment of a child with a psychosocial problem or with physical symptoms having psychosocial implications are to:

(1) appraise the nature and severity of the problem;

(2) identify likely causes, be they social, familial, individual, physical, or, as is commonly the case, a combination of these;

(3) plan with the child and family members a form of effective treatment and management.

Most psychiatric appraisal occurs because a family wants help with a problem in their child. However, increasingly psychiatrists and other professionals are requested to undertake assessments for other reasons, for example as part of a statement of a child's special educational needs, to assist a juvenile court in coming to a decision concerning a child charged with an offence, or as part of a child protection investigation.

A secondary but important set of aims relates to the need to communicate sympathy and support for the child and other family members, to convey an understanding of their problems, and to provide for them some meaning and significance to their predicament.

Finally, for administrative purposes, the clinician increasingly needs to make a diagnosis using a recognized multi-axial classificatory system such as ICD-10 (World Health Organization 1990) or DSM III-R (American Psychiatric Association 1987).

### Factors influencing assessment procedure

Patterns of psychological development are assessed in a wide variety of situations and different settings call for different forms of assessment. There are, however, general principles that should guide assessment regardless of setting.

The form of assessment will be influenced by the following factors:

1. The physical resources available—space, privacy, equipment (e.g. to test sight and hearing).

2. The time available to the clinician.

3. The expectations of the person or people who initiated the contact. In primary care this is usually one or both parents. In secondary care any one of a number of agencies may refer and each will have different expectations.

4. The composition of the group that attends. In primary care usually only mother and child attend. In a psychiatric setting the whole family is likely to be involved.

5. The motivation of those who attend. The group may vary in motivation—parents, for example, being highly motivated, a child negative towards attendance. Some families only come because they are sent by other agencies.

6. The nature of the problem. Clearly some presenting problems will be more complex and require more intensive assessment than others.

7. The skill and orientation of the clinician.

8. The follow-up and treatment facilities available. If, for example, the type of treatment available is limited, for example, to brief counselling and/ or medication, then assessment will need to be modified accordingly.

### General principles of assessment

#### The physical setting

Ideally the room in which a child is seen should be big enough to take both parents, the referred child, and at least a couple of sibs. The clinician will achieve greater rapport if he does not sit behind a desk, but beside or away from it. Equipment should be available to carry out a physical examination, to test hearing and vision, etc. There should be a supply of drawing paper and coloured pencils available. Useful toys include three or four jigsaws of varying levels of difficulty, a post-box with different shapes, a set of small family dolls, and a set of farm animals. Toys available should be robust and carefully looked after. It is also useful to have three or four books of graded difficulty to check the child's reading level. It is an advantage if all these can be kept in a cupboard or drawer so that they can be brought out as required.

#### Preparing the referral

In primary care or a busy paediatric clinic, little or no preparation is likely to be possible. However, the staff of a child psychiatric team may achieve a more satisfactory assessment if careful preparation is undertaken. The letter to the family should make it clear that it is highly desirable for both parents to attend and that it is often helpful for sibs also to be present at the initial interview. Obtaining permission to request a school report prior to the assessment has the advantage that the clinician will then have most relevant information available by the end of the first interview. If the assessment is to take place under unusual conditions, for example with the use of a one-way screen or video equipment, or if it is intended that students should be present at the interview, then again the family should be warned beforehand.

#### Initial greeting

Whatever the setting, the clinician should first introduce himself if he has not met the family before. It is usually advisable to see together all family members who have attended in the first instance. This might be just the mother and child or both parents and four or five children. A clinician may prefer to see a referred adolescent before other members of the family, especially if he has prior knowledge that the teenager is resentful about the referral. In general, however, at least a brief interview with the whole family or all those members who have attended is desirable. An initial step should involve asking the children how they would like to be addressed. The question 'What do you like to be called?', if the child's first name has a common abbreviation, often reduces initial tension and anxiety. The clinician should make sure he introduces anyone else in the room and explains what they are doing here. If, as is the case in some psychiatric facilities, a one-way screen is used, the family should be introduced to those behind it, and reassurance concerning confidentiality should be given.

### Clarifying the reason for referral

It is always helpful to know why the attendance has occurred at this particular time. Whose idea was it? In most children seen in primary care, it will be the mother who attends with the child, and she will be the person motivated for the attendance. She may, however, be attending only because some particular crisis occurred, because the father insisted something be done about a long-standing problem, or because she had confided a worry to a relative, neighbour, friend, or teacher. Reasons for attendance at a paediatric or child psychiatric facility also need clarification. A request for an opinion from a general practitioner often masks the fact that it was the parents who requested further advice. Referrals to psychiatrists from school agencies may fail to make it clear that a head teacher has said a child will be suspended from school unless the family attend a clinic for advice. It is essential to understand the motivation behind a referral if the clinician is to help in alleviating the anxiety responsible for it.

### Obtaining factual information from parents and children

There are various approaches to obtaining factual information. Most practitioners use an informal style of questioning, aiming to cover a number of areas systematically, but concentrating on particular issues likely to be relevant to the child in question. Such semi-structured interviewing (Bird and Kestenbaum 1988) has the advantage that it is flexible and allows the clinician to adopt a conversational approach. However, psychiatrists and psychologists are increasingly using a variety of more standardized questionnaires for this purpose (Orvaschel 1988). Such questionnaires as the Diagnostic Interview Schedule for Children (Costello *et al.* 1984) provide generally reliable information (Costello 1989), but suffer the disadvantage that they may be perceived by the family as more mechanical and less tailored to their own needs. Highly structured interviews are more suitable for research than for clinical purposes, and, at this stage, it seems sensible for most professionals to use a more informal assessment but ensure that all relevant areas are covered.

Some experienced psychiatrists and psychotherapists hardly structure assessment interviewing at all and rely almost entirely on the spontaneous expressions of the family members. They are concerned that, if the clinician structures the interview, family members will be inhibited from expressing feelings. There is however evidence that semi-structured interviewing need not place constraints on emotional expression and ensures more reliable gathering of factual information (A. Cox *et al.* 1981). Such an informal approach can begin by listening to the spontaneous complaints and other information provided by family members, and then by asking direct questions. Open-ended questions, for example 'In what ways have you tried to deal with his . . .?' produce more information than closed questions answerable with a yes/no, for example . 'Have you tried rewarding him for good behaviour?' It is never a good idea to submit a parent or child to a barrage of questioning without leaving a significant amount of time for the spontaneous expression of anxiety. In obtaining information about particular symptoms or problems, it is important to obtain an idea of the frequency of their occurrence, the details of what actually happened, provoking events, and the behaviour of parents and others that seem to make it better or worse. It is wise to concentrate on recent examples of the behaviour in question. Problematic behaviour is often situation-specific, i.e. it occurs in some settings and not in others. This often gives clues to effective management, so it is important to obtain as full information as possible.

### Eliciting feelings and attitudes from parents

Often these are the crucial variables in assessment. Positive feelings (such as warmth, tolerance, acceptance) and negative feelings (such as criticism, hostility, and rejection) may be elicited by appropriate questioning, by suggesting the presence of feeling states and observing parental reactions, and by reinforcing expressions of feeling when these occur. One can ask questions such as 'How does that make you feel? Do you feel that gets on your nerves?' Suggestions such as 'That must make you feel very angry?' or 'It sounds as though you quite enjoy it when he does that sort of thing?' sometimes elicit strong emotional responses. Showing interest and sympathy when emotion is spontaneously expressed may have a similar effect.

### Building family confidence

Parents who bring a child to a family practitioner, paediatrician, or psychiatrist for a psychosocial or psychosomatic problem not infrequently suffer from a deep sense of failure at what they imagine to be their own incompetence in dealing with the problem themselves. It is therefore particularly important to avoid conveying a critical attitude, and to take every opportunity to comment on the positive coping behaviour the parents are showing. One should assume, and in the great majority of cases this assumption is correct, that parents are doing their very best for their children. *An assessment interview should be perceived by parents as a supportive, not an undermining experience.*

### Interviewing children separately

Adolescents should always be seen separately at some point, and children from about 6 or 7 years to puberty if time permits. It is usually not desirable to attempt to see children younger than this without their parents.

An entirely verbal approach can be used in an individual interview with older children and adolescents. After setting the child at ease by talking about leisure activities and favourite television programmes, the interviewer can go on to broach the presenting problem and the child's view of its occurrence. Questioning can then go on to more general anxieties, depression of mood, suicidal thoughts and feelings, etc. 'I see a lot of children who get worried and anxious about things. What sort of things do you feel anxious about?' 'What sort of things make you angry at school?' 'What sort of things get you upset at home?' 'A lot of boys/girls I see worry something terrible is going to happen to their mum. Do you ever feel like that?' are all useful probes. Very unresponsive or mute children can be reassured—'I see a lot of children who have difficulty talking. I wonder why it is hard for you. It could be because you are feeling angry, ... or mixed up, ... or just because you can't think of anything to say?' If the child has come because of some misdemeanour, the clinician should avoid either condemnation or approval, but try to find out about the circumstances in which the wrongdoing occurred and the child's feelings about it in a morally neutral way. The interviewer should conclude by asking the child what else he or she would like to talk about.

For younger children, the supply of toys described above, especially family figures and farm animals, may well be a more effective trigger to elicit informative fantasies or anxieties the child may be experiencing. It is, however, surprising how many children of 6–10 years can communicate verbally without the use of toys and drawings.

Interviews with adolescents may need to include a number of topics not covered with younger children. Sexual concepts and behaviour and the possible use of drugs are examples. Adolescents may also be expected to elaborate to a greater degree than younger children on their mood states, relationships with peers, and special interests.

Before concluding any interview with a child alone, if it is thought desirable to raise any of the matters the child has discussed with the parents, then the child's agreement should be sought beforehand. If the child refuses (a rare event), then this should be respected unless the child would be put in danger as a result. The clinician may, however, say to parents that the child has expressed some private anxieties he would prefer not to communicate to them, and stress the normality of children sometimes having private thoughts they do not want parents to know about.

*Physical examination*
Children with physical symptoms will always require a physical examination to identify organic disease. General practitioners and paediatricians would naturally carry out such an examination themselves. Psychiatrists vary in whether they carry out physical examinations or refer to others for this purpose. Most psychiatrists would feel it desirable to examine the child themselves on at least some occasions, though not necessarily to check all bodily systems. A screening neurological examination alone can, for example, both rule out significant neurological disorder and identify developmental delays in motor co-ordination, visual perception, etc.

*Feedback*
The clinician should spend a significant proportion of time discussing his view of the problem with family members at the end of an assessment interview. It is a mistake to spend virtually all one's time asking questions and leave only a minute or two for feedback. In general the feedback of an opinion on the nature of a problem, its likely causes, and suggested treatment approach should be tentative and preceded by an enquiry concerning views the parents already hold. It is unhelpful, for example, to make confident suggestions about a treatment approach, only to be told by parents this is something they have already unsuccessfully tried.

*Planning future action*
Finally, at the end of an assessment, the clinician should make clear how he plans to proceed and to whom he would like to communicate his findings. If he is uncertain whether he should communicate certain confidential information, he should check with parents whether what he wants to do is acceptable. If he is not offering further appointments, he should make clear how the family can contact him again if this becomes necessary. If he is offering treatment, he should describe the likely length of treatment, frequency of sessions, purpose of treatment, and likelihood of change occurring either as a result of treatment or for other reasons.

### 1.2.2 FRAMEWORK FOR ASSESSMENT OF A PSYCHIATRIC PROBLEM

The type and amount of information that will be obtained during an assessment will vary depending on the nature of the problem. For example, a detailed account of the psychiatric state of the male cousins of the child might be very relevant if a sex-linked recessive condition, such as the fragile X syndrome is suspected, but would not otherwise be enquired about.

Attitudes to body appearance would be a main focus if the patient was anorexic, but not otherwise.

The following is a listing of information that would be relevant for a clinician making a comprehensive assessment of a child with a behaviour or emotional disorder. It is suggested that a clinician with little time available, making a less intensive assessment, should concentrate on obtaining information printed in heavy type.

1. **Nature and severity of presenting problem(s). Frequency. Situations it occurs. Provoking and ameliorating factors. Stresses thought by parents to be important.**
2. Presence of other current problems or complaints.
   - (a) Physical. Headaches, stomach aches. Hearing, vision. Fits, faints, or other types of attacks.
   - (b) Eating, sleeping, or elimination problems.
   - (c) **Relationship with parents and sibs. Affection, compliance.**
   - (d) Relationships with other children. Special friends.
   - (e) Level of activity, attention span, concentration.
   - (f) Mood, energy level, sadness, misery, depression, suicidal feelings. General anxiety level, specific fears.
   - (g) Response to frustration. Temper tantrums.
   - (h) Antisocial behaviour. Aggression, stealing, truancy.
   - (i) **Education. Attainments, attitude to school attendance.**
   - (j) Sexual interests and behaviour.
   - (k) Any other symptoms, tics, etc.
3. Current level of development.
   - (a) Language: comprehension, complexity of speech.
   - (b) Spatial ability.
   - (c) Motor co-ordination, clumsiness.
4. Family structure.
   - (a) Parents. Ages, occupations. **Current physical and emotional state.** History of physical or psychiatric disorder. Whereabouts of grandparents.
   - (b) Sibs. Ages, presence of problems.
   - (c) Home circumstances: sleeping arrangements.
5. Family function.
   - (a) **Quality of parental relationship. Mutual affection. Capacity to communicate about and resolve problems. Sharing of attitudes over child's problems.**
   - (b) **Quality of parent–child relationship. Positive interaction: mutual enjoyment. Parental level of criticism, hostility, rejection.**
   - (c) Sib relationships.
   - (d) Overall pattern of family relationships. Alliance, communication. Exclusions, scapegoating. Intergenerational confusion.
6. Personal history

(a) Pregnancy—complications. Medication. Infectious fevers.

(b) Delivery and state of birth. Birth-weight and gestation. Need for special care after birth.

(c) Early mother–child relationship. Postpartum maternal depression. Early feeding patterns.

(d) Early temperamental characteristics. Easy or difficult, irregular, restless baby and toddler.

(e) Milestones. Obtain exact details only if outside range of normal.

(f) **Past illnesses and injuries. Hospitalizations.**

(g) Separations lasting a week or more. Nature of substitute care.

(h) Schooling history. Ease of attendance. Educational progress.

7. Observation of child's behaviour and emotional state.

  (a) **Appearance. Signs of dysmorphism. Nutritional state. Evidence of neglect, bruising, etc.**

  (b) **Activity level. Involuntary movements. Capacity to concentrate.**

  (c) **Mood. Expression or signs of sadness, misery, anxiety, tension.**

  (d) **Rapport. Capacity to relate to clinician. Eye contact. Spontaneous talk. Inhibition and disinhibition.**

  (e) **Relationship with parents. Affection shown. Resentment. Ease of separation.**

  (f) Habits and mannerisms.

  (g) Presence of delusions, hallucinations, thought disorders.

  (h) Level of awareness. Evidence of minor epilepsy.

8. Observation of family relationships.

  (a) Patterns of interaction—alliances, scapegoating.

  (b) Clarity of boundaries between generations: enmeshment.

  (c) Ease of communication between family members.

  (d) Emotional atmosphere of family. Mutual warmth. Tension, criticism.

9. Physical examination of child.

A. Screening neurological examination.

  (a) Note any facial asymmetry.

  (b) Eye movements. Ask the child to follow a moving finger and observe eye movement for jerkiness, incoordination.

  (c) Finger–thumb apposition. Ask child to press the tip of each finger against the thumb in rapid succession. Observe clumsiness, weakness.

  (d) Copying patterns. Drawing a man.

  (e) Observe grip and dexterity in drawing.

  (f) Observe visual competence when drawing.

  (g) Jumping up and down on the spot.

  (h) Hopping.

  (i) Hearing. Capacity of child to repeat numbers whispered two metres behind him.

B. **Further medical examination (if relevant).** The clinician may either carry out such further examination himself or refer for specialist advice if this seems indicated. Referrals for more systematic visual and auditory testing are not uncommonly required.

10. Physical investigations. The assessment may reveal the need for physical tests including blood count, urinalysis, plasma chemistry, skull radiography, electroencephalography (EEG), brain scan, or magnetic resonance imaging (MRI).

11. Assessment by other professionals. A psychiatric appraisal may uncover further assessment needs that would best be undertaken by a colleague, if available. In particular, learning difficulties may require assessment from a psychologist and child protection issues may indicate the need for referral to a social worker.

## Formulation

Following an assessment along these lines, the clinician is in a position to formulate the problem. A formulation should consist of a statement of the main presenting features, a diagnosis (see Section 1.1.2), an indication of probable causes, an opinion concerning desirable management, and a view on outcome. An example of a formulation might be 'John is a 9-year-old boy who has been refusing to go to school for reasons of anxiety for the past month. He is also enuretic. The underlying problem is an over-close relationship with his mother, with father often away at work. A viral infection may have acted as precipitant. The diagnosis is school refusal. Family counselling sessions with a rapid but planned return to school are indicated. The outlook for return to school is good, but he will probably remain an anxious boy.' Ideally such a formulation should be written down, but, in primary care, if time does not permit this, the clinician will nevertheless find it helpful at least to establish a formulation in his own mind before proceeding further.

### 1.2.3 ASSESSMENT OF LEARNING DIFFICULTIES

## Introduction

In this section an account is given of a psychological approach to mental retardation and learning disabilities with emphasis on the value and limitations of testing undertaken by child psychologists.

Assessing and advising about learning problems form a significant part of the work undertaken by child psychologists, especially educational psychologists. The psychologist appraises the nature and severity of the problem, the circumstances in which it occurs, and the measures currently being taken to deal with it. The administration of standardized psychological tests, if carried out at all, is only part of the appraisal of the nature and severity of the problem. Usually, more time is spent by psychologists in working directly with teachers, in setting up individualized targets for children to achieve, in devising interventions to help them achieve these targets, and in monitoring the effect of the interventions on the individual child (Leach 1980). Nevertheless, tests continue to have a valuable place in the psychologist's repertoire. They are helpful in monitoring progress or deterioration in children with conditions that might affect mental function (e.g. a cerebral degenerative disorder or leukaemia) or when treatment might be expected to produce intellectual change. They also remain useful as research tools. Consequently it is important for non-psychologists working with children (e.g. clinical medical officers, paediatricians, social workers, and psychiatrists) to be aware of their uses and limitations.

## Types of test

*Skill to be assessed.* Standardized tests vary according to the functions they are intended to measure. Such functions include:

(1) general intelligence;

(2) special skills, for example language, perceptual ability, motor ability;

(3) educational attainment, for example in reading or mathematics.

Those involved in systematically assessing children's performance in school tend to use three main types of test (Assessment of Performance Unit 1978).

i. Comparative tests to discover where an individual child's performance stands in relation to that of other children. Appropriate normative data on the general population of children must be available if a test is being used for this purpose.

ii. Diagnostic tests to identify strengths and weaknesses.

iii. Curriculum related procedures to determine whether something taught has been learned (Cameron 1990).

In general, comparative tests are likely to be most useful where groups of children are being assessed, for example, to see whether a modification in diet has produced an improvement in overall intelligence in children with phenylketonuria, or whether a local educational authority is putting its remedial resources into schools with the highest rates of children having educational problems.

In contrast, diagnostic tests and curriculum-related procedures are most helpful when devising an educational programme for an individual child. The 'curriculum' may involve teaching a very basic skill such as doing up buttons, or acquiring a much more advanced set of information for a public examination. The principles are the same. One valuable 'curriculum-related' check-list approach to preschool children with serious learning difficulties is the *Portage guide to early education in the first six years of life* (Bluma *et al.* 1976).

In carrying out a diagnostic test on a child failing in school, the psychologist will not only note the level of the child's performance and the particular cognitive problems such as deficits in attention and concentration that the child may be experiencing, but will also observe whether performance is impaired by undue anxiety or other emotional states.

Learning is a process that proceeds by stages (Haring *et al.* 1987), and children with emotional problems may be impaired in the initial acquisition of a task or skill, in achieving fluency in its performance, in maintaining such fluency, in generalizing beyond the setting in which the skill has been learned, or in applying the skill appropriately.

**Constructions of tests**

Those involved in constructing standardized, norm-referenced tests of intelligence and attainment first identify the skills they wish to assess and the age range over which they wish the tests to apply. They then identify tasks that can be given to children that measure these skills and have high powers of discrimination, i.e. there is a suitably wide range of variation of achievement among children of the same age when given the task. They then develop criteria for scoring children's performance on the task. The reliability of the testing is then assessed by ensuring that children given the task twice perform in very nearly the same way on the two occasions (test–retest reliability) and that two observers watching the child perform the task rate the child in the same way (inter-rater reliability). They then examine the *validity* of the test, by using external criteria to determine whether the test does actually measure the skills intended. This can be done in a variety of ways, for example by seeing whether children who score highly on the task later achieve better in examinations than those who score poorly (predictive validity), or whether there is a correlation between the new test and another well-tried test assessing similar functions (concurrent validity). Finally, the test is standardized on a large representative population of children spanning the ages for which the test is to be applied.

**Tests of development in infancy and early childhood**

1. *Bayley Scales of Infant Development* (Bayley 1969). This American test covers the age from 2 months to 2½ years. A range of items is assessed so that the child can be scored on a Mental Developmental Index and a Psychomotor Developmental Index. The scale takes about 45 minutes to administer and, given a skilled and experienced psychologist, provides a reliable, comprehensive record of the child's level of functioning.

2. *Griffiths Mental Development Scale* (Griffiths 1954). This British test is designed for children aged up to 2 years. There is an extension available

for use above this age, but this is less satisfactory. The test has five sections: locomotor, personal–social, hearing and speech, eye–hand, and performance. It takes about an hour to administer. The test can only be administered by professionals who have been specifically trained in the use of the scale. It is currently being revised.

3. *Denver Developmental Screening Test (DDST)* (Frankenburg *et al.* 1975). This is a screening instrument for identifying children aged from birth to six years at risk for developmental delays and anomalies, so that they can be selected for further assessment. Four areas of development are assessed: personal–social, fine motor, language, and gross motor. The test takes about 20 minutes to administer. It can be applied by relatively inexperienced personnel, but is less accurate than the Bayley or Griffiths test.

## Tests of general intelligence

1. *The Stanford–Binet Intelligence Scale* (Thorndike 1986) is a test devised for use with childen aged 2 up to adulthood. Testing takes place in four areas: verbal reasoning, abstract/visual reasoning, quantitative reasoning, and short-term memory. The child's performance is assessed on a variety of tasks, depending on his age, the test taking about an hour to complete. Subsequently, an intelligence quotient can be calculated from data obtained from children in the general population.

2. *The Wechsler Pre-school and Primary School Intelligence Scale* (Wechsler 1989) is a test applicable to children aged from 4–6½ years. In the full form there are six verbal and six non-verbal tests, taking about an hour to administer, but a short form is available. Verbal IQ, performance IQ, and full-scale IQ scores can be computed. The test is reliable and well validated and provides a useful profile of the young child's pattern of abilities. However, as with the WISC (see below) the reliability of individual subtests is not high, and too much weight should not be put on isolated subtest results.

3. *Wechsler Intelligence Scale for Children—revised form (WISC-R)* (Wechsler 1974). This is a test of general intelligence applicable to children aged 6–14 years. There are 12 subtests, measuring different aspects of the child's abilities. Six of these (information, similarities, arithmetic, vocabulary, comprehension, and digit span) are used to obtain a 'verbal' IQ, and the other six (picture completion, picture arrangement, block design, object assembly, coding, and mazes) are used to obtain a 'non-verbal' score. A 'full-scale' score is obtained on the basis of the results of the 12 tests. The test takes about an hour to complete, but a standardized, reasonably reliable short form of the test can be carried out in about 20–25 minutes. The child's scores on this can be 'prorated' to achieve a rough equivalent to a score obtained on the full test.

Scoring on the test has been computed to ensure that the mean is 100 with a standard deviation of 15. About 15 per cent of the population therefore score below 85 and 2.5 per cent below 70. Results on tests such as the WISC-R are used in defining levels of mental retardation (Section 2.6.2).

4. *British Ability Scale* (Elliott *et al.* 1983). This is a test of general intelligence suitable for use with children aged 2–17 years. Testing takes about an hour, but shorter forms of the test can be used. A total of 24 scales are employed, and these assess speed of working, reasoning, spatial imagery, perceptual matching, short-term memory, retrieval, and application of knowledge. An advantage of the test is that it covers a wide age range, so that, if longitudinal assessment is required, its use, at least to some degree, overcomes the problems of relating results on one test to those obtained on another. Norms are so far only available for British children.

5. *The Goodenough–Harris Drawing Test (Draw-a-Man Test)* (D. B. Harris 1963). This is a brief test to assess non-verbal intelligence in children aged 3–10 years. The child is asked to draw the best picture of a man he can, and the drawing is scored against carefully defined criteria. The test is reasonably reliable, but provides only a modest indication of the child's general intelligence as assessed in other ways.

**Tests of educational attainment**

1. *Word Reading Subtest of the British Ability Scale* (Elliott *et al.* 1983). In this test, suitable for children aged 6 years and upwards, the child is asked to read a series of words of increasing difficulty until he has made a number of consecutive errors. A reading age, mainly reflecting accuracy of reading, is then obtained. The test takes about 10 minutes to administer.

2. *Neale Analysis of Reading Ability, Revised British Edition* (Neale 1988). Children aged 6 years and upwards are asked to read aloud a small number of stories, and are then asked questions about them. The child can be scored for accuracy, speed and comprehension, and reading levels for each of these computed. The test takes about 20 to 30 minutes to administer, but provides a broad, useful assessment of the child's reading abilities.

3. *Schonell Graded Word Spelling Test* (Schonell and Schonell 1950). This test is also suitable for children over the age of 6 years, and is similar in form to the Word Reading Test of the British Ability Scale.

4. *The Wide Range Achievement Test (WRAT)* (Jastak and Jastak 1978) is a test suitable for children aged 5 years and over. There are two forms, one suitable for children up to 12 years and the other for older children. Reading, spelling, and arithmetic skills are assessed. It takes 20–30 minutes to administer. The test is useful and reliable, but norms are only available for North American children, so that the test is of limited applicability elsewhere.

There is no very satisfactory standardized test of mathematical ability available for use in the UK, but the arithmetic subtests of the WPPSI, WISC-R, and British Ability Scale are sometimes used for this purpose.

## Adaptive behaviour

*Vineland Adaptive Behaviour Scale* (Sparrow *et al.* 1984). This is a useful measure of a child's level of independence and self-help. It can be used to assess children from birth to adolescence. A parent, usually the mother, is asked a series of questions about the child's social and self-help skills and the information scored accordingly. A social age and a 'social quotient' can be obtained to define the child's level of independent functioning.

## Specific skills

A variety of other tests, such as the Reynell Developmental Language Scales (Section 2.5.1), the Lowe–Costello Symbolic Play Test (Section 2.5.1), and the Bruininks–Oseretsky Test of Motor Proficiency (Section 2.4.2), are used to assess specific skills and abilities. These are discussed in the sections dealing with the relevant aspects of development.

## Application of tests

*Indications*

There is no doubt that psychological testing can play a most valuable part in the assessment of a child's mental functioning. In particular, results of testing will be helpful when:

(1) a child is failing educationally in relation to his peers;

(2) a child, while not failing educationally, appears to be functioning well below his level of ability;

(3) there is reason to believe that a child is suffering from a condition likely to affect intelligence or educational attainment, and it is desirable to obtain a record of progress or deterioration.

The results of tests can be used, in conjunction with observations of the child in the testing situation (e.g. by noting his approach to tasks, distractability, etc.), in the classroom, discussion with parents and teachers, and an appraisal of the child's responsiveness to different remedial approaches, for a variety of purposes. A specific programme of remedial instruction geared to the child's educational needs may be instituted. This may or may not involve a change in the child's class or school. The results of psychological testing may be one factor pointing to the need for a change in a child's treatment (e.g. anticonvulsant medication in a child with epilepsy).

*Risks and dangers of testing*

These can be classified as follows:

1. *Inappropriate test used.* The test employed may be for older or younger children, not relevant to the purpose for which it was administered,

or not standardized on the population from which the child was drawn. Thus it is inappropriate to test children from developing countries on unmodified tests standardized in Western countries.

2. *Unsatisfactory test conditions.* The child may be tested in noisy, distracting circumstances, the tester inexperienced, or the child unmotivated or over-anxious. The child may not speak fluently the language in which the test is administered, or may be hampered by hearing or visual impairment or other causes of incomprehension. A child with poor attention may be difficult to test unless the test conditions are good and the examiner extremely patient.

3. *Over-generalization of conclusions.* There is an inherent unreliability in all tests. Inevitably children given a test on one day will not score the same if given it even a few weeks later. The tester should be aware of the margin of error involved in the test he is using.

The tester should not draw inappropriate conclusions concerning the child's future. Especially in children under the age of 4 or 5 years, results of IQ testing are poor guides to later functioning unless the child's level is particularly low, and even then caution should be exercised in providing a prognosis (Section 2.6).

Sometimes variations in functioning lead to inappropriate conclusions. Thus a child who is reading somewhat below what one would expect from his level of intelligence is not necessarily underachieving (Section 2.5.5). Variation in test score cannot be used to diagnose brain damage, though tests have sometimes been used for this purpose. Certainly, however, results of testing can indicate that further physical investigation is necessary to establish whether indeed brain function or structure is impaired.

4. *Labelling.* The risks of children being labelled on the basis of psychological test results as dull and expected to achieve only poorly are real enough. It is the responsibility of everyone involved in administering tests and talking about the results with parents and teachers, to explain their limited predictive value, so as to ensure that expectations about the child are not inappropriately fixed at a low level.

## 1.3 Family influences and parenting disturbances

### 1.3.1 FAMILY INFLUENCES

#### Family structure

The conventional family (two married parents, two children) is no longer the norm in most developed countries. In both the USA and the UK, by the age of 16 years, about one in four children are no longer living with their two biological parents. Fortunately, healthy psychosocial development is compatible with a wide range of types of family structure. Survey data suggest that family structure can have an influence on development, even though, in most cases, that influence is not as great as the quality of relationships.

1. Children of single parents are somewhat more likely to show behavioural and emotional problems, especially if there is a lack of support from grandparents, friends, etc. Isolated, single parents often face serious financial disadvantage and, if they develop even minor illness, the lack of alternative care for their children may pose problems (Ferri 1976). However, there are few differences between children of single parents and children with two parents living in similar financial circumstances.

2. Children living in societies where there is a pattern of extended family care (with much reliance on grandparents, aunts and uncles, etc. for child care) have about the same rate of psychological disorders as do children living in nuclear families with two parents and a small number of children (Cederblad 1968). The strains and stresses of living in an extended family are probably different in nature, but no less severe than those occurring in nuclear families.

3. Anomalous or unusual family structures are not, in general, harmful to children unless the quality of relationships within the family is poor. Thus, for example, children brought up by two homosexual women do not, at least in their prepubertal years, appear to have an unduly high rate of problems (Golombok *et al.* 1983). Some anomalous family structures, such as communes, are almost inevitably unstable, and thus are likely to be unsettling for children brought up in them.

4. Large family size is associated with somewhat lowered intelligence, and with higher rates of educational and behavioural problems (Davie *et al.* 1972). Having a large number of brothers and sisters is, of course, quite compatible with normal development, but a large sibship does predispose to psychological difficulties. Large families are more likely to suffer economic hardship, and parental stimulation is likely to be less easily available. By contrast, only children appear to have a small advantage as far as intellectual development is concerned.

5. Ordinal position or position in the family is of little psychological significance as far as risk of behaviour disturbance is concerned, though oldest and only children have a slight advantage in intellectual development (Davie *et al.* 1972). Twins are somewhat slower to develop language, and this is probably because of a mixture of divided parental attention and a reduced need for communication in twins who have each other for company.

## Family social circumstances

The social status of a family is usually assessed in population surveys according to the employment of the father or the main breadwinner in the home. Intelligence and educational attainment are related to social class assessed in this way, with children of professional and other middle-class parents performing better than the skilled working-class and much better than children with unskilled working-class parents. Behavioural and emotional disorders are, however, found to be only weakly related to these measures of social class (Rutter *et al.* 1970a). About a third of UK and about a half of US mothers of preschool children go out to work. Whether a mother works or not has a major influence on the life of her children (P. Graham 1990). However, there is little evidence that maternal employment is, in itself, a significant advantage or disadvantage for children as far as their behavioural or cognitive development is concerned. Of greatest importance is the quality of substitute care available, and whether the mother feels comfortable and satisfied with her decision about work. The quality of mother–child attachment is closely related to whether a mother is satisfied with her employment status whatever decision about work she has made.

By contrast, more detailed investigations of the circumstances in which children are living suggested that adverse social factors are of major importance in the development of psychiatric problems. In particular, pre-school children are more likely to be disturbed if they are brought up in cramped housing conditions with inadequate play-space. Tower block or high-rise accommodation seems particularly unsuitable (N. Richman 1977a). The presence of serious financial hardship also predisposes to disorder, and, in teenagers, chronic unemployment is a significant stress factor producing emotional problems (Banks and Jackson 1982). Probably relative deprivation (the degree to which family members perceive themselves to be disadvantaged in comparison with others) is more important than absolute deprivation (a standard of living which is low on objectively defined criteria) (Runciman 1972).

### Family functioning

The influence of family function on children's behaviour and development can be viewed in two main ways. The family can be seen as a single unit, or one can consider the ways in which two individuals in the family (e.g. mother and child or father and mother) interact together.

*The family as a unit*
1. The *internal structure of relationships* in the family may be relevant. Thus in some families (Minuchin 1974) there is a clear generation boundary between parents and children. In others, the boundary is blurred: parents often act like the children, and children sometimes act like the parents. Special alliances between particular family members may exclude others from feeling they can relate to those forming such alliances. Parents may become over-involved or 'enmeshed' in their children's feelings and

lives, so that the children hardly know when they have feelings that are their own and when they are feeling what their parents want them to.

2. Families also may develop characteristic *strategies* or techniques for communication or problem-solving. Some families operate according to particular rules (with, for example, certain important subjects never mentioned), and this may lead to one or more of the children developing behavioural disorders.

3. Family units can also be seen as functioning on a *homeostatic* basis. If a change in one important central relationship occurs, this can inevitably lead to a reshuffling in the whole family, and a particular child may be adversely affected or suddenly released from stress. Parents with unexpressed marital conflict may 'need' to have a child with a problem whom they can argue about, and thus create a scapegoat. When that child leaves home, they may find they have to face their real relationship problems.

All these ways of looking at whole family functioning can be helpful in understanding and treating individual children's disturbances (see Section 6.2.4). It is also useful to examine the effects of separate influences of parents on children, of the marital relationship, and of the effects of one sib upon another.

### Parental marital relationship

The quality of the parental marriage is an important factor in whether a child develops an emotional or behavioural disorder. Children of parents whose marriages are characterized by quarrelling, tension, mutual dissatisfaction, criticism, hostility, and lack of warmth are more likely to become disturbed. The fact that children of divorced parents have a distinctly higher rate of disturbance (Hetherington *et al.* 1982), whereas children of parents separated by death, although suffering initial bereavement reactions (van Eerdewegh *et al.* 1985), have only a slightly raised rate of disturbance later on, suggests that it is parental disharmony rather than separation from parents that is the crucial factor.

In fact although their rate of disorder is distinctly higher than children in an intact family, only a minority of children of divorced parents become disturbed. Protective factors include the degree to which parents can make amicable arrangements for access, the quality of their own relationship with the child, the availability of other good relationships within the home (with sibs), and the child's temperament. The protective effect of a good relationship with an adult (usually a grandparent) outside the nuclear family has also been demonstrated in children whose parents have disharmonious marriages (J. M. Jenkins and Smith 1990).

### Quality of parental care and affection

Emotionally warm, continuous, sensitive care from parents or their substitutes is the main, but by no means the only necessary precondition for healthy psychological development. Its presence goes far to ensure the development of 'attachment' of the child to the parent (Section 2.1.2). This quality to the parent–child relationship results in the young child developing a whole range of adaptive forms of behaviour, including confident exploration of the environment and the appropriate seeking of parental protection when danger threatens.

Emphasis has been put on the first few hours and day of a baby's life as crucial in the development of attachment (M. Klaus and Kennell 1976). In fact, although of course early parent–child contact is highly desirable, continuing contact over the first years of the child's life is just as important (Herbert *et al.* 1982), and there is no scientific reason to think that the first few hours are by any means necessarily of such enduring significance. As the child gets older, so long as he remains in the company of familiar figures, he will be able to tolerate longer periods of separation and, indeed, he needs to achieve this in order to cope with independent living later on. Factors affecting response to separation are discussed further in Section 4.6 in relation to hospitalization.

Linked to this need for parental warmth and sensitivity is a requirement for parents to provide adequate consistent control and structure to the child's life. As the child gets older, so he will achieve greater autonomy, and parental capacity to allow the child independence in decision-making becomes important. The forms of reward and punishment (e.g. whether or not there is smacking or material rewards are given) are less important than the consistency of parental behaviour, agreement between parents, and maintenance of acceptance and responsiveness to the child's needs even when the child is being difficult.

### Parental stimulation

The quality of verbal and non-verbal stimulation provided by parents has a significant influence on intellectual development and academic attainment. The amount that parents talk to their children may be less important than the degree to which they succeed in maintaining two-way conversation (Tizard and Hughes 1984). The availability of toys is probably less important than the degree to which the child, with parental help, can be encouraged to explore objects and spatial relationships in his natural environment. The availability of books in the home, and the degree to which parents are able to help their children learn to read by listening to their efforts are factors of importance in the speed with which skill in reading is acquired (Hewison and Tizard 1980). Parental stimulation of these various types will be more effective if it takes place in the context of a warm, loving relationship.

### Parental mental illness

As in the rest of the adult population, mental health problems in parents are common. Mostly the disorders they experience are characterized by depression and anxiety, and are related to adverse social circumstances (poor marital relationships, inadequate housing, financial hardship, etc.). However a significant number of parents show other disorders such as schizophrenia, alcoholism, and aggressive personality disorders. The impact of mental ill health in parents on children has been extensively studied (Rutter and Quinton 1984).

A significant number of parents involved in non-accidental injury or sexual abuse (see Sections 1.3.2 and 1.3.3) have mental health problems, although serious mental disorder is uncommon among them. All types of mental disorders are linked to a raised rate of behaviour and emotional problems in children. Children of parents with psychoses such as schizophrenia and manic-depressive psychosis are less prone to show such problems than children of parents with depressive and anxiety states, alcoholism, and personality disorders.

Except in the children of parents with schizophrenia and possibly major depressive disorder, genetic factors are probably of minor importance in the transmission of mental health problems; other mechanisms are probably of more relevance. Many forms of mental ill health directly impair the quality of parental care. In particular, a depressed mother is likely to be less responsive to the needs of her child, less able to provide consistent discipline, and less likely to initiate interaction (Weissman and Paykel 1974; Mills *et al.* 1984). Morbidly anxious mothers often transmit their anxieties to their children by example. For example, a mother who is frightened of lifts or travelling on trains will often communicate fear of travelling to her children. Less usual forms of direct impact of illness occur when a child is involved in obsessional rituals or implicated in paranoid delusions.

Factors frequently associated with parental mental disorder are probably of greater importance than psychiatric symptoms themselves. It has been shown (Rutter and Quinton 1984) that childhood disturbance is more closely linked to parental marital discord and generally disharmonious marital relationships than to parental mental ill health. Parents with chronically aggressive personalities are particularly likely to have marital difficulties, so that their children are especially predisposed to develop behaviour and emotional problems themselves. Various processes are involved in the production of childhood disturbance in this situation. The child may imitate or identify with violent behaviour. Impulsive parents are likely to be inconsistent and frustrating to their children. Both parental separations and deprivation will occur frequently. Actual violence or threats of violence within the family will provoke anxiety. All these factors will predispose the child to serious disturbance.

Factors modifying the child's risk of developing disturbance if a parent has a psychiatric disorder include the mental health of the other parent, the quality of the relationship with the other parent, and the temperament of the child (Section 2.14.1).

### Parental physical illness

There is a slight but definite tendency for the children of parents with chronic physical ill health to have a raised rate of psychiatric problems. Such children may be involved in looking after the sick parent to a degree that isolates them from other children of their own age, and imposes responsibilities they are too immature to face. There may be poor communication within the family about the illness, so that the child is more uncertain and anxious about the future than is necessary. Finally, illness may cause or be associated with disturbed family relationships that are upsetting for the child.

### Sib relationships

There has been rather little study of the effects of brothers and sisters on each other's behaviour, but some studies have examined the factors affecting the quality of these relationships. Such relationships are often ambivalent and marked by both strong positive protective feelings and a sense of rivalry. The reactions of older children to the birth of younger siblings depend, for example, on the temperament of the older child, on whether the younger child is of the same or different sex, on the behaviour of the mother with the new baby, and on the quality of the mother–older child relationship before the younger child is born (Dunn 1988).

### 1.3.2 NON-ACCIDENTAL INJURY (NAI)

### Definition

NAI occurs when an adult inflicts a physical injury on a child more severe than that which is culturally acceptable.

Corporal punishment has only recently been abolished in British schools, and remains tolerated in many American states. Punishment by beating in the home is still socially acceptable in many cultures. There are other important cultural differences in what constitutes physical abuse, with some ethnic minorities viewing quite severe physical punishment as within the range of normal disciplinary practice, and others regarding almost any physical restraint as undesirable (Giovannoni and Becerra 1979). The use of physical punishment in itself cannot therefore be regarded as evidence of NAI unless it involves very young children, is frequent, or produces physical changes more marked than skin redness and mild transient bruising. This situation may change as countries follow the example of Sweden and introduce legislation to forbid the use of physical punishment by parents. The severity of physical injury is, however, only one indication of ill treatment. A child showing only minor physical changes might be severely neglected or emotionally abused, and this, in itself, would obviously be a matter of very serious concern.

## Prevalence

This will depend on the criteria used. A study from north-east Wiltshire, England (Baldwin and Oliver 1975), suggested that the annual prevalence rate of severe physical abuse was 1 per 1000 in children under the age of 4 years. Severe physical abuse was defined as that which resulted in bone fractures, bleeding into or around the brain, other severe internal injuries, repeated mutilation requiring medical attention, or death. Mortality from NAI in this study was 1 per 10 000 children. The prevalence of less severe physical abuse is uncertain, as it often does not come to medical attention. A national study of 3- to 17-year-olds in the United States (Straus *et al.* 1980) found that 14 per cent had been kicked, bitten, punched, hit with an object, beaten up, or assaulted with a weapon by one or both of their parents. Few of these children would have come to professional attention because of child abuse. A. H. Cohn (1983) has estimated that about one million children are physically abused in the US each year, with 2000– 5000 abuse-related child deaths and 2 per cent of all families involved in child abuse. There is some evidence that, in the UK, serious physical abuse resulting in death or physical disability reduced in prevalence during the late 1970s (J. Jenkins and Gray 1983), and this may be due to earlier identification. The rate of milder abuse increased, perhaps due to improved reporting procedures.

## Aetiology

Particular social, family, parental, and child factors are strongly linked to child abuse (P. Graham *et al.* 1985). *Cultural factors* include exposure of the population to violence through the mass media and acceptance of violence within the family and school as normal. *Neighbourhood factors* include the support and help people in the locality or housing estate give each other, the general level of affluence and employment, standard of housing, and availability of play space. *Family attributes* of relevance include family size, quality of the marital relationship, quality of parenting behaviour, and level of communication regarding family problems and feelings. *Parental factors* include parental age (very young parents are more likely to abuse), parental personality, and mental health. Only about 10 per cent of abusing parents have clear-cut forms of mental illness such as schizophrenia or depressive psychosis, but a high proportion of the remainder have relevant personality characteristics, such as impulsiveness, proneness to anger and irritability, and difficulty in forming relationships. Parental childhood experiences of early neglect and severe physical punishment are often present in the background of abusing parents themselves. Parental cognitive factors may also be of importance. Some parents regard their children in an egoistical way, merely as extensions of themselves, while others are able to accept that even their very young children have complex feelings of their own that deserve respect. The former are much more likely to abuse (Newberger and White 1989). *Child factors* include premature birth and early separation, need for special care in the neonatal period, congenital malformations, chronic illness, adverse temperamental characteristics, and style of obtaining attention. Most children subject to abuse are living in poor socio-economic circumstances, reared by young, inexperienced parents with personality problems. A minority, however, are living in well-to-do or comfortable circumstances and the main reason for abuse here lies in parental psychopathology.

In addition to these background factors, there are often triggering events such as an acute illness in the child or sib, a financial crisis in the family, or a marital dispute. It is rare to identify simple cause—effect mechanisms in child abuse. Personal environmental and triggering factors interact with each other to produce a situation in which abuse may seem an inevitable outcome.

### Clinical features

The child may present to a variety of agencies. Parents may bring the child to the general practitioner or accident and emergency department either with the injury or with some other physical or behaviour problem. The complaint may be that the child will not stop crying or just does not seem well. Alternatively, neighbours, a relative, day nursery worker, teacher, or other person in the community may become concerned and report the matter to the police, NSPCC, or social worker.

The most common forms of non-accidental injury result in multiple bruising, burns, abrasions, bites, torn upper lip, bone fractures and subdural haematomata, and retinal haemorrhages. Other less common forms of physical child abuse result in death by drowning or poisoning. The sudden infant death syndrome (cot death) (Section 5.2) is very occasionally produced by a deliberate parental act such as smothering.

Where physical injury has occurred, suspicion should be aroused if the parental reactions seem incongruous (unduly upset over mild injury or lack of emotion over severe injury), where the explanation is unconvincing or inconsistent, where there has been a delay in seeking medical advice, and where there is a history of suspicious injury in the child or a sib, or an earlier unsatisfactorily explained death of a sib. Certain types of explanation should be regarded with particular suspicion. Young babies, for example, do not injure themselves.

Admission to hospital may reveal other supportive evidence. Insensitive parental handling of the child may be observed, or a child may appear frightened of his parents or unduly 'watchful' of strangers. The physical appearance of the child, who may appear neglected and poorly thriving, is also relevant.

It should not be forgotten that injuries may have been caused by sibs or visitors to the home such as baby-sitters, and it is important to know from the parents who has been in recent contact with the child.

## Management

Where it is the family doctor or clinical medical officer who suspects abuse, he should refer the child to hospital, informing either the paediatrician or casualty officer as well as the Social Services Department of his suspicion. Children who present first to casualty departments and in whom non-accidental injury is suspected should, in all cases, be admitted to hospital for investigation. Parents should be told this is a necessary step so that further investigations can be carried out. If parents refuse, an Emergency Protection Order (Section 7.3.3) should be applied for. The patient's general practitioner and local Social Services Department should be contacted (if they are not already involved) and relevant information such as previous injury, placement of the child or sibs on the local 'Child Protection Register', or periods of time spent 'in care' should be sought.

A thorough assessment and complete record of details is particularly important in suspected cases of NAI, as records may well have to be produced in court at a later stage. An account of the circumstances surrounding the injury should be taken from the parents, together with a full medical history and details of the social background. A record of the nature of the injuries should be made with diagrams and photographs.

Physical investigations will be necessary both to document the full extent of the injuries and to exclude conditions that may mimic non-accidental injury (C. E. Cooper 1974). A full radiological survey (looking for old injuries and to exclude bony abnormalities such as osteogenesis imperfecta) should be performed. In some cases CT scanning of the head and ultrasound examination of the abdomen may be appropriate. The child's coagulation factors should be checked. Observations made by nursing and play staff of the parents, of the child, and of parent–child interaction are often very helpful.

Once all the relevant information has been collected, the consultant paediatrician or other senior doctor should, if NAI seems the most likely explanation of injury, talk with the parents. The extent and nature of the injuries should be described, and the parents given an opportunity to explain how the injuries have occurred.

The Social Services Department should be notified that the child is thought to be the subject of NAI. The decision whether to inform the Social Services Department is often difficult in cases where NAI is suspected, but the evidence is not strong, and medical staff are reluctant to involve themselves in the prolonged and sometimes painful procedures that an NAI case may entail. However, if in doubt, it is better to inform the Social Services Department, and thus share the responsibility for the child's future safety. Further, if the parents at this, or any previous stage, indicate they might remove the child from hospital, then a social worker should immediately make an application to a magistrate for an Emergency Protection Order.

### Case conferences

Once informed, the Social Services Department will organize a case conference about the problem. This will be attended by representatives of hospital and community medical, nursing, and social work staff, a local authority social worker, the general practitioner, and a member of the police (usually an officer designated to this area of work). Other professionals who have had contact with the family, such as playgroup leaders, child or adult psychiatrists, NSPCC officers, should also be present. If the child is of school age, a representative of the education service (usually an education welfare officer) will be present. Parents are usually informed of the timing of the conference, but not invited to attend, though the conclusions should be discussed with them as soon as possible afterwards. The aim of a case conference is to consider and share all the information available about the child and family, to make an agreed plan to ensure the child's safety and well-being in the future, to identify a 'key worker' to coordinate implementation of the plan, and to arrange a date to review progress (Department of Health and Social Security 1988).

Various decisions need to be made. Should the child's name be placed on the 'Child Protection Register'? If so, this will require a member of the Social Services Department to visit and monitor the situation regularly. Should the child be formally removed from parental care on a temporary and perhaps later on a permanent basis? If so, an application for a care order under Section 8 of the 1989 Children Act must be made to a court. Is the medical evidence of the cause of injury strong enough to stand up in court? Is it safe for the child to go home? If so, what level of supervision needs to be provided and who is to have responsibility for this? Are resources adequate to provide this level of supervision? How is the person taking main responsibility going to keep in touch with other professionals with whom the family may be in touch? Do other professionals, for example a child or adult psychiatrist, now need to be involved in further assessment and care? How are parents to be informed of decisions that have been reached? Usually a 'key worker' is nominated by the conference to ensure that decisions made are carried out.

### Removal from parental care

In assessing whether the child needs to be removed, or whether the risk of returning the child home under supervision can be taken, a number of considerations are relevant (Oates 1984). These include especially evidence of recurrent trauma to the child and sibs, the child's present physical state, the understandability of the act of violence (did it occur when the family was under unusual stress that is unlikely to be repeated?), the general level of parental care quite apart from the abusive episode, the presence of parental mental illness or personality disorder, the personality qualities of the parents and their apparent willingness to work with professionals and others who might be involved in helping them, the degree of responsibility the parents accept for what has happened, and the availability of social work resources to ensure adequate monitoring. In the end, the responsibility for the decision has to be taken by the Social Services Department, but in practice the decision is usually a consensus one.

In a minority of cases it will be apparent at an early stage that the best interests of the child will not be served by a return to the family, as the risk of repeated abuse and serious neglect is too great. In this case, court proceedings are likely to be necessary. The paediatrician will need to assist the court by describing the injuries and the reasons for believing they were not accidentally caused. The social workers and, on occasions, other professionals such as a general practitioner or psychiatrist will provide evidence on the social background, caring capacities, and mental state of the parents. If it is decided the child needs to be permanently removed from parental care, then arrangements will need to be made for long-term fostering or adoption.

## Return to the family

Assuming a child is returned home to the care of parents, either from hospital or after a period of temporary foster care, a programme of management or treatment needs to be agreed. The child's name will need to be placed on the 'Child Protection Register', so that the case automatically receives a certain level of attention and, if the family moves, the Social Services Department in the new area is automatically informed.

The key element to subsequent care will be the monitoring role played by a social worker, who will not only maintain regular contact, but will also attempt to help the family cope with material and other stresses and be available if, at any time, the parents feel they cannot cope and are reaching a situation where violence may occur again. Periods of temporary foster care may need to be arranged.

In addition, if resources are available, other types of intervention, perhaps requiring referral to a child psychiatric department, may be helpful. The parents may be given the opportunity to express their ambivalent feelings towards their child, and to understand how their own childhood experiences have failed to meet their own needs for nurture, and so made it difficult for them to care for their child. They may be helped by behavioural methods to develop skills enabling them to cope with their children's demanding behaviour (Gambrill 1983). Attendance at a day nursery, child psychiatric day centre, or other preschool facility may be arranged.

## Outcome

Children that have been the subject of abuse are at risk for a whole range of subsequent problems. The risk of re-abuse is between 10 and 30 per cent. In some cases this will be fatal. Abused children also subsequently have a high rate of physical problems, including accidents and cerebral palsy. Many will be developmentally delayed and have learning difficulties (Hensey *et al.* 1983; Lynch and Roberts 1982). There is a high rate of subsequent behavioural and emotional problems in childhood and adulthood, and children who have been abused may well have difficulties rearing their own children. Most of these later problems are likely to be the result of chronic emotional deprivation and lack of appropriate stimulation present in the abusive family, rather than of the injuries themselves.

### 1.3.3 CHILD SEXUAL ABUSE

### Definition

Sexual abuse has been defined as 'the involvement of dependent, developmentally immature children and adolescents in sexual activities that they do not fully comprehend, are unable to give informed consent to, and that violate the social taboos of family roles' (M. D. Schechter and Roberge 1976)

Such abuse can be classified according to the nature of the sexual activity:

1. The child may have been assaulted or mutilated in the genital area. This can be seen as a form of non-accidental injury and is dealt with elsewhere in this book (Section 1.3.2).

2. The child may have been subjected to sexual stimulation or intercourse, or been persuaded to stimulate an adult sexually. Such inappropriate sexual activity may take place:
(a) accompanied by violence (rape); or
(b) with the apparent consent of the child. In these circumstances, even though apparently consenting, the child may well comply because of fear.

3. The child may be involved in other sexual activities, for example posing for pornographic photographs or films.

The occurrence of abuse must be seen against the background of normal erotic behaviour and attitudes to nudity and sexuality within families. Studies in both the USA (Rosenfeld *et al.* 1984) and the UK (M. Smith and Grocke 1991) have revealed a very wide range of attitudes and behaviour.

It is important that definitions of abuse do not affect normal, warm physical feelings that parents may have for their children. Further, cultural factors need consideration in defining abuse. For example, ritual circumcision in females is the norm in some cultures and is regarded with abhorrence in others.

## Background

The prevalence of sexual abuse is unknown because it so frequently never comes to light. In the USA it has been estimated that about 19 per cent of girls and 9 per cent of boys are subjected to stressful sexual experiences during their childhood and early adolescence (Finkelhor 1979). In the UK it has been estimated from a survey of late teenagers that about 9 per cent of children have been sexually abused and about 1 per cent have actually had an incestuous sexual experience (Baker and Duncan 1985).

The subjects of sexual abuse are mainly adolescent boys and girls in their early teens. However, when a case involving a family member comes to light in a child of this age, it is often apparent that the sexual activity has been going on for several years. A significant number of younger children are involved, and in a small minority of cases children of preschool age are the victims. Abuse occurs within the family and with strangers in about equal proportions.

Abuse may occur in all socio-economic groups, but is more likely to appear in socially deprived families living in cramped circumstances. Depression in mothers and a history of antisocial behaviour or heavy drinking in fathers is commonly present.

In many cases, when such abuse has occurred, it is between father or stepfather and daughter. However, other family members are not infrequently involved. A minority of cases of sexual abuse involve rape by people from outside the family. Where boys are the subject of sexual activity, this usually takes the form of homosexual assaults by father or stepfather. Boys are less commonly involved in heterosexual activity with their mothers. Older siblings may be involved in sexually abusing their younger brothers or sisters and in some families a child may, at the same time, be both suffering abuse and perpetrating abuse.

It has been suggested (Bentovim 1988) that sexual abuse within the family can best be seen by viewing the family as a system in which each relationship has an impact on others. For example, sexual activity between father and daughter may begin after there has been considerable strain in the marriage, when the father turns away from his wife and looks to his daughter, not only for sexual gratification but also for emotional support. The wife colludes with the development of this alternative arrangement, and may choose to ignore evidence of the sexual activity as it emerges. She may be emotionally distant from her daughter, and communication in the family about feelings is likely to be generally poor. Such families are often rigid and moralistic with a strong taboo on the open discussion of sexual matters. Consequently, they are slow to change. However, the sexual attraction to young children that perpetrators experience may also be understood by considering intra-psychic mechanisms. A perpetrator's erotic or painful experiences of being abused as a child may create a dissociated mental component which may be reactivated in adulthood under particular circumstances.

### Clinical features

Sexual abuse may come to light because:

1. An allegation is made by the victim or someone who knows the victim. For example, the child may tell another child who informs a teacher. A family member, friend, or neighbour may become suspicious and
contact the social services department.

2. The child presents with a behaviour problem or learning difficulty which raises suspicion of sexual abuse by professionals who are consulted. Any type of stress reaction may be present: physical symptoms such as abdominal pain with no organic cause, emotional withdrawal, aggressive behaviour, or difficulties in concentration. These non-specific symptoms should raise suspicion of sexual abuse if no other obvious stress is present. Certain types of behaviour, such as over-sexualized talk or play, compulsive masturbatory activity, an unexplained suicidal attempt, or running away from home, should raise particular anxieties. Refusal to communicate (mutism) and anorexia nervosa can also occur as a response to abuse.

3. Physical signs suggestive or pathognomonic of abuse are present. There may be unexplained scratches, bruising or bleeding in the genital area or a foreign body in the vagina. Recurrent urinary infection may be a sign of sexual abuse. Sexually transmitted diseases such as chlamydia and gonorrhoea are virtually pathognomonic of abuse when occurring in the pre-pubertal child.

### Assessment

This will depend on the way in which the case has come to light. In particular, a doctor, especially a general practitioner or paediatrician, may be the first person to suspect that abuse has occurred, or he might be requested by another agency to confirm or refute the suggestion. In either event it is useful to observe the following principles:

1. A high index of suspicion should be maintained. It is probably unusual for children to lie or fantasize about being sexually abused. Professionals are understandably reluctant to take seriously suggestions that sexual abuse has occurred in an apparently normal family, but this attitude is unhelpful.

2. The comprehensiveness of assessment for CSA should depend on the

level of suspicion. For non-specific problems such as bed-wetting or sleep disturbance, it is not appropriate to undertake specific investigations. On the other hand, when serious suspicion is aroused because of specific concerns (see above), physical investigation together with an interview with the child and parents will be required.

3. Concerns about CSA should always be shared with other professionals. Anything that is more than a fleeting suspicion should be communicated to a colleague so that responsibility for deciding on the next step can be shared. Moderate or serious suspicion needs to be communicated to the social services or social welfare department.

4. Detailed investigations, whether physical or psychological, into CSA are usually unpleasant for the child and family and may be traumatic. In children who turn out not to have been abused, such investigations may come to be seen as a form of 'secondary abuse' and even in children who have been abused, investigation is a serious ordeal. Such investigations therefore need to be carried out by skilful professionals who are experienced in the techniques they are using, the child needs adequate preparation and continuing explanation, and the assessment needs to be carried out in the least-stressful circumstances that are possible.

5. Physical examination of the genitalia. A brief examination of the external genitalia for bruising, scratches, forms part of an ordinary general physical examination. A more thorough investigation to determine the patency of the vulva and anus, with collection of specimens for forensic purposes, needs to be carried out by trained paediatricians or police surgeons in a paediatric setting with the mother or someone else of the child's choice present.

It is important to remember that most sexual abuse, for example licking or fondling of the child's genitalia, or forcing the child to masturbate the perpetrator, will not produce any positive physical findings.

6. Interviewing the child. All family doctors, paediatricians, and child psychiatrists should feel competent to make general enquiries concerning the possibility of CSA by encouraging the child to talk about events that worry them and suggesting there might be problems the child may find it difficult to discuss. However, if responses to such questioning are positive, or if there are other strong grounds for believing the child has been sexually abused, then someone specially skilled in interview techniques will need to be involved.

Skilled interviewing in this area requires probing of the numerous ways in which the child might have been abused without asking leading questions or putting ideas into the child's mind. The child will require reassurance that he/she is doing the right thing by talking about the experiences and will not get into trouble because of it. It is vital to ensure that a complete record is kept of the interview, either by video-recording or by taking down what the child says verbatim.

## Management

Immediate management of a case in which a child has been sexually abused will follow along similar lines to that occurring in physical abuse (Section 1.3.2). In some cases, where the suspected perpetrator is in the home, the child will be admitted to hospital, though this will usually not be necessary if the perpetrator is not living in the home. The Social Service Department should be informed, and a case conference or a co-ordinating meeting called. For success in management, it is vital there should be a co-ordinated plan made between the health personnel (health visitor, general practitioner, paediatrician, clinical medical officer), the Social Services Department, and the police. It is widespread current practice in the UK for offenders to be charged. Minor and first offences are usually dealt with by cautioning, but more serious offences usually result in custodial sentences.

Continuing management will involve an attempt, as far as possible, to support the child within the family (Furniss *et al.* 1984). In a minority of cases, where care has been and promises to be generally inadequate and neglectful, the child will need to be removed from the home to the care of the Social Services Department. In any event the child should be given the opportunity to continue to communicate her feelings about the experience. Depression and guilt are commonly experienced by girls who have been the victims of rape or incest.

If the perpetrator is the father or stepfather, family breakdown is common, but by no means an inevitable outcome. A treatment approach involving initially separate interviews with the family members, then family interviewing, first with father absent, and then with father present, has been claimed to have a positive influence in a significant number of cases. Group therapy for girls who have shared this experience may provide valuable emotional support (Section 6.2.5). Parents whose children have been the victims of sexual assault from someone outside the family also need the opportunity to express their feelings and to be helped to understand their child's subsequent behaviour. Family interviews may be particularly useful in this respect.

## Outcome

No reliable figures are available for personality or marital outcome of children who have been the subject of sexual abuse. There are numerous anecdotal accounts of women who explain their adult sexual difficulties on the basis of such earlier inappropriate experiences. It does seem highly likely that sexual abuse in childhood frequently leads to later psychosocial problems, though there is no reason to think that this is an inevitable outcome (Tsai *et al.* 1979). The opportunity to communicate about the experience and a later relationship with a supportive and understanding adult partner are both likely to be helpful influences.

### 1.3.4 EMOTIONAL ABUSE

**Definition**

Many parents criticize the behaviour, appearance, and other attributes of their children from time to time in a manner that is unnecessarily hurtful. A minority are persistently hostile and rejecting to a degree that impairs the child's emotional development. Such children are said to be subject to emotional abuse. Emotional abuse has been defined in a Department of Health circular as present in 'children under the age of 17 years whose behaviour and emotional development have been severely affected and where medical and social assessments find evidence of either persistent or severe neglect or rejection'. Although the definition refers only to neglect and rejection, emotional abuse can also be said to occur clinically when children are grossly overprotected, and prevented from achieving normal independence.

**Clinical features**

Emotional abuse is often an accompanying feature of other forms of abuse (physical and sexual) and non-organic failure to thrive. A chronic physical illness or handicap may be present. It is also sometimes seen in children with psychiatric conditions, particularly conduct and emotional disorders, and faecal soiling. Very occasionally, it becomes clear clinically that a child without any symptomatology (physical or psychological) is being emotionally abused. Such children do not, strictly speaking, come within the legal definition of emotional abuse, but they nevertheless constitute a group seriously at risk for the development of disorder.

Emotional abuse can be recognized by the constant belittling, pejorative remarks that parents make about the child. There are no positive remarks and, if the child does something praiseworthy, the parents merely make this an excuse for further criticism—'You see, he *can* do it if he tries. The trouble is he never does.' Parents may show cutting humour, especially sarcasm, at the child's expense. Physical punishment may reach worrying proportions, but may fall short of physical abuse. The parents are likely to show impairment of other relationships, often getting on poorly with their own extended families. They may show signs of personality disorder, or other mental health problems such as schizophrenia. Finally they may show unusual beliefs (possibly culturally determined or resulting from mental illness) that damage the child emotionally—for example that the child is possessed by the devil.

The child, if of preschool age, is likely to show poor physical growth and impaired mental development. He may be clinging and show an insecure form of attachment with inability to explore away from his parents. There may be obvious fear of the parents. Older children may show the above features, but are also likely to show some more definite form of psychiatric disorder. Conduct disorders are more common than emotional disorders in this situation. The child's social behaviour may be disinhibited, so that he goes up to people in the street in an inappropriate manner. The child's self-esteem is often low, and he is likely to behave as though his own welfare and well-being are of little significance.

### Assessment

Emotional abuse is likely to be most evident when family members are seen together as a unit. It will be noted that one or both parents is openly and persistently critical of the child, perhaps with constant sarcasm and hostile comments. The child may be scapegoated, with other children in the family having their misdemeanours ignored or made light of. He may appear more disturbed in the presence of his parents than when he is away from them. It is often painful for clinicians to sit and listen while children are being discussed in their presence in such a negative way, but it should be remembered that the process is going on at home the whole time, and seeing the problem enacted before one is a necessary part of assessment.

Parents should also be seen separately, partly to establish the reasons for such negative attitudes, and partly to determine whether they are suffering from any form of mental illness. Parents may disfavour a particular child for a variety of reasons. They may have strongly wished for a child of the opposite sex, the child may be intellectually dull, physically unattractive, or deformed, or may remind the parents of another member of the family by whom they themselves were seriously rejected. For example a boy may remind his mother of her own father who sexually abused her, of an uncle with a criminal record, or, in the case of parental divorce and remarriage, of the biological father whom she now detests. It is important in addition to take a comprehensive history and, in particular, to determine whether the child has a significant emotional or behavioural problem.

Children over the age of 7 or 8 years should, with parental consent, also be seen alone. They can be reassured at the onset of the interview that what they say is confidential (Section 1.2.1). After time spent in discussion of neutral topics, they can be asked about parental behaviour as they see it with such questions as: 'I wonder if you feel dad gets on at you more than he does the others? What sort of things is it that he does that upset you most? How does it make you feel when that happens?'

### Treatment

This will depend particularly on the presence and nature of any psychiatric disorder present in the child or parent, and on the motivation of the parents to receive help. Many severely rejecting parents are unwilling to receive treatment for themselves and other family members. If emotional abuse is present, i.e. the child can be seen to be suffering a serious impairment of development as a result of constant rejection or neglect, then the case should be referred to the local Social Services Department. As in the case of non-accidental injury (Section 1.3.2) this will lead to a social evaluation, and case conferences involving the various professionals concerned, at the end of which the child is likely to be put on the Child Protection Register for regular surveillance or, in extreme cases, taken into care following legal proceedings. Magistrates are, however, often reluctant to make care orders where no sign of physical abuse exists. The decision of a doctor to refer such a case to the Social Services Department is rarely easy, as referral will almost inevitably result in loss of the goodwill of the parents. However, when the child has been clearly suffering over a prolonged period already, and treatment is refused, a firm decision to refer should be taken.

If treatment is accepted, family interviews may clarify the nature of the problem and help the parents to gain insight into their behaviour, with resultant behaviour change (Section 6.2.4). It is often useful to get parents to keep a chart of the positive aspects of their child's behaviour over a period of time, perhaps with a built-in reward system for the child (Section 6.2.2). Specific problems in the child and failure to thrive may require specific therapies. It is particularly important to give attention to the ways of improving the child's self-esteem both within the family and at school, as this may have been severely affected.

### Outcome

This will depend largely on the nature of the child's psychiatric problems, the presence of personality disorder or mental illness in the parents, and the motivation of the family to receive help. In general, families where parents are motivated to receive help do better, not necessarily because of treatment, but because perceiving a need for help is itself a good prognostic sign. Children showing conduct disorders in a strongly rejecting family are likely to do particularly poorly.

### 1.3.5 MUNCHAUSEN'S SYNDROME BY PROXY (ALSO KNOWN AS POLLE SYNDROME, MEADOW'S SYNDROME, FACTITIOUS ILLNESS)

### Definition

This is a condition in which symptoms or signs of illness in children are deliberately fabricated or induced, nearly always by the mother. There may be no underlying physical condition or, alternatively, a physical illness may be present but its manifestations grossly exaggerated (Meadow 1982). Minor exaggerations of illness by parents are common. This condition can only be regarded as present if parental behaviour is persistent over time and results in social and/or physical disability in the child.

### Clinical features

Children involved are usually of preschool age, though occasionally a little older. The child may be reported to have symptoms that in fact never occur. Epileptic fits and recurrent drowsiness are common examples. Alternatively, signs of illness may be artificially induced. Blood may be put in the child's urine or faeces by the mother using her own blood obtained by finger-prick or from menstrual products. The child may be covertly administered hypnotics, tranquillizers, or other medication and become periodically drowsy or comatose. Administration of salt or diuretic medication has produced bizarre biochemical syndromes. Artificially induced rashes and fevers have been reported. Some children are put on highly restrictive diets sufficient to produce malnutrition for imagined overactivity or other behaviour problems.

The conditions produced in this way naturally present considerable diagnostic difficulties. Extensive investigations on an in-patient basis usually take place, and rare diagnoses may be suggested. Eventually, it is realized that the symptoms only occur when the mother (or occasionally the father) is present. It will also be noted that the mother hardly ever appears to leave the child's side, and resents attempts to encourage her to do so. At this stage there is often great reluctance on the part of medical and nursing staff to accept the possibility that the illness is factitious. Considerable ingenuity may be required to confirm the diagnosis by observing the mother produce the symptoms, and indeed this may turn out to be impossible, in which case the diagnosis remains presumptive.

A number of case reports have appeared. The following features appear characteristic:

1. The perpetrator is nearly always the mother, but very occasionally fathers have been involved.

2. A previous child may have suffered or even died from an undiagnosed or very unusual condition.

3. The mother has often obtained medical information earlier in her life. Usually this involves a period of nurse's training.

4. Mothers do not usually have overt psychiatric illness, although a minority show personality disorders with widespread maladaptation in a number of areas of their lives. Usually, however, their abnormal behaviour towards the child appears to be an isolated psychopathological feature.

### Assessment

This will largely depend on the manner of presentation, but there are certain common features (Meadow 1985).

1. A high level of suspicion is required whenever a condition presents

which is unusual, only occurs when the mother is present, or for which the only evidence is based on maternal report.

2. Once suspicion has been aroused, there should be no further complex physical investigations. These may in themselves have an element of danger.

3. Initial emphasis should be placed on careful recording of symptoms and checking of signs in relation to the mother's presence, and thought should be given to possible mechanisms whereby the symptoms might be artificially produced. Efforts should be made to check the mother's account by obtaining independent evidence from father, other relatives, nursery school staff, general practitioner, or health visitor. This should be done tactfully, without raising the mother's suspicions. The mother may also be fabricating about other areas of her life.

4. An attempt should be made to separate the mother from the child to assess the effect on the child's symptoms. This may be resolutely refused by the mother, but is sometimes possible.

5. A psychiatric assessment may be helpful. This can sometimes be introduced to the parents as necessary to investigate possible behaviour or emotional problems in the child that may indeed be present. Such assessment may reveal the presence of parental psychopathology, and may also uncover the reasons for the mother's behaviour. One common pattern is that the mother has herself suffered chronic deprivation in childhood, and now seeks from caring nurses the sympathy and interest that having a sick child elicits.

6. In some circumstances, forensic pathological laboratory investigations or even police surveillance may be indicated. Open or covert observation of the mother's behaviour may be required.

### Treatment

**Confrontation.** As soon as the diagnosis has been sufficiently established such that court proceedings to protect the safety of the child could be successfully contested, the mother or both parents should be confronted with the diagnosis and the evidence to support it. A sympathetic rather than accusatory attitude should be taken up by the doctor (preferably the senior paediatrician involved). At the time of confrontation, help should be offered to the family. The father should at this point be involved in the discussions if he has not been previously involved.

**Immediate protection of the child.** In all circumstances the local Social Services Department should be informed that the child is at risk. The family may already be known to this Department. If the parents attempt to remove the child from hospital, an Emergency Protection Order should be obtained. In any event, a case conference attended by the various professionals involved will be needed in order to plan for the future safety and care of the child, and allocate responsibility for further supervision of the family.

**Further care.** This will follow lines similar to that described in cases of non-accidental injury (Section 1.3.2). A decision will have to be taken whether the child is to be allowed home to the care of his parents. Similar criteria should be employed as described with non-accidental injury. Family interviews, parental counselling, individual psychotherapy, and behavioural methods may all have a part to play (Nicol and Eccles 1985). Consideration should be given to the child's perception of his condition and the way it has arisen. As the child gets older he will need to have the opportunity to express his feelings and beliefs about his condition, and, in some circumstances, though probably not in all, he may need to be given a full explanation.

### Outcome

Systematic follow-up studies have not been reported. It is likely that there exist a number of undetected cases, in which a child is permanently handicapped by a belief that he is physically ill, when, in fact, he has no physical cause for his symptomatology. In diagnosed cases a better outcome is likely where the parents are prepared to acknowledge their responsibility and are motivated to receive help. A poor outcome is probable where there are insufficient grounds to remove the child from the home, but parental care is inadequate and the pathological mother–child relationship remains unresolved.

## Further reading

Bentovim, A., Elton, A., Hildebrand, J., Tranter, M., and Vizard, E. (ed.) (1988). *Child sexual abuse within the family*. Wright, London.

Cox, A., Hopkinson, K., and Rutter, M. (1981). Psychiatric interviewing techniques. II. Naturalistic study: eliciting factual information. *British Journal of Psychiatry* **138**, 283–91.

Graham, P. (1979). Epidemiological approaches to child mental health in developing countries. In *Psychopathology and youth: a cross-cultural perspective* (ed. E. F. Purcell), pp. 28–45. Josiah Macey Jr. Foundation, New York.

Graham, P. (1986). Behavioural and intellectual development. *British Medical Bulletin* **42**, 155–62.

Kempe, C. H. and Helfer, R. E. (ed.) (1980). *The battered child*, 3rd edn. University of Chicago Press, Chicago.

Rutter, M. (1989). Isle of Wight revisited: 25 years of child psychiatric epidemiology. *Journal of the American Academy of Child and Adolescent Psychiatry* **28**, 633–53.

Sattler, J. M. (1989). *Assessment of children's intelligence and special abilities*, 3rd edn. Allyn and Bacon, New York.

# 2

# Development and its disorders

## 2.1 Fetal and infant development

### 2.1.1 PREGNANCY, DELIVERY, AND THE PERINATAL PERIOD

**Introduction**

The satisfactory growth and development of the fetus depends on the presence of:

1. Normal genetic potential in the fetus.
2. A uterine environment that provides the nutritional and other physiological requirements of the fetus.
3. Absence of harmful agents or circumstances that might affect the fetus, for example infection, toxins, trauma, interruption to the blood supply, etc.

During the pregnancy, parents, and particularly the mother, experience complex psychological change. Adaptation to parenthood can be seen as a 'transitional experience' in which men and women come to see themselves taking on new responsibilities, as well as developing love and attachment to their new baby. Changes in the marital relationship commonly occur, with women sometimes becoming preoccupied with their baby-to-be and withdrawing from their husbands. A high proportion of pregnancies are unwanted in the early months, but most mothers are looking forward to their baby by the time of birth, though even at this time ambivalent feelings are not at all uncommon.

Psychosocial factors impinging on the mother-to-be may have a major indirect effect on fetal development because they may result in the fetus being exposed to harmful physical influences. The existence of direct effects on the fetus occurring as a result of physiological changes caused by stress on the mother is less certain. Abnormalities occurring in the perinatal period may have a direct effect on the brain functioning of the fetus. They may also produce long-lasting indirect effects by their impact on parental attitudes—for example, by making mothers much more anxious about their children and overprotective towards them, perhaps over many years.

**Psychosocial factors with indirect effects on fetal development**

*The emotional state of the pregnant mother-to-be*
Persistent negative attitudes to the pregnancy may be associated with high rates of certain complications of pregnancy. Essential hypertension occurring

during pregnancy has been linked to the occurrence of depression and ambivalent feelings in the pregnant woman (Wolkind 1981*a*). By contrast, hypertension occurring in later pregnancy is not associated with these emotional problems and is more likely to be physiologically determined. Vomiting in early pregnancy is not associated with neurotic difficulties, whereas persistent vomiting throughout pregnancy appears to occur more frequently in women living in deprived social circumstances with inadequate social support. A review of studies of nausea and vomiting in pregnancy (Macy 1986) concluded that causation was often uncertain, but that an ambivalent attitude to the pregnancy was the most common causative factor.

Minor psychological and psychosomatic symptoms are common in pregnant women and their partners. Although some women experience an improvement in their sense of well-being in pregnancy, the reverse is more common. In one study, Condon (1987) only 15 per cent of women felt better than usual during pregnancy, whereas 50 per cent felt worse. Prospective fathers also commonly experience psychological change during their partner's pregnancy. In the same study 10 per cent reported an improvement in well-being compared to 20 per cent who sensed a deterioration.

*Psychiatric disorders in pregnancy* consist mainly of anxiety and depressive states. Between 5 per cent (J. L. Cox *et al.* 1982) and 25 per cent (Kumar and Robson 1978) of pregnant women have been found to show such conditions. Palpitations, free-floating anxiety, chronic feelings of tension, and emotional lability at a level not amounting to psychiatric disorder are even more common. Mothers who are anxious during their pregnancies may be likely to have babies who, in the first few weeks of life, are irritable and cry easily: the mechanism is uncertain (Ottinger and Simmons 1964). Depression in pregnancy may occur as an isolated episode, or merely reflect predisposition to mood disorder before the pregnancy that is likely to persist after it.

Pregnant women who are depressed are likely to neglect their physical condition, eating irregularly and perhaps becoming overweight. They are less likely to attend antenatal check-ups regularly. Consequently their babies are more likely to be at risk for perinatal hazards, that may themselves produce psychological sequelae in the children later on.

### Smoking, alcohol consumption, and drug addiction in pregnancy

These substances are serious hazards to pregnancy, especially if usage is heavy. Babies born to moderate or heavy cigarette smokers are, on average, about 200 grams lighter than those born to non-smokers, after social differences have been taken into account (Butler and Alberman 1969). Working-class women are more likely to smoke as are women with psychiatric disorders. However, smoking has a clear, independent effect on birthweight even after socio-economic factors and psychosocial stress have been taken into account (Brooke *et al.* 1985). There are effects of smoking on later development, with babies of mothers who smoke having lower developmental quotients. This is probably mediated by effects on fetal blood supply affecting nutrition and brain development.

Heavy alcohol consumption in pregnancy results in babies born with the so-called fetal alcohol syndrome (Cooper 1987), a condition in which there is pre-natal and/or post-natal growth retardation, central nervous system involvement evidenced by developmental delay and characteristic facial dysmorphology, especially microcephaly, microphthalmia, or short, palpebral fissures, a poorly developed philtrum (median groove between the upper lip and nose), and a thin upper lip (Rosett *et al.* 1983). The prevalence of fetal alcohol syndrome is thought to be around 1 per 1000. Average reported IQ is around 68, but there are no specific cognitive deficits. Short attention span and learning difficulties are prominent in the school situation. It is likely that the mental retardation and attentional difficulties are at least partly organically determined, although the fact that many of these children live in deprived circumstances often makes it difficult to disentangle causative influences. The effect on the fetus of lesser amounts of alcohol consumption is uncertain. At this stage of knowledge, it seems improbable that consumption that involves a glass of wine a day or less is harmful (Smithells and Smith 1984), but higher levels of intake, especially before the third trimester may produce intrauterine growth retardation (Rosett *et al.* 1983).

A variety of drugs administered to the mother during pregnancy may affect the fetus (Johnstone and Forfar 1984). In particular, lithium can produce congenital cardiac abnormalities. It is however sensible to avoid the use of all inessential psychotropic medication in the first three months of pregnancy. While sporadic reports of fetal malformations occurring with use in early pregnancy of a variety of psychotropic agents are of only anecdotal value, it does not seem reasonable to take risks by using drugs with uncertain therapeutic benefit. In a woman planning a pregnancy, it also seems sensible to avoid the use of medication in the preconceptual period as soon as she ceases to use birth control.

Withdrawal effects in the newborn baby may occur if the mother is addicted to amphetamines, barbiturates, heroin, methadone, and morphia. Withdrawal effects from alcohol are less common. Irritability and hypertonicity are the most prominent symptoms in the neonate born to an addicted mother. The management of withdrawal effects in the addicted neonate requires the skills available in a neonatal unit.

## Direct effects of physiological responses to maternal stress on the fetus

Fetal heart rate increases if pregnant women experience fear, but there is no good evidence that stress in itself affects fetal development. Indirect effects through an influence on smoking and alcohol intake are much more likely (Newton 1988). There is also no evidence that mothers experiencing stresses in pregnancy are more likely to produce babies with malformations, nor that major life events can, except indirectly, trigger premature births. There may, however, be an effect on the later development of behavioural problems in the children. Maternal anxiety has also been linked to the later recurrence of pyloric stenosis (Dodge 1972). The evidence linking stress in pregnancy to later problems in children does not suggest that major delayed effects are at all common.

## Effects of low birthweight on later psychological development

There have been many follow-up studies of low-birthweight babies to assess whether later impaired psychological development is linked to the low birthweight itself, or is produced by premature birth or intrauterine growth retardation. The results are difficult to interpret for a variety of reasons:

1. Low-birthweight babies come from low social class groups more often than one would expect by chance. Consequently lowering of intelligence may be due to social rather than physical factors.
2. The low birthweight may itself have been caused by an abnormality in the baby, and this abnormality may be responsible for poorer intellectual performance later on.
3. The quality of care for babies born too small is improving year by year. Consequently, results obtained a decade or more ago may not be accurate reflections of the present situation.

Despite these problems, some reasonably firm conclusions can be drawn.

1. If the birth is otherwise uncomplicated, and treatment for respiratory problems or convulsions has not been required in the neonatal period, then the outcome of low-birthweight babies is as good as normal-birthweight babies from similar social class backgrounds. The presence of respiratory distress (often associated with haemorrhage into the periventricular region) indicates a poor outcome in those babies who survive.
2. The outcome of very low-birthweight babies weighing less than 1.5 kg (or in some studies 1.25 kg) at birth, although also generally good, is somewhat more problematic. They are more likely to show minor neurological signs, lower intelligence quotients, and more behavioural difficulties than matched controls when examined shortly after school entry (Marlow *et al.* 1988). However, the mean IQ of these children is well within the average range. Although it is possible to predict the development of later neurological abnormalities from ultrasound investigation using technology available in the early 1980s, it is not possible to predict cognitive abnormalities (Costello *et al.* 1989).

3. Babies suffering intrauterine growth retardation and born at a weight significantly lower than that predicted for their gestation (i.e. those babies who are 'light for dates') have a somewhat poorer outcome in terms of later behaviour and learning problems than those who are born early but at a weight appropriate for their gestation (Neligan *et al.* 1976). However, the evidence for this is not clear-cut (McBurney and Eaves 1986).

4. The great majority of babies born even at very low birthweight do very well subsequently from a psychological point of view, providing they do not show evidence of cerebral palsy, blindness, or other gross physical defect.

## Effects of perinatal complications other than low birthweight on later psychological development

The impact of other complications of pregnancy and delivery on the psychological development of the child is often difficult to evaluate. In general, a history of a single adverse event, whether it be haemorrhage in early pregnancy, toxaemia of pregnancy, a breech delivery, or a caesarean section, is usually of little significance in the appraisal of a child with a learning or behavioural problem (Davie *et al.* 1972). If, however, multiple complications have occurred, or a single complication (such as traumatic birth) followed by a major neonatal problem (convulsions, respiratory distress, etc.), then this may well be of significance as an indicator of underlying brain dysfunction. Single pregnancy complications (such as early haemorrhage) may, however, be of considerable significance if associated with intrauterine growth retardation and a baby born small-for-dates.

Babies with multiple perinatal complications born into families of low socio-economic status are more likely to show psychological deficits later on than babies with similar obstetric problems born into more socially advantaged families. Thus there is an interaction between the effects of birth complications and social factors (Werner *et al.* 1967).

## Psychosocial aspects of the neonatal period

Psychologically, the first month of life is often a period of great emotional intensity for the parents, especially for the mother. The establishment of a satisfactory feeding pattern, the development of a mutually rewarding relationship with the new baby, and adjustment to a changed sleeping pattern are all tasks that must be mastered over a relatively short period of time.

Problems in the neonatal period may occur as a result of factors affecting the baby or the mother. Unsatisfactory and insensitive professional care may also produce problems. These factors may interact so that, for example, depression in the mother may elicit unsympathetic or negative attitudes from professional staff who would otherwise have been more positive in their behaviour. Cumulative disadvantage may thus occur.

*Factors affecting the baby*
**Physical illness.** The baby may suffer from a variety of physical conditions, including apnoeic attacks, respiratory disorders, vomiting, jaundice, bleeding conditions, convulsions, and metabolic disorders. Any of these will require specialist assessment and treatment. The management even of those conditions with a benign outlook will inevitably interfere at least to some degree with the development of the mother–child relationship.

**Neurobehavioural factors.** Hyperexcitability and irritability, as well as apathy and somnolence in the newborn baby are common after a traumatic birth, or a birth in which the mother has been heavily sedated during labour. They may also occur for other reasons, presumably physiological, which prove impossible to identify. In any event, such behavioural patterns may make the establishment of the early mother–child relationship more difficult.

*Factors affecting the mother*
**Physical illness** in the mother in the immediate postpartum period is not common, but may occur. Breast abscess, gynaecological complications of the birth, and exacerbations of problems associated with pre-existing physical illnesses such as hypertension and diabetes mellitus are perhaps those most commonly experienced. If the mother needs specialist treatment in hospital, this will threaten the early development of her relationship with her baby.

**Emotional problems.** Transitory mood disturbances (postnatal 'blues') affect about 50 per cent of women in the immediate postpartum period. Crying, emotional lability, and feelings of ambivalence towards the baby lasting two or three days may occur. Puerperal depressive states occur in about one in eight mothers. These are likely to occur in the first month or so after the birth (J. L. Cox *et al.* 1982), especially in mothers who are under stress for other reasons. There is, in fact, no evidence of a rise in non-psychotic psychiatric disorder in women in the year after childbirth (P. J. Cooper *et al.* 1988), but when such disorders do occur, symptomatology often focuses around the baby. Excessive anxiety about the baby, self-blame, irritability, sleep problems, suicidal thoughts, and fear of harming the baby are prominent features. Such conditions usually respond fairly quickly to counselling and antidepressant medication and, if reasonable social support is available from husband, neighbours, and friends, separation from the baby is not usually necessary. Untreated, these disorders are likely to last several months.

Puerperal psychoses of schizophrenic, schizoaffective, and affective type, in which there are delusions, often involving the baby as well as paranoid and other bizarre ideas, are often associated with inappropriate mood. They occur in about 1 in 500 women in the postpartum period, when there is a rise in the rate, though not a change in the type of psychotic disorders (Kendell *et al.* 1981). Separation is often necessary for the safety of the baby, but this may be avoided if strict supervision is available in a specialist unit for mentally ill mothers and their babies. Successful intensive community nursing care at home has also been described, and found to be suitable for all but actively suicidal or infanticidal women, provided they are not living alone and are within easy reach of hospital (Oates 1988). Prognosis for the acute phase of puerperal psychotic disorders is good, but about one in seven mothers relapse following a further pregnancy. Psychotropic medication for mothers who are breast-feeding needs careful monitoring for its effects on the baby. Chloral, chlorpromazine, diazepam, and lithium prescribed to nursing mothers, have all been detected in breast milk. If the baby appears drowsy, then the mother's medication should, in any case, be reviewed.

*Factors in obstetric and neonatal units*
The policies of obstetric units vary widely in the degree to which they foster or hamper positive development of the early mother–child relationship. Encouragement of early mother–infant contact, facilitation of breast-feeding, awareness of and sympathy for maternal anxiety, and a willingness to appreciate that normal babies vary widely in their early behaviour are all factors likely to promote good early relationships.

The development of attachment is discussed in more detail when early social relationships are considered (Section 2.1.2). The first few days of the child's life have sometimes been suggested to be a vital period in the process of early 'bonding' between mother and child, upon which the success of later attachment depends (M. Klaus and Kennell 1976). In fact, attachment is a continuous process, and there is no good evidence that babies who, for one reason or another, are deprived of satisfactory early opportunities to develop relationships with their mothers in their first few weeks, cannot develop good relationships later on both with their mothers and with other adults and children (Sluckin *et al.* 1983). Nevertheless, there seems every reason why policies in obstetric and neonatal units should be carefully considered to encourage mutually rewarding mother–infant contacts in the first few days of life (see below).

## Psychosocial aspects of health care in pregnancy and the neonatal period

*Care in pregnancy*
Pregnancy represents the start of life for one individual and a transitional phase in the life of others, especially the prospective parents and sibs. A problematic pregnancy can set the scene for later psychosocial difficulties. However, there is much scope for preventive activity by professionals (Newton 1988).

Women are more likely to produce healthy babies if, during their pregnancy, they receive regular and sympathetic antenatal care. **Continuity of care** is a considerable advantage—mothers who see different doctors and midwives at each visit to a clinic are unlikely to receive very personalized attention.

Virtually all mothers have anxieties about whether they will produce a normal baby. **Sympathetic handling of anxiety**, with regular, prompt feedback of information concerning the progress of the pregnancy, especially after a complication or an investigation such as a scan, will reduce unnecessary worrying.

Advice concerning smoking and alcohol intake will alert mothers-to-be of the dangers of heavy smoking and alcohol consumption. Such advice should, however, be linked to practical help, especially to reduce smoking (Section 3.3.2). Early identification of heavy illicit drug use (often difficult because of poor antenatal attendance) will allow preparation for treatment of withdrawal effects in the newborn baby.

**Identification of depressive states** in pregnant women is of importance so that appropriate treatment can be provided to reduce distress. Women who have suffered such states before the pregnancy began are more likely to show continuing problems after the baby is born, so these should receive special attention. Practitioners should be aware that the presence of a depressive state is one reason for non-attendance at antenatal clinics. Careful follow-up of non-attenders, if necessary by domiciliary visits by the primary health care team, is therefore indicated.

Women may be vulnerable for reasons other than poor mental health. They may be socially isolated, in conflict over continuing their careers, in marital conflict, or fragile following a previous perinatal loss. Women who have previously suffered a loss of this type are understandably unusually anxious about their present pregnancy, though not generally anxious or depressed (Theut *et al.* 1988). The provision of extra social support for vulnerable women in pregnancy may well improve the psychosocial outlook and there is some evidence that it may improve birthweight (Oakley 1988).

Identification of mothers-to-be who are unlikely to be able to care for their babies after birth, because of severe mental illness or personality problems should also be undertaken. If this is a real possibility, early contact with the Social Services Department needs to be made, so that discussion with the parents can, when this is necessary, helpfully lead to rapid permanent foster or adoptive placement of the baby shortly after birth.

*Postnatal care*
The policies of neonatal units caring for mothers and babies in the immediate postnatal period should encourage early mother–child contact, while

taking account of the mother's need to rest and recuperate after a physically and emotionally exhausting experience. Arrangements for caring for babies of low birthweight or with other physical complications should not involve more separation than is absolutely necessary. Mothers of sick neonates interact less with their babies than mothers of normal children, unless special measures are taken to stimulate interaction (Minde *et al.* 1983). If the baby has to stay in a special care unit, then mothers should be encouraged to room-in and visit regularly, and to feel they have a valuable part to play in the care of the baby. Wolke (1987) has suggested a series of measures that can be taken to improve the environment for the sick infant in a special-care baby unit. These include careful observation of the individual infant to detect reactions to noise, light, handling, and attempts at social interaction. There should be an individual care plan with readiness to change the pattern of care as the baby matures. In general, parents are able to cope with the uncertainties about their fragile baby if they are kept well informed of changes in the baby's state and expectations for the baby's future as they occur. Other measures that can help parents include encouraging them to use soothing techniques to calm their baby when agitated, to keep records of their baby's progress, and to have access to a picture gallery of previous graduates of the unit with before and after photographs.

Many mothers, especially those with first babies, feel isolated and incompetent in the first few months of their baby's life. Regular visiting from health visitors, with emotional support and practical help, can reduce distress. In addition, mothers often feel benefit from attendance at groups run by voluntary bodies such as the National Childbirth Trust, or at general practitioner health centres.

Although the occurrence of transitory mood states is of only trivial significance, there is a real need for early identification of puerperal depressive disorders, so that appropriate treatment can be provided (see above). Identification of the early signs of puerperal psychosis should lead to immediate psychiatric referral.

Babies at risk for later abnormality by virtue of their perinatal history should be carefully followed up. In the absence of overt signs of brain damage, most such babies will develop normally, but follow-up will result in early identification of those with a less satisfactory outcome. Parents are, in any case, likely to be anxious about the development of such babies over the first couple of years, and regular monitoring is likely to reduce rather than increase their anxiety.

Finally, those involved in perinatal care should identify babies at serious risk for later abuse. In the perinatal period, it is possible to predict with some accuracy the occurrence of later maltreatment (Leventhal 1988). However, the modest sensitivity and specificity of such prediction means that the impact of professional intervention to 'high risk' babies on the frequency of later abuse is rather slight. Nevertheless, it is clear that young, inexperienced parents who have premature babies or babies with other early problems in development are at particular risk, especially if they have a history of psychiatric disorder or antisocial behaviour. In these circumstances, assuming the parents are agreeable, involvement of the Social Services Department will be indicated. If the baby is thought to be at definite risk, the Social Services Department should, in any case, be informed and a conference arranged (Section 1.3.2).

## 2.1.2 ATTACHMENT

The development of relationships can best be understood using attachment theory (Section 1.1.1). In this section the processes whereby attachment occurs are described.

### The attachment of family members

Parents, older sibs, and grandparents, especially if they are living with the family, will develop fantasies about the sex, appearance, and personality of the unborn child during the pregnancy. The appearance of the child and his early behaviour will dispel many of these fantasies, which will be replaced by perceptions based on reality. Intensity of feelings will, however, be heightened. In general, most family members, especially the mother who has had so much more close and physical contact with the child than the rest of the family, and who has undergone the usually intensely painful experience of giving birth, will rapidly develop strong unconditional positive warm feelings towards the child. These are sometimes mixed with ambivalence or even negative feelings, though normally positive feelings predominate.

Linked to these positive feelings are a number of other phenomena, all of which are most likely to be most strongly experienced by the mother, especially in the first few months of life. Family members are likely to show:

1. *Strong protective feelings towards the child*, especially if the child has been in danger during pregnancy, has been born after a long period of infertility, has been born prematurely, or has suffered physical illness.

2. *A need for proximity to the child*. This is especially likely to show itself at night, when, in the early months, the parents may wish to have the child sleeping in the same room.

3. *Exclusion of other relationships*. Not only may the mother pay less attention to the father and to other children, but older sibs may themselves wish to play less with their friends while the new baby is very young.

4. *Empathic feelings with the child*. A tense, anxious child is, for example, likely to make other family members feel the same way. This is not just a reaction to the child but a sharing of the child's feelings.

These feelings, linked to the attachment process, and experienced by family members, will be present in varying degrees of intensity throughout the child's early years. As the child gets older and takes his place in the family, they are likely to remain present but to a diminished degree.

## The attachment of the child

The process of attachment of the child to other family members is governed by the child's level of perceptual and other abilities, and the way in which these develop. The child's contribution to the development of social interaction is of great significance. These abilities develop at a much faster rate than has previously been realized. For example, in the first few days of life the child will distinguish his mother from other people by smell and by the sound of her voice.

Evidence that the child is becoming attached to other family members is shown by:

1. *Recognition of other family members as special people.* Recognition of family members as different from each other and different from outsiders is accompanied by differences in the child's behaviour towards the various people concerned. In particular, in the first 6 months of life, the child may smile differentially and may cry when some people, but not others, pick him up. Subsequently, from about 7 or 8 months to 3 years, the child usually goes through a phase of acute anxiety over separation from familiar figures. This anxiety will gradually diminish in intensity, but most children continue to experience unpleasant effects when they are separated from their families and, of course, separation from loved figures remains difficult for most people throughout their adult lives.

2. *Expression of especially intense feelings towards family members.* This is likely to be shown by demands for kisses and cuddling, but also by the fact that when the child is frustrated by family members, protest is likely to be more intense than with other people.

3. *Expectation that the family members will meet all needs.* The child will, in the first few months of life, turn automatically to his mother for food, warmth, and comfort in distress. In the second and third year of life, the child will often 'test out' the capacity of members of his family to be all-providing by showing difficult behaviour. Thus the child learns that parents will continue to provide for him even when under testing circumstances.

4. *Empathy with the feelings of other family members.* Especially in the first few months of life, just as the parents, especially the mother, are likely to be directly responsive to the mood of the child, so the child is likely to be directly responsive to the mood of parents. Children can be observed, for example, to become tense when their mothers are tense.

## Attachment interaction between child and family members

Attachment is shown less by the individual behaviour of the child and other family members than by the development of interactions between them. Characteristically these are likely to occur especially between mother and child in the first few months of the child's life. These interactions can be summarized as the establishment of:

1. *Mutually satisfactory biological rhythms.* In the early days and weeks of the baby's life, its sleep needs will probably differ from those of the mother. Normally babies of a week old sleep about 17 hours a day, but waking occurs at 3- to 4-hourly intervals. By three months, most babies are sleeping for six continuous hours at night. Mother and baby (if the child is breast-fed) and parents and baby (for the bottle-fed) will by this time usually have adjusted their sleep patterns to each other's needs, though dissonance of sleep rhythms may mean that this continues to be an exhausting time for parents. If the baby is breast-fed, again a rhythm is set up which meets the needs of both mother and baby for frequency and quantity of feeds.

2. *Bodily interplay.* Family members and the baby will gradually establish mutually satisfactory patterns of behaviour, involving holding and cuddling, to which even the young baby, by moulding his body and later by putting his arms up for holding, will contribute. Both parents and baby engage in mutual *en face* gazing and smiling at each other, again from the early days of life.

3. *Communicative interplay* (Brazelton *et al.* 1974). Observational studies using split-screen video techniques have established the way in which, from the first few weeks of life, vocalizations of mothers and babies complement each other. Although formed expressive single-word utterances do not normally occur much before the age of one year, parents and babies will normally have established a system of preverbal communication long before this time. The baby will be communicating its needs for being held, for food, warmth, and dryness, and the parents their own needs for body contact, talking, and indeed gossiping with the baby.

## Factors affecting the development of attachment

### Factors within the child

1. *The developmental maturity of the child.* In the absence of specific defects such as autism (Section 2.8.1), the child's level of social maturity is likely to be roughly similar to his development in other respects, so that mentally retarded children will be socially immature.

2. *Temperament.* A number of specific temperamental characteristics (Section 2.14.1) are of relevance. A temperamentally anxious child will maintain close proximity to the parent. This may result in the parent

becoming irritated and rejecting. A child with strongly positive emotional responses—a ready smile and a sunny disposition—may elicit strongly positive emotional responses in return. Twin studies reveal evidence of genetic factors in the infant's capacity to attune to its mother's affect (Szajnberg *et al.* 1989).

3. The presence of *sensory defects* (such as impairment of hearing or vision) or *other physical problems* affecting the child's capacity to respond to signals put out by the parents.

*Factors within family members, especially parents*

1. *The wish for the child in the first place.* The birth of many children is unplanned, and the discovery of pregnancy greeted with dismay. In most cases, as the pregnancy proceeds, much more positive feelings are experienced. In a minority this does not happen. It has been shown that when mothers are refused abortion and such pregnancies are carried to term, a high proportion of the children develop behaviour and learning problems (Forssman and Thuwe 1966).

2. *Parental personality, physical and mental health.* The presence of a warm, outgoing personality, good physical health, and the absence of parental mental health problems such as depression and anxiety, will facilitate attachment. Infants are acutely sensitive to the affective behaviour of their mothers (J. F. Cohn and Tronick 1989).

3. *Behaviour of older brothers and sisters.* This can facilitate or impair attachment in a variety of ways. An unduly jealous older child may, for example, distract parents from their younger developing baby. The behaviour of older sibs at the birth of a baby is dependent on a variety of factors (Section 1.3.1).

4. *Quality of family relationships.* Parents who find each other's company mutually rewarding will find it easier to relate to their children.

5. *Living conditions.* Attachment is more likely to occur when parents are not pressed by overcrowded living conditions and financial stress. However, very satisfactory attachment can occur in families living in materially deprived circumstances, providing that there are good quality family relationships.

Characteristic patterns of secure and insecure attachment shown by infants and young children are outlined in Sections 1.1.1 and 2.1.3, where attachment theory is described.

## 2.1.3 ATTACHMENT DISORDERS

These can be classified as:

1. Primary attachment disorder (Section 2.1.4):
   (i) reactive attachment disorder;
   (ii) disinhibited attachment disorder.

2. Other disorders in which disturbances of attachment play a major role:

   (i) preschool behavioural adjustment disorder (Section 2.1.5);
   (ii) conduct disorders (Section 2.10.2);
   (iii) preschool anxiety and depressive disorders (Section 2.9);
   (iv) elective mutism (Section 2.5.4);
   (v) school refusal (Section 2.9.3).

Disturbed social relationships are prominent in many other psychiatric disorders, but are of central relevance in those listed above.

## 2.1.4 REACTIVE AND DISINHIBITED ATTACHMENT DISORDERS

### Definition

Conditions in which the child, despite adequate intelligence, has failed to make specific attachments to a small number of caretakers, and instead shows either a complete lack of interest in initiating or responding to other children and adults, or indiscriminate but superficial attachment to any adults with whom he comes in contact.

### Classification

In the ICD-10 (World Health Organization 1990) classification, these two types of attachment disorder are classified separately as two forms of disorder of social functioning with onset specific to childhood and adolescence. In the DSM III-R classification, both types are termed 'reactive attachment disorder of infancy or early childhood' and form part of a group of miscellaneous conditions. In this scheme, the presence of grossly pathogenic care is a necessary condition for diagnosis.

### Clinical features

In the reactive type, the child looks miserable and frightened. Self-injury is common, and the child may refuse to eat, resulting in failure to thrive. The child shows no interest in making age-appropriate attachments to adults or other children.

In the disinhibited form of attachment disorder, the child is unconcerned who is looking after him and approaches other adults and children in an unduly friendly way. The relationships so formed are superficial, and the child does not become sad or worried if his caretakers change. Such children are often very popular on hospital wards, where they make friends with nurses, and appear to cope with brief admissions extremely well.

Overactivity, short attention span, emotional lability, poor tolerance of frustration, and aggressive behaviour are common accompaniments. A sharp distinction between the hyperkinetic syndrome (attention deficit disorder), conduct disorder, and this condition is often not possible. Learning difficulties often arise which are usually at least partly secondary to poor concentration.

The child has usually been reared in his or her early years in a series of institutions and/or foster families, or has been seriously neglected in his own home. There may be a history of recent abuse, and this is more likely if the child is hypervigilant and watchful. Very occasionally, the clinical features occur in children reared in normal families who have sustained brain damage usually as a result of head trauma or cerebral infection.

## Assessment

Direct observation of the child's behaviour towards the clinician is often sufficient to suggest, if not to confirm, the diagnosis. Assessment should include a consideration of the quality of care provided by the present caretakers.

Attachment disorders must be distinguished from pervasive developmental disorders in which the child has little capacity for social reciprocity, fails to improve significantly when placed in a normal social environment, and shows associated characteristic language disorder, as well as stereotypies or mannerisms.

## Treatment

The primary aim of treatment should be to establish a settled home for the child with a limited number of caretakers (preferably not more than two). Some socially disinhibited children, however, fail to cope if placed in foster or prospective adoptive families where there is an expectation of warmth and conforming behaviour, and fare better in small, structured children's homes.

Behavioural treatment to enhance social skills (Section 6.2.2) may be helpful in enabling the child to approach familiar and strange people with more discrimination.

## Outcome

Children placed with adopted families can do well even after a period of institutionalization lasting up to the age of 4 years. Their lack of generalized behavioural disturbance is likely to be reduced after placement (see Tizard 1977), but the capacity of the children to develop appropriate social inhibitions and to make and sustain relationships may continue to be impaired. Long-term follow-up of children placed out of institutions into permanent homes suggests that such improvement may be maintained at least into mid-adolescence (Hodges and Tizard 1989).

## 2.1.5 PRESCHOOL BEHAVIOURAL ADJUSTMENT DISORDERS

### Definition

These are disorders occurring in the 0–5 age group in which there is a widespread disturbance of behaviour, especially in the areas of activity, control, and aggression. The behaviour problems are often more prominent in some situations than in others, and may be more apparent with one parent than with the other. There may or may not be underlying anxiety. This type of disorder is not separately categorized in either of the two main classificatory systems, but it does form a reasonably discrete entity. Attempts to classify the non-specific disorders shown by very young children into categories mainly intended for older children, adolescents, and adults are neither sensible nor rewarding.

### Clinical features

1. *Activity*. The child is often reported to have been restless and to have had difficulty settling from birth or shortly afterwards. Feeding may have been difficult as a consequence. Subsequently, in the third and fourth year of life, there is a lack of concentration. The level of activity will, however, be variable, and when seen in the consulting room, the child may appear to concentrate reasonably well.

2. *Control*. The child seems unusually intractable and difficult to manage. Foci of conflict in the third and fourth year of life include bedtime and mealtimes. Food is refused and may be thrown around the room, and there are frequent temper tantrums. Frustration is poorly tolerated. The parents often say they cannot take the child out shopping because the child shouts and screams when refused anything. Prolonged tantrums may occur several times a day.

3. *Aggression*. Once the child is able to speak, verbal aggression is almost invariably present. The child may also physically hit out, scratch or bite. This pattern of behaviour may well result in the child being found unacceptable in and excluded from a playgroup or nursery school.

Inconstant features include:

4. *Anxiety*. Although the child may be apparently rejecting parental control, it may be obvious that there is in fact a good deal of underlying insecurity. This may be shown by the child's unwillingness to be left with friends or familiar relatives even for a few minutes, or by his incapacity to let his mother out of his sight without a panic reaction.

5. *Breath-holding attacks*. These are often a response to frustration, in which the child, usually aged between 18 months and 4 years, holds his breath at the culmination of a temper tantrum. In other children, breath-

holding occurs as an immediate response to pain or an anxiety-provoking experience. In a minority of children, breath-holding continues until the child goes blue before he takes his next breath, and, in a tiny number, the anoxia produces a brief convulsion.

6. *Adverse parental behaviour.* It is often difficult to tell whether parental behaviour is of primary significance or secondary to the child's difficulties. Usually it is both. Parents may be unduly critical or rejecting of their child and give the impression they have never given the child appropriate warmth and affection. Their expectation of the child's behaviour may be so rigid that what the child does is almost inevitably unacceptable. Their disciplining of the child may be inconsistent and arbitrary. Some parents show unduly passive, unstimulating behaviour, so that it appears as if the child's management problems arise from a need to produce strong parental reactions where these are lacking.

Such adverse parental behaviour may be present, but it is by no means invariable. Sometimes parents of children with this type of problem retain warmth and affection for their children and appear no more than ordinarily irritated or exasperated.

## Prevalence

About 20 per cent of preschool children have a significant behavioural or emotional problem. About 7 per cent have a disorder of moderate or severe intensity (N. Richman *et al.* 1975; S. Jenkins *et al.* 1980). Approximately half of these are likely to show predominantly a disorder of behavioural adjustment. N. Richman *et al.* (1982) in their study of 3-year-olds in the general population found that 11 per cent were definitely difficult to control, 9 per cent were attention-seeking, 13 per cent were active and restless, and 5 per cent had frequent tempers (three or more a day or lasting more than 15 minutes). These problems are often associated. They occur more commonly in boys, especially in their more severe form. There is little association with social class.

## Aetiology

**Social factors.** Although social class measured by paternal occupation is unrelated, these children are more likely to be living in overcrowded circumstances, in high-rise buildings with inadequate play space. There is no clear-cut relationship with maternal employment. Some preschool children benefit from a period apart from their mothers and others become more insecure. If the mother does go to work, the quality of substitute care is a major influence.

**Child factors.** Adverse temperamental characteristics (Section 2.14.1) may have been present from birth. Indeed it may be difficult to draw a distinction between these characteristics and the behavioural disturbance itself. The child has a definitely increased likelihood of developmental delay such as language retardation and night wetting beyond the expected age.

**Parental factors** (see above). As this is an 'adjustment' reaction, parental behaviour can be seen as part of the clinical picture rather than as an aetiological factor. As indicated, parents may, as a primary feature or a secondary reaction to a difficult child, be critical, rejecting, lacking in warmth, passive, and unstimulating. Mothers, in particular, may show high levels of depression and anxiety, and family relationships, especially the marital relationship, may be seriously disharmonious. Some parents, however, may show none of these adverse characteristics.

The **mechanisms** whereby these factors combine to result in a child with a severe behavioural adjustment problem are various. In some cases, adverse parental behaviour is obviously primary, in others it is mainly reactive. More commonly, a child with adverse temperamental character-istics and some developmental delay has been born into a family whose members are unprepared for social or personality reasons to cope with an unusually difficult child, though they might have managed well with a more placid, compliant one. As the child becomes increasingly rejected, more insecure, and thus more difficult to handle, so a vicious cycle may become established.

### Assessment

A description should be obtained of the **nature and severity** of the various problems of which the parents complain, as well as other possible problem areas. It is necessary to establish the **situation specificity** of the various difficulties. With whom are they present, and in what circumstances? This will establish whether the problems are widespread or monosymptomatic, and often also whether the child's problems are related to underlying anxiety.

The level of disturbance in the **parental personality** and **handling of the child** should be assessed. How far are they fixed in their behaviour, and how far open to possible changes?

Finally, the child and parents should be observed together, to form a further impression of the **parent–child relationship.** Observation of the child is necessary to exclude other conditions such as the hyperkinetic syndrome and childhood autism, and to indicate the child's developmental level. The presence of physical symptoms, such as headache or stomach-ache may suggest the need for a physical examination.

The condition may be differentiated from conduct disorders by the presence of overactivity and feeding and sleeping disturbances, and from the hyperkinetic syndrome or attention deficit–hyperactivity disorder, by the lesser prominence of overactivity and poor concentration, the presence of significant feeding and sleep problems, and by the importance of the parental contribution to the behavioural difficulties.

## Management

Various approaches may be successful, and there is little satisfactory evidence to indicate which is preferable. Treatment of isolated feeding and sleep disturbances is described elsewhere (Sections 2.2, 2.3).

**Family interviews** are indicated where parental communication is poor and distorted family relationships appear of major importance. Family sessions may reveal that the child's behaviour forms a focus of conflict because the parents are unable to deal with other areas of disagreement or frustration in their own relationship.

**Behaviour modification** (Section 6.2.2) is likely to be successful where the management problems are reasonably well-defined and the parental personalities are not widely disturbed. The presence of parental disturbance should not, however, exclude the use of behavioural techniques. Regular charting of the child's behaviour, and the application of a behavioural management programme may provide a structure to the parents' behaviour which may in itself reduce the degree of parental inconsistency. Parental counselling may accompany the delivery of such a programme. Parents may need to be helped to see how some of their behaviour, threatening to leave home if the child is naughty, for example, may increase insecurity.

**Social measures** involving an attempt to support parents in applications for more appropriate housing or to arrange a placement in a preschool facility may be valuable.

**Medication and individual psychotherapy** for the child have little or no part to play in the management of these conditions.

## Outcome

It was once thought that behavioural disturbances occurring in the preschool period had an excellent outcome. It has now been shown (N. Richman *et al.* 1982) that about two-thirds of children with definite preschool behaviour problems remain disturbed at least into the early school years. Boys in whom early restlessness is prominent appear to have a less good outcome than others. An amelioration of social or family circumstances does not necessarily produce an improvement in behaviour, though of course it is desirable for other reasons.

# 2.2 Feeding and eating control disorders

## 2.2.1 ESTABLISHMENT OF FEEDING

The establishment of a satisfactory feeding relationship between a baby and its mother in the first few days of life depends largely on instinctively determined behaviour, based on physiological mechanisms. It is gradually shaped, however, by the experience of the feeding couple, whether this be mutually satisfying or unsatisfying, enjoyable or unenjoyable.

In breast-feeding, the secretion of milk into the breast is under hormonal control, and ejection is stimulated by the baby's sucking, which triggers the milk-flow reflex. Generally, after a few days, a rhythm is established between mother and baby so that there is a balance between the baby's requirements and the mother's flow and ejection of milk. Similarly, in artificially fed babies, a pattern of feeding times is established with gradual lengthening of the period between the late night and early morning feeds over the first few weeks or months. There are a number of good descriptions of the physiology of infant nutrition, the composition of breast and artificial milks, with discussion of early feeding patterns and problems (e.g. C. B. S. Wood and Walker-Smith 1981).

Satisfactory establishment of early feeding depends on:

1. *Factors in the child*. An intact and well-functioning feeding apparatus and gastrointestinal tract, as well as good general health are required. Intact neurophysiological mechanisms are also necessary, and babies who have suffered perinatal trauma to the brain may show temporary or more permanent feeding problems. The use of intrapartum medication may impair the baby's capacity to suck after birth. Finally, feeding provides an early opportunity for the baby to demonstrate individual temperamental characteristics (Section 2.14.1). Some babies are placid and relaxed; others fussy and irritable.

2. *Factors in the mother*. The mother's capacity to supply sufficient breast milk or appropriate artificial feeds is obviously relevant. This may be impaired by local problems, affecting the nipple or breast, or by impairment of physical health. More commonly, feeding will be adversely affected by the mother's emotional state, especially by undue depression or anxiety (see p. 60). The mother may be generally anxious and have a specific anxiety about feeding. Such emotional changes may be hormonally or psychologically induced.

### Breast- or bottle-feeding: psychosocial considerations

There are advantages and disadvantages to both breast- and bottle-feeding. In developed countries, breast-fed babies are less likely to develop obesity, have a slightly lower incidence of asthma, eczema, and cot deaths, and, if they develop gastroenteritis, have a lower mortality rate. (In developing countries breast-fed babies have considerable added advantages in relation to infection and growth.) Psychological satisfaction also appears greater in mothers who breast-feed, though it is possible that this is because relaxed, contented women find it easier to breast-feed in the first place.

By contrast, bottle-feeding allows the feeding to be shared by other family members, and this makes it more feasible for a mother to return to work earlier. This will be a major consideration if there is economic pressure on the mother and, as is regrettably often the case, breast-feeding at work is not feasible. Many mothers also find breast-feeding embarrassing.

There is no evidence that breast- or bottle-feeding influences the rate of subsequent behavioural or emotional problems in the child (Orlansky 1949; Sewell and Mussen 1952). As there are small, but clear physical advantages to breast-feeding, professionals should obviously encourage this practice. It is, however, neither kind nor rational to induce guilt in mothers who, for economic or social reasons, decide to feed their babies artificially. An increasing number of working mothers do, however, find it possible, once breast-feeding is well established, to breast-feed their babies morning and evening, while the baby receives artificial feeds during the day.

## 2.2.2 FEEDING PROBLEMS IN EARLY INFANCY (0–3 MONTHS)

Minor, transient problems and set-backs in early feeding are very common. Persistent, more severe difficulties are less normal and require careful assessment. Physical disorders may be present and, even if they are not, an unsatisfactory early feeding relationship may be a prelude to serious behavioural difficulties later on.

1. *Rejection of the bottle or breast.* If a baby is active and otherwise well, this is usually only a transient problem occurring in the first few days of life. It is most likely to be due to faulty technique, and to be amenable to practical advice from an experienced health visitor or family doctor. A faulty feeding position, too large or too small a hole in the teat, may be responsible.

2. *Failure to suck.* A baby showing general apathy and lack of interest in sucking may be suffering from neurological dysfunction following a traumatic birth, the late effects of intrapartum medication, a neuro-developmental disorder, or a generalized physical disorder such as an infection. It may be associated with nausea related to other chronic disease. Very occasionally, failure to suck appears unassociated with any identifiable physical condition. This problem is difficult to explain. It is improbable, but not impossible, that even at this early age, the baby has developed unpleasant fantasies about the breast or bottle that deter sucking. More probably, the baby has had an adverse early experience in sucking, and sucking avoidance has become a conditioned response.

3. *Crying.* Normally babies only cry for relatively brief periods before and occasionally after feeds. Longer periods of crying may be due to hunger, discomfort from the cold or a wet nappy, or lack of sufficient attention. Crying due to pain is discussed below.

4. *Diarrhoea.* Persistent passage of loose stools in infants may be due to overfeeding, infections (especially gastrointestinal or systemic infections),

cow's milk allergy, drugs, or the malabsorption syndrome. Older children sometimes go through a phase of passing some undigested food—'toddlers' diarrhoea'. Psychosocial factors may be of significance in this form of disorder (Dutton *et al.* 1985).

5. *Constipation*. Most commonly this is due to underfeeding, and especially inadequate fluid in the diet. Unusual physical causes include cystic fibrosis (although this more commonly presents with diarrhoea) and Hirschsprung's disease. Occasionally babies withhold faeces if potting is attempted in the early months, though most babies do not appear to object to this futile practice.

6. *Vomiting*. Slight regurgitation, often of no significance or due to overfeeding, needs to be distinguished from actual vomiting of whole feeds. The latter may be due to infectious conditions, pyloric stenosis, cow's milk allergy, or hiatus hernia.

7. *Abdominal pain*. If unaccompanied by vomiting or failure to thrive, this may be a distressing symptom, but one of no organic pathological significance. The characteristic pattern of 'three month colic' is that the baby cries with pain and is unsettled in the early evening over the first few months of life, but then gradually improves. Usually, no physical cause is identifiable, and it must be assumed that physiological factors, probably affecting the motility of the gastrointestinal tract, are producing the prolonged pain-induced crying. Psychosocial factors may, however, be relevant (see below) in causation and, even if they are not, parents can become very distressed by the constant crying and require support through a difficult time.

### Management of feeding problems in early infancy

Management of these early difficulties first involves an assessment of the likely cause, together with treatment of any underlying physical condition. Necessary procedures in the assessment and management of physically determined problems are well described elsewhere (C. B. S. Wood and Walker-Smith 1981). In many cases, however, such physical factors are absent or insufficient to account for the problem. In these circumstances, psychosocial factors may well be of relevance. These include:

1. *Adverse living conditions*. Mothers living in crowded, insanitary conditions will inevitably have greater difficulty in feeding their babies adequately and safely, especially if they are bottle-feeding.

2. *Problems in social relationships*. Among the most common of these are unsupportive husbands or boyfriends, and intrusive, critical grandparents. The inexperienced mother needs all the sympathy and support she can obtain in the first few weeks of her new baby's life, and this may be lacking.

3. *Generalized maternal anxiety or depression*. Postpuerperal affective

states are relatively common (Section 2.1.1). Irritability, apathy, and agitation are frequent features and will be shown in behaviour outside the feeding situation.

4. *Anxiety focused on the baby*. Exaggeration of normal, maternal preoccupation with the baby may result in concern being expressed about quite ordinary aspects of the baby's feeding behaviour.

A common outcome of any of these psychosocial problems is that the mother is so anxious in her handling of the baby at feeding times that the baby becomes unsettled, cries excessively, refuses to suck, and eventually develops behaviour that reinforces the mother's anxiety so that a vicious cycle is established.

### Psychosocial aspects of management of feeding problems in early infancy

This will obviously depend on the circumstances, and, in particular, on the mother's social and material conditions and emotional state. It is necessary to exclude physical factors in the mother as well as the baby, but, even if these are present, psychosocial factors may be relevant. It should be emphasized that most such problems will be dealt with by primary care professionals, especially health visitors and family doctors. Only occasionally, where difficulties are serious and persistent, will advice from a paediatrician be sought, and psychiatric advice will be taken even more rarely.

1. *Practical help*. Observation of the feed may immediately reveal the nature of the problem. An inappropriate feeding position is relatively common. Alternatively, the feeding position may be satisfactory, but the mother so tense and anxious that the baby is unsettled. Practical advice on the way the nipple is presented or on the size of the hole in the teat may be invaluable. An anxious mother may rapidly become more relaxed when, as a result of a simple tip, she becomes successful in her feeding. More subtle problems which direct observation may reveal include a lack of sensitivity to the baby's feeding behaviour—for example, an insistence that the baby continues to suck, when obviously it wants to take a break. Mothers can be helped by suggestions about altering the timing of feeds, so that the baby is not too hungry at the time the feed commences.

2. *Mobilizing social support*. It may be useful to discuss a mother's feelings of isolation and insecurity with, if possible, the husband or boyfriend present. A health visitor may be able to encourage visiting from neighbours to provide company for an isolated mother. A self-help group of mothers at the local clinic or health centre can perform a similar function. Especially for a young and inexperienced mother, a health visitor may provide the nurturance the mother herself may need while trying to give it to her baby.

3. *Counselling*. Some mothers, especially if seriously depressed and

anxious, may benefit from the opportunity to express their ambivalent feelings towards their difficult baby to a sympathetic listener. Wider issues, such as concerns over loss of femininity and attractiveness, can be discussed. The mother may have previously expressed regrets at giving up her work or worries about what is happening at her place of employment while she is on maternity leave. Such counselling also allows identification of circumstances which might lead to the possibility of non-accidental injury.

4. *Psychiatric referral*. Mothers who show persistent disturbances of mood that fail to respond to counselling should be referred for psychiatric assessment and management. It is important, however, that the primary health care team should remain in touch with the family as well as the psychiatrist involved, as the type of help provided for the individual needs of the mother may need to be supplemented by continuing monitoring of the feeding relationship.

### 2.2.3 FAILURE TO THRIVE IN LATER INFANCY AND EARLY CHILDHOOD (3 MONTHS TO 3 YEARS)

**Definition**

This involves a failure to grow and gain weight either from birth or after a period of normal physical development. Standardized centile charts are available by which an individual child's growth can be compared with others of his age. Those who are consistently below the third centile in weight, who have documented weight loss over time, or who have a reduction in growth velocity (as shown, for example, by the crossing of two or more major centiles on a growth chart), should be regarded as requiring serious concern and investigation. Some who are of low birthweight, but continue to grow, albeit at a slow rate, may be regarded as 'small normal' children. These have constitutional or genetic reasons for their size, and suffer neither from disease nor from environmental deprivation—the two major reasons for failure to thrive.

**Clinical features**

If organic disease is present, the presenting features will depend on its nature. The failing-to-thrive child appearing for a routine appointment at a follow-up clinic for low-birthweight babies, may present with breathlessness due to a cardiac or other physical condition, or with symptoms referable to the gastrointestinal tract such as chronic diarrhoea and vomiting.

In the majority of children failing to thrive, who do not have organic causes, the presentation usually occurs from 2–3 months of age up to the third year of life. Feeding problems are common, and there may be regurgitation of food, vomiting, or chronic non-specific diarrhoea. Alternatively there may be no complaint at all from the mother, but the child may be found on a routine check to be falling below the third centile or faltering in growth velocity. The dietary intake of the child may be described as adequate, although careful investigation may reveal the child is, in fact, not getting enough to eat, and this is the commonest cause of failure to thrive. Nutritional inadequacy is often caused by both a failure of adequate provision of food and by inadequate intake. The child is likely to be mildly or moderately delayed in cognitive development, especially in language comprehension and expression.

On examination, if physical disease is present, the child may show characteristic abnormalities. Failure to thrive for other reasons is not associated with particular behavioural characteristics, although developmental immaturity may well be present.

### Prevalence

The prevalence of failure to thrive will depend on the criteria used. In theory there will be exactly 3 per cent of children at or below the third centile, together with a number of other children who, although taller and heavier than this, are failing to achieve their expected growth. In practice, there are some children (small normals) a little below the third centile, whose parents are of short stature and whose birthweight suggests that they are just born to be small. These will not be a cause for concern. Mitchell *et al.* (1980), using multiple criteria, found that nearly 10 per cent of children under the age of 5 years attending a primary health care centre in the United States, showed failure to thrive. About 5 per cent of paediatric admissions in the UK are for failure to thrive, and the problem also frequently presents to general practitioners.

### Aetiology

Traditionally, causes of failure to thrive have been classified as organic and non-organic. This terminology is misleading (Frank and Zeisel 1988). All failure to thrive is produced by inadequate food intake or malnutrition, and is therefore organically determined. However, in a minority of cases observed in primary and secondary care (perhaps between 5 and 10 per cent), the malnutrition is related to physical disease such as infection, heart failure, cystic fibrosis, food allergies or the malabsorption syndrome. In the remainder, physical disease is absent and the causes of malnutrition lie in the behavioural or developmental characteristics of the child, in the environment, or perhaps most commonly in the interaction between the child and the mother, an interaction that may be specifically maladaptive to the feeding situation (D. Skuse 1989).

1. The social background of young children who present with non-organic failure to thrive is often, though by no means always, socially deprived. The families are living in unsatisfactory accommodation, perhaps on housing estates with a bad local reputation. They may be in financial

difficulties made worse by the burden of another child. In some cases, the social circumstances of the families may provide a large part of the explanation of the problem.

2. In other families, living in more fortunate social circumstances, the main causes may lie in the *mother's mental state* or in the *quality of parenting*—the feelings of the mother and the way in which these are shown in her behaviour when feeding the child. A mother's difficulties in being warm and sensitive to her child's needs while it is feeding may be part of a general problem she has in making relationships, they may be part of a specific difficulty in child-rearing, or they may only relate to the child in question. Some mothers become depressed because, although living in materially privileged circumstances, they are socially isolated—alone all day in a house or flat while the father is away at work. General personality problems are most likely to be explained by deprivation in the mother's own early background, often compounded by a lack of good current support from her husband or her own family. Specific problems with the child may be related to the fact that the baby was and perhaps is unwanted. More generally, relationships within the family may be disturbed by tension, and there may be an overt family conflict or a more covert, mutually unsupportive relationship between the parents.

3. The child's characteristics may be of greatest importance. The exclusion of physical disease does *not* mean that the primary pathology lies in the family or wider environment: non-disease factors in the child may be of great significance. These include adverse temperamental characteristics: the child may be constitutionally lethargic, or unusually active or restless. Systematic studies suggest, however, that temperamental differences are only infrequently a significant causative influence. Of greater importance may be minor degrees of oral-motor dysfunction leading to difficulties in chewing and swallowing.

4. Often the circumstances in which feeding takes place provide the best clues as to the real nature of the problem, and these may be best identified by direct observation in the home setting. The mother may sit the child in an inappropriate position (semi-supine rather than sitting upright). The child may be constantly distracted during feeding, for example, by television, pets, or other children. Feeding may be an event the mother regards as something to be concluded as quickly as possible, so that an attempt is made to press the child to eat at a quite inappropriate pace. Conversely, the child may require more encouragement to feed than the mother is prepared to provide.

It is most common for social circumstances, parental and child difficulties to be present in combination rather than in isolation.

### Assessment

The assessment of the problem will begin with an attempt to construct a growth line for the child from birth to the present time so that a judgement can be made concerning whether there has been a fall-off from expected growth. The birthweight and child health clinic records will be invaluable for this purpose. A significant minority of children who fail to thrive have been born prematurely or show intra-uterine growth retardation. Where this is present, consideration should be given to these factors in assessing the child's growth pattern.

Although the findings are likely to be negative, a physical examination of the child is essential. Particular attention should be given to the presence of dysmorphic features, signs of infection, and dyspnoea, all of which may indicate an organic explanation. Physical investigations should be limited unless there are positive indications. Normally, a routine blood count and urine examination will be adequate, though in areas where lead poisoning is endemic, lead levels should be estimated.

Enquiries should be made into the circumstances in which the family are living. Overcrowding, severe financial constraints and lack of social support may all be relevant. Enquiries should be made about the mother's own mental state. Questioning about the mother's mood can most tactfully be achieved by asking whether the child's problem (or anything else) is making the mother feel low, unhappy, or anxious. Maternal mood may indeed be secondary rather than primary in these circumstances.

A dietary history should be obtained, if possible with the help of a dietitian. In some children a dietary history will make it clear that the child is not obtaining sufficient to eat. In some cases this will simply be due to ignorance, but in others enquiries concerning the mother's feelings about the child and the feeding situation will clarify the reasons. Useful information may be obtained by requesting the mother, with the help of a dietitian, to keep a record of the child's dietary intake over a week or fortnight. In mild cases of non-organic failure to thrive a sympathetic explanation with advice on diet given in the consulting room or out-patient department will, on follow-up, turn out to have been sufficient intervention.

Observation of feeding, ideally in the home, but if not in the consulting room or clinic, is often the most illuminating part of the assessment. The child may be seated inappropriately, subjected to major distractions, feed too rapidly or too slowly. The food may be unattractively prepared or presented. The child may be observed to have difficulties chewing or swallowing, even though the food is appropriately prepared and presented, and this may be an indication of the presence of oral-motor dysfunction.

Where a child is more severely affected or there is a suspicion of un-diagnosed organic disease, despite primary care and out-patient investigation, either because of the symptomatology or because of the absence of evidence of a psychosocial explanation, then the child should be admitted to hospital with the mother. Intensive physical investigations should, however, be avoided until it has been established that the child will not gain weight even though getting enough to eat.

*Assessment of weight gain in hospital.* Most children admitted to hospital will gain weight on a normal diet. Children who fail to gain weight despite normal intake have an organic cause for their poor growth. Children who do not receive an adequate intake may have an organic explanation, such as a neurological disorder, affecting their sucking or chewing capacity, or a chronic infection impairing appetite, but they are more likely to be showing food refusal for psychological reasons. The single most important investigation in the assessment of 'failure to thrive' is a trial of feeding, and other investigations should be held in abeyance until it is established whether the child is able to gain weight on an adequate intake.

*Assessment of feeding behaviour of mother and child.* Where a physical explanation remains the most likely cause, a range of further physical investigations will need to be carried out in hospital. Even in these, however, positive organic findings may not emerge. Berwick *et al.* (1982) noted an average of 40 laboratory tests carried out in infants admitted to hospital for failure to thrive. In only 0.8 per cent was relevant diagnostic information obtained. A hospital assessment of physical state and response to adequate diet should therefore be accompanied by a further evaluation of the mother–child relationship, the feeding behaviour of the mother, and the behaviour of the child during feeding. In some cases the child will not feed with the mother but will feed with nursing staff. The mother may be noted to be stiff, uncomfortable, and perhaps uninterested or even disgusted by feeding her baby. Alternatively, she may be overconcerned, anxious, and intrusive.

### Management

Most cases of failure to thrive can be managed in the primary care setting, and indeed the opportunity for home visits will often make primary care management more rewarding than treatment delivered from hospital.

Assuming physical disease has been excluded, management will centre on those features of the family circumstances, parental relationships, attitudes and behaviour, or, most commonly the feeding situation, which assessment has revealed to be most relevant.

Where growth has fallen seriously behind, in order to achieve catch-up growth, the child will require 1.5–2 times the expected calorie intake for age (Frank and Zeisel 1988), and this will often not be easy to achieve.

Parents may need educating about the amount of food their child needs to grow properly and, in some cases, this will in itself be sufficient intervention. More commonly, more active intervention is required, and this may call for the combined professional input of the health visitor, family doctor, and, where available, dietitian, social worker, psychologist and psychiatrist.

The most rewarding form of intervention is likely to involve direct impact on the feeding situation. Simple behavioural measures including establishing a good feeding position, as far as possible excluding distraction when the child is feeding, ensuring the food is appropriately prepared and presented, making allowances for any difficulties the child has in chewing or swallowing, may all lead to improvement. A useful behavioural approach to the treatment of feeding problems in failure-to-thrive children has been provided by Iwaniec *et al.* (1985).

In children admitted to hospital, a plan of management will also require the combined skills of those with medical, social work, nursing, psychological, and dietetic expertise (Fig. 1). The mother is likely to get especially valuable help from supportive nursing staff. If the baby will not feed for her, but will feed for the nursing staff, the situation should be dealt with sympathetically and not critically. It should be explained that it is natural for babies to be more difficult with their mothers, because they are more emotionally involved with them. A mother who is deprived herself and consequently finds it difficult to care for her child will need a nurturant and not a critical relationship with nursing staff and social worker so that her self-esteem can grow and she can feel she has more to contribute to her baby's care than she thought. Practical advice from nursing staff in the feeding situation, if offered tactfully, can often be taken up by the mother

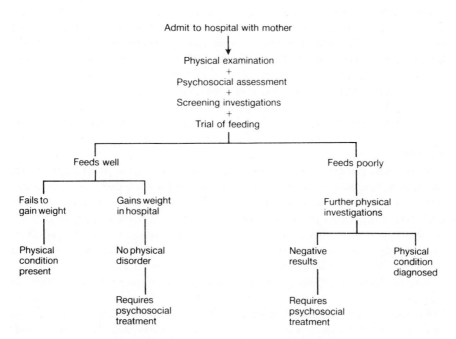

**Fig. 1.** Failure to thrive, failure to respond to out-patient care—plan of management. (Adapted from A. G. M. Campbell 1984.)

positively. The establishment of feeding programmes with very gradual introduction of solids in small, frequent feeds, can be useful. Help may also need to be given to ensure that the young child is adequately stimulated between feeds. Many children with failure to thrive show mild-to-moderate developmental delay, probably partly because of inadequate stimulation. It has been shown that malnourished children gain weight faster if active play and language stimulation programmes are instituted, and if the mother can be helped to undertake this type of activity she may be able to continue once the child leaves hospital.

Whether management is conducted by the primary care or hospital team, depression in the mother is an indication for an evaluation of the family and other circumstances that may be involved in lowering mood and energy. Counselling from social worker or psychiatrist and possibly anti-depressant medication may need to be arranged. The social worker may also be able to arrange for the alleviation of some of the social stresses acting on the mother by recruiting support from other family members, advising on financial benefits, etc. In cases where it is thought that the child is in serious danger of severe chronic neglect, the procedures described earlier for dealing with non-accidental injury need to be invoked.

In brief, the treatment of severe failure to thrive occurring for psycho-social reasons requires the combined skills of a multidisciplinary team. Follow-up should ensure the child is indeed now thriving physically and is receiving adequate nutrition. The frequent occurrence of developmental delay and behaviour problems in children who fail to thrive means that attention in management should not be solely focused on the child's physical growth. Cognitive development should also be monitored and, where necessary, additional stimulation provided either at home or in a preschool facility. Counselling should also be provided for any behaviour problems that may be present, and referral to a psychiatrist or psychologist for more intensive treatment arranged if this is not effective.

## Outcome

Non-organic failure to thrive is likely to improve slowly with appropriate treatment, but most children will remain small for their age. This is particularly likely to be the case where growth faltering has occurred early in the child's life, especially in the 3–6 months period, perhaps because of the rate of brain growth at this time and/or because maladaptive feeding patterns, once established, are particularly difficult to alter. A high proportion of children with short stature in middle childhood will have first presented with failure to thrive in infancy. Further, perhaps due to subnutrition or family adversity, there is a high rate of developmental delay with subsequent behavioural and learning problems in children who have, in infancy, failed to thrive. There is surprisingly high continuity between early feeding problems and eating control difficulties in adolescence (Marchi and Cohen 1990).

### 2.2.4 PICA

The mouthing and chewing of inappropriate objects is an uncommon presenting problem, although toddlers and young children often put small toys, buttons, pencils, and anything else available into their mouths. Persistent mouthing of earth and scavenging occurs in young, deprived children. Older children who continue to put objects in their mouth are usually, but not always mentally retarded. The dangers of pica include acute and chronic accidental poisoning. Children with pica are more likely to ingest tablets, bleaches, etc. if they are left lying around. They are also more likely to show high levels of body lead concentrations, although pica is now a very rare cause of lead poisoning, at least in the UK (M. Smith *et al.* 1983).

The management of children presenting with pica will depend on the nature of the problem. The removal of dangerous substances from the child's whereabouts is an important precaution. If the pica is a sign of chronic deprivation, then attention should be given to improving the social conditions in which the child is living. Behavioural methods (Section 6.2.2) are sometimes effective in treatment of pica in older mentally handicapped children.

### 2.2.5 OBESITY

**Definition**

Both objective and subjective criteria are important in the definition of overweight. Using standardized weight and height charts (Tanner *et al.* 1966) it can readily be ascertained whether a child is over the 97th centile for weight (or 20 per cent over expected weight for given height and sex). Eighty per cent over expected weight for given height and sex is sometimes taken as an objective criterion for serious concern. Similarly, a child's degree of adiposity can be assessed using skinfold callipers (Tanner and Whitehouse 1975) and compared with standardized tables. There is a reasonably good correlation between an observer's subjective comparison of a child's degree of obesity and these standard measures. The subjective judgement of the child and family may, however, be very different. Older girls may regard adiposity well within the 'normal' range as objectionable, and many parents of children with worrying objective evidence of obesity do not perceive their child as having a problem.

**Clinical features**

There are four main modes of presentation of obesity in children, each of which has somewhat different implications for management. The child may be noted to be overweight during a routine procedure such as immunization, or when presenting with an acute physical disorder such as a respiratory infection. Second, the child may be noted to be overweight by teachers or others outside the family. Third, the parents may become concerned about the rate at which the child is gaining weight. Finally, in older children, and especially in adolescent girls, the child herself may express concern. A presentation in which the family or child have themselves been motivated to seek help is more likely to respond to attempts at treatment than where only others have been worried.

The degree of overweight is rather poorly correlated with the amount of concern expressed by family members. Some grossly overweight children are never brought to attention by their parents, while other parents express considerable anxiety over rather minor degrees of overweight. Motivation to attend is often enhanced by a child's unhappiness at school because of teasing.

## Prevalence

Obesity is the most common nutritional disorder in developed countries. Because of varying standards, precise prevalence figures cannot be stated, but estimates vary between 2 and 15 per cent (Lloyd and Wolff 1976). In fact, probably the lower figure is more realistic if considering those children whose health is definitely at risk because of obesity. On the other hand, the higher figure is an underestimate if one considers all those teenage girls who consider themselves overweight.

Objectively defined obesity is more common in girls than in boys and, probably because of their greater intake of cheap starchy foods, the condition is more common in working-class than in middle-class children (P. W. Wilkinson 1975).

## Aetiology

Clear-cut single-factor organic causation is rare, though obesity does occur as part of the Prader Willi and Laurence–Moon–Biedl syndromes, both associated with mental retardation. Hypothalamic syndromes affecting the appetite mechanism may also result in obesity, but again these occur very rarely, and other signs of hypothalamic and pituitary malfunction are obvious.

Most obesity is **multifactorially determined** (Woolston 1987). Obesity certainly runs in families. Genetic, as well as environmental factors, are important. Although the evidence from twin studies is conflicting, a large investigation of Danish adoptees (Stunkard *et al.* 1986) clearly pointed to the importance of genetic factors. Obesity was related to the weight of biological parents (especially the mother), but not to the weight of adoptive parents. Nevertheless it is likely that environmental factors play a part in determining which genetically vulnerable children become obese.

Obesity occurs when there is an **excess of calorie intake** over energy output. By the time obese children are professionally assessed, their dietary intake is often not unusually high, though it is sufficient to maintain the degree of overweight. In this case, either the intake at some previous point in time must have been excessive, or the child is predisposed from an early age to put on excessive weight with a normal intake. Artificially fed babies are more likely to be overweight, both currently and in mid-childhood than breast-fed babies. In many children, the dietary intake is excessive at the time of presentation.

Obese children **exercise less** than the non-obese, but the differences do not appear great enough to suggest that low energy output is largely responsible for obesity and, in any case, obesity is quite as likely to be a cause of taking little exercise as the other way round. It is, nevertheless, interesting that the prevalence of obesity increases according to the number of hours of television viewed by the child (Dietz and Gortmaker 1985).

**Inappropriate patterns of eating in the family** are probably of significance. The artificial feeding of babies and excessive intake of cheap, starchy foods in an unbalanced diet occur more commonly in working-class families. More specifically, the intake of bread, potatoes, and between-meal snacks of crisps, biscuits, and sweets is encouraged in some families, and restricted in others. Family patterns of overeating are probably the single most important cause of obesity.

**Family and individual psychopathology** is thought important by many clinicians, but good evidence is lacking. It has been suggested that mothers who feel guilty about some aspect of their own behaviour may overprotect their children and fail to allow them autonomy. The child is never allowed to feel hungry, and so loses the capacity to determine its own need for food. There is no evidence that obese children are unusually aggressive as one might expect if their autonomy was severely restricted. Certainly some obese children, like some obese adults, do overeat when they are miserable, and pre-pubertal, obese children are indeed more depressed than age-matched controls.

Many very overweight children have good cause to be miserable. They are often derided by other children, teased, and called names to a degree that leads to a serious degree of social isolation (Tobias and Gordon 1980). Sociometric studies have shown that obese children are less likely to be wanted as friends than children with other physical disabilities, even those with conspicuous deformities. Obese adults have been shown to be self-denigrating, anxious, and sensitive to social criticism, and the same is probably true of overweight children.

### Assessment

An adequate history should attempt to clarify the onset of obesity. If the onset is recent, was it associated with any particular event? Details of the child's intake at mealtimes and between meals should be obtained. The intake of other family members, and their own degree of overweight will be relevant. The presence of any behavioural or emotional problems should be established and, in particular, it is helpful to know whether the child's relationships with other children have been affected by his or her appearance. Finally, it is useful to know what dietary measures, if any, have been tried to date, and how motivated the child and family are to achieve loss of weight.

Physical examination should establish accurately the height and weight of the child for baseline purposes. The presence of any endocrine disorder can usually be excluded purely on physical examination.

### Management

This will depend particularly on the degree of obesity and on the motivation of the child and family. In all children over the 97th centile, it is reasonable to point out the health hazards of remaining overweight, but in many cases this, in itself, will not be a sufficient motivating factor. In those who are motivated, usually either because of teasing at school or because the degree of obesity has reached even more alarming proportions, a positive treatment approach is indicated.

1. *Dietary restrictions.* An 800–1000 kcal diet taking into account the child's food preferences should be recommended. This is most unlikely to be a sufficient treatment measure in itself, and even for short-term success a variety of other measures will need to be implemented.

2. *Behavioural measures* directed toward modification of the eating patterns of the child along lines suggested by Stuart (1967) may be helpful. This approach, which has proved suitable for adolescents (I. Gross *et al.* 1976) involves obtaining a clear account of the stimuli that normally precede eating, so that alternative responses can be achieved. When, for example, the child feels miserable or has reached a time when he would normally take a snack, he is given some alternative activity to eating, which is then reinforced. In adults, behavioural methods have been shown to have more long-lasting effects than appetite-suppressant drugs (Stunkard *et al.* 1980). Focused treatment programmes involving behavioural control of eating patterns, education of the child and family concerning calorie intake, and encouragement of exercise do appear to have at least a short-term effect.

3. *Family and group approaches.* An attempt should also be made to change family patterns of eating behaviour, for a child is unlikely to reduce his intake if the behaviour of the rest of the family is unchanged. Family sessions, in which behaviour at mealtimes is discussed, may be helpful in this respect, particularly by stimulating involvement if one of the parents appears unconcerned and indifferent to the problem. Some older children and adolescents will also benefit from attendance at groups run on 'weight-watcher' lines.

4. *Hospitalization.* In-patient treatment is indicated for a variety of reasons. First, when there is scepticism among family members that excess intake is responsible, an admission may demonstrate that weight loss is indeed possible without any more complicated procedures. Second, it may be necessary for life-saving purposes when the very seriously obese child is showing hypoventilation (Pickwick syndrome), and useful results can be obtained by admitting the child. Third, if family patterns of eating are inappropriate, a period away from home may allow an older child to develop autonomy in the realm of eating behaviour so that, when he returns home, he has the capacity to refuse food when it is pressed upon him.

More drastic forms of treatment applied in adults to the very severely obese include jaw-wiring and the ileal bypass operation. These have also been used in older adolescents with reasonably successful results. Clearly, however, these can only be contemplated when the obesity has become life-threatening.

### Outcome

Obese children have a high risk of remaining obese in adulthood. Thus Stark *et al.* (1981) found that four out of 10 obese children became obese adults. Long-term obesity is associated with hypertension, cardiac disease, diabetes, emphysema, varicose veins, and poor operative risk. These adverse outcomes are, however, only common in the severely obese. Most forms of treatment produce immediate short-term improvement, but, except for individual case reports, most follow-up studies reveal that such improvement is usually short-lived.

### 2.2.6 ANOREXIA NERVOSA

### Definition

This is a condition defined by the presence of significant, self-induced weight loss (or in the case of pre-pubertal children, failure to gain weight), body-image distortion and a widespread endocrine disorder involving the hypothalamic–pituitary–gonadal axis resulting in delayed menarche or secondary amenorrhoea (World Health Organization 1990). The DSM III-R criteria comprise (i) refusal to maintain body weight, (ii) intense fear of gaining weight, (iii) disturbance in the experience of body weight, size, or shape, and (iv) primary or secondary amenorrhoea (American Psychiatric Association 1987).

### Clinical features

**The age of onset.** This may be anything from 6 to 60 years, but the great majority of cases begin between 14 and 19 years of age. In one study of early-onset anorexia nervosa, age of onset varied from 7.7–13.7 years

(Fosson *et al.* 1987). Dieting and loss of weight are the central, characteristic features of the condition. Onset of dieting behaviour may be sudden or gradual extending over weeks or months. There may be an obvious precipitating factor—commonly a group of girls has decided to lose weight together; most fail to lose any but a minority do lose weight and then stop dieting. One goes on to develop anorexia. Alternatively, a rather solitary, introverted girl may develop the condition following a chance remark about her appearance and need to lose weight.

**Dieting.** Initially this may be hidden from the rest of the family or it may be overt. After a time it becomes apparent that there has been a marked reduction in food intake, and this leads to family arguments at meals. Frequently parents will press food on to the girl, and she will respond by eating less and less, cutting food up into small pieces, spending an apparently interminable time eating tiny quantities. Girls may enhance the effects of reduced food intake by taking drugs such as purgatives, appetite suppressants, or diuretics, or by self-induced vomiting, though this is unusual in the younger age group. Some girls also exercise in order to lose further weight.

Excessive dieting may alternate with episodes of binge-eating (bulimia). The bingeing may involve eating vast quantities of food, or quantities that the girl perceives to be enormous but are really quite normal. Again this is not a common feature in younger patients, and bulimia nervosa, in which binge-eating is the main feature of the eating control disorder, is unusual in children and early adolescence.

The dieting behaviour is based on a morbid fear of becoming fat. Usually there are particular areas of the body the girl is particularly worried about, and these include especially the thighs, buttocks, breasts, and abdomen. The reasons why the girl has such an excessive fear of fatness are often, at least partly, unconscious, but occasionally a girl may attribute her desire to be thin to her wish to be attractive and popular. It is most unusual for girls to admit they are trying to avoid the biological changes of adolescence or womanhood.

**Bodily misperception.** Accompanying the morbid fear of fat, there is also frequently a misperception of bodily appearance—the girl overestimating the size of her thighs, stomach, etc. in comparison with other girls. As she genuinely perceives herself to be bigger than she really is, this misperception leads to further family conflict and arguments. It is also common for girls to misperceive the quantities of food they are given to eat—believing that their mothers are giving them excessive quantities, perhaps in order to make them fat.

The **referral** of the patient may occur for reasons other than loss of weight. It is quite common for girls to present to their family doctor with other symptoms such as constipation or amenorrhoea.

**Atypical syndromes.** As well as the classical form of the condition that is described above, atypical forms also occur. In one of these (bulimia nervosa) the overeating or bingeing is a more prominent feature than undereating, but periodic dieting behaviour does occur, and the central psychopathological feature remains a fear of becoming fat (G. F. M. Russell 1979). Another characteristic picture, more commonly seen by primary care practitioners, is a marked preoccupation with physical appearance and constant ineffectual dieting, unaccompanied by amenorrhoea or even excessive loss of weight. Nevertheless, the condition is different from normal dieting in that the preoccupation with food and appearance 'takes over' the patient's life and causes a great deal of suffering.

**Physical examination.** In moderate or severe classical cases, this is usually negative, apart from the presence of signs of undernourishment. Parts of the body may be covered with lanugo or fine downy hair, but elsewhere the skin may be coarse and rough. Blood pressure may be reduced, and in severe cases hypothermia may be present. Laboratory investigations are usually also negative, though there may be a reduction of gonadotrophin levels, with low blood levels of oestrogens, LH, and FSH. In contrast, there may be elevated levels or cortisol and growth hormone. The hormonal disturbance is of hypothalamic or suprahypothalamic type. Gonadotrophin levels are also reduced in males. It is unusual to find either abnormal physical signs or abnormal results of laboratory investigations in younger patients. In adolescents who are severely affected, osteoporosis may occur, and cardiovascular effects include bradycardia and arrhythmias.

## Background information

**Prevalence.** The condition occurs classically in about 1 per cent of 15–19-year-old girls (Crisp *et al.* 1976). However, atypical mild forms occur more commonly than this in the age group at particular risk. The condition is probably increasing in frequency, though moderate and severe forms remain uncommon. It is extremely unusual in developing countries, though it is just beginning to occur among girls in Westernized families living in large cities. The condition is more common among certain groups selected for thinness, such as ballet dancers.

**Family factors.** There is a familial tendency to the condition, and sibs are affected more commonly than one would expect by chance. Twin studies suggest that there is a genetically determined element to the condition. It occurs more commonly in middle-class families, and often in those with an apparently particularly high moral standard of behaviour, strongly influenced by the 'work ethic'. Characteristic patterns of family functioning have been described in which there is overprotectiveness, linked to inadequate patterns of communication about feelings. Consequently, the overprotected girl is unable to establish her own individuality and autonomy, and, because of poor communication, she cannot protest and make her views and feelings known. Establishing control, at least over her food intake, is her means of achieving autonomy, but she only does this at tremendous cost to her physical health and social relationships. This pattern of family relationships, while certainly sometimes found, is not universal, and there is a wide variation of family pattern.

**Personality.** Many, but not all, affected girls, have a history of unusually conforming behaviour. They tend to be quiet and somewhat compulsively altruistic, often wanting to train to be nurses or air hostesses or they are attracted to other occupations in which service to others is a prominent feature. Again this is not universally found. What does seem more common in the psychopathology of patients is a lack of desire to enter the adult world of womanhood with its implications for independence, sexuality, and motherhood. The primary fear of fatness can, at least sometimes, be seen as a wish to avoid the rounded contours of female adolescence.

**Societal factors.** Probably societal attitudes to female appearance are of relevance. The condition is less common in societies where a rounded female figure is the desirable norm. In Western industrialized countries, the 'perfect' figure (as evidenced by the vital statistics of pin-ups on magazine covers) has, over the 1970s and 1980s, steadily approximated to a drain-pipe shape.

The endocrine changes that occur in the condition, such as the reduction in gonadotrophins, are probably secondary, although, once established, they may inhibit recovery. As discussed below, psychological improvement sometimes occurs rather rapidly once the physical changes produced by malnutrition are reversed. Physical factors may, therefore, be of little importance in primary causation but of major significance in the maintenance of the condition. The lack of primary importance of endocrine factors is confirmed by the occasional presence of the condition in classical form (apart, of course, from amenorrhoea) in hormonally normal males. Males are affected 10–15 times less commonly than females. However, in pre-pubertal children, sex differences in prevalence are much less marked with a female preponderance of only three or four to one (Fosson *et al.* 1987).

**Neurochemical factors.** Although most of the endocrine and metabolic changes are probably secondary to starvation, evidence for a primary neurochemical abnormality also exists (Herzog and Copeland 1985). Increased endogenous opioid activity has been detected and decreased levels of cerebrospinal fluid norepinephrine persist after weight recovery.

## Assessment

The form of the initial assessment will depend on whether this takes place in the general practitioner's consulting room, in a paediatric or general medical out-patient department, or in a psychiatric setting. In all settings, however, once the diagnosis is suspected, it is desirable to see both parents and the girl together so that the problems (including fear of fatness, dieting, binge-eating, over-exercising, etc.) can be identified within the family as openly as possible. Especially relevant are a description of mealtimes and the parental attitude to the girl's growing independence. Her previous personality and the means by which the family communicates and resolves problems more generally are also informative. Many girls are secretive about their dieting and exercise habits, and their statements cannot always be accepted at face value.

Some time should be spent with the girl alone, in order to attempt to establish her own view of the situation and attitudes to adolescence, sexuality, and womanhood. Other aspects of special relevance include her mood, the presence of any actual bodily delusions, as well as her level of weight and motivation for change. It is helpful to ask questions such as 'What would your ideal weight be?' and 'Are there any parts of your body you feel especially concerned about?' Physical examination should include obtaining an accurate height and weight (wearing only underwear), and noting pubertal status, and signs of malnutrition. If the latter are present, or a significant amount of weight has been lost, then further physical investigations will be required.

There are both physical and psychiatric conditions with which anorexia may be confused. Physical conditions include malabsorption syndromes and endocrine disorders such as hypopituitarism. A careful history and physical examination should allow exclusion of such conditions. It is important to remember that anorexia nervosa may coexist with physical syndromes. In particular, a link has been described with Turner's syndrome and diabetes mellitus.

Psychiatric states that may be confused include depressive disorders and, much less commonly, schizophrenia. Many girls with anorexia nervosa do not suffer from a depressive reaction, and indeed their capacity to remain cheerful in the face of their apparently appalling predicament is rather characteristic. Nevertheless, a significant number of girls are seriously depressed, and here the problem is to know whether the depression is primary with secondary loss of appetite and weight, or vice versa. The presence of other depressive symptomatology and attitudes to bodily appearance will usually clarify the diagnosis, but mixed pictures are seen especially in the younger age group. Depressed girls usually have a reduced appetite (normal in anorexia nervosa) and they usually wish to gain weight normally. Their activity level tends to be reduced. The presence of strongly delusional ideas related to bodily appearance may raise the possibility of schizophrenia. A small minority of girls presenting with apparently classical anorexia nervosa do develop unmistakable signs of schizophrenia. Males with anorexia nervosa share most of the features of the classical condition, except of course the amenorrhoea, but they are less likely to come from middle-class families.

## Treatment

Once the diagnosis has been made, the nature of the condition should be explained to both the girl and her parents, if possible together. Mild variants of the condition can be treated on an out-patient basis, but loss of more than a few pounds in weight is an indication for in-patient assessment and treatment. A trial of out-patient treatment lasting a month or two can be followed by in-patient admission if unsuccessful. Earlier admission would be indicated if the weight loss worsened, there was evidence of definite metabolic disturbance or hypotension, or there was severe depression with risk of suicide.

Medication has only a small part to play in management. Appetite stimulants are unhelpful. The presence of severe anxiety and tension calls for the use of an anxiolytic agent, such as chlorpromazine. Tricyclic antidepressant medication may be used on a short-term basis if depressive features are prominent.

The mainstay of treatment, whether on out- or in-patient basis, is the use of a firm approach to weight gain followed by supportive and sometimes interpretive psychotherapy to the patient and family. Few girls accept admission to hospital easily. In most cases a family interview is necessary to point out to both the parents and the girl that anorexia nervosa is a life-threatening illness, and that, even if survival is not in question, the girl cannot be healthy or lead a normal adult life later on unless she is well nourished. By the time of presentation it is common for the patient to be dominating the family and to attempt to refuse admission. In these circumstances, the clinician should point out the inappropriateness of the child being in control, affirm the seriousness of the condition and encourage the parents to assert their authority in the girls' best interest. With this approach, compulsory admission is very rarely necessary. Mild cases may be managed on an out-patient basis.

A target weight range should be established between 90 and 110 per cent of the mean weight for the patient's height and age. The patient and family should be told that full health cannot be achieved until weight is maintained within this range. The girl should, in consultation with the dietitian, be given a diet on which she might expect to gain 3–4 lb (1.5–2 kg) a week. One or two food dislikes may be respected, but, in general, the girl should be expected to eat a full range of foods. Quarter-or half-portions may be presented for the first week or two, and the food should be presented in as palatable a way as possible. If weight loss has been severe, the patient should be nursed in bed until an agreed intermediate weight is reached, at which point limited, and then full, activity can be allowed. Tube feeding is very rarely required, and should not be employed except as a life-saving procedure. Twice-weekly weighing on accurate scales in underclothes is adequate to assess progress. Many patients are worried about becoming overweight, and these should be reassured that, at least while in hospital, nursing staff will see this does not happen. There are advantages if the criterion for progress is weight gain and not food intake, as the latter may be subject to distortion by the girl. Some authorities do, however, use food intake as a criterion of satisfactory progress. Improvement in food intake and weight gain may be accompanied by difficult behaviour, and it should be explaining to parents beforehand that this may occur and is indeed an encouraging sign.

Once the target range of weight is reached, it must be maintained, and family meals in the in-patient unit, followed by weekends home allow assessment of progress in this respect. During the period of weight gain and immediately afterwards, it is usually unwise to comment favourably on the girl's improved appearance. She may not regard herself as looking better, and such remarks can precipitate relapse. During this period of rehabilitation family interviews are particularly helpful, not just to allow exploration of feelings about the girl's eating habits, but also to discuss other issues concerning autonomy, decision-making, and communication in the family. Marital problems between the parents sometimes emerge at this point and may require separate sessions. If they are prepared to come, the presence of any sibs at these interviews can be additionally helpful. Once acceptable weight gain has been achieved, individual psychotherapeutic interviews also become more worthwhile, and it is often then possible for the first time to explore concerns about the meaning of adolescent bodily changes, sexuality, and adulthood with the girl. The girl can be helped to achieve more autonomy in decision-making, and to feel more in control of her life generally. It has, however, been demonstrated that in younger sufferers from this condition family therapy is superior to individual therapy (G. F. M. Russell *et al.* 1987). Even after full recovery in a prepubertal girl, height gain can be disappointingly slow. If periods fail to start when the girl has reached normal weight, clomiphene can be helpful to stimulate menstruation.

After discharge from hospital, contact with the girl and her family should be maintained to monitor weight and review general progress. A point of stability is usually reached after a few months in which it becomes clear that further intervention is not likely to achieve further change.

## Outcome

With these measures, a proportion of girls, perhaps about a third, appear to make a full recovery (Hsu *et al.* 1979). Their attitude to their bodily appearance appears normal, and they no longer seem to have concerns about adolescent development. About a third retain a reasonable body weight, but they remain morbidly preoccupied with their appearance and often have other personality problems affecting especially their relationships with the opposite sex. Finally, a third, after perhaps failing to reach satisfactory weight, or after reaching it only with very great difficulty, fail to maintain it and may either be chronically underweight or develop periodic bingeing with consequent weight fluctuations. Girls in this group who have developed the condition before puberty may remain of short stature and never menstruate. Personality problems remain severe in this group. Suicidal attempts are not uncommon and about 5 per cent do commit suicide. The mothering capacity of the intermediate group may be impaired and the more severely affected will rarely marry. Full recovery is, however, usually followed by a normal fertility rate and mothering capacity (Isager 1985). Poor prognostic factors in pre-pubertal children include early age of onset, depression during the illness and disturbed family relationships (Bryant-Waugh *et al.* 1988). Severe cases requiring re-admission often do particularly poorly.

## 2.3 Sleep and its disorders

### 2.3.1 NORMAL SLEEP PATTERNS

The newborn baby sleeps on average 16–17 hours a day, though there is rather wide variation. Cycles of sleep and waking are relatively brief. Over the first 3 months of life the total amount of sleep declines little, but the cycles are longer, so that, by the age of 3 months, 70 per cent of babies are sleeping right through the night. By 6 months 85 per cent of babies sleep through the night, but there are still 10 per cent waking every night by the age of 1 year. By this age the great majority of babies have established a stable pattern of sleep and wakefulness with a long sleep at night and a nap in the morning and afternoon. Social class and the parity or experience of the mother are not related to wakefulness, but breast-fed babies are more likely to wake at night than bottle-fed (Eaton-Evans and Dugdale 1988).

### Developmental neurophysiology of sleep

The neonate has a characteristic EEG sleep pattern that gradually approaches but does not attain the adult pattern by the age of 1 year. Adults show two types of neurophysiological activity during sleep—quiet sleep and rapid-eye-movement (REM) sleep. So-called quiet sleep has four stages, proceeding from stage 1 with low-amplitude, fast-frequency waves, to stage 4 with high-amplitude, low-frequency waves. For about 20 per cent of sleep time adults show REM sleep in which the EEG shows low-amplitude fast-frequency waves, associated with decrease of muscle tone, variations in pulse rate, respiration rate, and blood pressure, as well as bursts of rapid eye movements. A sleep cycle from the beginning of one phase of REM sleep to the beginning of the next lasts about 1½ hours. By contrast, neonates show nearly 50 per cent REM sleep and a good deal of sleep that is poorly differentiated in EEG terms. Cycles are of short duration, lasting about 45 minutes. The adult, when falling asleep, goes through a prolonged phase of quiet sleep before a REM phase, while the neonate goes straight from wakefulness to a REM phase. By 3 months, however, the infant, on going to sleep, does pass through a phase of non-REM sleep. From 3 months to 1 year the infant's sleep EEG pattern also becomes much more clearly differentiated with increasingly prolonged periods of slow-wave quiet sleep.

Infant sleep in the first year of life gradually comes to be increasingly modified by environmental influences. By 6–8 months, a change in sleeping arrangements, anxiety or depression in the mother, or an alteration in the bedtime ritual may result in an increase in wakefulness. Disrupted quiet sleep is commoner in babies who have suffered pregnancy and birth complications (Blurton-Jones *et al.* 1978) and in babies born to diabetic mothers.

At about the age of 18 months, and perhaps earlier, infants show signs of dreaming. Dreams occur largely in REM sleep. When woken from REM sleep, adults will report a dream on about 80 per cent of occasions, whereas in non-REM sleep they report dreams on only about 10 per cent of occasions.

## Influences on sleep in early childhood

In the second and third year of life, normal children often show difficulties in getting off to sleep. During the night, wakeful periods occur and a high proportion of children do not go back to sleep without calling for their parents. In one population study (Crowell *et al.* 1987) more than a third of children aged 18–36 months woke their parents during the night at least once or twice a week. Unwillingness to go to sleep in this phase of life may be related to a variety of developmental factors. In order to go to sleep the child must stop paying attention to the world outside him, withdraw into himself, and think his own thoughts in his internal world. Many children go through a phase when they are unwilling or frightened to give up the external world. As they go to sleep they may wish to carry an image of their mother or some other familiar person with them. But their capacity to retain an image in their minds (their capacity for object constancy) is limited, so they may need to recall their mother for a brief reminder. The presence of a doll or bit of blanket, a transitional object (Winnicott 1953), may ease the journey over the bridge from wakefulness to sleep. Wakefulness is also commoner in babies and toddlers with adverse temperamental characteristics—those who, in the daytime, are more negative and intense in their expression of mood and more irregular in their habits.

Older children may fear a loss of control during sleep. They may, for example, be frightened they will pass faeces or urine and thus disgrace themselves. Others will be upset at the thought of missing television, curious about what their parents might get up to in the middle of the night, or determined to show their power to keep their parents running from one bedroom to another in the early hours of the morning. Lying in the dark or semidarkness, perhaps with the wind blowing the curtains through a half-open window, is often sufficient to stimulate anxieties in the older child of burglars, ghosts, giants, or other frightening phenomena.

In the fourth and fifth years of life, the need for naps is usually gradually reduced and then eliminated. Children are more likely to sleep through the night, though they may well wake earlier than their parents wish. Problems in getting off to sleep and wakefulness in the night sometimes persist, though nearly all children will sleep through the night once they start infant school. Bad dreams or nightmares become commoner and are often associated with a frightening experience, such as a television programme, seen on the previous day.

## 2.3.2 SLEEP PROBLEMS

Sleep problems are common among children. Simonds and Parraga (1982) surveyed a population aged 5–20 years and found restless sleep (28 per cent), sleep talking (13 per cent), fear of the dark (10 per cent), and bed-time rituals (8 per cent) to be among the most frequently occurring difficulties.

### Wakefulness at night

This is by far the commonest sleep difficulty, with a peak frequency between 12 and 24 months of age (N. Richman 1981). About 20 per cent of children of this age either take at least an hour to get to sleep or wake frequently or for prolonged periods during the night. About 10 per cent of children show waking at night at 3 years; by 8 years the prevalence has dropped to 4 per cent (N. Richman *et al.* 1982). In the young child, waking at night may be associated with adverse temperamental characteristics, maternal depression and anxiety, as well as stress in the family (N. Richman 1981). In most children with sleep problems, however, none of these factors is present, and the undue wakefulness cannot readily be explained by them.

### *Management*

The management of difficulty in getting off to sleep and wakefulness at night (N. Richman *et al.* 1985) should begin with an assessment of the problem. An account should be obtained of the frequency and duration of the child's waking episodes over the previous week or fortnight. Enquiries should be made about the rituals at bedtime and what the parents actually do when, for example, the child wakes and comes into their room. Details of other possible areas of disturbance in the child—his social relationships, any feeding difficulties, etc.—should be obtained. The presence of depression and anxiety in either parent, and the social circumstances of the family should be noted. Types of treatment already received, especially the previous use of hypnotics, should be identified, for, by 18 months, nearly a quarter of UK children have received hypnotic medication.

When, as is usually the case, the wakefulness is an isolated problem, parents can be reassured that the problem will pass and is of no serious import. Some sleep problems are, however, so exhausting for parents to deal with that they present a real threat to the stability of family life. Children with such problems can be treated.

The treatment of choice is behavioural management (described below). Hypnotic medication, for example trimeprazine tartrate 1–2 teaspoons *nocte*, may sometimes provide some temporary improvement, but is unlikely to result in lasting benefit.

A course of behavioural management should start with the parents charting the child's sleep pattern and their own response to it over a period of 10–14 days to a fortnight. The record thus obtained is very likely to show that parents are inappropriately rewarding the child for wakeful behaviour and failing to provide the best conditions for helping a child to get back to sleep after waking in the middle of the night. The father and mother may be spending up to a couple of hours in the child's room with the child calling out or crying if the parent attempts to leave. The child may be given quantities of sweet drinks if he wakes in the night and demands them. The parents may be alternately shouting at the child or taking him into their bed if he wakes.

Alterations in the management should then be discussed with parents. They should first be asked their own views on how they might change their behaviour. Cutting down on the length of bedtime rituals, replacing sweet drinks with water, resisting demands to get into the parents' bed would all prove helpful on occasions, provided that the parents can be consistent. If the child is old enough, a star chart marking nights when the child sleeps through and does not wake his parents can be kept. A minority of parents will be prepared to let the child 'cry it out', but many will not wish to do this. Some parents when faced with behavioural advice will turn out not really to be sufficiently motivated to change. Others will reveal levels of guilt or anxiety not originally detected and these may require more intensive counselling or psychotherapy.

Once a programme has been agreed, the parents should be encouraged to continue charting the child's sleep pattern and their own behaviour. A further two or three sessions to advise on continuing changes in management are usually sufficient to produce definite improvement, though often the child will still show minor sleep problems. It should be emphasized that this behavioural management approach, though effective in the majority of young children, is not likely to succeed where there are widespread emotional or behavioural difficulties, where parental anxieties or personality problems are thought to be a primary cause, or where social circumstances make it impossible for parents to institute a consistent programme.

The outcome of monosymptomatic sleep problems is good, and most do not betoken disturbance in the future. If they form part of a widespread preschool behavioural disturbance, then they are much more likely to be followed by significant psychological problems at least in the middle school years (Section 2.1.5).

### Nightmares

These are frightening dreams that occur in the rapid-eye-movement (REM) phase of sleep. The child will wake in an anxious state with slightly raised heart and respiratory rate, and then be fully orientated and able to recount his dream. The pattern of the nightmare and its underlying neurophysiological basis is, therefore, very different from a night terror (see below). Virtually all children experience at least occasional nightmares. Peak frequency is at about the age of 5–6 years, after which there is a decline in rate, though most people continue to experience them occasionally throughout childhood and in adult life.

A child experiencing frequent nightmares, occurring perhaps more often than once or twice a week, is a cause for concern. Most nightmares are stimulated by frightening experiences the previous day, such as watching a horrifying television programme. Parents whose children are having frequent nightmares will usually prohibit the watching of such programmes. A child having nightmares for some other reason is probably going through an unusually anxious stage and requires investigation. Enquiries should proceed along the lines described for other anxiety states (Section 2.9.2).

### Night terrors, sleep-walking

These are described together because they have a common neurophysiological basis and genetic origin.

**Night terrors** are very upsetting for parents to watch. Usually about 2 hours after going to sleep the child sits up and appears terrified. He is likely to scream, shout, and moan, and sometimes tries to get out of bed. He may appear hallucinated and even deluded. The child does not respond when spoken to or attempts are made to calm him. There is a considerable rise in heart and respiratory rate. After a few minutes the child slowly settles and appears to go back to sleep. In fact, he has never really awakened.

**A sleep-walking episode** usually starts with the child sitting up 2 or 3 hours after going to sleep with a glazed expression on his face. He gets out of bed and walks mechanically, usually to his parents' bedroom or to the lavatory. Occasionally, he may wander around the house and, very rarely indeed, out into the street. He avoids objects and thus appears, at least to some degree, to be in contact with his surroundings. However, he does not respond to questions and is very difficult to wake. If his arm is taken and he is guided back to bed, he will usually comply and go back to sleep. Often he will put himself back to bed within a few minutes.

Night terrors and sleep-walking each occur in about 3 per cent of children, though one report has suggested that as many as 15 per cent of 5–12-year-old children have had at least one sleep-walking episode. The peak age for night terrors is 4–7 years and for sleep-walking somewhat older than this. The episodes occur in a period of arousal from EEG stages 3 and 4 sleep, so that the child is in fact quite deeply asleep when showing the behaviour. As many as 50 per cent of sleep-walkers and children showing night terrors have a close family member who has shown one or other of these phenomena (Kales *et al.* 1980). The condition may be inherited in an autosomal dominant mode. Both conditions are likely to occur if, during the previous day, the child has experienced a stressful event, but, if they occur monosymptomatically, they are not in themselves an indication of emotional disturbance.

*Management*
Management of both sleep-walking and night terrors is similar. An account should be taken from the parents of the frequency and nature of the episodes, stressful events that might be upsetting the child, and the way the parents react when the episodes occur. Parents should be advised that it is unnecessary to wake the child. They should stay with the child while he is showing the behaviour, but beyond guiding the sleep-walking child back to bed, they need not take any other active measures. If parents are worried, and wish to take active measures, Lask (1988) has described an apparently very effective procedure. This involves the parents keeping a record of the times at night when the child experiences night terrors, and then systematic- ally waking the child up about 15 minutes before an attack is anticipated. If episodes remain frequent and persistent, occurring more often than once a week, but there is no evidence of unusual daytime anxiety, then it may be reasonable to try the effect of diazepam 2–5 mg *nocte*. This drug alters the proportion of stages 3 and 4 sleep, and has been reported to be helpful. Parents of sleep-walking children should be advised to secure external doors and windows, as, rarely, fatal falls from windows and dangerous wandering out into the street have been reported.

The outcome of both night terrors and sleep-walking is good, and usually they remit well before adolescence. A small minority of people continue into adult life.

## Hypersomnias

These are disorders in which children or adolescents suffer from patho- logical excess of sleep. They are rare and of two types: narcolepsy and the Kleine–Levin syndrome.

**Narcolepsy.** The patient experiences an irresistible desire to go to sleep, sometimes in the daytime. The sensation comes on very rapidly and usually lasts from 5 to 30 minutes. After the episode the child is usually fully alert within seconds. Usually attacks occur out of the blue, but sometimes they may be precipitated by unusual excitement. The subject may also at times experience cataplexy (sudden loss of muscular tone with a fall to the floor) or sleep paralysis (inhibition of muscle tone as the subject goes to sleep). About 10 per cent of patients suffer from epilepsy, implying that the condition is probably due to some form of brain dysfunction. The condition is probably largely neurophysiologically determined, but experience of an emotional or stressful event can trigger attacks. A high proportion of adults with this condition have associated psychiatric disorders.

Narcolepsy is rare before the age of 10 years and more commonly occurs in late adolescence.

**Kleine–Levin syndrome.** This is also a rare disorder occurring mainly in adolescence. Subjects show excessive somnolence during the day associated with markedly increased appetite.

*Management* (Guilleminault *et al.* 1974)
Both these types of hypersomnia need to be distinguished from normal behaviour. It is, after all, common for people to take a nap in the afternoon, and many teenagers eat voraciously and seem, to their parents, to spend an excessive amount of time dozing in their beds. The distinction from normal behaviour lies mainly in the episodic nature of the disturbance, together with (in the case of narcolepsy) the rapid onset and recovery. The presence of epilepsy also needs exclusion. The main diagnostic feature, available only when a sleep laboratory is accessible, is that the patient, when showing sleeping behaviour, can be demonstrated to have gone from the waking state straight to REM sleep. Normal sleep involves an initial period of non-REM sleep. Other psychiatric disorders, especially hysterical reactions, should also be considered in differential diagnosis.

Treatment should primarily involve the identification of causative factors such as emotional stress. The life of the child or adolescent should be altered to take account of the attacks. Teachers of a teenager with narcolepsy should be informed of the condition and encouraged to let their pupil withdraw from a class if there is warning that an attack is coming on. Older teenagers with narcolepsy should not be allowed to drive a car while attacks are still occurring.

**Medication.** Methylphenidate 5 mg *mane* and lunchtime, increasing in older children to 10 mg *mane* and lunchtime, is sometimes helpful in both these conditions. In older teenagers there is a danger of drug abuse with this medication, and, if prescribed, the drug should only be administered under strict control. Clomipramine 10 mg t.d.s. to 25 mg t.d.s. has been reported to reduce the frequency of narcoleptic attacks. The associated psychiatric disorders that may well be present will require psychological treatment, usually individual psychotherapy or family therapy.

## 2.4  Motor development and disorders of movement

### 2.4.1  NORMAL MOTOR DEVELOPMENT

Development of motor skills allows a child to gain greater control over his environment. Children who are keenly motivated to achieve mastery over the world around them achieve such skills more rapidly. Further, in order to master a skill of any degree of complexity, it is important that the child has a plan of action in his mind as he tackles the task. A child who wants to drink from a cup has to execute a series of coordinated movements involving his hands, his eyes, and his lips. The amount of tilt of the cup, the coordination of the movements required to lift the cup without spilling the contents, the neck movement required to ensure that the cup meets the mouth and not the nose—all these components of the task require planning and cognitive or intellectual ability without which the child will fail. An intact visual pathway and perceptual skills are also necessary.

Table 2 sets out information concerning:

1.  The average age when children are able to perform a particular motor skill.
2.  The age at which, if the child cannot perform the task, there is a need for concern, further assessment, and investigation. Such investigation is necessary because the child is at high risk, either for the presence of pathology or for the presence of a maturational delay carrying significance for later functioning. In some cases, however, the delay discovered will only be a temporary lag. In a survey of children not walking unaided by the age of 18 months, Chaplais and Macfarlane (1984) found, for example, that only 32 per cent had a pathological condition.

Performance of the tasks in Table 2 does not just depend on muscular power and coordination. A child with a visual defect who cannot locate a rattle in space will be unable to grasp it. A child with a perceptual deficit may be unable to copy a square, not because he lacks fine motor control, but because he is unable to plan to reproduce the shape he perceives visually.

### Influences on normal motor development

The sequence in which different aspects of motor development occur is largely under genetic control influencing the rate of CNS myelination. Thus, no matter what the social circumstances or cultural context, children are likely to go through similar sequences in which head control precedes

## Table 2

| Movement | Average age of performance (months) | Need for concern if not performed by age (months): |
|---|---|---|
| **Gross motor movements** | | |
| Sits with head steady | 3 | 5 |
| Sits without support | 7 | 10 |
| Stands holding on | 10 | 15 |
| Takes 2–3 steps unaided | 14 | 18 |
| Walks up steps holding rail | 20 | 30 |
| Jumps up and down on 2 feet | 30 | 42 |
| Hops on 1 foot | 48 | 72 |
| **Hand skills (Fine motor skills)** | | |
| Grasps a rattle | 3 | 5 |
| Transfers a cube from hand to hand | 7 | 10 |
| Uses 'pincer' (finger–thumb) grip for small object | 9 | 12 |
| Drinks from beaker | 14 | 20 |
| Builds tower of 4 cubes | 20 | 33 |
| Copies circle | 36 | 42 |
| Threads beads | 40 | 48 |
| Copies + sign | 48 | 54 |
| Copies square | 54 | 66 |

the ability to stand unsupported, which itself precedes skilled finger–thumb movements and so on. In the absence of deprivation or cruelty, the acquisition of gross motor skills (sitting up, walking alone, etc.) is largely independent of the social circumstances in which the child is reared. Twin studies support the importance of genetic factors. Suggestions of racial differences in the speed of motor development, with children of African origin developing skills on average somewhat earlier than children of Caucasian origin (Capute *et al.* 1985) are based on evidence that is difficult to evaluate.

Fine motor development, by contrast, is much more sensitive to social influences and to the opportunities available to the child for exploration, play, and stimulating experience. If the child has no, or very limited opportunity to use a cup or spoon, or build towers with bricks, then these skills will not be acquired and, when the child is exposed to these objects, he will appear delayed in development.

### Assessment

A rapid assessment of motor development in a baby or young child can be made with a small amount of equipment, a rattle, half a dozen plastic or wooden cubes, a piece of fine cord and some wooden beads, coloured pencils, and paper. It is particularly important to distinguish loss of motor skills from failure to attain them, as the former is likely to be related to the presence of a significant physical disorder. Fuller accounts of technique and interpretation of developmental assessment are available elsewhere (e.g. Bax *et al.* 1990).

### Causes of delay in motor development

#### General developmental delay

Most cases of delay in motor development are associated with delay in other spheres, especially language and social development. General mental retardation therefore remains the most common reason for motor delay. However, it is important to note that many children with general mental retardation show little or no delay in the development of gross motor movements. Gross motor development is therefore a poor predictor of mental retardation.

#### Specific motor delay

Delays of motor development in the absence of general developmental delay may be produced by:

(1) cerebral palsy;
(2) other abnormalities of the central nervous system and muscle (especially hypotonic disorders), that may be progressive or non-progressive;
(3) specific motor dyspraxia (clumsiness).

Psychosocial aspects of cerebral palsy, and other abnormalities of the central nervous system are described in Section 5.7.

### 2.4.2 CLUMSINESS (SPECIFIC MOTOR DYSPRAXIA)

Most children who are clumsy are also delayed in other aspects of their development, but a minority are specifically uncoordinated, sometimes to a remarkable degree (Gordon and McKinlay 1980). Typically, clumsy children, though normal in their language development, are somewhat slow to walk, and subsequently have difficulty feeding themselves and are very delayed in being able to undress and dress. They have particular difficulties in doing up buttons, tying shoelaces, holding a pencil, and using scissors. As they get older, their problems show themselves particularly in handwriting and sports activities. They may develop secondary educational problems (Henderson and Hall 1982), and their self-esteem may suffer because they are teased for their ungainliness and ineptitude at games. Consequently they can become depressed or unduly aggressive.

The cause of specific clumsiness is unknown, but genetic factors are probably often of importance. Intrauterine growth retardation and birth complications may be of significance in the individual case. Some children are given inadequate opportunity to learn fine motor skills.

**Assessment** can be carried out informally by observing the child build a tower with cubes, use a pencil, jump up and down on the spot, hop, carry out the finger–thumb apposition test, and copy a pattern. An experienced observer will be able to identify a specifically clumsy child by noting disparity between the way the child carries out these tasks and his verbal ability. More formal assessment to confirm the diagnosis (especially to rule out general mental retardation) and develop a plan for remediation can be carried out by a psychologist. The Bruininks–Oseretsky Test of Motor Proficiency (Bruininks 1978) is a test suitable for use with children aged 4½–14 years of age. Gross and fine motor function in skills requiring, for example, balance, agility, dexterity, and strength are assessed in a comprehensive manner. The test takes about 45 minutes to an hour to administer, and is useful for identifying children with specific deficits in motor co-ordination. Formal neurological examination and EEG are usually non-contributory, except in cases where clumsiness has developed in a child with previously normal dexterity.

**Management** will depend on the nature and severity of the problem. It may be sufficient to explain to parents and teacher that the child, although of at least average ability, has a specific difficulty and is not being lazy. He can be expected to improve with time, although he will probably remain more uncoordinated than other children. Clumsy children can be helped to gain confidence and improve their skills by physiotherapists, and occupational and speech therapists with a specific interest in their condition. Practice of movements the child finds difficult and discovering new ways of circumventing the child's motor problems will also be helpful (Gordon and McKinlay 1980).

### 2.4.3 STEREOTYPIES AND HABIT DISORDERS

These are intentional, but pointless movements, carried out in a repetitive manner. They may form part of a wider disturbance, such as autism, or be monosymptomatic. They include head-banging, rocking, and hair pulling as well as usually rather trivial habits such as nail-biting and nose-picking.

### Head-banging and rocking

These are commonest in infancy, beginning at about the age of 6 months to a year, but occasionally persisting into later childhood. They are usually present at night when the child has been settled in a cot and left to go to sleep, but may be present in the day when it is likely they occur in response to frustration. Although usually best seen as comforting habits, not indicative of wider disorder, they may provide a clue to severe neglect and lack of proper care. In either event, protective padding to the cot and a re-arrangement of the child's night-time routines may be helpful, but if care is inadequate, then improving it will be of major importance.

### Hair-pulling (trichitillomania)

This is an unusual problem, which may be isolated or part of a general disturbance. The child may pluck his or her hairs out to a degree that a considerable patch of baldness is created. If the hair is swallowed, as it sometimes is, then a hairball (trichobezoar) may be created, which may produce an intestinal obstruction requiring surgical intervention.

Management depends on the nature of the problem. If an isolated phenomenon, behavioural management is likely to be effective. If part of a general anxiety state, then attention to the source of anxiety, with family or individual psychotherapy, will be helpful. Hair-pulling may also be a sign of chronic social deprivation, in which case ways should be found to improve the quality of care more generally.

### 2.4.4 TICS AND TOURETTE SYNDROME

### Definition

Tics involve rapid, involuntary, repetitive muscular movements. They can be transient or persistent, local or widely generalized. In their mild, transient, or 'simple' form they are of very little significance; where severe, generalized, and 'complex' they can represent one of the most handicapping forms of childhood disorder.

### Clinical features

Tics usually begin between the ages of 5 and 7 years. In **simple form** the child may have involuntary blinking movements of the eyes together with frowning and twitching of the cheek muscles. Often the movements have begun at a time of stress for the child—for example, entering a class with a teacher the child does not like, or an illness in the child or other family member. The tics persist over a few weeks or months and then usually disappear. They often do not come to medical attention, and, if they do, they respond to reassurance and encouragement to put less pressure on the child.

When they occur in more **severe 'complex' form,** tics begin at about the same age (Torup 1962), though they may appear as early as infancy or as late as adolescence. They usually rapidly spread to other parts of the body, including especially the shoulders, limbs, and trunk. There may be complicated movements present such as jumping and kicking, as well as simple twitches. If, in addition, vocal tics are present, the condition is usually described as a form of Gilles de la Tourette syndrome or (more briefly) Tourette syndrome (Shapiro *et al.* 1973).

Tourette syndrome is characterized classically by convulsive muscular jerking, inarticulate cries, together with coprolalia (emission of obscene utterances), and echolalia (automatic repetition of sounds heard). By convention, however, the condition is now regarded as present if both motor and vocal tics are present, even if there is an absence of coprolalia.

Associated behavioural and emotional symptomatology is found in a large number of severely affected ticquers, though it is much less common in the mildly affected. Encopresis, comforting habits, obsessional and hypochondrial symptoms are particularly prominent. There may also be signs of developmental delay, such as specific speech retardation.

## Prevalence

Mild tics occur in about 10 per cent of children at some stage of their development (Macfarlane *et al.* 1954; Lapouse and Monk 1968). In severe tics, boys are more commonly affected than girls at all grades of severity in a ratio of 3 to 1. There are no significant social class associations. The prevalence of Tourette's syndrome is uncertain, but it is much less common, occurring in about 0.5 per cent of adolescents.

## Aetiology

In the individual case, the cause is usually uncertain, but there are various pointers.

1. *Genetic.* A number of cases have been described in which both one parent and a child have been severely affected. No clear-cut pattern of inheritance has been demonstrated but family studies suggest a dominant major gene is implicated (Price *et al.* 1988). A high proportion (perhaps as many as 50 per cent of first-degree male relatives) show chronic tics or Tourette's syndrome. First-degree relatives also show a raised rate of obsessive-compulsive disorder.

2. *Neurological.* A high rate of non-specific EEG abnormalities has been reported. Post-mortem and animal studies suggest that there may be a defect in the striopallidal connections of the basal ganglia.

3. *Neurochemical.* The effectiveness of the butyrphenone drugs and the similarity to syndromes produced by L-dopa suggest an abnormality of dopamine metabolism. An excess of dopamine concentration in the corpus striatum may be present.

4. *Familial and other forms of stress.* In mild cases parents are sometimes over-pressurizing, and the child may be subjected to other forms of stress, at school or elsewhere. In more severe cases, it is usually impossible to disentangle primary familial factors from the family disturbance that is produced by the presence of the child's extremely disturbing pattern of behaviour. When children are convulsed by frequent twitching and screeching, it is only natural that parents, often told there is nothing physically

wrong with the child, should become exasperated and upset. A punitive approach is relatively common.

5. *Individual psychopathology.* The obscene nature of the explosive utterances has naturally given rise to a speculation that the condition is due to repressed aggressive and sexual feelings. It is likely that these are indeed sometimes of importance, as occasionally children are relieved of tics, but become, for example, more generally aggressive. In severe cases, individual psychopathology is more likely to be secondary. The nature of the utterances may relate more to their social unacceptability than to individual sexual or aggressive psychopathology. For example a case was described in the 1970s from the Peoples' Republic of China in which the explosive utterance was a counter-revolutionary slogan.

In brief, in mild cases, the condition is probably a developmental phenomenon, brought on by interaction between stress and genetic predisposition. Severe cases, including Tourette syndrome, are probably largely organically determined by genetically determined dysfunction of the striopallidal connections and interference with dopamine metabolism.

### Assessment

As well as obtaining an account of the onset of the condition, and factors affecting improvement and deterioration, it is particularly important to identify family reactions to the presence of the movements. A check should be made for the presence of other behavioural problems or developmental delay. Physical examination is usually negative. On interviewing the child, the movements are often more marked than in ordinary situations because of the particular stress of the clinical examination. The child and parents can be asked how the severity of the tics during the consultation compares to that in everyday life. It is useful to observe the parental reactions to the tics, and whether the severity of the tic fluctuates according to the topic being discussed.

The differential diagnosis is usually not difficult. The movements may, however, be confused with those of choreic or choreiform type. These may be present in various forms of cerebral palsy, Huntington's chorea, and (in countries where this still occurs) rheumatic chorea. In these conditions, the movements are non-repetitive, and a variety of different muscle groups are involved, whereas in tics, single muscle groups are affected. Movements in myoclonic epilepsy are repetitive, but usually involve larger muscle groups and other epileptic phenomena are present. Mannerisms, stereotypies, and comforting habits may be confused with tics, but are purposive in nature. They may coexist with tics.

### Treatment

This will depend very considerably on the severity of the condition.

*Mild tics*
These respond to reassurance. It is useful to consider with parents and child
the presence of possible stresses and to remove these if at all possible.
Parents should be encouraged to ignore the tic as much as possible. With
this form of management the tics usually disappear within a few weeks or
months.

*Severe tics, including Tourette syndrome*
A combined approach is required, involving behaviour modification, individual and family counselling, and (probably most effective) medication.

**Behaviour modification.** Principles described in Section 6.2.2 should be
followed. Parents and/or child should be asked to chart the tics to provide
a baseline of frequency of their occurrence. The period of charting might be
relatively short (half an hour or an hour a day) or might involve the whole
day. Once a baseline has been established, appropriate behaviour modification techniques can be introduced. These might involve a system of stars or
other rewards for periods of time in which the tics do not occur. Massed
practice is also sometimes temporarily effective (Section 6.2.2). This consists of getting the child to produce the tics voluntarily, as frequently as
possible, over a defined period of time, say 5–10 minutes twice a day. This
form of practice sometimes enables the child to achieve greater control over
the tics at other times.

**Family and individual counselling.** Children with severe tics and their
parents can be helped by sympathetic discussion of ways in which the tics
might be reduced and means by which the impact on the child's social life
can be mitigated. As far as is compatible with the behaviour modification
programme, the parents should be encouraged not to mention the tics or to
tell the child to stop them. Some parents find this virtually impossible, and
they can sometimes be helped by behaviour modification programmes, the
child rewarding the parents for not mentioning the movements. Most
children co-operate with this form of treatment enthusiastically. Individual
and family sessions can also be helpful for exploring family and internal
conflicts that may be relevant to the maintenance of the tics.

**Medication.** Haloperidol is the treatment of choice, and may be given in
doses increasing from 0.02 mg/kg bodyweight daily in two doses to 0.05
mg/kg bodyweight daily in two or three divided doses. Unfortunately
side-effects are common. The drug should not be prescribed to a level
which makes the child sleepy, but sometimes it is worth accepting a mild
level of sedation for a few days to see whether the child habituates, but
retains the therapeutic benefit. Dystonic reactions may occur, in which case
the child should be withdrawn for two or three days before recommencing
at a lower dose, combined with orphenadrine 50 mg t.d.s. Most children

with severe tics benefit from haloperidol, but only a small minority become symptom-free.

If haloperidol is ineffective, a dopamine-blocking agent, pimozide 1 mg b.d. increasing by 2 mg increments weekly to a maximum of 8 mg can be prescribed (Gillies and Forsyth 1984). Clonidine at an initial dose of 3 $\mu$g/kg may also be prescribed, but is less effective. Beneficial effects can sometimes be obtained with doses of about 0.05 mg t.d.s. Some clinicians would regard regular monitoring of blood pressure as necessary if this drug is prescribed. Stimulant medication should not be used as it may aggravate the symptomatology.

### Outcome

This varies according to the severity of the tics, the presence of associated psychiatric problems, and the degree of family disturbance. Mild tics, confined to the face and occurring in the younger age group, generally tend to disappear over a few months. Recurrences may occur, but these tend to be short-lived.

The prognosis in those children with severe widespread tics and Tourette syndrome is much less favourable, but nevertheless a significant proportion (about 50 per cent) do improve markedly before they reach adulthood. A number of these improve only in later adolescence. Good prognostic signs include an absence of associated psychiatric problems, harmonious family relationships, IQ average or above, and lack of signs of developmental immaturity or brain dysfunction. Haloperidol, although usually producing worthwhile temporary symptomatic relief, probably does not shorten the total length of the disorder. In a small minority of cases, deterioration of personality with schizophrenic features occurs.

## 2.5 Language and its disorders

### 2.5.1 NORMAL LANGUAGE AND SPEECH DEVELOPMENT

Communication is the process of transmission and reception of information. There are many different modes of communication, non-verbal as well as verbal. Language is the sum of the skills required to communicate verbally. Speech is the oral production of language. Communication begins immediately after birth. Indeed, some would suggest that the mother's belief that she knows how her baby is feeling while she is pregnant implies that it begins earlier than this.

In the first few hours of life babies learn to distinguish their own mother's voice. From the first few days of life, the cry of pain can be distinguished from the cry of hunger. From 2 months of age, babies can be seen to move their lips in response to their mothers talking to them. By 3–4 months early babbling is usually occurring, and this gradually becomes more complex and speech-like. At this time, mother and baby can be observed to be involved in conversational 'turn-taking' in which the child will respond by babbling when the mother stops speaking to him. The baby's oral productions become increasingly imitative. A phase in which the baby automatically 'echoes' adult speech and appears to 'play with' sounds is common. At about the age of 12 months, the first words with meaning, such as car, 'bicky' (biscuit), Mama, or Dada usually occur. By 18 months the child is usually combining words and, shortly after this, symbolic (make-believe) play can be observed. There is therefore a characteristic development of communicative ability before the onset of speech.

After the use of first words with meaning, there is rapid increase in the quality and quantity of speech, especially over the subsequent two years. Expression usually lags somewhat behind comprehension, the child understanding more than he can say. Vocabulary increases, as does the complexity of grammatical utterance. Table 3 summarizes information concerning language development.

The fact that a child has not acquired a particular skill by the suggested age does not mean that the child's development will necessarily be impaired subsequently. It does mean that the child has significant language delay, is at risk for subsequent impairment, and requires assessment.

### Theories of language development

Various theories have been developed to explain the process of language development. B. F. Skinner proposed a behavioural model, children learning language largely by imitation and selective reward or reinforcement for

**Table 3** Language development

| Skill | Average age acquired (months) | Concern if not acquired by age (months) |
|---|---|---|
| Vocalizes using vowel sounds | 3 | 5 |
| Two way vocalization | 4 | 6 |
| Repetitive babble | 8 | 11 |
| Waves bye-bye to spoken request only | 13 | 20 |
| Symbolic play with large doll | 16 | 24 |
| Uses three words other than dada, mama | 20 | 26 |
| Selects 3 or 4 common objects on request | 20 | 30 |
| Combines words to express 2 ideas, e.g. 'daddy work' | 27 | 33 |
| Understands a request with 3 components, e.g. put the little spoon in the box | 36 | 48 |

correct utterances. Noam Chomsky has put emphasis on innate mechanisms such as a 'language acquisition device' to help explain why children produce rule-based grammatical utterances (e.g. 'he bringed the money') that they cannot have learned by imitation. Jean Piaget and Jerome Bruner have attempted to explain language development in terms of more general cognitive processes and symbolic thought. None of these theories is fully explanatory. It is clear that imitation is of minor significance, as children's language is so different from that of the adults they hear, but innate mechanisms can hardly explain the wide variations in language observable in children brought up in different cultures. More than most aspects of child development, the process whereby language is acquired remains mysterious, although there is now a considerable body of information about the factors associated with and sometimes causing language problems (Bishop 1987).

Psycholinguistic studies suggest that it is useful to consider different components of language development. These include the phonology of language (the sound system), grammar or syntax, semantics (the representation of meaning in language), and pragmatics or how language is used. Thus in place of the medical diagnoses in traditional use (autism, mental retardation, etc.), speech therapists are increasingly using terms such as phonological deficits or semantic pragmatic disorders to describe speech and language problems.

## Influences on normal language development

**Genetic factors** are of particular importance in establishing the sequence with which different stages are reached, but are also important in determining the pace of development.

**Physical factors.** Children who have suffered from intrauterine growth retardation, or where there has been a prolonged second-stage labour, are at risk for slower than average language development. Recurrent otitis media, occurring in the first and second year of life, is a common cause of mild language delay.

**The quality of the social environment** is the other factor of significance (Fundudis *et al.* 1979). The pace and quality of language and speech acquisition are linked to:

1. *Social class.* Middle-class children being more advanced, especially after about the age of 3 years.
2. *Size of family.* Children in large families are, on average, slower to speak, probably because adults have less time to talk and listen to them.
3. *Multiple births.* Twins speak a little later than singletons, perhaps because twins often communicate a good deal together, and do not need to communicate with adults so much.
4. *Sex.* Girls' early language development is slightly more advanced than

boys, perhaps because girls are more responsive, or because mothers converse with their daughters a little more than they do with their sons in the first year or two of life.

5. *Quality of stimulation*. Children reared in isolation or in institutions where they are severely neglected show marked, but often reversible delay in language development. The quality of language environment in family-reared children is also important. What seems to matter is not so much the quantity of language stimulation, but the quality of language interaction between parents and child. Children with parents who listen, react, and draw out their children conversationally are likely to be more linguistically advanced than those whose parents fail to acknowledge their attempts to communicate or just talk at rather than with them.

6. *Bilingual households*. Children reared in families where two languages are spoken from the outset, show no disadvantage and usually acquire both languages well. Needing to speak two languages may, however, be a disadvantage for children language-delayed for other reasons.

### Assessment of language

It is usually possible to gauge the rough level of an infant and young child's speech and language reasonably accurately by asking the mother questions with the child present about this aspect of development, and by observing the way the child uses language during history taking. It is a good idea *not* to talk to preschool children directly to begin with as the intrusion of a stranger may only serve to make the child shy, and sometimes such shyness lasts longer than would otherwise be the case. Where there is a discrepancy (usually with the child not showing on observation such an advanced level as the mother reports), this is often because the child is inhibited in a new situation. It is helpful to have available appropriate toys or puzzles, and to ask the mother to use them with the child. This will give some idea of the quality of language interaction. If it is wished to assess particular aspects of language, it is again best to ask the mother to provide the stimulus. It may be necessary to make precise requests such as asking her to get the child to pick out a colour or an object without pointing. Direct interaction with preschool children to assess language development is usually best achieved through imaginative play. Pointing to a picture in a book may also stimulate language production. In older children, engaging the child in conversation about a recent experience or a television programme will often elicit a response useful for language assessment.

After assessment of language ability in this way, one should have a reasonable idea of the child's capacity for comprehension, expression, use of inner language (best shown in capacity for make-believe play), and response to gesture.

A rough estimate of the child's hearing can be obtained in children up to the age of one year by observing the child's ability to localize sounds, and later by making simple requests, either in front of the child or (with older children) by minimal voicing standing behind the child. The availability of tests of hearing (see Section 2.11.2) which assess the child's ability to hear over a range of sounds of different tones, has meant that most children are now competently assessed for hearing ability in surveillance examinations carried out by general practitioners or clinical medical officers. However, hearing loss can be episodic, and it is unwise to assume that a child's hearing is currently unimpaired because a normal hearing test has previously been carried out, especially if there is a history of recurrent otitis media.

Children in whom there is an apparently significant delay in language development or in hearing ability should be referred for more expert assessment. There are various excellent tests of the language ability of the preschool child, including the Reynell Developmental Language Scales (Reynell 1969). The Illinois Test of Psycholinguistic Ability (Kirk *et al.* 1968) and the Test for Reception of Grammar (Bishop 1983) can be used with older children, and are specific tests of language ability. For an assessment of inner language, a test of symbolic play (Lowe and Costello 1976) can be helpful.

## 2.5.2 LANGUAGE DELAY IN YOUNG CHILDREN

A child who is not talking or who is unusually slow to speak may be suffering from any of the following, either alone or (and this is not unusual) in combination:

(1) mental retardation; Section 2.6.2;
(2) specific developmental language delay and aphasia;
(3) infantile autism—see Section 2.8.1;
(4) hearing impairment; Section 2.11.2;
(5) elective mutism—see Section 2.5.4.

### Mental retardation

If language delay is associated with delay in other aspects of development to roughly the same degree, then the child is likely to be showing some degree of mental retardation. This is the most common cause of significant language delay, and is discussed in more detail in Section 2.6.2.

### Specific developmental language delay and aphasia

In both the DSM III-R (American Psychiatric Association 1987) and the ICD-10 (World Health Organization 1990) classifications, specific developmental disorders are defined as disorders characterized by inadequate development not due to demonstrable physical or neurological disease. Both classifications divide developmental speech and language problems into articulation, expressive language and receptive language disorders, with ICD-10 including an additional category of acquired aphasia with epilepsy.

In most children with specific language delay, articulation, comprehension and expression are all immature to some degree. Most are also generally mentally retarded, at least to a mild degree, although because their lack of language development is markedly below their non-verbal ability, they must be regarded as showing specific language delay. A minority are of average or above-average general ability. The child cannot understand simple requests at a level expected for his age, and has difficulty in naming common objects. The condition is commonly associated with immature speech production. It should be noted that small discrepancies between verbal and non-verbal ability are common in the first 2 or 3 years of life, and such discrepancies are usually of little significance.

Severe forms of specific language delay, with particularly marked discrepancy between non-verbal and verbal ability, are classified more appropriately as forms of aphasia especially if the child is of average or above-average general ability. Such aphasias may be mainly sensory or mainly expressive in nature. In sensory aphasias, the child will not understand spoken speech despite the presence of normal hearing (as assessed by special investigations), and shows well developed ability in non-verbal skills. Spontaneous speech is usually very limited. In expressive aphasia, the child's hearing and understanding of speech is normal or near normal, but his ability to express himself is severely deficient.

### Background factors

Significant specific language delay with or without associated mental retardation is present in about 6–8 per cent of three-year-old children (Silva *et al.* 1987). Developmental expressive language delay not associated with mental retardation occurs in about 5–6 per 1000 children. It is about twice as common in boys as in girls, and there is a strong association with family size, social class, and the presence of behavioural and emotional disorders (Stevenson and Richman 1976). There is also often a family history of similar delay. Severe, pure receptive aphasia occurs in a much smaller proportion (about 1 per 10 000 children). Other forms of aphasia are also rare. There is no specific association between aphasia and social class.

### Aetiology

Specific developmental language delay is usually caused by a combination of polygenic inheritance and a poor quality of language environment.

Aphasias may be congenital (present from birth) or acquired. Congenital forms of aphasia are usually constitutionally or genetically determined (in which case a family history will often be obtained), but may be secondary to perinatal birth trauma or severe hearing loss. Acquired aphasias arise from brain trauma (usually sustained in a road traffic accident), cerebral infection, or cardiovascular accidents. Chronic receptive aphasia may also arise following, or in association with, a cluster of epileptic seizures—the Landau–Kleffner syndrome (Bishop 1985).

*Management*
Children with developmental language delay often show multiple signs of deprivation and behavioural disturbance. Action to improve the social circumstances of the family should be taken if this is possible. Improvements in housing conditions and establishment of the right to financial benefit may result in a better environment for language stimulation.

If the child appears to be relatively neglected at home, other measures to improve the quality of care should be taken. Advice to parents on ways of opening up conversation with their children may be helpful. In particular, parents can be helped to respond to their children's comments in ways that open rather than close the way to further communication. Labelling and talking about everyday activities should be encouraged. A special time should, if possible, be put aside for play and parents should be encouraged to let the child help with domestic activities, and to talk with the child while these are going on. Parents need to be particularly patient in trying to understand a child with language delay, and helping him cope with the frustration of not being understood. Sibs and peers will often not be so patient, and this may lead the child to develop behavioural problems arising from his inability to communicate. Placement in a playgroup or nursery school will be helpful if the child is ready to separate.

More severely language-delayed children will benefit from advice and sometimes treatment from a speech therapist. Individual or group sessions may be helpful, but probably of at least equal value is the specific instruction that therapists can give to parents with advice on how to practise the sounds with which the child is having particular difficulty.

A small number of severely language-handicapped children, the 'aphasic' group, may benefit from special schooling specifically designed for the education of the poorly communicating child of average or above-average ability. The need for such schooling should be apparent in the preschool years, but may only surface in middle childhood when less specialized educational measures have proved ineffective. Educational units for young non-communicating children, often taking a mixture of autistic, developmentally retarded, and specifically language-delayed children, can play a most valuable role. All language-delayed children need repeated examination of their hearing to identify remediable hearing defects.

*Prognosis*
The child with a specific language delay of mild severity will probably catch up. However, more severely delayed children, for example those speaking

only a few single words by 3 years, are at high risk for later behavioural and educational difficulties (N. Richman *et al.* 1982). Many will require special education for the sequelae of such developmental delay. These include continuing language delay, specific reading difficulties, and behavioural disorders. Later language development in children diagnosed as showing aphasia is variable. Many children will be chronically handicapped by their language delay, and later show reading and writing difficulties. The outcome for some children with acquired aphasias is, however, better than this, and marked improvement may occur in the first two years after an insult to the brain is sustained.

**Hearing impairment** (Section 2.11.2)

Hearing impairment arising from recurrent otitis media is a particularly common cause of mild language delay and immaturity of speech.

**Differential diagnosis**

The differential diagnosis of language delay therefore involves these five conditions (mental retardation, specific developmental language delay, hearing impairment, infantile autism, and elective mutism), that sometimes occur in combination. In determining which is present, assessment should involve careful history-taking and observation with particular attention given to the level of non-verbal ability, response to sounds, the use of gesture, the presence of neologisms and echolalia, the variability in language usage, and the child's speech intonation (see Table 4).

**Table 4**

|  | Mental handicap | Specific developmental language delay | Deafness | Autism | Elective mutism |
|---|---|---|---|---|---|
| Non-verbal ability | Poor | Average or above | Average or above | Average or below | Average |
| Response to sounds | Normal | Normal | Absent or poor | Variable | Normal |
| Use of gesture | Present | Present | Markedly present | Absent | Present |
| Neologisms and echolalia | Absent or transient | Absent or transient | Absent | Present | Absent |
| Variability in language use | Absent | Absent | Absent | Present | Present |
| Speech intonation | Normal or immature | Often immature | Abnormal | Abnormal | Normal |

## 2.5.3 DISORDERS OF SPEECH PRODUCTION

### Dyslalia

This involves immaturity of speech production and is characteristic of children who also have developmental language delay. The child frequently omits and sometimes substitutes consonants, and consequently is difficult to understand, though his parents and sibs may be reasonably good at understanding him. Some consonants, such as 'b', 'd', and 'm' are characteristically well pronounced early, while others, such as 's' and 'f' are developed later. Lisping is a form of dyslalia including substitution of 'th' for 's'. If linked to specific language delay, assessment and management of dyslalia should proceed along the lines described below. If occurring in isolation, the condition has a good prognosis and requires no special treatment, though an opinion and advice from a speech therapist will be reassuring.

### Stammering (stuttering)

This involves explosive repetition of consonants and word-blocking. It is relatively common in the preschool period and, almost invariably starts around the age of 3–4 years. The prevalence is around 3 per cent, and it is about twice as common in boys as in girls. Mild hesitancies of speech remain common into middle childhood, when the prevalence is still found to be around 3 per cent.

Genetic factors are probably of importance in causation, for it tends to run in families in a manner that suggests polygenic inheritance, and no specific adverse child-rearing practices have been clearly identified. It is sometimes said to be linked to poorly established cerebral dominance, and to attempts parents make to convert left-handed children to right-handedness, but there is no good evidence for this. Further, in some countries, where all left-handed children are persuaded to use their right hands, there does not seem to be an increased rate. It is not particularly associated with emotional disorder, though sometimes, in severe forms, stammering does lead to great anxiety and distress.

### *Management*

Parents with young children with mild or moderate forms of stammering should be strongly encouraged to ignore it, and to avoid telling the child not to do it or punishing the child. This will be easier for parents to achieve than for older sibs and other children who may tease the stammering child mercilessly. Self-consciousness about the stammer is the main cause of its perpetuation. As with other forms of explosive motor behaviour, like hiccoughs and tics (Section 2.4.4), stopping thinking about the problem helps it to go away. This advice to ignore the condition should not, incidentally, be applied to emotional and behaviour problems from which the child may coincidentally be suffering. Dealing with these while ignoring the stammering may be additionally helpful.

In older children, in whom stammering remains persistent over the age of 5 years, is accompanied by associated movements of the face and other parts of the body, and is at least moderately severe, special techniques applied by psychologists or speech therapists involving the use of relaxation and timed, syllabic speech have been found to improve, though often not to eradicate the condition. Haloperidol (Section 6.2.6) has been reported to produce some symptomatic improvement in adolescents who stammer.

## 2.5.4 ELECTIVE MUTISM

### Definition

The condition is defined in ICD-10 (World Health Organization 1990) as characterized by a marked, emotionally determined selectivity in speaking, such that the child demonstrates language competence in some situations but fails to speak in others. Diagnostic criteria in DSM III-R (American Psychiatric Association 1987) comprise (a) persistent refusal to talk in one or more major social situations and (b) ability to comprehend spoken language and to speak.

### Clinical features

There is often a mild or moderate delay in the early development of language in the first 2 or 3 years. The condition usually comes to attention when the child enters playgroup, nursery school, or infant school. Refusal or inability to talk becomes evident at this point, and is usually (and in many ways appropriately), attributed to unusual shyness. As the child progresses through school and still refuses to talk, it becomes apparent that a more serious disorder is present. Very occasionally the mutism occurs following a traumatic event in middle childhood following which the child refuses to speak. Children with this condition usually talk normally in the home, although sometimes there is a degree of selectivity in whom they will speak to even at home. Associated features may include:

1. Unusual personality features in the child. Most children with elective mutism have been persistently obstinate, self-willed, and manipulative from their early years.

2. Adverse family factors. The parents, especially the mother, are likely to have been somewhat overprotective in the early years. A history of mental ill-health in one or both parents is often present.

### Aetiology

The **temperament** or early personality of the child is probably of major importance. Children with elective mutism are often excessively shy and withdrawing from strangers in their early years. The importance of **developmental immaturity** is suggested by the early history of language delay. These child characteristics compounded with **parental overprotection** result in the child having little experience of the external world. There may be an almost symbiotic relationship beween mother and child (Hayden 1980). Subsequently, the parents put pressure on the child to speak, but are met with persistent obstinacy and refusal. There is also a tendency for members of some of the families of children with elective mutism to opt out of conflict by entering into silence. Mutism has also been seen as a fear-reducing form of learned behaviour (Reed 1963).

### Prevalence

Excessive shyness on school entry is relatively common. In one study (J. B. Brown and Lloyd 1975) about seven per 1000 children entering school at the age of 5 years had not uttered a single word after 8 weeks. By the end of the school year nearly all were speaking normally. The prevalence of clinically significant mutism is probably around one per 1000 (Kolvin and Fundudis 1981). There is a roughly equal sex ratio, and no particular link with social class. Children with this condition are usually of average ability or a little below.

### Assessment

An account that a child speaks normally or nearly so at home, and has not done so at all at school for a year or more is pathognomonic of the condition. Milder levels of non-speaking should be classed as pathological shyness rather than elective mutism. The fact that the child shows affectionate social relationships in the home and uses gestures to express needs in school excludes the possibility of autism. Deafness is usually ruled out by the normal responses within the home.

Evaluation of the temperament of the child and of the quality of family relationships form a necessary component of assessment.

### Treatment

It is often helpful to begin treatment with a small number of **family interviews** to clarify patterns of communication within the family and identify the presence of special alliances (especially between the mother and the identified child). Family interviews may also be helpful in bringing to the fore the often considerable anger experienced by both parents and child.

Subsequently, **behavioural methods** (Section 6.2.2) are likely to have most success, although these by no means always produce improvement. A behavioural contract can be set up with positive reinforcement for gradually increasing use of verbal interchange. The imaginative use of tape-recorders on which the child can play back his own voice in different settings may be helpful.

Individual psychotherapy with mute children is not usually effective, and medication has no place in the treatment of this condition.

## Outcome

A 5–10 year follow-up of children with elective mutism suggested that slightly less than half improved (Kolvin and Fundudis 1981). Failure to improve by the age of 10 years was thought to be a poor prognostic sign, but otherwise there were no clear indicators of outcome. Longer-term follow-up of this condition into adult life does not seem to have been carried out, but one would expect a further proportion to improve on leaving school.

### 2.5.5 READING AND READING RETARDATION

The process of reading involves understanding the meaning of written or printed symbols. Most children are able to read single, short words by the age of 6 or 7 years and by 10 or 12 years have acquired sufficient reading ability to cope with everyday tasks such as reading a tabloid newspaper. The acquisition of reading ability is a necessary requirement for full participation in a literate society.

## The development of reading ability

In order to achieve a reasonable level of reading ability, a child must be able to:

1. Make perceptual discriminations, for example to discriminate between m and n, p and q, and also between many other similar shapes.
2. Discriminate between sequences, for example to distinguish between BAD and DAB.
3. Achieve transfer of information from one sensory modality to another—visual symbols must be transformed into auditory internal or spoken speech.
4. Blend sound satisfactorily, for example articulating the word CRAMPS involves five blending tasks.
5. Comprehend language at the level of the written material.
6. Integrate information acquired from new reading material into an already existing knowledge store.

The acquisition of reading ability probably arises in two main ways, so that there are two routes to reading. The first is phonological and involves the slow and painful deciphering of the single letters or common combinations of letters (e.g. th, ing) in words along the lines described above. The second or lexical route involves the establishment of the meaning of a passage by using a variety of other higher-order skills. Whole words rather than single letters are perceived. Some words less essential to understanding meaning are either very briefly scanned or not scanned at all. Most children begin by using the phonological route, but rapidly adopt the lexical route, only reverting to the phonological when they have difficulty in deciphering an essential word, phrase, or longer passage.

## Types of reading difficulty

Standardized tests of reading ability (Section 1.2.3) such as the Neale (1958), the Schonell and Schonell (1950), and the Wide Range Achievement Test [WRAT] (Jastak and Jastak 1978) allow the establishment of a child's reading age. This may be above or below his chronological age.

## General reading backwardness

Many children who have difficulty reading are behindhand in all aspects of their development—the majority of these are mildly mentally retarded. There is a reasonably high but imperfect correlation between reading ability and intelligence ($r = 6$) so that quite a number of retarded children are found to read somewhat above or below the level one would expect from their general ability.

## Specific reading retardation (SRR)

Specific reading disorder is defined in ICD-10 (World Health Organization 1990) in terms of a specific and significant impairment in the development of reading skills that is not solely accounted for by mental age, visual acuity problems, or inadequate schooling. Diagnostic criteria for developmental reading disorder in DSM III-R (American Psychiatric Association 1987) comprise (a) reading achievement, as measured by a standardized, individually administered test that is markedly below the expected level given the person's schooling and intellectual capacity (as determined by an individually administered IQ test); (b) interference with academic achievement or activities of daily living requiring reading skills; and (c) absence of a defect in hearing or visual acuity and neurological disorder.

Using a regression equation, it is possible to calculate a predicted reading age from a knowledge of the level of the child's non-verbal intelligence. Thus, if a child of 10 years has a non-verbal IQ of 100 one can predict that there is a high probability that he will be reading at an average level for his age. If the child has a reading age of about 13–14 years, a statistical calculation is likely to show that his reading age is about two standard deviations above predicted, and if he is reading at about the 6–7-year level, this is likely to mean that he is about two standard deviations below that predicted. Specific reading retardation is said to exist when a child is reading at an arbitrary level (usually 1½ or 2 standard deviations) below that expected from his non-verbal ability.

Attempts have been made to define a syndrome of **dyslexia** from among the group of children with SRR. Dyslexia is said to exist when, with a background of adequate instruction and a good social background, genetic or constitutional factors are responsible for neuropsychological deficits leading to reading failure. There is a lack of evidence from general population studies to suggest that it is possible to distinguish at all clearly a group of dyslexics from other children with specific reading retardation, but some clinicians nevertheless still find the concept useful in individual cases.

### Prevalence

The rate of SRR (defined as two standard deviations below non-verbal IQ, was found to be 4 per cent in Isle of Wight 10- and 11-year-old children, and approximately 10 per cent in an inner London borough (Berger *et al.* 1975). The rate is higher in boys (3:1) and in children from working-class families. Large family size and overcrowding is also commoner in children with specific reading retardation, and indeed in children with general reading backwardness (Rutter *et al.* 1970*a*).

### Background features

It is especially difficult in this condition to distinguish between those background factors that are causative and those that are merely associated. It should not therefore be assumed that all the factors listed below have aetiological significance, though a number probably do.

**Familial factors.** The rate of SRR is raised in family members of affected children. This is likely to be due to a mixture of social and genetic factors: in some affected children the family environment is of obvious importance, but in others a reasonably clear-cut pattern of genetic inheritance can be detected. Recent studies of the genetics of dyslexia (Pennington 1990) have confirmed the importance of genetic influences and indeed single gene effects may very occasionally be demonstrable in producing reading problems. Genetic factors may be of special importance in children with persistent spelling difficulties (Stevenson *et al.* 1987).

**Social factors.** As already indicated, children with SRR tend to come from large families, living in overcrowded circumstances. A lack of books in the home and low interest in reading are also often of importance. It has been demonstrated (Hewison and Tizard 1980) that if parents of infant school-aged children can be encouraged to listen to their children reading on a regular basis, this has a definite preventive effect on the development of reading problems.

**Neuropsychological deficits.** These fall into two main categories:

1. *Perceptual deficits.* Reading-retarded children often have a poorly developed visuospatial ability, and have difficulty in telling left from right. Their capacity to make both fine auditory and visual discriminations may be impaired. It is not surprising that children who find difficulty in telling the difference between similar but not identical sounds, and similar but not

identical visual patterns, should also often have reading problems. These perceptual problems may arise from a delay in the establishment of cerebral dominance, as there is a suggestion that children showing little preference for the use of one hand are more likely to be reading retarded. Mixed eye–handedness and mixed hand–footedness are not, however, particularly associated with reading problems, nor do left-handed children have a particularly high rate.

2. *Language deficits.* Reading-retarded children are often slow to speak, and subsequently their language development may continue to be delayed. Both comprehension and expression are likely to be affected.

**Sensory deficits.** Episodic partial hearing loss, usually due to recurrent ear infections, is sometimes associated with delay in reading development. The relevance of chronic secretory otitis media ('glue-ear') in early childhood to later language delay and reading problems is uncertain, some believing it to be a widely prevalent cause, and others that it is of much less significance. Visual defects, such as uncorrected errors of refraction, are also occasionally of significance.

**Behavioural disorders.** Children with conduct disorders, characterized by antisocial behaviour are particularly likely to show specific reading problems (Rutter *et al.* 1970*a*). Further, children with reading problems have a particularly high rate of conduct disorders. The reasons for this close association are unclear. It seems likely that a number of children who fail educationally become discouraged and turn to antisocial behaviour in frustration. It is also possible that children who are impulsive and lack concentration as a result either of their early experience or of inherited personality traits, may develop both antisocial behaviour and learning problems for these reasons.

**Educational opportunity.** Teachers, like parents, vary in their ability to teach children to read. Monitoring and encouraging the reading progress of a group of 30–35 young children is not an easy task. Children, exposed to many changes of teacher, or to teachers who find it hard to maintain discipline and a positive approach to learning in their pupils, are more likely to have reading disabilities. Absence from school alone, unless it is considerably prolonged, is unlikely to make too much difference to a child's educational progress in the early years of schooling. Educational catch-up is the rule rather than the exception.

## Clinical features
**Early identification.** Children who are slow to develop language, are clumsy, or have difficulty with spatial tasks such as jigsaws are at risk for developing later learning difficulties. However, a significant number of children will 'catch up'. It is doubtful whether there are any satisfactory tests that can be applied in the preschool period to predict later reading failure with any useful degree of accuracy, though this is sometimes claimed (Baird and Hall

1985). However, children with language delay and poor concentration, drawing skills, and motor coordination are certainly at greater risk.

**Reading retardation.** Children may be brought to health professionals because of failure to learn to read or because of very slow progress. The purpose of such a consultation is usually to rule out an underlying medical cause, such as sensory impairment. Alternatively, with less justification, the doctor may be asked to consider the possibility that the reading problem is itself a medical condition such as dyslexia. As indicated above, it is usually not possible to distinguish dyslexic specific reading retardation from non-dyslexic reading retardation.

**Associated or secondary disorders.** Children with severe reading problems not infrequently come to attention because of somatic symptoms such as abdominal pain or headaches produced by tension or anxiety. Even more commonly, children presenting with some medical complication of antisocial behaviour such as a physical condition produced by glue-sniffing (Section 3.3.2) will be found to show a serious degree of educational retardation.

### Management

The identification, assessment, and remediation of specific reading retardation is largely an educational matter, but health professionals are appropriately involved in a number of ways:

1. *Early identification.* Children with developmental problems such as early language delay may be identified during surveillance procedures. It is important that information about such delays is passed on to school teachers, so that extra attention can be given to the child from an early age.

2. *Later identification.* Children brought to doctors for physical or emotional problems may turn out to show serious educational failure. Again it is important to ensure that teachers are alerted to the learning difficulty if they are not already aware of it. Referral to an educational psychologist can then be made if this appears necessary.

3. *Assessing medical and psychiatric causes of learning failure.* Children may be referred for this reason, and indeed, if the retardation is moderate or severe, a medical assessment is indicated. Vision and hearing should be carefully checked, and referral for treatment made if necessary. Tests of motor clumsiness and visuospatial difficulties may be undertaken, although usually this will take place more systematically during assessment by the educational psychologist. Attention to stressful factors in the home, neighbourhood, or school may indicate that a child's learning difficulties and lack of concentration are related to anxiety. The presence of an associated psychiatric disorder should lead to treatment or appropriate referral.

4. *Remediation.* The treatment of specific reading retardation is an educational matter, unless there are associated medical or psychiatric

problems. The most frequently used approach is small-group tuition with teaching methods based on an understanding of the child's perceptual and linguistic strengths and deficits. Such understanding is likely to emerge most clearly from a systematic assessment by a specialist teacher or educational psychologist. If the child is discouraged and unmotivated, the teacher may need to spend a number of sessions building a positive relationship with the child. A number of specific programmes have been devised over the years, but there is no evidence that any one approach will be suitable for all children with this problem. Children probably benefit most from intensive, personalized teaching in as small a group as possible, and indeed, individual one-to-one tuition is probably best of all.

5. *Advocacy.* Doctors or other health professionals may be approached because of a lack of educational progress or because there is a lack of adequate remedial facilities locally. In these circumstances a doctor can make enquiries about the local availability of specialist educational help, so that he can either reassure parents that all possible steps are being taken, or put them in touch with educational facilities that may be able to offer a greater amount of help.

*Outcome*
Children aged 10 years or over who are seriously and specifically retarded in reading are unlikely to make very rapid progress, even with the best educational help available. However, given appropriate help they will make slow steps forward and, although most will continue to find reading difficult, a minority will make substantial progress. Spelling difficulties in adolescence and adulthood are very common in children who have had difficulties in learning to read.

One major aim of parents and professionals concerned with children having difficulty reading must be in the prevention of secondary antisocial problems, by encouraging the child in skills such as sports and carpentry, in which verbal ability is not of prime importance. Loss of self-esteem with consequent delinquency is a definite risk among children who are failing educationally.

# 2.6 Intelligence and learning disorders

## 2.6.1 INTRODUCTION

### Definition

Intelligence can be considered as the sum of those cognitive abilities that underlie adaptation to the environment. Like other biological characteristics, intelligence is distributed normally throughout the population. Its development is under both genetic and environmental control, though the relative contribution of each of these is a subject of controversy (Madge and Tizard 1980). In Western, developed countries, twin and adoption studies suggest that polygenic factors contribute between 30 and 60 per cent to the variance of intelligence. In developing countries, where environmental variations are more extreme, the genetic contribution may be smaller.

The **continuity** of intellectual development from early infancy to childhood, and from childhood to adulthood is variable. Except for those at the extreme lower end of the scale, later childhood and adult intelligence cannot be predicted with any useful accuracy from testing carried out under the age of about 5 years. Continuity is considerably greater from 5 years upwards, predictive strength increasing with age. This may be as much a result of continuity of test material and environmental conditions as to genetic factors. Continuity of intelligence test scores is, however, only a group phenomenon, and the individual child may vary considerably in IQ test results from age to age, without this having any pathological significance. Thus Honzik *et al.* (1948) found, in children repeatedly tested between 2 and 18 years, that 58 per cent varied by 15 points or more on different testing occasions.

The **relationship between IQ, educational attainment, and later occupational achievement** is positive, but tenuous. Many highly intelligent people perform disappointingly, and others, with very average IQs, may do remarkably well. This is because factors other than intelligence, such as drive, persistence, attention, inherited wealth, useful social contacts, and highly developed social skills, are of major importance in the achievement of worldly success.

The measurement of intelligence by the use of intelligence tests is discussed in Section 1.2.3.

### 2.6.2 MENTAL RETARDATION

#### Definition

Slowness in intellectual development may be widespread and affect all aspects of cognition. Only rarely will a child's functioning be retarded to the same degree over the entire range of skills, but where such skills are almost all significantly impaired, it is reasonable to think of the child as showing general mental retardation. This term is not universally accepted, and there are various synonyms.

Previously mental retardation was referred to as mental subnormality. It is also known as mental handicap or general learning difficulty. Preschool mentally retarded children may be referred to as generally developmentally delayed. In the UK, the Education Act 1981 has abolished such administrative categories. Schools for mildly retarded children (see below) are now referred to as schools for children with moderate learning difficulties. Schools for children with severe learning difficulties cater for more seriously retarded children.

**Table 5** Classification of mental retardation by IQ

|  | ICD-10 | DSM III-R |
| --- | --- | --- |
| Mild | 50–69 | 50–55 to 70 |
| Moderate | 35–49 | 35–40 to 50–55 |
| Severe | 20–34 | 20–25 to 35–40 |
| Profound | Below 20 | Below 20 or 25 |

N.B. Concurrent deficits in adaptive functioning must also be present.

| Education categories and IQ | |
| --- | --- |
| Moderate learning difficulties (MLD) | 50–70 |
| Severe learning difficulties (SLD) | Below 50 |

N.B. IQ is only one consideration in educational categorization.

For a diagnosis of mental retardation or handicap to be appropriate, there are two important requirements. The child's IQ should be less than about 70, and the child should be functionally impaired in his everyday life. The severity of retardation is classified slightly differently in the two major classification systems (World Health Organization 1990; American Psychiatric Association 1987). Terminology is further complicated by the fact that, in the UK, as well as in other countries, educationists use different criteria in classifying severity (see Table 5). Mild retardation differs from more severe forms of retardation in presentation, aetiology, associated features, prevalence, appropriate management, and outcome, so the distinction by severity is vital.

## Clinical features

Mental handicap may first be identified as probably or certainly present when a condition such as Down's syndrome, known to be strongly associated with it, is first diagnosed. Conditions associated with mental retardation may be diagnosed antenatally (e.g. by biopsy or amniocentesis), at birth (e.g. by the presence of congenital abnormalities), or shortly after birth (e.g. by biochemical screening or the detection of cerebral palsy). Alternatively, the first indication of mental handicap may be the presence of general developmental delay in a child hitherto thought to be normal.

When mental handicap is diagnosed in a child previously regarded as normal, the mode of presentation will usually depend on the severity of the condition. Moderately and severely retarded children will often present with delay in their motor milestones in the first few months of life. The child will be slow to obtain head control, sit unsupported, etc. A large number of moderately retarded children, however, show normal motor development and present for the first time with language delay. The child may be thought to be deaf because he fails to take notice of sounds, or shows lack of single words or word combinations at the appropriate age (Section 2.5). Mildly retarded children may not be detected until they enter school when failure of educational progress may be found to be due to a general slowness of development rather than to a specific learning disability. Usually, however, it will be found that the early development of the mildly retarded, especially their early language development, has been slow. Occasionally, mental retardation arises as a result of some postnatal event, such as a head injury or cerebral infection. In these cases the time course of the condition will, of course, be different.

Once diagnosed, the clinical features of children with mental retardation will depend more especially on:

(1) the severity of the condition;
(2) the presence of associated physical and psychiatric conditions;
(3) the quality of care and education the child receives.

These related factors are discussed in more detail below.

### Prevalence

Mild mental retardation (IQ 50–70) occurs in 2–3 per cent of the population. However, this prevalence figure is not very meaningful, as IQ tests are standardized to ensure that roughly this proportion of children score in this range. If the intelligence of the entire population rose, tests would be restandardized and the rate of children scoring at IQ level 50–70 would remain the same. Moderate to profound mental retardation (IQ less than 50) occurs in about 3 per 1000 children. Boys and girls are roughly equally affected, though, because of their higher rate of associated disorders, especially behavioural problems, there are usually more boys than girls in special schools for the mentally retarded. In the UK, mild mental retardation occurs most commonly in children from low socio-economic groups. Middle class children are under-represented in schools for children with moderate learning difficulties. In countries with less socio-economic inequality, such as Sweden, mild mental retardation has been found to show little link with social class (Hagberg *et al.* 1981). Moderate and severe mental retardation is not currently associated with social class, though as more conditions producing it become preventable (e.g. by abortion following amniocentesis), one may predict that, because of differential access to the best health care, the proportion of severely retarded children from low socio-economic groups will rise. There is already some indication from West Germany that this is occurring (Liepmann 1979).

## Aetiology

### Mild retardation (IQ 50–70)

**Polygenic influences and multiple deprivation.** Most children with mild retardation come from deprived family backgrounds in which the quality of parental care provided has been poor. There is a strong link between mild retardation and low socio-economic status, large family size, and overcrowded housing. Because the emotional, intellectual, and financial resources of the parents have been stretched, they have been unable to provide the 'language environment' which promotes language development. Such children have been little listened to, talked to, or played with.

Parental intelligence is usually below average, though only a minority of parents of mildly retarded children are retarded themselves. Family and twin studies suggest that polygenic influences are also of importance in aetiology.

In addition, all specific causes of moderate, profound, or severe retardation (see below) may, on occasion, produce mild retardation if the insult to the brain is less serious. As social conditions in developed countries improve, it is likely that more mild retardation will be produced in this way, and a lower percentage will be related to polygenic influences and environmental deprivation. There is already evidence this is occurring in Scandinavian countries (Hagberg *et al.* 1981) and the UK (Lamont and Dennis 1988). The latter found that 42 per cent of children attending schools for children with moderate learning difficulties (mild mental retardation) had probable specific medical reasons for their retardation. This figure is probably an overestimate of the medical contribution, in that mildly retarded children not attending special schools were not included in the survey, and the aetiological significance of the perinatal abnormalities they recorded was sometimes doubtful. Nevertheless it is clear that there is indeed now a sizeable number of mildly retarded children with specific medical (in contrast to polygenic and environmental) reasons for their slow development, of which chromosomal abnormalities (especially Down and fragile X syndromes—see Section 5.3) and perinatal abnormalities are the most common.

Other physical causes of mild retardation include

1. **Malnutrition.** The nutritional state of the mother during pregnancy affects fetal development, including development of the fetal brain. After birth, malnutrition may also be a cause of psychological deficit or abnormal behaviour even in developed countries. Children who fail to thrive because of inadequate calorie intake, especially in the first year of life, frequently show mild learning difficulties when they enter school, and investigation of the link suggests that inadequate nutrition to the brain may be a significant causal factor (D. Skuse 1989). In developing countries, malnutrition usually co-exists with infection and severe environmental

deprivation. It probably affects mental functioning both directly and in an indirect way. The malnourished infant and young child is often lethargic and slow to respond to stimulation. He is prone to infection, and therefore more likely to suffer cerebral damage with effects on psychological functioning. Even in developing countries, early malnutrition probably affects later performance as a result of interaction between physical and environmental factors (Stein and Susser 1985).

2. **Toxins.** There is controversy as to whether exposure to the lower levels of lead in the atmosphere and in dust created by vehicle exhausts is a hazard to children's mental development. Some epidemiological studies carried out in the UK suggest the effect is non-existent (I. Smith *et al.* 1983) or small (Thomson *et al.* 1989). Others carried out in US cities where lead pollution is heavier suggest that effects may be more sizable (Needleman *et al.* 1979). Alcohol intake in pregnancy producing the fetal alcohol syndrome (Section 2.1.1) is another example of a toxin causing mild mental retardation.

3. **Specific trauma.** Some children with normal intellectual potential suffer a specific trauma to brain function sufficient to impair intelligence to some degree, but not to such severity to cause moderate or even more serious retardation. Some children with cerebral palsy, post-encephalitic states, or trauma to the head fall into this category.

4. **Idiopathic.** Finally there is a small number of cases of mild mental retardation in children who come from apparently non-deprived backgrounds with no obvious physical condition present. Here the question arises whether there has been 'hidden' deprivation perhaps with the child being neglected for long periods in an affluent household by au pairs, nannies, etc. or whether there is a physical cause of unknown aetiology. Sensitive history-taking is necessary in these situations as the distinction is important for future management, but sometimes uncertainty remains even after a careful assessment.

*Moderate to profound retardation (IQ less than 50); Fig. 2*
Although this level of retardation may be produced by gross deprivation, the great majority of children functioning at this level of intelligence have organic brain pathology accounting for their retardation (Gustavson *et al.* 1977; Hagberg 1978; McQueen *et al.* 1986). Occasionally the effects of an organic lesion are compounded by coexisting neglect.

**Chromosomal defects** (Section 5.3) now account for about 40 per cent of the moderately or more severely mentally retarded.

1. *Down syndrome* (trisomy 21) accounts for about three-quarters of this 40 per cent, i.e. nearly one-third of all cases of moderate to profound retardation.

2. *Fragile X syndrome.* This syndrome accounts for about 10 per cent of

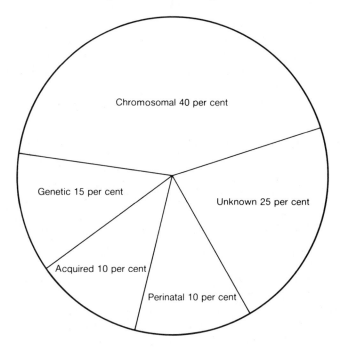

**Fig. 2.** Causes of moderate, severe, and profound mental retardation in childhood.

moderate or more severe retardation in boys, and occasionally for retardation in girls (Section 5.3).

3. *Sex chromosome abnormalities.* Children with Turner's syndrome (XO), Klinefelter's (XXY), and other abnormalities of the sex chromosomes although usually of intelligence in the normal range, sometimes show general mild or moderate mental retardation, or, more commonly, specific cognitive deficits.

4. *Other autosomal abnormalities.* These include disorders in which there is a chromosomal deficit (e.g. *cri du chat* syndrome) or excess (other trisomies).

**Genetic defects.** Single gene defects account for about 15 per cent of moderate to profound retardation. These are mainly metabolic disorders such as galactosaemia and phenylketonuria exerting their effects by altering the metabolism of amino acids, lipids, carbohydrates, and, rarely, other bodily constituents. Some endocrine disorders as well as a range of other rare genetic disorders are included in this category.

**Abnormalities of pregnancy and the perinatal period** account for approximately 10 per cent of cases:

(1) infections of pregnancy, for example rubella, toxoplasmosis, cytomegalo-virus, acquired immune deficiency syndrome (AIDS);

(2) alcohol or drug abuse in pregnancy;
(3) maternal phenylketonuria;
(4) perinatal abnormalities, including birth trauma and postnatal anoxia;
(5) neonatal disorders—including infective and metabolic conditions such as hypoglycaemia.

**Postnatal causes.** These include head injury (accidental or non-accidental), infantile spasms, and cerebral or meningeal infections and exposure to toxins, such as large quantities of ingested lead producing encephalopathy. Mental retardation also very occasionally follows prolonged anoxia of the brain caused by cardiac arrest or obstruction to the respiratory tract, as in near-miss drowning or sudden infant death syndrome.

**Other causes.** In about 25 per cent of cases of moderate to profound retardation no cause is identifiable. In a proportion of such children the presence of other signs of developmental abnormality such as deformities or organ malformations makes it likely that the mental retardation has arisen as a result either of a single gene or chromosomal disorder, or as a failure of early fetal development. In others the presence of epilepsy makes it clear there is an undiagnosed underlying cerebral abnormality. Some children do not show such stigmata and indeed look perfectly normal. In the absence of a history of gross deprivation or non-accidental injury, however, it seems reasonable to assume that such children are suffering from an unidentified organic disorder. Any unjustified assumption that such cases might be caused by covert parental neglect is likely to increase the already serious emotional burden on the family.

Table 6 provides information on the medical features of some different causes of mental retardation.

### Assessment

There are four components to assessment of the retarded child:

(1) determination of the level of functioning;
(2) identification of associated physical and psychiatric problems;
(3) investigation of the cause of the retardation;
(4) assessment of family care, expectations, and coping capacity.

*Determination of level of functioning*
Techniques are available that can be used to assess the level of the child's functioning in different aspects, general intelligence, language, gross motor and fine motor development, and social and personal development. The generally retarded child will, of course, require assessment in all these different areas.

A clinician with experience and a small amount of available equipment will be able to make a reasonably accurate assessment of a child's developmental level in each of these areas. A more precise estimate can be obtained

**Table 6** Notes on some causes of mental retardation

| Syndrome | Aetiology | Clinical features | Comments |
|---|---|---|---|
| **Chromosome abnormalities** (For Downs syndrome and X-linked retardation see text) | | | |
| Triple X | Trisomy X | No characteristic feature | Mild retardation |
| Cri du chat | Deletion in chromosome 5 | Microcephaly, hypertelorism, typical cat-like cry, failure to thrive | |
| **Inborn errors of metabolism** | | | |
| Phenylketonuria | Autosomal recessive causing lack of liver phenylalanine hydroxylase. Commonest inborn error of metabolism | Lack of pigment (fair hair, blue eyes) Retarded growth. Associated epilepsy, microcephaly, eczema, and hyperactivity | Detectable by post-natal screening of blood or urine. Treated by exclusion of phenylalanine from the diet during early years of life |
| Homocystinuria | Autosomal recessive causing lack of cystathione synthetases | Ectopia lentis, fine and fair hair, joint enlargement, skeletal abnormalities similar to Marfan's syndrome. Associated with thromboembolic episodes | Retardation variable. Sometimes treatable by methionine restriction |
| Galactosaemia | Autosomal recessive causing lack of galactose 1-phosphate uridyl transferase | Presents following introduction of milk into diet. Failure to thrive, hepatosplenomegaly, cataracts | Detectable by post-natal screening for the enzymic defect. Treatable by galactose-free diet. Toluidine blue test on urine |
| Tay–Sachs disease | Autosomal recessive resulting in increased lipid storage (the earliest form of the cerebro-macular degenerations) | Progressive loss of vision and hearing. Spastic paralysis. Cherry-red spot at macula of retina. Epilepsy | Death at 2–4 years |
| Hurler's syndrome (gargoylism) | Autosomal recessive affecting mucopolysaccharide storage | Grotesque features. Protuberant abdomen. Hepatosplenomegaly. Associated cardiac abnormalities | Death before adolescence |

**Table 6** (*continued*)

| Syndrome | Aetiology | Clinical features | Comments |
|---|---|---|---|
| Lesch–Nyhan syndrome | X-linked recessive leading to enzyme defect affecting purine metabolism. Excessive uric acid production and excretion | Normal at birth. Development of choreoathetoid movements, scissoring position of legs and self-mutilation | Can be diagnosed prenatally by culture of amniotic fluid and estimation of relevant enzyme. Post-natal diagnosis by enzyme estimation in a single hair root. Death in second or third decade from infection or renal failure. Self-mutilation may be reduced by treatment with hydroxytryptophan |
| **Other inherited disorders** | | | |
| Neurofibromatosis (Von Recklinghausen's syndrome) | Autosomal dominant inheritance | Neurofibromata, *café au lait* spots, vitiligo. Associated with symptoms determined by site of neurofibromata. Astrocytomas, menigioma | Retardation in a minority |
| Tuberose sclerosis (Epiloia) | Autosomal dominant (very variable penetrance) | Epilepsy, adenoma sebaceum on face, white skin patches, shagreen skin, retinal phakoma, subungual fibromata. Associated multiple tumours in kidney, spleen, and lungs | Retardation in about 70 per cent |
| Lawrence–Moon–Biedl syndrome | Autosomal recessive | Retinitis pigmentosa, polydactyly, sometimes with obesity and impaired genital function | Retardation usually not severe |
| **Infection** | | | |
| Rubella embryopathy | Viral infection of mother in first trimester | Cataract, microphthalmia, deafness, microcephaly, congenital heart disease | If mother infected in first trimester, 10–15 per cent infants are affected (infection may be subclinical) |
| Toxoplasmosis | Protozoal infection of mother | Hydrocephaly, microcephaly, intra-cerebral calcification, retinal damage, hepatosplenomegaly, jaundice, epilepsy | Wide variation in severity |

| | Aetiology | Clinical features | Course / treatment |
|---|---|---|---|
| Cytomegalovirus | Virus infection of mother | Brain damage. Only severe cases are apparent at birth | |
| Congenital syphilis | Syphilitic infection of mother | Many die at birth. Variable neurological signs. 'Stigmata' (Hutchinson teeth and rhagades often absent) | Uncommon since routine testing of pregnant women. Infant's WR positive at first but may become negative |
| **Cranial malformations** | | | |
| Hydrocephalus | Sex-limited recessive Inherited developmental abnormality, e.g. atresia of aqueduct, Arnold–Chiari malformation. Meningitis. Spina bifida | Rapid enlargement of head. In early infancy, symptoms of raised CSF pressure. Other features depend on aetiology | Mild cases may arrest spontaneously. May be symptomatically treated by CSF shunt. Intelligence can be normal |
| Microcephaly | Recessive inheritance, irradiation in pregnancy, maternal infections | Features depend upon aetiology | Evident in up to a fifth of institutionalized mentally retarded patients |
| **Miscellaneous** | | | |
| Spina bifida | Aetiology multiple and complex | Failure of vertebral fusion *Spina bifida cystica* is associated with meningocele or, in 15–20%, myelomeningocele. Latter causes spinal cord damage, with lower limb paralysis, incontinence, etc. | Hydrocephalus in four-fifths of those with mylomeningocele. Retardation in this group |
| Cerebral palsy | Perinatal brain damage. Strong association with prematurity | Spastic (commonest), athetoid and ataxic types. Variable in severity | Majority are below average intelligence. Athetoic are more likely to be of normal IQ |
| Hypothyroidism (cretinism) | Iodine deficiency or (rarely) atrophic thyroid | Appearance normal at birth. Abnormalities appear at 6 months. Growth failure, puffy skin, lethargy | Now rare in Britain. Responds to early replacement treatment |
| Hyperbilirubinaemia | Haemolysis, rhesus incompatibility, and prematurity | Kernicterus (choreoathetosis), opisthotonus, spasticity, convulsions | Prevention by anti-Rhesus globulin. Neonatal treatment by exchange transfusion |

Source: Gelder *et al.* (1989)

by the application of a standardized developmental or psychological test, or one of the instruments more specifically designed for this purpose, for example the Vineland Adaptive Behaviour Scales (Sparrow *et al.* 1984) or the AAMD Adaptive Behaviour Scale (Foster and Nihira 1969) can provide invaluable information.

Which children should a clinician refer to a psychologist? This will largely depend on local practice and the availability of psychological time. In theory, any child who might benefit from specialist advice on education or the promotion of development should be referred. In practice, it is desirable to refer all children in need of, or likely to need, special schooling. As a rough guide this would include all 2-year-olds functioning at less than a 15-month level, all 3-year-olds functioning at less than a 2-year level, all 4-year-olds functioning at less than a 2½-year level, and all 5-year-olds functioning at less than a 3½-year level. Some of these will not, in fact, require special schooling, but there is a high enough risk to justify referral.

At least as important as an assessment of the child's ability on a standardized assessment is an estimate of the child's level of functioning in everyday life. Some children who appear very slow on testing are capable of a much higher level of functioning and vice versa. It is useful to ask parents about what their child can and cannot do, using one of the standardized developmental assessment sheets such as the Denver Developmental Screening Test (Frankenburg *et al.* 1975).

Assessment is not a once-for-all exercise. Retarded children will require regular monitoring of their developmental progress and the aetiology of the condition may give clues as to the pattern of development that may be expected.

*Identification of associated physical and psychiatric problems*
The most common physical problems associated with mental retardation are sensory disorders, especially hearing and visual defects. In all cases of suspected mental retardation, the child's sight and hearing should therefore be carefully checked. Checking should be repeated at regular intervals as such defects may arise during the child's development or become detectable when previously this was not possible. In addition, a general physical examination is necessary in order to identify deformities or other stigmata that might give a clue to aetiology, as well as require treatment in their own right. Cerebral palsy is present in about a quarter of the moderately or more severely retarded. The presence of attack phenomena should be enquired about, as a similar proportion (about 25 per cent) of the moderately or more severely retarded suffer from epilepsy (Corbett *et al.* 1975).

Behavioural and emotional problems are particularly common in retarded children, especially in the more seriously retarded (Corbett 1979). Systematic questioning to elicit the presence of such problems is therefore especially necessary in this group. The hyperkinetic syndrome (attention deficit disorder) and infantile autism are much more common in severely retarded children than in the general population. About 10 per cent of moderately or more severely retarded children show the hyperkinetic syndrome, and roughly the same proportion are autistic (about 300 times the rate in the general population). About the same proportion (10–15 per cent) of severely retarded children and adults in institutions show self-injurious behaviour, this problem being particularly common in the Lesch–Nyhan and de Lange syndromes (Murphy 1985). Self-injurious behaviour is less common (around 2 per cent) in the non-institutionalized mentally retarded population (Rojahn 1986).

The presence of social problems in the family should also be established. The clinician should try to determine whether the child is neglected or inadequately stimulated, and, if in doubt, can refer to a social worker.

### Establishing a cause

Identifying a cause for mental retardation is not by any means always possible. The advantages of making a diagnosis are that:

1. The genetic implications will be clearer.
2. Very occasionally a treatable cause may be found.
3. Parents will, in general, find it a relief to have this source of uncertainty removed.

**Moderate to profound retardation.** The level of functioning will provide an important clue to aetiology. Mild retardation often does not have an identifiable physical cause; moderate or more severe retardation usually does. Some causes, for example phenylketonuria, may be suspected before birth because of a previously affected sib. A condition sometimes linked to mental retardation such as spina bifida may have been diagnosed on the basis of amniocentesis during pregnancy. Physical examination at birth may reveal the presence of an obvious chromosomal abnormality such as Down syndrome. A history of traumatic birth, postnatal anoxia, or hypoglycaemia may alert the clinician to the possibility that subsequent development will be slow. The results of routine biochemical investigations carried out after birth will reveal the presence of disturbances of amino acid metabolism, such as phenylketonuria. Observation of hypotonic postural abnormalities or asymmetrical movements shortly after birth will alert the clinician to the possibility of cerebral palsy with associated mental retardation. Most causes of severe retardation will therefore be established by the time the baby is a few weeks old on the basis of routine history-taking, physical examination, and biochemical investigation.

The remaining cases of severe retardation will be identified as a result of unusually slow development. This is likely to be more marked in the sphere of personal–social and language development than in motor development. Affected babies are likely to be passive, unusually 'good' babies in that they may rarely cry for attention. They take little notice of their surroundings and may first be suspected of deafness or blindness (which may indeed be present in association with retardation).

All children found to be functioning at a moderately or more severely retarded level, for whom history-taking and clinical examination have failed to reveal a cause, require further investigation. To rule out the most common causes, this should include plasma and urinary amino acids, skull X-ray, computer-assisted tomography (if available), and chromosomal analysis (including, where available, an investigation of fragile sites). The conduct of further investigations will depend on the results of history-taking and clinical examination as well as the availability of resources, but, if the investigations named above are negative, even the most thorough and expensive subsequent testing will uncover aetiology in only a small minority of cases.

**Mild retardation.** The establishment of a cause for mild retardation normally only begins much later, perhaps when the child is at school and failing educationally. In a minority of cases the history and physical examination will provide some clues to a specific physical cause, and further investigations can be requested accordingly. If no physical cause is suspected, especially if there are obvious social factors likely to be responsible, investigation should be very limited. Plasma and urine amino acids can be tested, and the conduct of this investigation may be reassuring to parents. Chromosomal investigation, including investigation for a fragile site, should also be carried out. In the absence of multiple social causes, further physical investigation along the lines described for more severe retardation may be justified, especially if the child is at the lower end of the mildly retarded group (i.e. IQ 50–60).

*Family care, expectations, and coping capacity*
The manner in which the history is given, the way in which the parents speak of the child, dress and undress it, answer questions about the way the child responds to being talked to and played with, these will all provide useful information about the quality of care the parents are able to provide.

It should also be possible to assess the sort of expectations the parents have of the future development of their child by asking questions about how they think the child is doing compared to other children of similar age, what they think will happen when the child gets to school age, etc.

Finally, the parents' coping capacities are of major importance in relation to future management. At an initial assessment these will often be quite uncertain. It is a brave person who attempts to predict how he himself would cope, for example, with the lifelong burden of a severely retarded child, and perhaps only the foolhardy would attempt such a prediction for someone else. Nevertheless, as time goes on, it does become much more possible to assess the capacity of parents to cope with their retarded child.

The most vulnerable families are likely to be those experiencing multiple stresses, or where there have been relationship difficulties prior to the birth of the affected child (Byrne and Cunningham 1985). The reader is also referred to Chapters 4 and 5 in which psychosocial aspects of chronic physical disorders are discussed. Much of those chapters is relevant to the families of children with a retarded child.

## Management

The various components of management to be discussed will be:

1. Breaking the news.
2. Counselling on the promotion of development.
3. Dealing with associated deficits and behaviour problems.
4. Advising on appropriate education.
5. Genetic counselling.
6. Providing social and emotional support.

### Breaking the news

Imparting information to parents about the nature, likely causes, and further development of a mentally retarded child is one of the most difficult tasks for the clinician. Even allowing for faulty memory, parental reports obtained after the event suggest that it is often badly done but that a thoughtful, systematic approach can markedly reduce the rate of dissatisfaction (Cunningham *et al.* 1984). If the condition is diagnosed shortly after birth, it is good practice for the news to be broken as soon as possible to both parents, with the affected baby present in the room. This task should be undertaken by the consultant paediatrician, accompanied by the health visitor who will be involved in follow-up. The interview should be conducted in privacy and there should be a subsequent interview with the paediatrician a day or so after the initial disclosure. The health visitor should be available for further support and advice during the following months.

In counselling parents where retardation is identified at a later stage, a number of principles need to be borne in mind, especially at the lower end of the mildly retarded range, or the more severely affected.

1. The doctor will often be mainly interested in identifying a cause. If he fails to diagnose an organic condition, he may suggest that there is 'nothing wrong' with the child. It is hard to imagine a more misleading suggestion.

2. All parents have some idea what is wrong with their child before they are seen. It is always helpful to discover 'where they are at' in their understanding and expectations of their child's condition. It is, for example, useful to ask them what sort of age they think their child has reached as far as his functioning is concerned. They may have known a similar child, read a relevant book, etc.

3. Information is poorly absorbed when parents are anxious and upset. They need repeated opportunities to ask questions and check out their own level of understanding.

4. Communication that 'nothing can be done' for a retarded child is misleading. There may indeed be no useful medical treatment, but parents should be encouraged to believe that there is a great deal that can be done. It is useful to have some clear ideas of the positive steps that parents, often with the help of teachers and other professionals can take to promote development.

5. Many people may be involved in imparting information about a retarded child. They may include the health visitor, paediatrician, general practitioner, social worker, clinical medical officer, and teacher. Whoever is involved does well to check out with parents what they have previously been told before imparting their own views. If a professional provides different information from that previously given, it is always a good idea to indicate that it is not unusual for professionals to hold varying views and, in the end, the parents, who are the only real experts in their own child's development, have to work out for themselves what the real nature of the problem is. Free communication between professionals will do much to prevent different information being provided unintentionally. Incidentally, many parents are confused by the complexity of the organizations providing services, and there are considerable advantages if the professionals involved can agree on a 'link-person' who can take responsibility for coordinating services and explaining their delivery to the family.

6. Parents will want to share the information provided with relatives and friends. They may lack the skills to do this, and it is often helpful to ask if they feel confident about it, and, if not, whether they would like to rehearse it. This also provides an opportunity to check the accuracy of their understanding.

7. Parents faced with painful knowledge react in a variety of ways. Most are able to absorb the knowledge, cope with the loss of the normal child they had hoped to bring up, and get on with rearing the child they do have as best they can. A minority are persistently angry, depressed, unrealistic, inappropriately guilty, or show other, more complex reactions. The management of such reactions is discussed in more detail in connection with parents of the physically handicapped in Chapter 4.

8. All parents want to know what is ultimately going to happen with their retarded child. As the child gets older, the future becomes more certain, but, in the preschool period, unless the child is profoundly retarded, it is often quite uncertain how well a child will develop later. In general, and there will be exceptions, parents should be encouraged to look no more than 3 or 4 years ahead at a time until the child reaches teenage, when planning for a longer span of postschool life becomes appropriate.

9. In considering outcome, parents often want to know whether a child

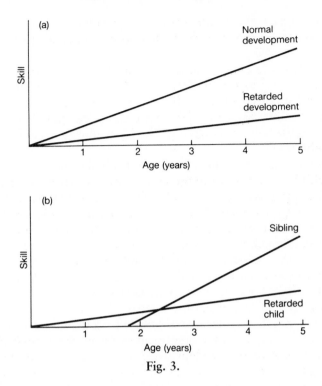

Fig. 3.

will 'catch up'. Figure 3(a), although very simple, may be found helpful in explaining why it is that children appear to fall further and further behind normal children, even though they acquire new skills each year. If a retarded child has a sib, say a couple of years younger, the modification shown in Fig. 3(b) may be helpful.

If the diagram in Fig. 3 is drawn, it is important to point out that the line drawn for the retarded child will be affected to some degree by his particular pattern of development, and by the quality of home care and education he gets. The lines drawn are only a rough guide to development. Further, it is important to say that development can be expected to continue, albeit at a slow rate, into the late teenage period and early adulthood.

10. If the retarded child is present, it should not be forgotten that he may be capable of understanding much that is being discussed. In these circumstances, it is useful to include the child in the consultation, though at a level the child can comprehend.

*Counselling on the promotion of development*
Retarded children take longer to learn new material and, once they have learned something new, they usually forget more easily than do ordinary children. Consequently they need more help, and more systematic help

from parents, teachers, and others in the acquisition of skills. In particular, they often fail to learn by observation, and therefore need more structured teaching.

This help needs to be provided at an appropriate level for the child. It is worse than useless to try and teach skills too far ahead of the child's existing level, so parents should, for example, be discouraged from trying to teach their children to read when they cannot discriminate simple shapes or label common objects correctly. There are more important things for parents to be doing.

In the preschool period the main role for professionals such as teachers, speech therapists, etc. is in helping parents to find ways to stimulate their children's development. The availability of programmes such as the Portage scheme (Revill and Blunden 1979), which enable professionals to guide parents in setting realistic targets and choosing methods to achieve them, has done much to help parents feel they can carry out useful tasks with their children. Clinicians should therefore make sure that, as early as possible, parents of retarded children are put in touch with professionals such as teachers and psychologists who have access to this type of material.

*Dealing with associated deficits and behavioural problems*
As already indicated, common associated physical deficits include impairment of vision and hearing, epilepsy, and cerebral palsy. Commonly linked psychiatric deficits include the hyperkinetic syndrome and infantile autism. The management of each of these is discussed elsewhere in this book. Self-injurious behaviour can be treated with the use of protective devices, medication, and behavioural modification (Murphy 1985), though not always successfully. Physical protection, using a helmet or other device, is often a necessary first step, and may be required continuously. Medication, though widely used, is often ineffective and commonly produces undesirable side-effects, especially oversedation. However, neuroleptic medication does sometimes have a useful role to play in this condition (Singh and Millichamp 1985; Farber 1987). Medication needs careful monitoring if it is not to have prolonged side-effects. Particularly in institutionalized retarded children, but also in those living at home, whatever may have been the original reason for prescribing drugs, there are risks of children remaining on medication for years, sometimes with multiple drug combinations, long after there is evidence for their effectiveness. Behavioural therapies (Section 6.2.2), especially reinforcement of positive behaviour and extinction, have probably had greater success, but put more demands on staff.

The mentally handicapped are not only prone to the development of unusual psychiatric syndromes, such as autism, the hyperkinetic syndrome, self-injurious behaviour, stereotypies, faecal smearing, and pica, but also show an excess of the commoner behavioural and emotional problems. Assessment and treatment in these cases should follow similar lines to that in children with normal intelligence. Although some authorities have suggested that mental retardation is a contraindication for psychotherapeutic procedures, they can, in fact, be helpfully applied at least in the mildly and moderately retarded group. Psychotherapeutic techniques similar to those used with younger children of the same mental age can be readily applied to children with mental retardation.

*Advising on appopriate education (Section 7.4.2)*
Although decisions on appropriate education will be taken by teachers, educational psychologists, and other educational professionals, clinicians will very often need to provide advice to enable sensible decisions to be made. In the UK such advice will be formally sought when a full assessment is made for a child with special education needs under the 1981 Education Act or, in the USA, under Public Law 94-142. However, it is desirable to provide information before this stage is reached. The clinician will be expected to provide information on medical matters such as the need for medication, physiotherapy, hearing aids, spectacles, etc. In addition, however, the clinician's observation of the child's attention span, persistence, level of activity, capacity to form relationships, etc. will enable useful suggestions to be made about whether the child has a particular need for individual attention, or for a more or less structured setting.

A clinician should make clear to parents that he does not have responsibility for final decisions on education, but only for advising the education authority. It is usually unwise to say that a particular named school would be best for a child, but if a clinician has a view about this, then mention can be made of the school in the report to the authority as a possible placement. Medical reports to education authorities are made available for parents to see (Section 7.4.2) either in full or abstract form, so it is often sensible to tell parents what it is intended to put in a report before it is written.

Once the child is placed in a particular school, any change in the child's medical and psychiatric management needs to be made known to the school either directly to the head teacher or via the school doctor.

*Genetic counselling*
All parents should receive information about the risks involved if they have further children, even if it is improbable that they will. Further, if the retarded child has sibs, parents should be encouraged to discuss the genetic implications with their normal children for their own progeny once they reach their mid-teens. In nearly all cases such information can be reassuring and, if it is not provided, there is a likelihood that unnecessary anxiety will be experienced.

The risks involved in some common conditions for further full sibs of the

**Table 7**

| Condition | Risk (per cent) | |
|---|---|---|
| | Further full sibs of the retarded child | Children of full sibs of the retarded child |
| Down syndrome: trisomy 21 | 1 | As general population |
| Down syndrome: translocation type | 2–10 if parent carrier | As general population (if not a carrier) |
| Phenylketonuria | 25 | As general population |
| Fragile X | 50 (boys—if mother carrier) 50 (girls—carrier status) | 50 per cent if mother carrier.* Otherwise nil |
| Unidentified cause | 4 | As in general population |

* N.B. Risk of substantial learning difficulties if fragile site is inherited is about 80 per cent in boys and 30 per cent in girls.

retarded child and for the children of full sibs are shown in Table 7. The figures in Table 7 may need to be modified if there are special features present (e.g. if there is a further mentally handicapped person in the family, the child has an atypical form of the condition, etc.). In these cases, in particular, specialist genetic advice is desirable.

Risks involved in less common conditions and principles of genetic counselling including specialized techniques for antenatal diagnosis are described in detail elsewhere (Rodeck 1987).

*Providing social and emotional support*
Parents with handicapped children will usually obtain emotional support from a variety of sources, but especially from relations, friends, neighbours, health visitors, and, as the child gets older, teachers. Health professionals and those working with them such as hospital social workers can provide valuable support and, in some cases, the help they give in this respect can be crucial.

Parents may often feel a special relationship with the clinician who made the original diagnosis and wish to turn to him for help when subsequent crises arise. Some paediatricians, clinical medical officers, and psychiatrists make a practice of offering parents of severely mentally handicapped children regular review appointments on an annual basis to review the child's progress. They can offer encouragement despite the fact that often this has been painfully slow. Such a review, which may form part of a monitoring of the special education that the child is receiving, also provides an opportunity to check the possibility that further assessment and treatment of associated deficits (e.g. hearing, vision), are required. The child may have developed new problems (e.g. behavioural difficulties) or be entering a new phase of life (e.g. puberty) which raises fresh anxieties in the parents. Problems may arise not only in the handicapped child but also in other members of the family, such as the child's sibs (Gath 1989). Again review appointments offer an opportunity for counselling and, where necessary, referral.

Many of the principles mentioned in relation to the counselling of parents of children with chronic physical handicap (Section 4.5) are also relevant in connection with parents of the mentally handicapped.

### 2.6.3 GIFTED CHILDREN

**Definition**

Children with unusual intellectual gifts fall into three main categories:

1. Those with high measured intelligence. It is generally accepted that an IQ of over 140 on one of the standardized tests such as the WISC (Section 1.2.3) marks a child out as having unusual all-around ability.

2. Children of any level of general ability with an unusually well-developed isolated gift, for example in mathematics or music. Children with unusual athletic ability, who undergo prolonged periods of intensive training, share many features with this group.

3. Mentally handicapped children with one or two exceptional, isolated, capacities, for example for complicated computation. These are referred to as 'idiots savants' and some, but not all, are autistic (Section 2.8.1).

**Psychiatric aspects**

Contrary to much public belief, highly intelligent children (those with IQ 140+) have either the same or somewhat lower rates of behavioural and emotional problems than do children in the general population (J. Freeman 1983). There is no good evidence for the view that is sometimes stated that gifted children have particularly low needs for sleep and are prone to overactivity. They are likely to be active, inquisitive children with bright, intellectual parents, but not all fit this stereotype. When they do have problems, these may be related or unrelated to their giftedness. Related problems include boredom at school (sometimes leading to behavioural problems there), due to an inappropriate curriculum or teachers failing to recognize a child's special capacities. Parents who are told that their children are gifted may develop unrealistic expectations about their performance. Although a minority do become very high achievers, most gifted children do not become gifted adolescents or adults and this often leads to disappointment. Social isolation, arising because of the child being set apart from others less talented, may also be a problem.

Emotional problems unrelated to the giftedness are more common in this group. Parental marital disharmony and maternal depression are among the most common features.

The management of problems related to giftedness is largely educational. Most young gifted children can be well educated in ordinary schools, provided special attention is given to their curriculum and social needs. A variety of enrichment and accelerating educational strategies can be used for exceptionally able pupils (S. Cohn 1988). A minority thrive better in schools (e.g. music schools) in which their special capacities can be nurtured. Problems arising from factors other than the giftedness can be dealt with as in the non-gifted group with family or individual assessment and treatment.

# 2.7 Attention and attention-deficit disorder

## 2.7.1 DEVELOPMENT OF NORMAL ATTENTION

Studies of the development of attention have revealed a characteristic pattern throughout childhood. Newborn babies are equipped perhaps to a surprising degree to pay attention to certain kinds of stimuli rather than others. Thus they will selectively attend to moving objects more than stationary ones, to patterned sounds rather than monotonous tones, and, within the first few weeks, to a picture of a human face rather than a picture of a jumble of facial features. The development of attention is characterized by change along a number of dimensions.

**Exploration.** The attention of infants and toddlers is dominated by features of the environment. The young child is highly exploratory and, above all, his attention is captured by novel stimuli he has not met before. As he gets older, the nature of the child's environment and what is happening in it becomes less important, and the child's attention becomes more related to a preconceived internal plan or intention. Relatively undirected exploration is replaced by search.

**Search.** A child's tendency to search for certain features in his environment will be affected by his needs and motivation, therefore, becomes increasingly important in his span of attention. Success in searching will often be rewarded, so that his capacity to attend will be reinforced and thus lengthened.

**Distraction.** A feature of the young child's attention to novelty is that he remains open to distracting stimuli. As children get older and work more to a preconceived plan in their searching, distracting stimuli have to be excluded. The capacity to inhibit irrelevant stimuli is developmentally determined.

**Span of attention.** An increasing length of the span of attention is a characteristic feature of development. Older children will sustain attention for longer periods of time.

**Links with other cognitive skills.** Attention allows absorption of information and ensures better adaptation to the environment. An older child attends to a book because he enjoys reading it and wants to know how the story develops. If his short-term memory or his visuospatial ability (e.g. capacity to discriminate letter shapes) are poor, or if he has some degree of language delay, then attending to the book will be unrewarding for him, and his attention will not be sustained. The development of the capacity to attend is therefore dependent to some degree on the development of other abilities.

## 2.7.2 ATTENTION-DEFICIT HYPERACTIVITY DISORDER (HYPERKINETIC SYNDROME)

### Definition

This condition is characterized by an early onset, the combination of over-active, poorly modulated behaviour with marked inattention and lack of persistent task involvement, with pervasiveness of these characteristics over situations and their persistence over time (World Health Organization 1990). The DSM III-R classification criteria (American Psychiatric Association 1987) are similar, but the presence over a 6-month period or more of at least eight out of fourteen types of behaviour, including especially fidgetiness, difficulty in remaining seated, high distractability, and impulsivity, are required.

The term 'minimal brain dysfunction' (MBD) syndrome has also been used (Clements 1966), especially in the 1950s and 1960s to explain on physiological grounds (unspecified brain dysfunction) a behavioural syndrome very similar to or identical with the attention-deficit disorder/ hyperkinetic syndrome described below. Most clinicians now believe that it is undesirable to use the term MBD to describe this behavioural syndrome because it may imply an inaccurate or at least an incomplete view of aetiology (E. Taylor 1984).

### Clinical features

**Onset.** Children usually present to health services between the ages of 3 and 7 years though, sometimes in children who present with other problems later in life, it will be clear from the story that the child was over-active and lacking in concentration earlier on.

Complaints about the child's behaviour may come either from the parents or from playgroup organizers or teachers. This is partly because people vary in their capacity to tolerate difficult behaviour and partly because the child may show considerable variability from situation to situation in his activity level and attention span.

**Behavioural characteristics.** Characteristically the child will, from babyhood, have been a restless and difficult feeder, sometimes sleeping poorly and irregularly. As walking begins and the child becomes more mobile, he is noted to be unusually active and into everything. He may only be able to sit for a moment or two at a mealtime. This high level of activity means that the child is particularly prone to accidents. As the child is expected to spend more time on pursuits involving concentration, it is seen that he is highly distractable and cannot settle to any activity for more than a brief period. Puzzles, construction toys, and picture books do not engage his attention. Though he may be able to sit in front of the television for an indefinite period, it is notable that he is fidgety and seems to take little in.

Entry to a preschool facility may bring to light a problem that has been present for some time, because the child is suddenly expected to conform and to cope with more structured expectations. Many severely over-active children are unsually uninhibited in their response to strange people, failing to show normal stranger anxiety (Section 2.9.1). They may show associated developmental problems, especially language delay and other behaviour problems, particularly a tendency to aggressive behaviour.

As the child gets older, over-activity usually becomes less of a problem, but defects of attention and concentration may lead to learning problems, while impulsiveness and lack of inhibition may result in antisocial behaviour.

## Background information

**Prevalence:** This has been found to vary widely, especially between North America and the UK, with prevalence rates of children attending services with this diagnosis in the region of 0.06 per cent in the UK and 1.2 per cent in the USA, a twentyfold difference (E. Taylor 1986). Community studies suggest, however, a higher total prevalence rate in the region of 6 per cent (Szatmari *et al.* 1989). Possible artefactual reasons for this discrepancy include differences in parent and teacher perception of behaviour, differences in clinic referral rates and differences in diagnostic practice. Systematic differences in diagnostic practice due both to the classification systems used and to the training of clinicians have indeed been demonstrated (Prendergast *et al.* 1988). It is, however, also possible that there are real variations in the prevalence of ADD/hyperkinetic disorder between different countries.

Boys show hyperkinetic problems about three or four times as commonly as girls. There is no strong link between social class and the condition. Children showing the hyperkinetic syndrome who also show aggressive behaviour have high rates of family disharmony, parental mental illness, and other indices of family and social disruption in their background, but this is less marked in hyperkinetic children without antisocial features (Stewart *et al.* 1980).

## Aetiology

Causative variables thought to be of importance include:

1. *Genetic factors.* Adoptive studies (e.g. Cantwell 1975) as well as twin studies suggest that there is a strong genetic contribution, especially in boys (Goodman and Stevenson 1989). Monozygotic twins show greater concordance than dizygotic in activity levels, and the biological parents of affected children have a strong history of hyperkinesis in their own childhood. Genetic factors may operate by their effect on temperament and on neurochemical processes (see below).

2. *Brain dysfunction.* Children with evidence of brain damage, for example those with epilepsy and cerebral palsy, show high rates of the hyperkinetic syndrome. Rates are even higher in children with associated mental retardation (especially if this is severe), with around 10 per cent so affected (Corbett 1979). It is important, however, to stress that most children with the hyperkinetic syndrome do not show overt evidence of brain dysfunction.

3. *Neurochemical processes.* Brain dysfunction is also suggested by the presence in some studies, but not all, of a high rate of non-epileptic EEG anomalies. In addition, abnormalities of dopamine metabolism have been described in experimentally induced hyperactivity in rats, although this mechanism has not been confirmed in affected children. Dopamine is a neurotransmitter substance, found in considerable concentration in the activating system of the brain, and is therefore important in the control of the level of arousal. There have also been less well confirmed reports of abnormalities of serotonin and noradrenaline metabolism.

4. *Diet.* Hyperactivity has been related by a US allergist, Feingold (1975), to the presence of various additives and naturally occurring substances in the diet. Tartrazine and salicylates are said to be particularly noxious. Attempts to prove this theory by removing these substances from the diet and then re-introducing them in controlled clinical trials using a double-blind technique have not confirmed these claims (Goyette *et al.* 1978). Nevertheless, additive-free diets remain widely used. It is possible that their effectiveness relates more to the fact that their use requires parents to be firm and more consistent, at least over their children's food, than to any direct action arising from withdrawal of noxious substances. However, it is also possible that with a more systematic approach to identification of noxious substances, more children whose over-activity is diet-responsive may be identified (Egger *et al.* 1985; Kaplan *et al.* 1989) (see below).

5. *Lead.* Lead ingested or inhaled in high concentrations can produce severe brain damage. It has been claimed that lead in much lower concentrations, previously considered to be within the normal range, can produce hyperactivity (Needleman *et al.* 1979). The evidence for this claim has been called into question, and it is uncertain whether removal of lead from the environment would produce any lowering at all in the rates of hyperactivity (M. Smith *et al.* 1983). This is, of course, no reason for not minimizing children's exposure to lead as much as possible.

6. *Temperament.* Children who have been active from the first few months of life are likely to remain so at least during their childhood years (Thomas and Chess 1977). Continuity of behaviour that sometimes seems relatively impervious to outside influences has been taken as evidence that the child's temperament or personality plays an important part in the development of the syndrome.

7. *Psychological factors.* Psychological mechanisms implicated include inclination to seek immediate reinforcement, arousal modulation problems (children sometimes showing unduly low and sometimes unduly high arousal), poor inhibitory control, and basic impairments of attention.

8. *Social factors.* Young children are more likely to show behaviour disorders, including disorders characterized by high levels of activity, if they are living in poor social conditions, with inadequate housing and in families with stretched financial resources (N. Richman *et al.* 1982). It is likely that these factors operate by their influences on parental health and parental behaviour.

9. *Parental behaviour.* It has been observed clinically that mothers of children showing the condition are sometimes relatively unresponsive to their children's demands. It may, therefore, be that some of the child's over-active behaviour is undertaken in order to elicit a response in the caring adult. Parents of hyperkinetic children also sometimes show inability to provide consistent control for their children, and are occasionally passive and unusually accepting of difficult behaviour. However, the role the parents play in producing hyperactivity should be considered cautiously. It has been shown that parental behaviour becomes more appropriate when children are successfully treated with stimulants (Barkley 1989), suggesting that inappropriate parental behaviour may be secondary to the child's behaviour problems.

10. *Early childhood experiences.* Children separated from their parents in early life, then reared in institutions for two or three years, and finally settled in good adoptive families may continue to show behavioural problems that are sometimes of hyperkinetic type (Tizard and Hodges 1978). Further, some parents date the development of hyperkinetic problems from stressful events such as a hospitalization occurring in the first year of life. There is no systematically gathered evidence of the later effects of traumata of this type, but it is certainly possible that they occur.

### Interactional effects

These various causative influences are rather rarely seen in pure form. Much more commonly the child is suffering from a multiplicity of adverse influences. It is very common, for example, to see over-active children in whom adverse temperamental characteristics interact with parents under stress or with personality problems, who are therefore incapable of providing consistent control and affectionate care. The pattern of interaction may be even more complex with, for example, adverse experience producing neurophysiological dysfunction and problems in modulating and monitoring experience. The child may then develop learning difficulties that have a secondary disturbing effect on his behaviour.

## Assessment

As precise as possible an account of the level of activity, degree of inattention, and related problems should be obtained. Because of the situation-specific nature of much hyperactivity, it is particularly important to obtain separate accounts of behaviour from both parents and teachers, or, for example, playgroup leaders. It should be noted whether the child's behaviour varies within situations. At home, is he better when his father is around? At school, is he more difficult in the structured classroom than in the playground, or in less structured classroom activity? Has any response to dietary constituents been noted? It should be emphasized that a parental complaint that a child is over-active does not necessarily mean that the child's level of activity is unusually high. One or both parents may perceive quite ordinary levels of activity to be pathological.

An account of the child's past activity should be taken with particular reference to the possibility of brain damage—perhaps from a very difficult birth or a prolonged epileptic fit. Definite developmental delay, either general or specific (e.g. language delay or clumsiness), would support a biological component to the problem. In any event, the child's developmental level and learning ability would also require informal or more detailed assessment.

The family background, quality of family relationships, and way in which the parents respond to the problem will all be relevant. Observation of the interaction between parents and child when he is being difficult is also likely to be revealing. It is important to remember that child–parent interaction in a consulting room may not be representative of what goes on at home. The parents can be asked about this.

Finally, it is often helpful to ask the parents to keep a chart over a week or fortnight of a particularly troublesome type of behaviour that they have described.

## Treatment

Although specific therapies (medication and behaviour modification) are usually given most attention in discussions of management of this syndrome, equally important elements of treatment include helping the parents to see the problem in perspective, and environmental manipulations of one sort or another.

Counselling of parents should focus first on an acknowledgement of the extent of the problem, the probable importance of biological factors, and the way these interact with the reactions of those around the child. The fact that the child is not going to change rapidly, but that the long-term outcome may well be good providing one can avoid the child becoming discouraged and low in self-esteem, should also be emphasized.

**Environmental manipulations** that may be helpful if they are feasible include (in school) providing more opportunity to engage in free play, and ensuring that, in structured settings, the opportunities for the child to be disruptive are reduced. This can sometimes be achieved by careful attention to classroom seating arrangements, and by not expecting prolonged periods of sustained attention too soon. Praise should be given if the child's level of attention improves even slightly. At home it is often helpful to encourage parents not to overstimulate a hyperkinetic child. Having only one or two toys out at a time, and limiting the visits of friends to small numbers may be useful.

**Formal behaviour modification programmes** can take place either at home or at school. Whether they are feasible or not will largely depend on whether the parents are sufficiently motivated and organized and whether skilled psychologist time is available. An operant approach can be embarked upon which follows the principles described in Section 6.2.2. The child's behaviour in one or two especially troublesome respects, for example sitting at table, concentrating on a puzzle, should be charted by parents or teachers. Rewards agreed with the child should be instituted for appropriate behaviour, and inappropriate behaviour should be ignored. Even though inappropriate behaviour may demand attention, care should be taken to ensure that the child receives *more* attention at times when his behaviour is more appropriate.

**Medication** should be limited to cases where the child's behaviour is seriously affected both at home and school. The child should have been personally observed by the clinician to have high levels of inattention and over-activity and, although there may be adverse circumstances in the family, there should be a lack of evidence that these are the prime causes of this type of behaviour.

Stimulant medication was reported many years ago to have a beneficial effect on hyperkinetic children. It remains a well-evaluated form of treatment for the severely affected child. Methylphenidate (*Ritalin*) should be prescribed morning and lunchtime in doses increasing by small amounts every 3 or 4 days. It is sensible to begin with methylphenidate 2.5 mg *mane* and increase by 2.5 mg at these intervals while parents and teachers monitor the child's behaviour. A ceiling dose of 10 mg *mane* and 10 mg lunchtime would probably be appropriate for a 4- to 5-year-old child, with somewhat lower doses for younger and higher for older children. Some clinicians use doses two or three times higher than this.

Side-effects may include depression, irritability, twitchiness, reduction of appetite, insomnia, and a worsening of symptoms. Sometimes this worsening occurs over 2 or 3 days and is then followed by definite improvement, so it may be worth persisting with the drug even if it seems ineffective to begin with. Some of the side-effects may be mitigated by 'drug holidays' at weekends, or for more prolonged periods. The child's growth may be temporarily affected, but there appears to be catch-up growth once the drug is stopped (Roche *et al.* 1979).

With all these side-effects, it is often a matter of careful negotiation with parents to determine whether the drug should be prescribed or persisted with. Parental feelings on the matter are obviously of great importance. Nevertheless, a minority of children receive clear benefit from the use of stimulant medication. In most children, a period of medication lasting several months or a year or two is necessary. It is usually the child or the parents rather than the clinician who decides when the drug has outlived its usefulness. Parents can be reassured that addiction will not occur, but they should keep tablets out of the way of teenage sibs.

As indicated above, **diets**, especially additive-free diets, are very widely used by parents, even though the scientific evidence for their effectiveness is very limited (E. Taylor 1984). It is, however, unhelpful for professionals to be dismissive about this form of treatment and more appropriate to help parents to apply it rationally. Some guidelines are provided in this section. It must be emphasized that evidence for the effectiveness of stimulants (see above) is much stronger than that for dietary treatment. In applying an additive-free diet, all foods are restricted that contain artificial colouring and preservatives (E. Taylor 1985). Unless home-made or labelled as additive- and artificial colouring-free, the following foods are most likely to contain such products:

(1) processed ham, bacon, salami, sausages, ready-cooked and frozen meat, frozen fish;
(2) ice-cream, ice-lollies, and all forms of confectionery, flavoured crisps, sauces, etc.;
(3) fizzy drinks and fruit squash (except those specially labelled additive- and artificial colouring-free);
(4) most types of wrapped bread and breakfast cereals;
(5) shop-bought cakes, biscuits, desserts, jellies;
(6) margarine, processed or coloured cheese.

If the application of an additive- and artificial colouring-free diet, or the avoidance of one or two specific foods, such as orange squash, chocolate, sweets, lollies, or ice cream, appears to be helpful in improving a child's behaviour, it would not be sensible to discourage its use. Such a diet will result in a child's life being somewhat socially restricted, but will have no other ill-effects.

In considering initiating dietary treatment, the severity of the problem as well as the presence or absence of features suggestive of dietary aetiology should be taken into account. Other measures should be used in managing mild levels of hyperactivity.

Positive indications for dietary treatment include:

(1)  an allergic illness in the child in the past or currently;
(2)  a definitive allergic disorder in a first-degree relative;
(3)  episodic changes of behaviour, especially if these are associated with intake of specific foods, rather than the child being frustrated;
(4)  changes of behaviour that result in the child being in an episodic, excited, explosive, or tense state;
(5)  the observation, usually by the parent, of physical changes such as blotching of the skin or facial pallor, associated with behavioural changes;
(6)  the absence of obviously stressful factors in the background.

If an additive-free diet is ineffective, but the child is moderately or severely affected, then the use of a more systematic and drastic approach to dietary restriction of naturally occurring foods may be applicable (Egger *et al.* 1985). This treatment, which has also been used in the management of severe migraine, requires the services of a dietitian. The child is put on a small number of foods (perhaps one or two vegetables, one or two types of meat, and a source of carbohydrate such as rice) with added vitamins and calcium gluconate. The child is maintained on this restrictive diet for about 3 weeks, and, if improvement occurs, foods are reintroduced one by one at about weekly intervals, in order to identify positively activating items of diet. Foods most commonly identified as harmful using this approach are wheat, milk, and eggs.

This restriction/reintroduction dietary approach does, however, have dangers. Unless properly supervised, growth failure may occur as a result of malnutrition. Some parents have appeared to restrict diet quite unnecessarily on the grounds of imaginary hyperactivity—a form of Munchausen's syndrome by proxy (see Section 1.3.5). Dietary approaches may distract attention away from important social or familial factors affecting the child. Finally, reintroduction of foods after a long period of abstinence may, very occasionally, produce anaphylactic shock. Despite these drawbacks, the dietary approach may be useful in selected cases, although it must still be regarded as of unproven value.

## Outcome

A number of follow-up studies have been carried out and the prognosis is variable. Over-activity in the preschool years tends to predict conduct disorder in middle childhood (N. Richman *et al.* 1982). Mild, pure forms of the condition have a reasonably good outcome, though there may be persisting restlessness. Followed up to adolescence, children who have been diagnosed as severely hyperkinetic show impaired cognitive test performance and poor scholastic achievement. Later, in adult life, a high proportion develop antisocial behaviour disorders, but this is largely accounted for by the presence of aggressive behaviour in childhood (Gittelman-Klein and Mannuzza 1989). The hyperkinetic child not showing aggressive behaviour may have a better outlook, but the evidence for this is not clear-cut.

# 2.8 Pervasive developmental disorders

## 2.8.1 CHILDHOOD AUTISM

**Definition**

This is a condition characterized by delay and deviation in the development of social relationships and language. In addition, the child is likely to show mannerisms, resistance to change, attachment to unusual objects, and acute emotional reactions of excitement or anxiety in situations which do not usually provoke such responses.

Childhood autism is now classified (American Psychiatric Association 1987; World Health Organization 1990) as a pervasive developmental disorder rather than, as used to be the case, a psychotic disorder. The DSM III-R diagnosis requires a qualitative impairment in reciprocal, social interaction, qualitative impairment in verbal and non-verbal communication and in imaginative activity, and a markedly restricted repertoire of activities and interests, with onset during infancy or childhood. The ICD-10 diagnosis of childhood autism is based on similar criteria but, abnormal or impaired development must be present before the age of 3 years. In fact, the term 'pervasive developmental disorder' is somewhat misleading, as the deficits shown are, in fact, not fully pervasive. Nevertheless, the present classification systems represent significant advances.

The condition may first become apparent in the first few months of life and, in these 'no onset' cases, there is an assumption that it has been present from birth. In other cases there is a period of normal development usually lasting no longer than 3 years, followed by the gradual appearance, usually over a few weeks or months, of the classical features.

**Behavioural features**

**Social relationships.** In 'no onset' cases, the baby is 'different' from birth. He is slow to smile, appears generally unresponsive and passive, and does not seem to enjoy being picked up. He will not put his arms up to be cuddled, and may wriggle and squirm when taken into his mother's arms. He is likely to avoid eye contact. As he reaches 2 and 3 years, he will show a preference for his own company and, if he hurts himself, will not come to his parents for comfort. He will be slow to distinguish his parents from other people.

These social abnormalities change with age. As the child gets older, he may, much later than normal children, become very attached to his parents and be distressed by their absence. He is still, however, likely to show little interest in playing *with* other children, though he may play alongside them. As he moves into puberty, the child's social relationships are likely to remain unusual. There will be no imitation of other people's activities. He will probably still prefer his own company and, even if he has reasonable language, it is likely that his social relationships will be marked by lack of awareness of the feelings of other people, and by an incapacity to take on another person's point of view. He will make no real friendships with others his age.

**Speech and language development.** Impairment occurs in both verbal and non-verbal communication. Both comprehension and expression are usually markedly delayed, especially when considered in relation to other aspects of the child's development. About half the children diagnosed as autistic are not speaking single words by the age of 5 years and, of these, only a minority later develop any useful speech. When speech does occur, it shows various abnormalities including persistent echolalia (mechanical repetition of sounds or words) and neologisms (use of words the child has made up). The echolalia may be of delayed type, the child repeating a word that has been said to him some time previously over and over again. The echolalia may be partly responsible for pronominal reversal, the child saying, for example, 'you want' when he means 'I want'. There is often an associated abnormality of voice production. A source of particular distress to parents is the fact that the child's language usage is often highly sporadic. The child may, for example, utter one or two well-formed phrases, and then not speak again for several weeks or months. The child may make requests by taking an adult's hand and leading in the required direction to a cupboard or door. There is usually no use of gesture. The child shows no or very little make-believe play, and indeed there is usually a total absence of all imaginative activity, probably because of a deficit in symbolic thought processes.

As the child gets older, expression and comprehension will continue to improve, but usually at a disappointingly slow rate. Further, the child's intonation may remain mechanical, monotonous, stereotyped, and repetitive. It is impossible for the child to sustain a meaningful conversation for more than a sentence or so.

**Mannerisms.** These take a variety of forms, but particularly characteristic are rapid flapping movements of the arms and hands, occurring especially when the child is excited or upset. Twirling of the body and walking on tiptoe are also characteristic. There may be a whole range of other mannerisms involving, for example, speech and gait. The child may examine objects by sniffing them. Sometimes one set of mannerisms follows another in time.

**Resistance to change.** The child often develops a stereotyped pattern to his life and becomes very distressed if there is any deviation from it. Food has to be prepared in a particular way, the table must be laid just so, a walk to the shops must follow a particular route. Sometimes these rituals threaten to constrict family life to a serious degree and they result, of course, in considerable restriction in the child's range of activities.

**Attachment to particular objects.** Normal toddlers show attachments to cuddly objects they like to have around them at bedtime or when they are upset. Autistic children sometimes become attached to unusual non-cuddly objects, bits of wire, tin cans, etc., from which they hate to be separated at all times of the day.

**Acute emotional reactions.** There are frequent outbursts of excitement, anxiety, misery, or anger, and these may be precipitated by alterations in the environment that would not be expected to cause more than a slight ripple in the emotional life of an ordinary child.

**Other behavioural symptoms.** There may be a wide range of other symptoms including sleep problems, over-activity, distractability, impersistence, soiling, and wetting.

### Intellectual development

**General intellectual development.** As language is specifically delayed in this condition, this is best gauged by considering the non-verbal development of the child. Non-verbal development may be within the average range, but is much more likely to be retarded. About 70 per cent of autistic children have non-verbal IQs below 70, about 50 per cent below 50. Only about 5 per cent have IQs over 100.

**Specific deficits other than language.** The child is likely to be particularly poor in tasks involving symbolic thought, and in those that require him to perceive how other people are thinking or feeling.

**Special abilities.** Islets of ability have been described in autistic children. These usually involve mechanical tasks such as rote memory—learning things off by heart. In general these abilities only take the child into the normal range and are not exceptional, except in relation to the generally low level of ability. Very occasionally, however, some abilities are developed to a most unusual degree. These may be mathematical (e.g. the capacity to multiply two four-figure numbers together in the head, or to calculate the day of the week of a date several years ago) or they may just involve rote memory. Not all children with these unusual abilities (idiots savants) are autistic, but a proportion are (O'Connor and Hermelin 1984).

### Physical conditions

Usually the children have no identifiable physical abnormalities. However, they may show an underlying physical cerebral disorder (see below) and they may develop epilepsy. About 15 per cent of children with autism develop epilepsy, although often the fits are neither severe nor frequent.

## Background factors

The condition occurs in about 3–4 per 10 000 children and is about four times commoner in boys than in girls (Steffenburg and Gillberg 1986). However, the rate of mentally retarded children with atypical autism, but with a triad of similar social, language, and relationship impairments is much higher, increasing the rate to as many as 21 per 10 000 (Gillberg *et al.* 1986). It has been described as occurring in all countries throughout the world where it has been looked for. It was thought to be commoner in middle-class families, but the findings are inconsistent. The nature of the condition is a subject of considerable controversy. In particular, there is disagreement concerning the basic psychological deficits, to which the other aspects of the condition are secondary. At present, it seems most likely that the basic deficit is a cognitive one, perhaps with deficiencies of the perception of facial expression and of symbolic thought as central (Rutter 1983). Affected children have particular difficulty in accurately identifying the emotional expressions of other people (Hobson 1986). Such evidence implies that the problems in social relationships, mannerisms, acute emotional reactions, etc. are secondary to the child's difficulties in making sense of his environment. However, the unusual nature of these other deficits makes it unlikely that there is a simple or straightforward link between these and the cognitive deficits.

Although the basic psychological defect is uncertain, there is now more agreement that in most, if not all cases, some degree of brain dysfunction, rather than a reaction to adverse environmental influences, underlies the condition. The evidence from this view arises from knowledge of:

1. *Genetic factors.* The condition runs in families with 2–3 per cent of sibs affected. Sibs of autistic children also have a higher than expected rate of language delay. Concordance is much higher in monozygotic than in dizygotic twins (Le Couteur *et al.* 1989). A small number of cases have been described with chromosomal defects, especially the fragile X syndrome (Section 5.3), but probably this does not account for more than about 5 per cent of cases of autism (Wahlstrom *et al.* 1989).

2. *Underlying physical conditions.* Autistic children have been found to suffer from an excess of complications of pregnancy and delivery. In a minority of cases, the condition arises on the basis of a well-identified physical condition. This may be infective (fetal rubella, congenital syphilis), metabolic (phenylketonuria), genetic/developmental (tuberose sclerosis, fragile X syndrome), or of unknown aetiology (e.g. infantile spasms). Numerous, so far unsuccessful attempts have been made to identify a specific neuroanatomical, neurophysiological, or neurochemical cause for

the condition. Various anatomical abnormalities including hypoplasia of specific cerebellar lobules and widening of the fourth ventricle have been described. Abnormalities of catecholamine and endorphin metabolism have also been inconsistently reported.

3. *Parent and family factors.* There is no evidence that abnormal parental attitudes or behaviour are responsible for the development of autism, though this has been alleged in the past (Koegel *et al.* 1983). Clinical impressions that fathers of autistic children tend to be schizoid in personality have been confirmed by systematic study (Wolff *et al.* 1988), but the link is likely to be genetically determined.

On the other hand, the impact of having an autistic child on family life and parental health is often powerful. Autistic children are amongst the least rewarding to bring up because of their lack of affection and slow progress in social relationships and language development.

4. *Link with schizophrenia.* A small number of autistic children develop schizophrenia (and indeed other psychotic conditions) in later life (Petty *et al.* 1984). However, the distinctive clinical characteristics and family studies make it clear that autism and adult schizophrenia are genetically distinct syndromes (Kolvin 1971).

## Assessment

Most autistic children come to attention in the second or third year of life because of developmental delay—usually delay in language development. The diagnosis is made on clinical grounds using the criteria described above. Atypical cases are quite common, but it is unwise to make a diagnosis of autism unless both significant language delay and the characteristic disturbances of social relationships are present.

The condition needs to be differentiated particularly from Asperger's syndrome (autistic psychopathy), schizoid personality, and schizophrenia (in which language delay is either absent or an inconspicuous feature), as well as developmental aphasia (in which the disturbance of social relationships is absent) and deafness, in which the child responds to and uses gesture as well as relating well to his parents.

In girls who have progressed normally for the first 1–2 years, but then regressed with the development of autistic features, associated with jerking or writhing, involuntary movements, and abnormalities of muscular tone, a diagnosis of *Rett's syndrome* should be considered (Olsson and Rett 1987). Criteria include female sex, early behavioural, social, and psychomotor regression, loss of purposeful hand-skills with very characteristic hand-wringing or hand-washing movements, and a failure of skull growth. A similar picture of regression after a normal period of development in the first 1–2 years, followed by very slow subsequent mental development may also occur in children of both sexes in the absence of autistic features.

Disintegrative disorder is a term used to describe a condition in which there is a loss of cognitive function after a period of normal development. It is similar to autism, but occurs after a period of normal development lasting longer than 2 years (Evan-Jones and Rosenbloom 1978). The condition appears to have a prognosis similar to, or rather worse than autism or pervasive developmental disorder of later onset. It is uncertain whether the condition can be meaningfully distinguished from autism (Volkmar and Cohen 1989).

Most difficulty arises in the diagnosis of children who are both autistic and moderately or severely mentally retarded. Autism should be diagnosed only if there is a significant language deficit in relation to other disabilities. In children who are severely mentally retarded, the presence of autistic features is sometimes rather a minor matter when a child is so handicapped by serious learning deficits.

## Management

Management should begin with an explanation to the parents of the nature of the diagnosis, its probable cause, the ways in which the child can be helped, and the likely outcome. It is important to reassure parents that their behaviour has not produced the condition, though this does not mean that, in the future, there is not much they can do to help the child. In the preschool years it will not be possible to predict outcome with any degree of certainty, but it is important that parents should be prepared for a chronic and often seriously disabling course to the condition. The positive aspect of the child's development, and the fact that he will continue to make progress, although this may well be slow, also needs to be stressed. It is important to put aside sufficient time to discuss parents' questions about the condition.

Neither medical treatment (medication) nor psychotherapy directed towards the child have any significant place in management, though brief courses of medication may sometimes be useful to reduce the impact of associated behavioural problems. For example, both phenothiazines and butyrphenones, such as haloperidol, may sometimes be beneficial in reducing severely aggressive behaviour. Early reports of the value of fenfluramine have not been confirmed. In contrast, behavioural and educational methods have been demonstrated to be helpful. Rutter (1985) suggested that there should be four goals of such treatment—the fostering of normal development, reduction of the rigidity and stereotypy that often dominates the life of the child and family, the removal of maladaptive behaviours, and the alleviation of family distress. These goals are most likely to be met by a combination of counselling, behaviour modification, and special schooling (Howlin and Rutter 1987).

Normal development is fostered by encouraging parents to provide intensive, stimulating experiences more consciously geared to the child's developmental level than would be necessary with an ordinary child. For example, if parents take obvious pleasure in early vocalization, label objects verbally when they use them, communicate with their child in every possible way (with gesture and signs as well as words), the child's language will be given its best chance to develop. *Special schooling* with a favourable teacher–pupil ratio, and teachers experienced in teaching non-communicating children in a highly structured setting, plays a most important part in the encouragement of normal development (Rutter and Bartak 1973).

Both at home and at school the reduction of stereotypies and other maladaptive forms of behaviour can best be approached by **behavioural methods** (Section 6.2.2). Operant methods involving reward for appropriate behaviour or responses are effective in enhancing communication skills and in improving eating and sleeping patterns. The technique of 'graded change', introducing alterations in the child's environment in very small but firm steps, may be effective in reducing the rigidity with which some autistic children control their environment. Methods of this type need to be applied by parents in the home, as generalization of change from clinic or hospital to the home setting is often limited. Various more extreme behavioural approaches have been applied. For example, considerable success has been claimed for 'holding' techniques in which the child is forced into close physical bodily contact with a therapist for a prolonged period of time. This procedure is, of course, extremely upsetting and distressing for the child in its early stages. Although the technique may produce short-term improvement in social interaction, it is as yet uncertain whether the medium-term benefit justifies the distress it produces.

**Parental counselling** to alleviate distress and promote coping skills is particularly necessary in the first few months after the diagnosis is made, but remains important subsequently. Parents take a very variable amount of time to overcome their grief at 'losing' the normal child they expected to rear, and react in a variety of ways to their distress—especially with anger and depression. Anxiety about the future, often largely based on realistic concerns, sometimes becomes preoccupying and gets in the way of useful measures that could currently be taken. The mental health professional, with the capacity to understand and accept parental reactions, advise about ways of fostering normal development and tackling behavioural problems, and with knowledge of the likely outcome, can often play a helpful part in supporting parents of these children (Howlin and Rutter 1987).

### Outcome

The prognosis of classical infantile autism is poor. A high proportion of children so diagnosed remain dependent as adults and require continuing care at home or in hostel accommodation. Only about 10 per cent achieve a really significant degree of independence, and a somewhat smaller number are able to obtain a job on the open market (Lotter 1978). Personality problems involving impairment of social relationships are usually prominent.

About one in six children develop occasional attacks of epilepsy. A minority of children develop more serious aggressive problems or marked mood swings, sometimes amounting to manic-depressive psychosis during adolescence, but these difficulties usually abate in early adult life. The best predictor of outcome for independent living is the measured intelligence of the child—those with a non-verbal IQ over 70 performing best later on.

Three provisos need to be made to this bleak account of prognosis. First, some cases of classical but mild autism occurring in the preschool period which are associated with normal intelligence do abate in middle childhood. Second, although progress is very slow, in nearly all cases there is, in fact, continuing improvement throughout childhood and adolescence, so that each year the child gains some new skills. Third, the quality of parental care and education is important in outcome. Autistic children who are neglected will deteriorate and not make the gains of which they are capable.

### 2.8.2 ASPERGER'S SYNDROME

**Definition**

This is a relatively rare condition, present from early childhood, in which there is a characteristic lack of empathy and capacity for social relationships, pedantic speech, and preoccupation with a special interest about which the child may accumulate extensive information in a mechanical fashion (Wing 1981). The condition is sometimes described as schizoid personality or autistic psychopathy.

**Clinical features**

After what is usually a relatively normal early development, the child is usually noted to lack warmth and interest in social relationships round about the third year of life. Language is not usually delayed, but the quality of speech is often impaired with lack of variety in intonation. The combination of a monotonous delivery of speech, and often laborious emphasis on exactitude in the use of language gives a pedantic quality to the child's verbal output.

As the child gets older he is likely to be isolated socially both by the fact that other children find him odd, and by the fact that he appears little interested in making friends. Lack of social sensitivity is prominent, and this sometimes leads to the child giving offence when none is intended. The child is likely to spend more and more time reading about a special interest, such as railway timetables or characters in space-fiction epics. The children show a variety of other features. They are often clumsy in motor movement. They may show mannerisms, though these are usually not prominent. Physical health is generally good, and there are usually no specific learning disabilities.

## Prevalence

The condition occurs in about 1 or 2 per 10 000 children. It is about five to six times more common in boys than girls. There may be a slight tendency for the condition to be more common in middle-class families. Intelligence is usually within the normal range. Parents may show difficulties in forming social relationships themselves.

## Aetiology

There is controversy as to whether the condition is a variant of autism, and indeed in careful studies comparing children with the syndrome and autistic children, Gillberg (1989) and Szatmari *et al.* (1989) have been unable to identify any clear-cut differentiating features. As with autism, the aetiology is unknown, but it is suspected that genetic factors are of major importance. Occasionally Asperger's syndrome and autism may occur in the same family, suggesting a common aetiology.

## Assessment

Assessment of the condition involves taking a full developmental history and an account of the behavioural and emotional problems. Physical examination is usually negative, though clumsiness can often be detected. Physical investigations are not usually indicated.

As an indication, there is sometimes difficulty in distinguishing the condition from autism, in which language delay is prominent and mental retardation often present. It has been suggested that the term 'Asperger's syndrome' should be reserved for those children who would otherwise be diagnosed as autistic, but who use language freely, fail to make language adjustment to fit the social context, wish to be sociable but fail to make relationships with peers, are clumsy, develop idiosyncratic but engrossing interests, and have marked impairments of non-verbal expressiveness affecting their facial expression and gesture. Most clinicians do continue to find it useful to make a distinction between Asperger's syndrome and autism. In children with a mild variant of Asperger's syndrome, it may be in their interests not to make a daunting sounding diagnosis, but to consider them merely as having a somewhat unusual personality. The point at which one can make a distinction between a normal and abnormal personality is arbitrary. Children with schizophrenia differ in that they have a later onset, and are likely to show thought disorder with characteristic delusions and hallucinations.

## Treatment

There is no medical treatment for this pattern of personality. Parents and teachers can, however, protect children with such unusual personalities in a variety of ways. Acknowledging that the child's personality is unlikely to change very much over time and accepting the child for what he is are important first steps, Such children respond badly, with fearful reactions or aggression, to unpredictable events, so establishing a routine is helpful. Tactful explanation to other children in the neighbourhood or classroom about the child's vulnerability, with firm handling of any bullying, reduces the stress to which the child is exposed at school and home. Most such children will cope in an ordinary school, but special educational arrangements are occasionally necessary.

As children get older they themselves may be helped by supportive counselling about the nature of their condition, and ways in which they can improve their relationships with others.

### Outcome

This will depend on whether the child's personality is widely deviant or whether he merely shows some mild personality characteristics of the type described. Most individuals will be able to obtain employment in a fairly routine job, though they will not be able to cope with a complex work situation. Successful relationships with the opposite sex leading to marriage are uncommon. For more severely affected individuals, sheltered employment and residence in hostel accommodation may be necessary.

There is an increased tendency towards various forms of mental illness, including anxiety and depressive reactions and obsessional disorders. A group of children with 'schizoid personality' (a closely related condition), who were followed up into adult life, showed a similar persistence of their personality problems with some increased tendency to antisocial behaviour (Wolff and Chick 1980). The development of schizophrenia is unusual, but has been observed.

## 2.9 Emotional development and disorders of mood

Fear, anxiety, and depression develop in characteristic fashion during childhood. Their causes, the way in which they manifest themselves, and their adaptive function change as the child progresses from infancy, through childhood to adolescence. As most morbid or abnormal mood states are probably only extremes of normal mood, so the nature of mood disorders changes over this time.

In this section, fear and anxiety (both normal and abnormal) will be treated separately from depression, but it should be remembered that, in the individual child, there is often considerable overlap. The experience of one mood state is likely to be accompanied, at least to some degree by the presence of the other.

## 2.9.1 DEVELOPMENT OF NORMAL FEAR AND ANXIETY

### Definition

Fear and anxiety are terms used relatively interchangeably to describe unpleasant emotional states, accompanied by physiological changes (such as tachycardia) in which there is anticipation of an undesired outcome. In order to be sure that anxiety is really present, therefore, one must be able to tell both that a person is experiencing an emotion and what it is he is worried about. Obviously, infants and very young children cannot talk about their feelings, but their behaviour in certain situations often makes it clear that they are indeed anxious and why this is, even though they are themselves unable to describe their emotions.

### Anxiety in the early years

**Communicated anxiety.** From the first few days of life, infants can be seen to respond to the emotions of those who are holding them, usually their mothers. A baby held by an anxious mother will often appear tense and this empathic behaviour may, in itself, make feeding even more difficult for the worried mother.

**Stranger anxiety.** Anxiety in the presence of strangers is not an all-or-none phenomenon, and the factors affecting its presence are of importance to those providing health care for children in hospitals. There is individual variation in the age at which children show anxiety, and to some degree it is situation-specific (Waters *et al.* 1975). Thus children are less likely to show it in the presence of their mothers, or if the stranger does not come too close too quickly. Their previous experience of the stranger, and their ability to control the situation will also affect their level of distress.

By about the age of 4 months, using special observational techniques, infants can be seen to show distress with strangers. Wariness of strangers at this age is a somewhat delayed reaction, but by 7 or 8 months, the presence of a stranger or strange situation will make the child look obviously unhappy and distressed immediately. At about the same age the child develops 'separation anxiety', or distress at being separated from his mother or other familiar figure. Separation anxiety usually increases until about the age of 18 months and then slowly declines in intensity (Bowlby 1975).

Although separation anxiety and stranger anxiety are separate phenomena, and one can occur without the other (e.g. when a child being held by its mother shows fear of a stranger), their course runs closely parallel. Both emerge around 6–9 months, are at their peak around 12–18 months, and decline between 2 and 3 years.

### Later development of anxiety

As children get older, the range of possible outcomes that may make them feel anxious increases considerably. They will continue to be afraid of dangerous and unfamiliar situations and of separation, but may also be fearful of physical illness, mutilation, or death, of the experience of other unpleasant emotions such as shame and depression, of their own unaccept-able aggressive or (later) sexual feelings. Each of these factors may make a child anxious. Whether this occurs or not will depend on the individual characteristics of the child, his previous experience, and the degree of personal threat the future holds for him. Anxieties are also influenced by the level of the child's understanding. Apparently irrational anxiety may become readily explicable if one obtains a better picture of the child's view of a situation. For example, a child may become very anxious before an anaesthetic when he is told he is 'just going to have a little sleep' if this is what he was told the previous week about a pet animal that had to be put to death, and that he has not seen again.

## Normal defence mechanisms against anxiety

Anticipatory anxieties are experienced as in adults with an unpleasant set of mental or physical sensations (palpitations, a feeling of nausea, vomit-ing, stomach-ache, weakness of the legs, etc.) closely linked in the child's mind to the feared eventuality. The child may deal with his anxiety by the use of various 'defence mechanisms'. Although each of these would be abnormal if carried to extremes, most can be observed at least transiently in normal children.

1. *Regression.* The child may regress in the face of an anxiety-provoking situation, behaving like a normally fearful younger child, and therefore eliciting greater protection from those around him.

2. *Ritualization.* He may develop obsessional behaviour such as check-ing rituals. For example, a four-year-old worried about his mother going away for a holiday might start to lay all his toys out and count them endlessly.

3. *Somatization.* Some children 'somatize' their anxiety; they may de-velop aches and pains in their abdomen or head. These pains may be due to muscular tension, to undue sensitivity to normal abdominal sensations, or to attention-seeking.

When these types of behaviour or symptomatology occur therefore, anxiety may be a prominent underlying cause.

## Development of fear

Fears are states of acute anxiety, linked to the presence of particular objects, persons, or situations. They result in avoidance behaviour. Different stages of child development are associated with different types of fear (Marks 1987). Fear of heights begins at around 6–8 months and increases when the baby begins to walk. Novel stimuli (of which fear of strangers is the most powerful example) begin to be anxiety-provoking around 6 months of age, and reach their peak around 18 months to 2 years. Between 3 and 5 years, fear of the dark, of imaginary monsters, and of animals are very common. In middle childhood, from 6 to 11 years, the child begins to be afraid of ridicule and other shameful social situations. In adolescence, many normal teenagers are frightened of failure, and of large social gatherings such as parties. This is also the phase of life when fear of death and nuclear war emerge most strongly.

## 2.9.2 ANXIETY STATES AND SPECIFIC FEARS

### Classification

There is a separate category for 'emotional disorder specific to childhood' in ICD-10 (see page 10) under which are included separation anxiety disorder, phobic disorder, social sensitivity disorder, and sibling rivalry disorder. Adult-type phobic disorders including agoraphobia, panic disorder and generalized anxiety disorder are separately classified. In DSM III-R (see page 9) panic disorders are divided into three subtypes. Agoraphobia (fear of being in a place from which escape might be difficult), social phobia, and simple phobia. A panic attack is defined as present if at least four out of thirteen symptoms including palpitations, nausea, trembling, sweating, and fear of dying have been experienced. Generalized anxiety disorder is present if there is excessive anxiety or worry about two or more life circumstances, with six out of eighteen possible symptoms. These symptoms are grouped under the headings of motor tension, autonomic hyperactivity and undue vigilance.

### Clinical features

*Preschool (0–5 years)*

**Generalized anxiety and fearfulness.** Preschool children with undue general fearfulness and anxiety will often show problems in even brief separation from their parents, especially their mothers. They will tend to follow their mothers around and scream if separated, even if this is for a short time and with a relatively familiar person. They are often generally miserable, whining children, who cry easily. Their general irritability and tension results in a tendency to have frequent temper tantrums. Specific problems of management, such as separation difficulties at night, when going to sleep, are common. Attendance at playgroup is an unhappy experience—the children take days or weeks to settle down and may never do so and have to be removed. They tend to mix poorly with other children, and not to want to go to other people's houses.

**Specific fears.** Specific fears are very common in the preschool period. Indeed this is a time when they are most frequently experienced and are most intense (J. W. Macfarlane *et al.* 1954). In the first two years of life, the child has a special tendency to develop fears of stimuli discrepant with his normal experience. Thus, strange noises or an unfamiliar object in the room may produce fear. Fear of some strangers may be more pronounced than with others. Occasionally a child may develop a fear of a completely familiar person such as his or her father. This could be for an obvious reason that may be serious (the child has been hurt by his father), or trivial (the father has started to grow a moustache), or unaccountable. By 3 years, fear of animals (spiders, dogs, mice) is very common, as is fear of the dark and of ghosts. The abnormality of the fear can only be judged by its severity, persistence, and the social disability produced. Most normal fears in children of this age last at most only a few weeks or months, and do not interfere with the child's life.

### Middle childhood (5–12 years)
The child of this age with an anxiety state will often show:

1. *Tension and worry throughout the day.* The anxiety will often vary depending on the situation the child is expected to face. A new school term, going to a party, parents going away on holiday together without the child, a test at school or a change of teacher, a minor illness either in the child himself or in a close family member (especially the mother)—these are all likely to increase the child's anxiety.

2. *Physical complaints for which no organic cause can be found.* Headaches and stomach-aches are particularly common, but limb pains, recurrent cough, and other symptoms occur frequently. The mechanism whereby these symptoms occur is often obscure, but sometimes it appears that ordinary bodily sensations are misinterpreted.

3. *Separation fears.* As well as general tension and anxiety, there may also be fearfulness related to specific situations. Undue concern about separation from parents is particularly common. The anxious schoolchild is therefore especially likely to worry about admission to hospital, or going away on a school journey.

4. *Regression to immature behaviour.* Separation anxiety (see above) can be seen as a form of regression, but other manifestations of regression are common. Undue dependence on parents for self-care in washing and feeding may occur. The child may revert to a pattern of wetting and soiling. Waking during the night and insistence on sleeping with parents may occur.

5. *Manipulativeness.* Secondary to the presence of anxiety, the child may attempt to become very controlling of situations, and indeed, as a response to anxiety, may succeed in achieving a good deal of control within the family. Ensuring that parents do not go out in the evening, and managing to persuade parents to give notes for school absence when this is clearly not justified by physical illness are two examples.

6. *Obsessional behaviour.* Again, as a response to anxiety, the child may

develop rituals which form just one part of a general anxiety state rather than a full-blown obsessional disorder (Section 3.2.1). This occurs uncommonly, but a number of anxious children do become preoccupied, for example, with checking locks on doors or with the cleanliness of the toilets.

7. *Motor movements*. Tics are described elsewhere (Section 2.4.4). Occasionally anxious children show these rapid, repetitive movements as part of an anxiety state.

8. *Enuresis and encopresis*. Although both of these may occur for other reasons (and indeed enuresis is more usually a sign of developmental immaturity), they may also occur as part of a generalized anxiety state.

9. *Other mood changes*. Finally, anxiety states in childhood are often associated with other mood changes, especially irritability and depression.

10. *Specific fearfulness*. Apart from school refusal (Section 2.9.3), handicapping specific fearfulness is somewhat unusual in middle childhood. Most common are fears connected with sleep. Children may be unwilling to enter their bedrooms by themselves or be too frightened to stay in bed. Lying in bed alone, the child may experience fears of abandonment by his parents or fears of burglars, ghosts, etc. An open window with a flapping curtain, or a creaky staircase, may enhance fearfulness. Fear will often lead the child to refuse to go to bed or stay in bed alone. As children with such fears are often taken into the parental bedroom or allowed to watch TV until well past their bedtime, the child is likely to come to perceive advantages in fearfulness, and this may lead to manipulative behaviour—the child insisting on privileges and claiming fear when none is present. Manipulative behaviour and fearfulness are often commingled.

Other specific fears in this age group have usually arisen at a younger age. They include fear of dogs, other animals, and thunderstorms.

### Adolescence (12 + years)

Adolescent anxiety states have many features in common with those occurring at earlier ages, but there are some important points of difference. By this age, as well as general tension, salient features are more likely to include:

1. *Free-floating anxiety*. A sense of mental unease exists without attachment to any particular object, person, or situation. This is now likely to be accompanied by a sense of depersonalization (a sensation of uncertainty about personal identity and bodily integrity) and derealization (a more or less acute sense of unreality).

2. *Existential anxiety*. As well as an acute sense of depersonalization, a teenager may show anxious feelings related to an unusual concern for the reality of his own personal existence, his capacity to make real choices, and about the meaning of life.

3. *Hypochondriasis*. Headaches and stomach-aches with no organic

basis are, at this age, likely to be replaced by a more adult form of hypochondriasis, with unusual concern over bodily health. Minor complaints are likely to lead to unnecessarily prolonged absence from school, college, or work. There may also be worries about personal appearance. Girls may be chronically anxious, for example, about the size of their breasts and buttocks. Exaggerated concern over bodily appearance in girls may also be more specifically linked to partial or full anorexic syndromes (Section 2.2.6). Anxious boys may be unduly worried about their height, their acne, or the size or shape of their genitalia.

4. Teenagers with chronic anxiety are also prone to suffer from **specific fearfulness**, but this is likely to take different forms from that occurring at younger ages. Social anxieties about the loss of friendship and sexual unattractiveness, are likely to be more prominent. There may be unusually severe worries about the teenager's own personal future, about death of the teenager himself, his parents or others close to him, or more cosmic anxieties about the future of the world, nuclear warfare, etc. Such anxieties may, of course, be well founded and merit anxious concern. The morbidly anxious youngster is likely, however, to experience a chronic sense of unease, qualitatively different from that of a teenager who is concerned and politically aware, but not disabled by such preoccupations.

Specific fears, other than those connected with school refusal are rather unusual in adolescence. However, anxiety linked to social situations, such as parties or other occasions, may be so severe as to reach phobic proportions. Fear of leaving the home (agoraphobia) may begin in adolescence, occasionally as a sequel to school refusal (I. Berg *et al*. 1974). Agoraphobia may be linked to specific fears of different forms of travelling, in trains, buses, cars, lifts, the underground, or on escalators. The central feature is fear of being in a situation from which it would be difficult, embarrassing, or impossible to escape.

## Prevalence

General anxiety and specific fearfulness are amongst the common emotional problems occurring in childhood. In the preschool period, between 2 and 3 per cent of three-year-old children were found to have many different worries, or worries over certain things for long periods (N. Richman *et al*. 1982). Nine per cent of children of this age were found to have three or more marked fears or six or more fears altogether. Most common fears in this age group are fears of the dark, thunder, animals (especially large dogs), and insects. By 8 years fearfulness had reduced with only 2 per cent of children affected. Anxiety disorders accounted for about a quarter of psychiatric disorders seen in children identified as disturbed in the Isle of Wight study, so that about 2 per cent of the children were affected both at 10 and 11 years, and at 14 years (Rutter *et al*. 1970*a*). In adolescence, specific fearfulness is less common, and when it occurs is most likely to focus on social situations, or to take the form of school refusal (early adolescence) or agoraphobia (late adolescence).

In contrast to late adolescence and adulthood, when females are more commonly affected, younger boys and girls do not differ in their rates of anxiety and fearfulness. Rates also do not differ by social class as assessed by parental occupation, though children living in socially adverse circumstances may be more commonly affected (see below).

### Background factors

1. *Genetic and constitutional factors*. Although anxiety states have a familial tendency, it is unclear to what degree genetic factors are responsible.

2. *Temperament*. Children who, in the first year of life, are rather withdrawn, quiet, and 'slow to warm up' have a predisposition to develop anxiety reactions in later childhood (Thomas *et al.* 1968).

3. *Parental behaviour*. Characteristically, anxious children have parents who tend to be overprotective. Overprotectiveness has been defined (D. Levy 1943) as provision of excessive physical contact, prolongation of infantile care, prevention or discouragement of independent behaviour, and insistence on retaining control. Such overprotection may be explicable in terms of the parents' own insecure childhood, or of some feature of the child's early life. The child may be 'special' because it was conceived after a long period of infertility or, for example, because of a near-miss cot death or other serious illness (M. Green and Solnit 1964). However, other parental behaviour, especially inconsistency, excessive criticism, and punitiveness, may also make children anxious.

4. *Communication of anxiety*. Specific fearfulness and general anxiety are particularly likely to be communicated directly from parent to child. Parents who themselves, for example, show unusual fear of insects are likely to induce fear in their children by their own emotional behaviour (Eisenberg 1958). Modelling is an important process whereby fears are transmitted, most commonly from parent to child.

5. *Specific experiences*. Excessive fears are sometimes triggered by unusually frightening events. A boy's fear of thunderstorms may begin after a severe storm when he thought his parents were not in the house. Most specific fears, however, do not seem to have their origin in such an event.

6. *Social adversity*. Parents who are settled in their lives, without serious financial, housing, or marital difficulties, are more likely to be able to show sensitivity to their children's worries and anxieties, and thus help overcome them. Children living in homes where parents are pressurized by adversity are more likely to experience insecurity, and thus be predisposed to specific fears and anxiety.

## Causative mechanisms

**Biological–developmental.** Fear is an adaptive phenomenon. The experience of fear in the presence of danger encourages behaviour designed to achieve safety, especially proximity to protective figures. Young children living as they do in a state of greater vulnerability and need for protection, naturally show a marked tendency to fearfulness. This gradually reduces with age. In some individuals, excessive or prolonged anxiety reactions can be seen as an overdevelopment of an adaptive mechanism.

**Learning theory.** Children learn avoidance behaviour as a result of escaping from dangerous situations. The escape is associated with relief from the unpleasant sensation of anxiety. Avoidance is thus reinforced or rewarded and is more likely to recur.

**Psychodynamic theory.** Unacceptable aggressive or sexual impulses may be repressed so that the child is unaware of them. However, the focus of aggression may remain a stimulus for fear reactions, and the stimulus may become overgeneralized. Thus a boy with repressed aggression towards his father may show excessive fear not only of his father but also of other authority figures.

## Assessment

Information should be obtained on the duration of anxiety and fear reactions, and the degree to which they disable the child socially. It is important to establish the situations in which the anxiety is experienced, and the coping mechanisms the child uses to deal with the emotion when it occurs. The experience of anxiety reactions by other family members, especially the parents, is usually of considerable significance. Possible anxiety-provoking stresses both at home and school should be probed, even if they are not mentioned spontaneously.

## Treatment

### Anxiety states

The aims of treatment of general anxiety states should be to reduce or remove unnecessary stresses producing anxiety, to help the child and family to understand more clearly the nature and function of experienced anxiety, and to enhance existing coping mechanisms.

1. *Stress reduction and removal.* Where the child is responding to inevitable, ordinary stresses, it will not be appropriate or possible to remove them, but in some cases the child is in fact reacting to unnecessary stress. A child with a physical disorder may be fearful as a result of an unrealistic view of the likely outcome of his condition. A child fearful of sleeping alone can be moved temporarily into another bedroom shared with a sib. A child frightened that his mother is going to carry out a threat to leave home may be given reassurance.

2. *Improvement in understanding the nature of anxiety.* Brief, focused counselling sessions with the family may enable insight to be gained into the way anxieties are transmitted from one family member to another. Alternatively, family members may be helped to see how tensions in the family are resulting in a child experiencing chronic insecurity.

A *brief course of focused individual therapy* (Section 6.2.3) may, with children over the age of about eight years, enable insight to be gained into the way hitherto unexplained anxieties have arisen as a result of unacceptable impulses. In both family and individual sessions, a simple explanation of the way anxiety produces physical symptoms can be helpful.

3. *Enhancing coping mechanisms.* The child is probably already using partially effective coping techniques to deal with the experience of anxiety. Encouraging the child to avoid unnecessary stresses, but to deal more effectively with necessary stress can be helpful. *Relaxation techniques* (Section 6.2.2) are sometimes useful in children over the age of about 10 years. *Medication* is not usually effective in dealing with anxiety states in childhood, and is only slightly more helpful in adolescence. In severe, generalized anxiety states, a minor anxiolytic agent such as diazepam (Section 6.2.6) may produce some symptomatic relief. The use of such medication should be limited to 2–3 weeks.

### Phobic states or specific fears

Behavioural methods should be employed in these conditions. Desensitization of *in vivo* type (Section 6.2.2) is usually effective. The child is encouraged to produce a hierarchy of feared situations, and then achieve relaxation, first with the least anxiety-provoking and then, when this has been achieved, with more severely anxiety-provoking situations. Operant methods can be used to reinforce or reward behaviour that retains the child in gradually increased contact with the feared situation. Alternatively, or in addition, the child can be encouraged to practice relaxation while imagining the feared situation or facing it in real life. If the feared situation is social, then it is often useful to get the child or adolescent to practise beforehand with parents or a friend how he is going to cope when the real situation occurs. Such role rehearsal can be useful, for example, when a child is frightened to go into a shop to make a purchase. A child may also be helped by a 'modelling' procedure—watching an adult or another child go through, in a relaxed way, an experience of which he is afraid.

### Outcome

In general, both anxiety and phobic states have a reasonably good prognosis (P. Graham and Rutter 1973), with two-thirds remitting over 3–4 years. Severe anxiety reactions are often self-limiting, and are likely to last only a few weeks. A high proportion of the specific fears of childhood disappear within a few months. Although anxiety reactions are often short-lived, children who experience them are at risk for recurrence in adolescence and adult life, and they are likely to be somewhat shy and diffident in later life. Enhancement of coping mechanisms is likely to improve the prognosis, by giving the child a better capacity to deal with his anxieties in the future.

### 2.9.3 SCHOOL REFUSAL

#### Definition

In school refusal an irrational fear of school attendance is the core symptom. Fear may be based partly or fully on fear of separation from home or from one or both parents (Bowlby 1975). It may, however, also be specific to school attendance or some aspect of it, in which case school phobia is an appropriate term to use. The condition is distinct from truancy in which children conceal their absence from school.

#### Clinical features

The problem usually shows itself first at a time of change of school or after a period of absence for some other reason, such as a minor illness. The onset may be acute, but is more often gradual with absences building up over a few weeks. Unwillingness to go to school may be expressed openly—the child, for example, saying that he dislikes a particular teacher who has been unpleasant to him, or that he has been bullied at school. Alternatively, the refusal may present with physical symptoms of tension and anxiety, more or less obviously linked to school attendance. Abdominal pain, headache, nausea, limb pains, attacks of palpitations, and a range of more unusual symptoms may be present (Hersov 1960). On questioning, it becomes clear that the symptoms are either absent at weekends or during school holidays, or present only in mild form at these times. The child may be able to go out in the evenings to play with his friends.

   **Previous personality** The previous personality of the child is variable. Most have been rather 'good', quiet, conforming children who have never been in any trouble at school and are keeping up well with their work. A proportion, however, have been previously outgoing with many friends, and some have been having difficulties with their school work.

   **Family background.** In the family background there is characteristically a rather anxious and perhaps depressed mother who has, for some years, had an overly close relationship with the child in question (Skynner 1974). The child may be special because of being the last of a sibship (the 'Benjamin' syndrome) or for some other reason, such as a very premature birth. Father may be absent through death or separation, or he may be a somewhat passive person who has difficulty in establishing his presence.

   **School factors.** These have received little systematic examination. There may indeed be a stress at school which many of the children in the class are finding problematic—such as a very unsympathetic teacher. Alternatively, the school system may be rather lax in checking up on absences from school or there may be an overconcern regarding minor somatic illness amongst the school staff. Much more commonly, the teachers are aware of the irrational reasons for absence and have made sensible but unsuccessful attempts to get the child back to school.

**Prevalence.** The condition in its severe form is not a common one, occurring in about 1–2 per cent of cases seen in child psychiatric clinics (Chazan 1962). Community surveys suggest that severe school refusal occurs in less than 1 per 1000 children aged 10–11 years, although it is more common in 14-year-olds (Rutter *et al.* 1976*b*). Minor forms of school refusal with anxious reluctance to go to school for a few days, or chronic anxiety, in a form that never actually causes the child to miss school, are much more common.

**Age and sex.** The peak ages for occurrence are at the time of school transition, but there is also a peak in adolescence, so that 5 years, 11 years, and 14–15 years are the most common ages for presentation. There is no particular social class trend, and boys and girls are affected about equally.

## Assessment

If the presentation involves a physical symptom, any necessary investigations should be rapidly carried out to exclude the presence of a physical condition. Usually the absence of physical pathology is fairly obvious, but just occasionally this may not be so. School refusal has been called the 'masquerade syndrome' (Waller and Eisenberg 1980) because its presentation may mimic so many other conditions. These include chronic viral infections, gastroenteritis, peptic ulcer, migraine, and cerebral tumour. In the remainder of this section it is assumed that these and other conditions have been excluded. It should be pointed out that school refusal for reasons of anxiety may coexist with any of these physical conditions as well as others.

Once the nature of the problem has been established, it is useful for whoever is dealing with the problem—family doctor, paediatrician, educational psychologist, or child psychiatrist—to see the whole family, including father if he lives at home. The way in which the problem has arisen should be discussed, and particular attention should be given to the possibility of a special relationship between the child and the mother, and any difficulties father may have in asserting himself in the family. The attitudes of the family towards the real nature of any physical symptoms should be explored—are they seen as due to a physical condition or to anxiety? The nature of aggressive feelings that the child and mother might have for each other because of their mutual dependency might be examined. The child should be seen, at least briefly, separately, so that anxieties and worries that cannot be expressed in front of the family can be explored. It is also often helpful to see the parents together briefly, in case there are aspects of their own relationship that might be relevant. Finally, it is important for any health professional dealing with the problem to contact those in the school system—teacher, educational welfare officer, educational psychologist, who have been involved—to get their view of the situation.

At this stage the practitioner should have a view of the extent and severity of the problem, and the degree to which school or family relationship factors are relevant. Most commonly the problem can best be viewed as a combination of a disturbance of family relationships, and anxious personality traits in the child, sometimes associated with specific stresses in school.

**Differential diagnosis** includes truancy, depressive disorder in the child, an unusual and difficult-to-diagnose physical condition such as peptic ulcer (that may or may not itself be linked to stress or anxiety), and a stressful but well covered up situation in school, such as severe bullying, which might make any child reluctant to attend. Truants, unlike school refusers, disguise their absence from school, are usually involved in other forms of antisocial behaviour, generally come from more disrupted families, and are often failing in school subjects. In older children, however, the distinction between truancy and school refusal is sometimes not easy to make.

## Management

### Return to school

In mild cases and those of recent onset, or where physical symptoms have been present for some time, but it has only recently been clarified that there is no physical condition present, the main aim of treatment should be to get the child back to ordinary school as soon as possible (Kennedy 1965; Blagg and Yule 1984). The practitioner should explain to the family how the problem is viewed, usually with emphasis on how this is not just the child's problem, but that of the whole family. Father may need to be encouraged to be firm with the child about school attendance and other matters. It is often helpful to sympathize with the mother concerning her feelings of being separated from her child while he is in school. The school should be contacted and a day agreed when a planned return is to be made. The assistance of the educational welfare officer may be invaluable. Indeed most cases of school refusal are dealt with by educational psychologists and school welfare officers without any health professionals being involved at all. Some minor concessions may need to be made concerning attendance at assembly or physical education classes. The use of short-acting anxiolytic medication on the morning of return may occasionally be justified. Using this approach a high proportion of mild cases and severe, but acute recently diagnosed cases will be resolved.

### Chronic school refusal

Long-standing problems are not likely to recover so quickly. Referral to child psychiatric services is often indicated. The child may, for many months, have been enjoying a relaxed time at home with his mother during the school day, and it is not surprising that the prospect of a noisy school day produces overwhelming anxiety. The child may have a more deep-seated intrapsychic problem, or family relationships may have become seriously disturbed with little prospect of change. In some cases, parents develop an overvalued idea, almost amounting to a delusion, that their child has a physical problem making it impossible for him to go to school. It is especially important in these circumstances that there is good communication between health, education, and social services staff.

A variety of psychiatric and educational measures may be helpful in chronic school refusal. Some educational authorities run small units for disturbed children that can be attended first full-time, and then part-time while the child is integrated back into ordinary school. Pressure by parents to arrange home tuition should in general be resisted, as this merely results in the child being even more embedded in his home situation. Educational measures can be supplemented with individual psychotherapy for the child (Section 6.2.3) or a series of family interviews. Behavioural management, with techniques such as relaxation, and imaginal and *in vivo* desensitization (Section 6.2.2) can also be helpful.

In more intractable severe cases residential placement may be indicated, in a child psychiatry in-patient unit, in a residential school for children with emotional and behaviour problems, or in a boarding school taking a proportion of disturbed children. The admission needs to be carefully planned, but, if confidently handled, it can nearly always be successfully arranged. Once away from home, separation anxieties are often, at least temporarily, abated. A proportion of school-refusing children admitted to in-patient units will require long-term residential placement in schools for children with emotional and behaviour disorders.

### Outcome

Most mild and acute cases resolve rapidly without further problem. More severe cases, especially those occurring in older children with widespread personality problems, do not have such a good outcome (I. Berg 1970). A proportion, about a third, go on to have significant psychiatric disorders of neurotic type in adulthood. Occasionally, school refusal develops into work refusal at school-leaving age, with features of the so-called 'inadequate personality' present (I. Berg *et al.* 1976). In women, school phobia may emerge into agoraphobia or fear of leaving home. A significant proportion of women with agoraphobia give a history of school refusal in their childhood. Nevertheless, most such children, even with severe school refusal, grow into healthy adults without serious mental health problems.

### 2.9.4 DEVELOPMENT OF DEPRESSIVE FEELINGS, BEHAVIOUR, AND BELIEFS

The concept of depression embraces a number of different components. They include depressive feelings, depressive behaviour, depressive cognitions or beliefs, as well as depressive disorders. Although both in normal development and in psychiatric disorders these are often linked together, they require separate description.

**Depressive feelings**

It is only by about the age of 6 years that children begin to use the language of emotional affect in adult fashion, referring, for example, to themselves and others as sad or miserable. It is reasonable to infer that younger children are experiencing such emotions when they look unhappy, but the knowledge of normal depressive feelings in very young children is limited because of lack of introspective information. By 10 years of age, 11 per cent of boys and 13 per cent of girls living on the Isle of Wight were reported by their parents to 'often appear miserable, unhappy, tearful, or distressed', and teachers of these children reported only slightly lower proportions to be affected in this way (Rutter *et al.* 1970*a*). In the same population, by 14 years of age, 21 per cent of boys and 23 per cent of girls reported themselves as often feeling miserable or depressed on a questionnaire, but nearly twice this proportion reported feelings of misery at a psychiatric interview (Rutter *et al.* 1976*b*).

**Depressive behaviour**

Characteristic depressive behaviour includes both non-verbal elements, especially crying and a sad facial expression, and verbal elements—usually expression of feelings of despair, hopelessness, and various depressive beliefs (see below). Not all crying is, however, a sign of depression.

Newborn babies cry in the first moments after birth, and subsequently crying from distress is frequent throughout the early years of life. The quality of cry varies according to its cause. Thus a sensitive mother can distinguish between the cries of hunger, pain, and discomfort, and indeed these three different types of cry can be distinguished electrographically. Normal infants may behave in a sad manner from the age of a few months, especially when they are separated from their mothers or other caretakers. Such separation may be physical or psychological, as when, for example, the mother is depressed and less sensitive to the needs of her young child. On these occasions the infant or toddler may become temporarily apathetic, refuse food and generally lack vivacity and energy. Crying from pain and physical distress diminishes rapidly after the preschool period.

Normal crying associated with mental distress persists, though to a diminished extent, throughout childhood and adolescence. Sadness may be present when a child temporarily goes quiet, does not want to play with his friends or develops irritability in mildly frustrating situations that would normally be shrugged off.

Self-destructive threats and behaviour appear with increasing frequency throughout late childhood and adolescence. An American study found that 33 per cent of 12- and 13-year-olds had at least occasional thoughts of harming themselves, and 8 per cent of Isle of Wight 14-year-olds had had definite suicidal ideas over the previous year (Rutter *et al.* 1976*b*).

## Depressive cognitive beliefs

Unrealistically negative beliefs about oneself, one's future, and of the world in general are a prominent feature of depressive disorders (Beck 1967). Transient thoughts of this nature appear normally during childhood. By the age of 7 or 8 years children can experience a devalued view of themselves as worthless, and not worth bothering about. By early puberty they can have views that their future holds little for them, and by mid-puberty that the world around them is rotten. In children with depressive disorders, such beliefs are held intensely and persistently.

## Reaction to bereavement

Studies of children who have lost a parent by death reveal that characteristic reactions occur (Raphael 1984; van Eerdewegh *et al.* 1985). A relatively prolonged period of sadness, crying, and irritability lasting from a few days to several months may occur throughout childhood. Younger children are particularly likely to show bed-wetting and temper tantrums, and older children, especially girls, may show sleep disturbance and more clear-cut depressive reactions. School performance is often impaired, at least temporarily. Most children are not severely affected, and studies do not suggest that major psychopathology is at all common following early bereavement.

Reaction to bereavement is likely to be modified by the level of social support available to the family and the social and financial changes resulting from the death. The reaction of the surviving parent will be of major significance. The previous personality of the child and the coping mechanisms he employs will be relevant. A minority of young children blame themselves or the surviving parent for the parental death, but usually (though not always) communication within the family is sufficient to reduce the guilt and anger so produced. There is evidence that brief family interventions can improve the outcome in the bereaved child (Black and Urbanowicz 1987). There is a tendency for adults who have suffered bereavement in childhood to show slightly higher rates of psychiatric disorder, but the findings are somewhat inconsistent and, in particular, it is not clear whether the child's age at the time of parental death is of particular significance.

## 2.9.5 DEPRESSIVE DISORDERS

### Definition and classification

Depressive disorders are conditions in which persistent lowering of mood (dysphoria) and lack of a sense of pleasure in life (anhedonia) are prominent and persistent features. The DSM III-R classificatory system (see page 9) divides depressive disorders into major depression (which may have single or recurrent depressive episodes) and dysthymia. A depressive episode is made up of at least five out of nine symptoms of which either depressed mood or diminished interest or pleasure in everyday activities must be present. Others include significant weight loss or gain, insomnia, agitation or retardation, poor concentration, feelings of worthlessness, and recurrent thoughts of death. Dysthymia which, in children, must have been present for at least one year, involves depressed or irritable mood and two out of five symptoms including poor appetite, insomnia, low energy, low self-esteem, poor concentration, and feelings of hopelessness. The ICD-10 classification (see page 10) makes similar distinctions, categorizing depressive symptoms into depressive episodes, recurrent depressive disorder, and dysthymia. Depressive psychoses and bipolar affective disorders are discussed in Section 3.1.3.

### The meaning of depression

The classification systems described above reduce the meaning of the term 'depression' to a set of reasonably clearly defined symptoms. Although this is useful for research purposes, 'depression' has a set of wider and richer clinical meanings than the classifications suggest. The term can be used (Nurcombe *et al.* 1989) to describe an affect or mood ('I have felt really depressed over the last month'), a dynamic ('He was depressed by the loss of his father'), a syndrome (a cluster of behaviours and feelings demonstrably related to each other statistically), a disorder (a handicapping condition with a characteristic aetiology and course), and a disease (a disorder arising from scientifically defined pathophysiological processes leading to rational treatment).

Too little is known about depression to think of it as a disease. The classifications aim to describe a set of depressive disorders but, in fact, there is little evidence that, in childhood, these different subtypes of disorder can be clearly defined from each other. It is likely that in childhood so-called 'dysthymia' merges imperceptibly into ordinary sadness and misery. In adolescents and very occasionally in younger children, dysthymia emerges also, in its more severe form, into depressive syndromes more characteristic of adult-type depressive disorders of 'nuclear' type.

## Clinical features

### Preschool children

Although it is difficult or impossible to apply some of the features (such as low self-esteem, hopelessness, and recurrent thoughts of death) to classify depression in this age group, there is no doubt that babies, toddlers, and other preschool children can appear depressed over a significant period of time and that it is meaningful to think of their psychological problems in these terms. Characteristically, when depressed, infants and toddlers are apathetic and refuse food. They are miserable, unhappy, and irritable, and may spend a good deal of their time crying and rocking. Such children are often referred to paediatricians because of failure to thrive. Growth failure is often accompanied by non-specific diarrhoea. On examination, no physical abnormality is found to account for the symptoms, but the child is fretful, insecure and unhappy. There is often evidence of mild developmental retardation. The picture may occur as a result of gross and obvious neglect or there may be more subtle reasons why the child is so disturbed.

### Middle childhood

Depressive states at this age are often poorly differentiated from emotional disorders in which both anxiety and depressive features are prominent. It is unusual for children to be brought to the practitioner with depression of mood as the presenting feature. More commonly the depressed child presents with psychosomatic symptoms, especially headache or stomach-ache. Failure to make progress in school, with poor concentration and inattention, is also a common mode of presentation. Irritability, social withdrawal, and incapacity to cope with even minor frustration without a temper outburst may occur, but sleep and appetite disturbances are less common. Very occasionally, however, prepubertal children present with more classical features of depressive disorder—apathy, verbal and motor retardation, and loss of appetite. Feelings of low self-esteem can often be elicited by asking questions whether the child feels as good or worth as much as other children, or whether he thinks his parents like him as much as his sibs. Although deliberate suicide is extremely rare in this age group and even suicidal attempts are most unusual, depressed children do often wish they were dead and express this desire in response to questioning.

Children of this age may not admit to feelings of depression readily. They are more likely to complain of being 'bored' and of lack of interest in usual activities. Normal children tend to complain of boredom only in circumstances when they have nothing of interest to do, while depressed children are likely to say they feel bored most of the time. The depressed child nearly always looks miserable and unhappy and appears to lack vitality and energy.

As indicated above, attempts to distinguish major depressive disorder from dysthymia, and dysthymia from normal sadness and misery, although they may be necessary for administrative and research purposes, are not usually very helpful clinically in the pre-adolescent.

### Adolescence

Depressive states in adolescence are more similar to those seen in adulthood. The teenager will often complain of feeling sad and apathetic, as well as lack of energy. Appetite and sleep disturbances are more common. There may be impairment of food intake, or the child may overeat to comfort himself. Difficulty in getting off to sleep at night and waking during the night may leave the teenager exhausted during the day. Bodily preoccupations are common, and the child may be taken up by worries over his appearance or symptoms of minor ill health such as acne. A sense of futility and hopelessness may be experienced. Suicidal thoughts are relatively common, although suicide threats are much less usual.

### Prevalence

The prevalence of depression in the preschool period is uncertain, largely because a lack of agreed criteria appropriate for this age group. In the pre-adolescent, although as many as 10–15 per cent of children show depressed mood (Rutter *et al.* 1970*a*), only about 2 per cent are diagnosable as depressed using DSM-III criteria. There is a sharp rise in the rate of depressive disorders in adolescence (Rutter *et al.* 1976*b*).

There is little difference between pre-adolescent boys and girls in the rate of depressed mood and depressive disorder, but in adolescence depressed girls outnumber boys by about two to one. The reasons for this may lie in hormonal or psychological differences. It has been found, for example, that the sexes are similar in their levels of perceived self-competence in pre-adolescence, but diverge during and after puberty. Low self-esteem may then act as a mediating variable, differentiating the sexes in their predisposition to depression (see below). Similar mechanisms may be responsible for the fact that depressive symptoms have been found to be more common in United States population studies in young adolescents from minority race, of low socio-economic status, and with low school grades (Garrison *et al.* 1989).

### Aetiology

Causes of depression can be divided into predisposing or vulnerability factors, and triggering or precipitating factors. A significant number of studies have been carried out in adults with depression to elucidate aetiological factors using this model (e.g. G. W. Brown and Harris 1978), but there are few relevant studies in childhood depression. Nevertheless, this is a growing area and more information is now available.

Child Psychiatry →

depression p. 183 →

*Predisposing factors*

1. *Genetic.* Most investigators (e.g. Strober and Carlson 1982) suggest depressive symptoms are raised in family members of seriously depressed children. However, the picture is not clear-cut. The familial nature of depression leaves open whether the mechanism is genetic or environmental, and there are no clear-cut findings to elucidate this issue.

2. *Temperament.* Studies suggest that quiet children with regular habits, slow to adapt to new experiences, are predisposed to emotional disorders if they become disturbed (Thomas *et al.* 1968). It is uncertain whether such factors also predispose specifically to depressive disorders.

3. *Biological factors.* In adult depression there is some evidence that there are abnormalities of neurotransmitters, especially monoamine metabolism. There is also inconclusive evidence of cortisol hypersecretion. The evidence for neurotransmitter and endocrine abnormalities is weaker in children, but Foreman and Goodyer (1988) have found raised mean salivary cortisol levels in a group of depressed children.

4. *Chronic life adversity.* Babies and toddlers who are neglected and deprived of affection are at special risk for developing depressive states characterized by apathy and withdrawal. Older children with depressive conditions may also be living in circumstances where they are undervalued. The low valuation they place on themselves may be merely a reflection of the poor opinion held of them by members of their family, their teachers, and other children.

Depressed children are more likely to have mothers who have poor confiding relationships and high rates of distress themselves. They are more likely to come from broken homes. Parents rearing their children in conditions of serious social adversity, under financial pressure, poorly housed, or with marital problems, naturally find it difficult to provide sensitive, loving care for their children.

Various psychological mechanisms have been described, more especially in adults, to explain why these factors should predispose to depression. One helpful concept is that of learned helplessness (Seligman 1975): after several unsuccessful attempts have been made to escape from unpleasant circumstances, lack of motivation and apathy may supervene. A number of depressed children in what seem to them to be hopeless situations seem to fit this model well.

*Triggering or provoking events*

1. Undesirable life events occur in excess in the twelve months before the onset of disorder (Goodyer *et al.* 1987). Although in depressed adults such life events are more likely to cause depression in predisposed individuals, the evidence for this in childhood is less certain. Nevertheless, it is probable that threats to a child's well-being that arise because of an event

at home or at school will have a more powerful effect if the child is vulnerable because of his personality or because he has been sensitized by other previous adversities. It has been shown that child psychiatric attenders with depressed mood have more commonly previously sustained parental loss than disturbed children without such mood disturbance (M. G. Caplan and Douglas 1969). The onset of depression in children, as in adults, is commonly associated with the experience of loss, and a wide variety of experiences, often specific to the particular child affected, may be responsible.

2. *Viral illnesses.* Lack of energy and mild depression of mood are common after febrile illnesses, especially measles and infectious mononucleosis. In a minority of children such postviral reactions are severe and may last weeks or months. It is uncertain whether these psychological reactions occur as a result of physiological or environmental factors. There is some evidence that, in adults, these reactions are accompanied by immunological changes, especially abnormalities of T-cell function (Hamblin *et al.* 1983).

### Assessment

It is important to try to establish the nature and extent of depressive symptoms, and the degree of disability they are producing. Differential diagnosis of a depressive episode would involve exclusion of depressive psychosis, anorexia nervosa, a physical condition accompanied by mood change, and normal reactive feelings of sadness and unhappiness. Depressive symptoms may be present in a whole range of psychological disorders of childhood, and the diagnosis of a depressive condition should not rest merely on the presence of sadness of mood, but on the existence of a constellation of symptoms (the depressive syndrome) described above. The presence of 'masked depression' (depression underlying other, usually physical symptoms) should be considered cautiously. With careful questioning, depressive symptomatology can be elicited from children and their parents if it is present and, in the absence of such symptomatology, the condition should probably not be diagnosed (Carlson and Cantwell 1980). An attempt should be made to identify the time of onset and events of possibly causative significance immediately preceding this point in time. Events involving a possible experience of loss are likely to be relevant, but physical factors, such as viral illness, should also be noted. A family interview may reveal the fact that the entire family is burdened by stress with the identified child acting for some reason as the only family member 'allowed' to experience distress. An interview with the child may elicit specific stressors, occurring for example at school, of which the parents may be unaware.

### Treatment

**Relief of stress.** An attempt should be made to alleviate any stresses identified during the assessment procedure. Thus, the parents of a child depressed by a repeated failure in school can be encouraged to see if remedial help is available and to ensure that they and, if possible the child's teachers, praise the child for even a very small improvement on previous performance. Such measures will often not prove sufficient on their own, but they are always worth taking.

**Improvement of communication.** This can be considered in two main ways. The child may first be helped by individual counselling to become more aware of the factors responsible for depressed affect when this occurs. Some children, for example, become depressed whenever they are unable to express anger against those whom they feel are frustrating their wishes. Second, communication between family members can be improved in family interviews. The loss of a grandparent, for example, may have affected the whole family, but only the identified child may have showed prolonged symptoms. A family interview may clarify the degree to which *all* family members have suffered, thus allowing the identified child to share his burden of loss.

**Cognitive behavioural methods.** It has been pointed out that depressed individuals often have a distorted view of the way other people see them. Low self-esteem can be perpetuated if a child or adolescent misinterprets a remark as derogatory when, in fact, it was not intended to be so. Treatment of depression (Beck *et al.* 1979) can focus on encouraging patients to keep careful records of situations in which they feel their depressed view of themselves is confirmed. Using these techniques, in discussion of these situations with the therapist, a child may develop a more realistic view of the way he is seen by others.

Other methods of enhancing the child's self-esteem may also be useful. A child who feels himself disfavoured by his parents in comparison to his sibs may well be misinterpreting their behaviour. Yet it can be helpful to suggest to parents that their depressed child has a (hopefully temporary) need for more affection and comfort than his sibs.

**Physical methods of treatment.** Medication should be reserved for the more seriously depressed child as the risk of side-effects is likely to out-weigh benefits in the mildly depressed. Further, even in the seriously depressed, a trial of psychological therapy should precede the use of medi-cation. If psychological therapies fail, then imipramine (for the slowed-up depressed child) or amitriptyline (for the agitated depressed child) should be prescribed (Section 6.2.6). Improvement should be expected within 10 days, and if this does not occur, the drug should be discontinued after 3 weeks. If effective, the drug should be stopped after about 2 months to determine if it is still effective. A variety of newer anti-depressants are available, but have no clear advantages.

### Outcome

Most children (about two-thirds) with depressive disorders improve sub-stantially following referral to a medical agency. This is probably because referrals are often precipitated by temporary crises and worsening of symptoms during the course of a fluctuating condition. The great majority of referred boys with major depression and dysthymia improve, though full recovery may not occur for several months or even years, and there is an increased liability to further episodes of depression (Kovacs and Gatsonis 1989).

### 2.9.6 SUICIDE AND ATTEMPTED SUICIDE

#### Definition

Suicide involves a deliberate form of self-harm resulting in death, while attempted suicide is an act of self-injury with a non-fatal outcome. It is important to note that motivation to die does not enter into these definitions. Some successful suicides have not intended to kill themselves, while others who have not died have genuinely wished to do so. Kreitman (1977) has coined the term 'parasuicide' to describe deliberate acts mimicking suicide, but not resulting in a fatal outcome.

#### Epidemiology

**Suicide.** The rate of completed suicide in the 10–14-year-old age group in the UK is very low at about one per million annually, and suicide in children under the age of 12 years is very rare indeed. In the 15–19-year age group, however, the rate is higher, at about 15 per million per year, and has remained fairly steady over the last two decades (McClure 1988). Rates in the USA are two to five times higher (Schaffer and Fisher 1981). Completed suicide is more common in males by a factor of about 2:1. There is no particular social class distribution.

    **Attempted suicide.** This is a much more common phenomenon. In the 10–14-year age group, it is infrequent, but in 15–19-year-olds it occurs in about 400 per 100 000 per year. Females are affected about three times as commonly as males, and there is an excess in lower socio-economic groups. The rate of attempted suicide rose sharply to reach a peak in the 1970s, but remained steady or declined in the 1980s (McClure 1988).

#### Background features

Apart from the sex ratio, socio-economic status, and means of self-injury (see below), characteristics of those who commit suicide and those who deliberately harm but do not kill themselves are rather similar. These include:

    1. *Parental characteristics.* Parents often suffer from mental health problems especially depressive syndromes and personality disorders. Marital difficulties between the parents are often present.

    2. *Parental discipline.* This is likely to be inconsistent, but rigidly applied. Parents may be generally rather permissive, but then impose unreasonable restrictions.

3. *Family communication patterns*. There is likely to be rather poor communication both of feelings and information within the family. Parents are often unaware of the suicidal ideas experienced by their teenage children (Monck and Graham 1988).

4. *Social isolation of the child*. The child may be socially isolated and receive little social support from friends or family. Social support from friends may have been lost just before the suicide attempt. Some teenagers have run away from home before the attempt and are almost totally isolated.

5. *Psychiatric state of the child or adolescent*. There is a high rate of depression (about 50 per cent) among suicide attempters. Such depression is usually acute and reactive, but may be chronic. There may be a long history of disturbed behaviour including serious antisocial problems. In the older age group, drug and alcohol abuse are fairly commonly found.

6. *Physical health*. Children with chronic physical ill health are particularly at risk for attempted suicide (Hawton 1982). In girls, the possibility of pregnancy should always be considered, especially as the discovery of pregnancy may have precipitated the attempt.

7. *Contact with other suicide attempters*. There is a contagious component to attempted suicide. Youngsters engaging in this type of behaviour often have acquaintances or friends who have behaved similarly. Outbreaks of overdosing in schools have been described (Brent *et al.* 1988). This may explain why media programmes designed to reduce suicide in the young paradoxically appear to increase rather than diminish the phenomenon (Shaffer *et al.* 1988).

*Characteristics of suicide attempts*

1. The great majority of those who attempt suicide do so by taking overdoses, usually of non-opiate analgesics such as aspirin, but occasionally with psychotropic or other drugs. (Completed suicides involve a much wider range of methods with poisoning from vehicle exhaust fumes, hanging, shooting, and suffocation by other means as prominent causes.)

2. The attempt itself may be only a trivial threat to the health of the child (swallowing half a dozen aspirin tablets) or may require skilled intensive care in hospital over several days in order that the life of the child can be saved.

3. Attempts are usually precipitated by crises in relationships, either with family or friends. A dispute over discipline is common, as is a row with a boyfriend or girlfriend. Bullying at school may have occurred.

4. Attempts are often preceded by threats that have not been taken seriously. After the attempt, the teenager usually, but not always, communicates what he or she has done, often to the person towards whom most anger is directed.

5. About half those who attempt suicide have attended a medical agency,

usually the general practitioner, in the previous month. The visit is often connected in some way to the problem that leads to the attempt.

### Assessment

This should be carried out as soon as possible after the suicidal attempt has been identified, and, if this has been necessary, the teenager has been successfully resuscitated. It should take place before the patient is discharged home. The involvement of a psychiatrist in the assessment is desirable, but not essential. It is important, however, that it should be carried out by someone capable of assessing the risk of recurrence and the need for intervention on other grounds. For children under the age of 16 years it is preferable for assessment to be carried out on a paediatric ward. An appropriate form of assessment involves first an interview with the teenager, then an interview with the parent(s), and finally a family interview. Some professionals prefer to begin with a family interview.

Information of particular relevance that needs to be gathered includes:

1. Details of the attempt, number of tablets taken, intention of the teenager at the time, whether anyone was informed after the tablets were taken.
2. Precipitating circumstances.
3. Presence of emotional or behaviour problems in the months before the attempt.
4. Quality of family relationships, especially in communication of feelings.
5. Social network of the child.
6. Previous involvement of helping agencies, general practitioner, social worker, etc.
7. Current suicidal intentions of the child.
8. Present mental state of the child in other respects.
9. Degree to which the attempt has altered the circumstances that precipitated it.

With this information available, the clinician should have a reasonably clear understanding of the reasons for the suicidal behaviour and of the action that needs to be taken. Where reasons for the attempt remain obscure, the possibility of sexual abuse should be considered.

### Treatment

The risk of recurrence is greater when the attempt has been seriously life-threatening, the child is depressed, the circumstances precipitating the event are unresolved, or the child's social supports remain poor. Follow-up is required if any of these conditions pertain, but otherwise this is a matter of clinical judgement. In particular, follow-up is less necessary if the attempt has been relatively trivial, and the precipitating circumstances appear to have been resolved.

The nature and intensity of treatment will depend on the circumstances and resources available. A small minority of attempters will be found to have obvious psychiatric disorders requiring admission to an inpatient unit. A further small number will require a few days admission on the paediatric ward for observation. Most will be able to return home with follow-up appointments for treatment or to check on progress.

In cases where faulty communication has been responsible, family therapy sessions (Section 6.2.4) should be offered. Some teenagers will, however, benefit from individual counselling. Where school stresses have been significant, contact with the school should be made, together with an attempt to resolve the problems experienced there.

### Outcome

The outcome of adolescent suicide attempters as a group is unknown. It is likely (Hawton 1982) that those living in circumstances of chronic stress or who have shown other serious behaviour and emotional problems are those at greatest risk for recurrence of further attempts, completed suicide, and other psychiatric disorders.

## 2.10 Later social development

### 2.10.1 INTRODUCTION

The early development of social relationships and attachment, because of its major importance in psychological and psychiatric disorders occurring in the first five years of life, is discussed in Section 2.1. Later social development is considered much more briefly here.

As children enter the third and fourth year of life, relationships with others of their own age become more important. Experience with children from other families at home or in preschool facilities begins to take up more time and attention. Entry to a playgroup or nursery school usually provides the first experience of prolonged separation from the mother. Such experience involves the mastery of new skills in independent living, but also a significant loss of immediate maternal contact, and many children find this stressful. The way in which this transitional experience is handled by parents and playgroup or nursery school staff will largely determine the smoothness with which transition occurs, but the temperament of the child will also be of importance.

In older children, social experiences at school and in the immediate neighbourhood gradually come to attain as great or greater importance than experience within the family. The quality of the relationships found in middle childhood and adolescence will partly depend on the quality of early intrafamilial relationships, and partly on the current situation. The personality and interest of peers with whom the child forms special relationships will, for example, also play a large part.

## 2.10.2 CONDUCT DISORDERS

### Definition

Nearly all children at some time show antisocial behaviour, i.e. behaviour which contravenes social norms, or personal or property rights. Most, at some time or another, have brief episodes of stealing and lying. A substantial minority go through phases when they bully other children, are aggressive in other ways, or truant from school. The child with a conduct disorder is different in the extent and severity of his antisocial behaviour. There is therefore no sharp dividing line between normal and abnormal antisocial behaviour.

**Social norms** will affect judgement as to when antisocial behaviour is a cause for concern. In some cultures, fighting among boys in the school playground is regarded as a matter for serious concern: in others it is taken for granted this will occur. Inevitably the police, as well as professionals such as psychiatrists concerned with assessing and attempting to help antisocial children, will be affected in their judgements of abnormality by values held by others in their society. The perception of antisocial behaviour may also vary within a society. Many in our society would have a different reaction to reading about a group of undergraduates turning over cars following celebration of a rugby cup victory, to reading of a group of black youths engaged in similar behaviour after being refused entry to a dance hall.

**Delinquency.** Children who show antisocial behaviour may or may not be breaking the law. If they are, their behaviour may be considered a form of juvenile delinquency. Delinquency is defined by the law. Thus legislation protects property, so that virtually any form of stealing may be regarded as a delinquent act. By contrast, lying, by itself, is not delinquent unless it occurs in court under oath. The great majority of antisocial acts are not detected. Antisocial behaviour can therefore be divided into that which is non-delinquent, undetected delinquent, and delinquent. For understanding causation, assessment, and management, these legal distinctions are not particularly useful, but, of course, in determining what happens to a child who may or may not have broken the law, they may be crucial.

**Perspectives on antisocial behaviour.** Psychologists, psychiatrists, the police, and lawyers are likely to view antisocial behaviour from very different perspectives. For the courts, an antisocial act requires prosecution of the offender, so that, if appropriate, consequent punishment or other measures can follow. For a doctor or mental health professional, a child's antisocial behaviour requires assessment and understanding, so that action can be taken to help the child avoid recurrence. Inevitably, there are sometimes clashes between those taking these two perspectives. Nevertheless, most members of the police would acknowledge that, in some situations, more than punitive action is required to deal with an offender, and most psychiatrists or psychologists would accept the need to punish at least some delinquents.

## Classification

In ICD-10 (World Health Organization 1990), conduct disorders are divided into those confined to the family context and into unsocialized, socialized, and oppositional defiant types. The last category is mainly used for younger children. DSM III-R classifies conduct disorders and oppositional defiant disorders as two forms of disruptive behaviour disorders. Conduct disorders are divided into group, solitary aggressive and undifferentiated types. Diagnostic criteria for a conduct disorder include a pattern of behaviour lasting at least 6 months in which at least three out of sixteen listed forms of behaviour have been present. (These include stealing, lying, truanting, destruction of property, initiating fights, cruelty to animals, and fire-setting.)

## Clinical features

### Socialized and unsocialized antisocial activity
A distinction is commonly drawn between socialized and unsocialized aggressive or delinquent behaviour (R. L. Jenkins 1973). Socialized delinquents are those who commit offences in the company of others, who generally get on reasonably well with other children their age, and whose offences might be seen as acceptable behaviour by other family members and friends. They are more likely to show guilt over their misdemeanours and sympathy for their victims. Unsocialized delinquents, by contrast, offend alone and not in company, have poor peer relationships, and offend against their group norms as well as those of wider society. They show little guilt for their offences or indeed concern for other people in general. They are also said to show more neurotic traits. This distinction may be easily applied when looking at individuals falling at the extremes of the socialized/unsocialized continuum, but most antisocial children show features of both types. The distinction is probably most useful in adolescence.

### The preschool period
Conduct problems in the preschool period are most commonly shown by aggressive behaviour against other children both at home and in facilities such as play groups and nursery schools. Away from home, aggression is commonly triggered off by situations in which the child wants something another child has, or is dispossessed of something he wants by another child. Frustration is followed by an immediate attack. Some preschool children will physically attack their parents when frustrated but it is most unusual for other adults to be involved. Aggression towards parents is frequently triggered by conflicts over feeding, bedtime, or refusal of the parent to allow the child to have something he wants. Deliberate destructive

behaviour, tearing down wallpaper or breaking furniture, may occur in the home. It is important to distinguish between this type of deliberate destructiveness and the spoiling of toys or other objects which occurs when the child takes things to pieces to satisfy his curiosity.

At this age antisocial behaviour is commonly accompanied by at least some features of the hyperkinetic syndrome or attention-deficit disorder with hyperactivity (Section 2.7.2), including overactivity, short attention span, impersistence, and lack of concentration. Frequent temper tantrums are a common accompaniment. There is indeed some degree of diagnostic overlap between conduct disorder and attention-deficit disorder with hyperactivity, but children with conduct disorder tend to be older, have fewer developmental delays and are more psychosocially disadvantaged (Szatmari *et al.* 1989).

### Middle childhood

**Aggressive behaviour** characterized by physical attacks on other children and, to a lesser extent, parents, is still the most common form of antisocial activity. Verbal attacks, involving swearing and name-calling, are frequent accompaniments. Being thwarted by a parent or by another child is the most common precipitant of an aggressive outburst. However, at this age, more deliberate, premeditated aggression begins to occur. The child may plan to injure another either by himself or with a group. He may also be involved in deliberate cruelty to animals.

**Stealing.** Before the age of 5–6 years, the concept of personal property is likely to be poorly developed, so that taking other children's possessions cannot meaningfully be regarded as stealing. After this age, stealing in a real sense does become possible, though many 6–7-year-old children appropriate other children's belongings in what is often rather muddled territory between borrowing with permission, borrowing without permission, and deliberate stealing. By 7–8 years, children will sometimes be involved in stealing from shops, either alone or in groups. Stealing must be regarded as problematic if it occurs repeatedly, despite detection and reprimand, inside or outside the home. In younger children, stealing is most likely to result from a child's desire to comfort himself for feelings of deprivation. Older children's stealing is more likely to occur in gangs as part of marauding or 'self-proving' behaviour. Stealing is often followed by lying in order to cover up misdemeanours, and indeed the denial of stealing is probably the most common reason for lying in this age period. Whether stealing is followed by lying to cover up, often depends on whether the parent or teacher, who knows very well that a child has been stealing, insists on the child owning up. Adults who do not do this often save the child from compounding his offence, but, perhaps misguidedly, it is often thought that it is important to give a child the opportunity to admit an offence in order to mitigate the seriousness of his behaviour.

**Lying.** Less commonly, lying may take the form of spinning fantasies for no very obvious reason, but not to cover up offences. Children of this age may invent stories in which they have had accidents or illnesses that have in fact never occurred to them, or they may invent, for the benefit of their friends, a totally different set of family relationships from that existing in reality. Children who engage in fantasy production of this type may be seriously deprived of affection, but this is often not the case. They may lack social skills in making friends.

**Disruption.** Antisocial behaviour in school may also take the form of disruptive activity not involving physical or verbal aggression. Some children may disrupt classes by frequent disobedience, wandering around, throwing things, and generally irritating other children and the teacher. *Truancy* is relatively uncommon in this age period, but can occur before puberty.

**Fire-setting** is an uncommon form of antisocial behaviour, but does sometimes occur. It may arise as a result of experimentation with matches, or as a deliberate destructive activity. The latter occurs both as a group activity and in isolation. It is often accompanied by signs of other serious antisocial behaviour, and there is frequently a high level of disturbance in the home (Stewart and Culver 1982).

Other forms of antisocial activity in this age period include drug-taking, especially solvent abuse (Section 3.3).

### Adolescence
As children can be charged with an offence in the UK only once they reach the age of 10 years, inevitably the proportion of antisocial children that can be classified as either detected or undetected delinquents is higher in adolescence than it is in younger age groups. There is, however, no sharp dividing line in the frequency of the actual behaviour that occurs.

**Physical aggression** as a sign of antisocial activity is much less common in this age group though, when it occurs, because of greater physical strength, the outcome may be more serious. Gang-fighting in large cities is the most common setting in which physical aggression occurs. Some individual children, however, show seriously aggressive behaviour at home or in the classroom. When frustrated, they may lash out against the person who has thwarted them. Sometimes such aggression may occur in a 'blind rage' with apparent lack of awareness of the surroundings. Such 'episodic dyscontrol' (Nunn 1986) may require differentiation from an epileptic attack, though usually there is a little good evidence for the presence of epilepsy.

**Breaking into property and stealing** is the most common form of offence for which youngsters of this age are charged, and in fact interviewing potential victims makes it clear that there are many more offences of this type committed (probably mainly by teenagers) than come before the courts. Stealing also takes the form of taking and driving away cars and motor bicycles. Vandalism of public property is a common occurrence, especially in some council housing estates, and this is also an offence mainly committed by teenagers. All these types of offence most commonly occur as a form of group activity.

**Truancy.** During the final years of compulsory schooling, especially in inner-city schools, truancy becomes much more common. Persistent truancy is uncommon in primary school children. However, in the UK, there are as many as 20 per cent of children in their last year of school absent for unexplained reasons (Fogelman *et al.* 1980). Most truants spend the day out of school in groups, but in a minority it is an isolated activity.

Antisocial activity in adolescence may also take other forms, including drug-taking (Section 3.3), and sexual offences, including sexual abuse of younger children (Section 1.3.3). Sexual activity is common among teenagers in Western society, and there is no reason to regard it as a form of antisocial activity. However, some girls, and a small minority of boys, who become involved in prostitution, and others whose sexual promiscuity puts them at risk of exploitation, can reasonably be regarded as engaged in a form of antisocial behaviour.

*Associated problems*

Although classification of child psychiatric disorder tends to make a sharp distinction between antisocial disorders and emotional disorders, many children who show antisocial behaviour are also unduly tense, anxious, and prone to anxiety and depressive feelings. In adolescence they also have a tendency to adult-type neurotic disorders, including hysterical reactions. Some are prone to feelings of hopelessness and make suicide attempts.

**Background information**

**Prevalence.** It is particularly important to realize the arbitrary nature of the cut-off point between normal rebelliousness and conduct disorder when considering the prevalence rates of this condition. On the Isle of Wight, the one-year prevalence rate of antisocial disorder was found to be about 3 per cent, and in London the rate was found to be at least double this (Rutter *et al.* 1975*b*). In Ontario, Canada, approximately 5 per cent of boys and 2 per cent of girls were rated by teachers to show conduct disorders (D. Offord *et al.* 1989). Delinquency rates vary very widely from place to place and from time to time. In the late 1980s, around 3 per cent of 10–16-year-olds in England and Wales were cautioned or sentenced for an offence in any one year. Not all boys charged with offences need necessarily be regarded as showing conduct disorders. For example, a boy charged with an isolated offence of shoplifting might well be considered within the normal range for antisocial behaviour. Conversely, self-report studies reveal that teenagers never before the courts have often committed dozens of offences in which they have not been detected and for which they have not been charged (Belson 1975).

**Age.** The rate of physically aggressive behaviour declines with age after 8 or 9 years. Criminal activity, mainly offences against property, reaches a peak around 13–15 years and thereafter declines (Farrington 1981).

**Sex ratio.** Boys are charged with offences seven or eight times as commonly as girls. There is a less marked sex difference for conduct disorders in younger children. Indeed at 3 years there is little difference between the sexes: by 4 years the gap has begun to widen.

## Social and neighbourhood factors

**Social class.** Conduct disorders do occur more commonly in working-class than in middle-class children, but this is almost entirely due to differences between the social classes in family size, rates of overcrowding, and level of child supervision (Kazdin 1987). Similarly, once these factors have been taken into account, there is little or no difference between ethnic minorities and indigenous populations in their rates of conduct disorder. However, serious crimes, such as carrying offensive weapons, are much more likely to be committed by less-privileged children. Even so, the police practice of selectively charging working-class children results in biasing the delinquency statistics. When detected in minor crime, middle-class children are more likely to engage the services of a solicitor and be warned and cautioned by the police (Palmai *et al.* 1967).

**Ethnic status.** Currently in the UK, Asian children have somewhat lower, and West Indian children higher rates of delinquency than does the indigenous population. The rates among West Indian youths increased in the 1970s in line with unemployment rates, and this may well be a causative association (Rutter and Giller 1983).

**Schools.** Factors within schools can reduce delinquency rates even among children living in deprived urban neighbourhoods. Those schools where the staff are committed to educational values, set and mark homework consistently, and concentrate on rewarding good rather than punishing bad behaviour show lower rates of truancy and other forms of delinquency among their pupils (Rutter *et al.* 1979b).

**Other social factors.** Children living in inner cities, in poor overcrowded conditions, in neighbourhoods where there are high crime and unemployment rates, are at high risk for both conduct disorders and delinquency.

## Familial factors

**Communication.** Children living in families in which there is poor communication of feelings and a lack of awareness of and respect for the feelings of others, probably have higher rates of conduct disorder.

**Family size and structure.** Children from large families are at particular risk. Those living in homes broken by divorce and separation are also at high risk. If family relationships are discordant and disharmonious, this raises the risk of antisocial behaviour to a very significant extent (Rutter and Giller 1983).

**Parental personality and health.** There is a relatively strong link between the presence of an antisocial personality in one or both parents and similar behaviour in the child (West and Farrington 1973). However, most young children showing conduct disorders do not have an antisocial parent. Depression in the mother is also linked to antisocial behaviour.

**Parental child-rearing practices.** These effects are often difficult to distinguish from those of parental personality. Parents who are inconsistent in applying rules and regulations, disagree between themselves, and rely heavily on harsh punishment for bad behaviour rather than reward for good, are likely to have antisocial children (Patterson 1982).

**Television and comics.** Children with aggressive tendencies have a special predisposition to watch violent television programmes and read violent material in comics, so the fact that aggressive children watch more violence does not necessarily mean that one can blame the media for violent behaviour. There is, however, some evidence that violence on television is responsible for promoting some degree of violent behaviour among boys (Belson 1978).

### Personal factors

**Genetic.** Twin studies suggest that, while genetic factors are of definite importance in adult criminality, they are of less significance in conduct disorders. Among adopted boys, the risk of juvenile delinquency is greater if the adoptive father has antisocial tendencies, but is raised very little if the biological father has a criminal record but the adoptive father does not (Hutchings and Mednick 1974). Nevertheless, genetic factors probably do play some part in the causation of aggressive behaviour, especially in that minority of boys who continue to show aggressive behaviour in adult life. They probably exert this effect by their influence on temperament and personality (see below). There is now good evidence that genetic and adverse environmental factors interact to contribute to cause antisocial behaviour (Cadoret *et al.* 1983).

**Temperament and personality.** 'Difficult' babies are more likely to show aggressive behaviour problems later on. A baby inheriting characteristics resulting in his being irritable and negative in mood may elicit rejecting behaviour in his parents. The effect of this vicious cycle may result in delinquency later on. There is not now thought to be any close association between 'extroversion' or 'neuroticism' and aggressive behaviour but affected children do seem to be more impulsive, and less concerned about long-term than immediate gratification. Underlying these personality difficulties there may be specific patterns of autonomic reactivity, but these have not yet been identified.

**Physical health and body-build.** Children with epilepsy and other disorders of cerebral function are at increased risk for conduct as well as for emotional disorders (Rutter *et al.* 1970*b*). There is also a tendency for children with other physical disorders to show somewhat higher rates. It used to be thought that children with a stocky body-build were especially at risk, but the evidence for this is not strong.

**Educational retardation.** There is a strong link between conduct disorder and educational failure, especially specific retardation in reading ability. There has been much discussion why this link should exist. It is probable that many children who fail in school become discouraged and low in self-esteem, and take to delinquency to obtain rewards they are missing elsewhere. It is also possible that underlying personality factors such as impulsiveness and excitability both impede learning and predispose to aggressive and antisocial behaviour.

### Labelling

Once a child has been labelled as antisocial, perhaps following a court appearance, the risk of further antisocial activity is increased (Farrington 1977). Delinquent children who have appeared in court seem to have a slightly worse outcome than those of a similar level of antisocial behaviour who have avoided a court appearance.

### Assessment

Information should be obtained on the precise nature of the antisocial activity and the immediately precipitating circumstances. It is particularly important to know whether the behaviour was impulsive and whether it was preceded by a frustrating experience. It is also important to obtain systematic information on other aspects of the child's behaviour and emotional life, especially the presence of significant anxiety and depressive symptoms.

The developmental history, including experiences of illness and separation from either parent, will be relevant. Particularly important will be information on current family functioning, the quality of relationships in the family, the strength of positive and negative parental attitudes to the child, the way in which discipline is exercised, the attitude of other family members to the behaviour, and their view as to why it has occurred. Finally it is important to know how the child and the family see the future in terms of the likelihood of recurrence, and whether they are motivated to receive any psychological help or think this could be useful.

The assessment of children with conduct or antisocial disorders is made more difficult by the fact that the person involved in assessment, whether this be a psychologist, doctor, probation officer, or social worker, is likely to be seen as an authority figure by the child and other family members. As one of the problems may be an antisocial attitude, obtaining information is often a tense procedure. It is therefore particularly important that the assessment is carried out by someone who can convey a sympathetic and understanding attitude, in a non-condemnatory manner. It is worthwhile spending time getting to know the child and family before embarking on obtaining precise details of antisocial behaviour.

Information about young children with conduct disorders is often best obtained within the family setting, or at least with both parents and the child in question present. It is useful to acknowledge in the interview that the child would probably prefer not to be in the room hearing unpleasant facts about himself. The child may be asked right at the beginning whether in fact he wanted to come to the consultation, and a sympathetic but firm attitude to this taken from the outset. Very occasionally, parental attitudes may be so rejecting and negative it may be thought better to continue the interview with the child out of the room. However, if parental attitudes are of this nature, it is important to remember that the child is probably exposed to constant attack at home as well, so the experience of rejection is not likely to be a new one for him. Having seen the family members together, it is usually helpful to see the child separately and then the parents alone.

With older children and teenagers it will often be preferable to see the child by himself before seeing the parents, and to complete the assessment by conducting a family interview. If this is so, it is important not to transmit information the child gives in private back to the parents, without checking with the child that this is agreeable to him. The reverse is, of course, also the case.

Sometimes the child or one of the parents will not be willing to be interviewed at all. In many cases a useful service can be provided by seeing those family members who are motivated to receive advice and guidance. A situation that arises frequently is one where teachers are worried about a child's antisocial or aggressive behaviour, but the family shows no apparent interest in cooperating by going to a clinic or psychiatric department for an assessment. The options here are to conduct a school-based assessment, or to limit one's activities to obtaining further information from the teacher and providing advice on this basis alone.

## Management

The form of treatment that should be offered will depend especially on the motivation of the child and family, the nature of the conduct problems, and the presence or absence of significant neurotic symptoms, especially tension, anxiety, and depression.

**Counselling or psychotherapy** may be helpful in the minority of cases where, as well as the conduct problems, neurotic symptoms are prominent and the initial contact with the family provides evidence that its members are sufficiently 'psychologically minded' to achieve some insight. Such an approach may also be indicated by the presence of unusual antisocial symptoms suggesting a neurotic basis to the problem. A boy stealing underwear or showing unusual cruelty to animals would be an example of this. Such counselling may be provided at family or individual sessions. It is sensible to establish at the beginning how many sessions there are likely to be (maybe 6–12), and, in the first two to three sessions, to identify one or two foci or issues which seem of particular importance. In antisocial children, issues commonly identified for exploration would be anti-authority attitudes in family members, difficulties family members have in being firm and consistent about disciplinary matters, disputes between parents about rules and regulations, the presence of underlying marital problems as a primary source of conflict.

In a further minority of cases, family-based **behaviour modification approaches** will be helpful. Again it is important for the child and family to be motivated, to wish to change, and to be prepared to go to some trouble to achieve change. A parent management training approach is likely to be most successful in younger children where the main symptoms involve physically aggressive behaviour rather than signs of delinquency such as stealing. The principles of behaviour modification are discussed in Section 6.2.2. In children with conduct disorders, operant conditioning approaches are likely to be most effective. A focus for change such as outbursts of temper can be identified, and a functional analysis of the problem under-taken. The application of material and then social reinforcers for avoiding outbursts can sometimes achieve significant change. In older children, family-based interventions relying more heavily on increasing positive reciprocity, establishing clear communication, and the development of mutual negotiating skills may be more effective (N. C. Klein *et al.* 1977).

**Medication** has only a very small part to play in the management of children with conduct disorders. However, in a small minority of cases, where aggressive outbursts, though not epileptic in nature, appear to take the form of frequent 'blind rages' (Leicester 1982), haloperidol may be a useful treatment. A dose of 0.05 mg/kg body weight in two divided doses over a period of a few weeks or months may be helpful. The dose should be reduced if impairment of concentration, drowsiness, or dystonic symptoms occur.

As well as these specific techniques, management should include atten-tion to ways in which family relationships can be improved and the child's energies channelled in more positive directions. Marital counselling may be indicated for the parents. Advice may be given to fathers as to how they might spend more time with their sons, or how different children in the family may be given more individual attention. Such advice may be helpful, perhaps by improving the self-esteem of children whose antisocial be-haviour may be a sign of chronic discouragement and feeling of failure.

The management procedures described so far are likely to be helpful only where the child and family are motivated to receive help. In a substantial number of children this is not the case, and family members may even be unwilling to cooperate in assessment. Nevertheless, the lives of teachers, other pupils in the classroom, and other members of the community may be disrupted so that action may be indicated. In these circumstances a variety of interventions may be made, some by virtue of the powers held by juvenile court magistrates, others without such powers needing to be taken.

These measures include:

1. Advice to teachers on ways of interviewing parents or applying behaviour modification programmes in the classroom.

2. Supervision of the child and contact with the family by a social worker, probation officer, or education welfare officer (EWO). The EWO is likely to be of particular help in cases of truancy.

3. Court procedures such as the requirement for truants of regular reporting back to the magistrates on regularity of school attendance. Such procedures have been shown to be more effective in ensuring regular attendance than merely arranging for supervision without reporting back (I. Berg *et al.* 1978).

4. Compulsory attendance at community-based centres for what is now known as 'intermediate treatment'—a variety of activities geared to the abilities of the child and organized by Social Services Department staff.

5. Residential placement in special schools for children with emotional and behavioural problems, community homes with education on the premises (CHEs), or (for children at greater risk for violent or other serious antisocial behaviour) youth treatment centres.

Finally it is important to stress that, as the following section on outcome indicates, the prognosis for many children with severe conduct disorders, even with treatment, is not good. This means that it is important to place emphasis on preventive measures as described in Chapter 6.

### Court reports

When a child has been charged with an offence, psychiatrists and psychologists may be asked to make assessments for the benefit of the magistrates to help them to decide how the child should be dealt with. In this case, it is important to make clear to the family at the outset why they are being seen and that anything they say may be put into the report. The information can be obtained in a similar manner to that described above.

In writing the report (Gibbens 1974) it is usually helpful to describe the circumstances of the offence and any precipitating factors. Relevant background information about the child's development, physical health, and mental state should be provided as well as information about the child's family, including positive as well as negative factors. A statement should be made about the presence or absence of significant psychiatric disorder. Throughout the report one should carefully distinguish factual information from inferences. One should also distinguish information provided by the child, family, or other individual from that deriving from one's own observations. Jargon should be rigorously avoided, and any technical terms used should be explained. At the end of the report one should bring together the information one has obtained in a formulation (Section 1.2.2). A recommendation may then be made at the end of the report about what would be a preferable line of action. If any form of treatment is recommended, it is important to check both whether this is available, and whether the child and family are motivated to receive it.

### Outcome

The long-term outcome of children with conduct disorders varies considerably with the nature and extent of the antisocial behaviour (Robins 1978). In general, children whose antisocial behaviour is limited to minor delinquency or isolated delinquent acts and whose relationships with other children are good do reasonably well. By contrast, children with a range of antisocial problems, some of which they show in severe form, and who get on poorly with others, do not do well in adulthood. The outcome of children with antisocial problems as a group is therefore distinctly worse than those with emotional disorders, but there is a good deal of individual variation. The best predictors for outcome lie in the behaviour of the child and not in the social circumstances in which the child is living. In particular, social class is unrelated to outcome. There is a tendency for children who are 'labelled' antisocial by a court appearance to do slightly worse than children with problems of similar severity who are not so labelled (Farrington 1977).

The problems that antisocial children develop in later life are diverse. They have an increased risk not only of antisocial or aggressive personality disorder, but also of neurotic conditions and schizophrenia. Their social lives are more likely to be disrupted by frequent job changes, and they may also show difficulties in establishing long-lasting relationships. Divorce rates are therefore high.

Although continuity of antisocial behaviour from childhood to adulthood is not high (most antisocial children do *not* grow up to be antisocial adults), the reverse is not the case. Nearly all seriously aggressive, antisocial adults have shown this pattern of behaviour in childhood. This means that the possibilities for prevention of antisocial behaviour in adulthood by taking action in childhood are, at least theoretically, very considerable. Reports of successful community-based preventive programmes for youth at high risk are relevant in this connection (D. R. Offord *et al.* 1987).

## 2.11 Sensory development

### 2.11.1 DEVELOPMENT OF NORMAL HEARING

The development of hearing enables a child to respond to the world around him, and is especially vital in the development of language. At birth, hearing babies show a startle response to a loud sound, and by 1 week can discriminate the voice of their mother from other people. By about 6 weeks they will quieten or 'still' to the sound of the mother's voice. At 4–7 months, the infant will attend to weak auditory stimuli in his near environment. Between 10 and 15 months, the baby will be able to locate sounds above and below ear level without the aid of visual cues.

### Screening for deafness

Auditory responsiveness of newborns can now be tested, using an auditory response cradle. High-risk cases (e.g. when there is a family history of deafness) should be screened in this way. Early detection of deafness in the UK still mainly depends on screening procedures using a distraction test between 7 and 9 months (Hall 1989). The baby's capacity to localize sound by head turning to quiet sounds made just outside his visual field is assessed. Parental doubts about their child's hearing expressed before this age should, however, always be taken seriously. There is some controversy concerning whether screening for hearing impairment should continue between 1 and 5 years, but it is agreed that parents should always be asked if they suspect their child's hearing may be affected when the child is screened for other purposes during this time. Parental concern about hearing loss, the presence of language delay, and recurrent otitis media are indications for repeated testing. Children thought possibly to be suffering from hearing impairment require audiometric procedures, a description of which is beyond the scope of this book, but it should be noted that the sooner detection occurs and remedial measures instigated, the better the prognosis.

Auditory perception continues to mature throughout childhood. For example, the child's capacity to discriminate between consonants and different musical sounds increases up to the age of puberty.

### 2.11.2 HEARING IMPAIRMENT

Deafness can be classified according to its cause (conduction or sensorineural) or according to its severity (mild, 15–30 decibel loss; moderate, 30–45 decibel loss; severe, 65–95 decibel loss; or profound, greater than 95 decibel loss). Severity may vary depending on the pitch and tone of the sound. A particular pattern of loss, for example loss of high or low frequency, may make generalization from average loss misleading.

Mild conductive hearing loss is relatively common, and is usually caused by secretory otitis media. Conductive loss may also occur in association with certain syndromes, for example Down syndrome and cleft palate in which the middle ear is often affected.

More severe degrees of deafness commonly have a sensorineural cause. Sensorineural deafness occurs in about 1 per 2000 children (I. G. Taylor

1984). Genetically determined conditions and fetal infection (especially rubella and cytomegalovirus) are responsible for a high proportion. It is important to remember that congenital rubella infection may be re-activated after birth causing hearing impairment in a child whose hearing has previously been normal or less severely affected.

## Psychological development

In deaf children who do not have associated handicaps such as cerebral malformations, cerebral palsy, and visual impairment, non-verbal intelligence does not differ early on from that found in the general population, but may begin to decline even in the preschool years (Sonksen 1985). Early conceptual development of the hearing-impaired usually proceeds normally. Deaf children show normal make-believe play, thus confirming that language development is not a necessary requirement in the development of some forms of symbolic thought.

Language development is, however, likely to be impaired even in the mildly hearing-impaired child, and will be seriously affected in more severe forms of deafness.

The causes of language delay in the mildly deaf child are often complex. Recurrent ear infections are more likely to occur in understimulated children living in deprived and overcrowded circumstances. A child with mild hearing loss living in a chaotic household with the television or radio on all the time is much more likely to have difficulty in making sense of verbal messages and making himself understood than a child with a similar level of hearing loss living in more fortunate circumstances. The former child is not only more likely to show early language delay, but also difficulties in reading later on.

It has been frequently suggested that chronic secretory otitis media (CSOM) is a common cause of reading failure, but rigorously controlled studies (e.g. Webster *et al.* 1989) suggest that this is probably not the case. The adverse social conditions linked to CSOM are more likely to be responsible for educational problems. Nevertheless, CSOM is an unpleasant condition that impairs communication while it is present, and it certainly deserves early identification and treatment.

Severe deafness affects both comprehension and expression. The child's inability to hear diminishes his capacity to monitor his own vocalizations. His capacity to understand what is going on around him and communicate with others, especially members of his own family, is thus potentially seriously threatened.

## Early management of the severely deaf child

Parental suspicion of deafness often precedes diagnosis by several months. The intervening period is one of uncertainty for the family, although the diagnosis is less often accompanied by hopelessness and despair than is the case with visual impairment. Help from competent professionals (such as audiologists, speech therapists, and psychologists) will reduce secondary problems.

Assessment of the intellectual development of deaf children requires special skills. Psychologists will use intelligence tests, such as the Merrill Palmer (Section 1.2.3), which require little or no language ability.

Mild or moderate degrees of deafness can usually be compensated to some degree by hearing aids, so that communication with the family need not necessarily be seriously affected. With more severe degrees of deafness, especially in the preschool years, parents may need to be helped to understand and cope with behaviour problems due to impaired communication and lack of coping skills. There is some controversy as to whether parents should be encouraged themselves to learn one of the various sign languages, but most now agree that all forms of facilitating communication, both in the family and in the school setting, should be promoted. Unusual degrees of dependence may arise because the child cannot use auditory cues to locate his mother and so needs to be physically close to her. Temper tantrums arising from frustration because of inability to make their needs known are commoner in young deaf children. Special care will need to be taken to encourage the deaf child to mix with other children who, in the early years, are likely to ignore a child who cannot respond readily to requests. Efforts should be made to encourage communication and minimize forms of behaviour that may be off-putting to others. Hospitalization of the seriously deaf child also requires special preparation because of the child's impaired level of verbal understanding.

With modern aids, the education of all but the most severely deaf or multiply handicapped children can take place in normal school or in a special unit attached to a normal school. For a small minority of children, special residential education will be required. Wherever the child is educated the staff need to have a comprehensive appreciation of the hearing-impaired child's strengths and weaknesses as they interact with the academic and social demands of the school situation.

### Psychiatric disorders

The rate of emotional and behavioural difficulties in the hearing-impaired child is higher than that in the general population. No special problems have been identified—both emotional and conduct disorders are common. R. D. Freeman *et al.* (1975) found that 23 per cent of 5–15-year-old deaf children showed moderate or severe psychiatric disorders—a rate about three times that in the general population. Deaf children were particularly likely to be described as restless, possessive, and destructive. Causes consisted mainly of disturbances in family relationships and inappropriate child-rearing practices and attitudes. Such problems are however heightened by difficulties in communicating with the deaf child. Divorce is not more common in parents of deaf children, but when family conflict occurs it often focuses around problems of communication with the child in question.

Assessment and treatment of disturbance in deaf children poses unusual problems for the professional. It has been suggested (R. D. Freeman 1977) that mental health professionals who are involved should either be able to sign or to have a competent interpreter, that they should be sure to collect information from more than one source, and that they should be wary of wrongly interpreting deviant verbal expression as a sign either of thought disorder or mental retardation. In general, once communication problems with the deaf child or adolescent have been at least partly overcome, the principles of assessment and treatment need not vary greatly from those employed with hearing children.

## 2.11.3 DEVELOPMENT OF NORMAL VISION

The development of visual skills allows children to explore and learn about their environment, and to integrate information obtained from the other senses in a way that enormously increases their capacity to adapt to the world around them. It is not surprising that blindness is one of the handicaps most feared by parents for their children.

The pupillary reflex is present from about the thirty-first week of gestation. At birth, newborns will turn their head to a diffuse light source such as a window. The newborn is also capable of visual fixation, and by 4 weeks can be seen to watch his mother carefully when she speaks to him. Visual input during the first 3 months of life leads to the development of integrated circuits at cortical level which are revealed by behaviour such as reaching for an object in the field of vision, and turning to a sound source at 4 months, blinking to a threat at 5 months, and searching for a recently lost object at 8 months.

Visual perception also shows rapid maturation in the first year of life. At 2 months an infant will look more at a picture of a face than at a shape with 'scrambled' facial features on it. By 6–7 months, and often earlier, a baby can tell familiar from unfamiliar faces. Perceptual maturation continues at least during the first 10 years of life. As children get older they become more capable of distinguishing shapes such as letters that are similar but somewhat different from each other (Walk 1980).

Screening for visual defects, for example congenital cataract, should take place through careful inspection of the eyes after birth. Parents should be asked at all subsequent screening examinations (recommended in the UK at 6 weeks, 8 months, 21 months, and 39 months) whether they have any concern about the child's sight. Any parental anxieties about the child's vision should be taken seriously and would normally indicate referral to an ophthalmologist. Squints can often be detected at the 8-month examination. Visual acuity should be checked periodically by the school nurse throughout the school years in case the child has need of spectacles (Hall 1989).

## 2.11.4 VISUAL IMPAIRMENT

Visual defects to a degree requiring special education occur in about 4 per 10 000 children. They are caused by a wide variety of physical conditions. Optic atrophy, cataract, and diseases of the macula, retina, and choroid account for over 50 per cent of cases of registered blindness (Fine 1979). Most causes of blindness or partial sight are present at birth, and about half are genetically determined (Jay 1979). Visual difficulties may be caused by cortical visual impairment, a condition in which the ocular structures are normal, but the occipital lobes are damaged. The condition is difficult to diagnose as the visual impairment may vary considerably depending on the task in question.

Additional handicaps are common among children with visual defects. Rates of associated handicap depend on the severity of the visual defects involved, but in some series as many as 70 per cent of visually handicapped children have been found to be suffering from some degree of mental retardation. It is, however, often uncertain how far such retardation is organically determined (see below). Cerebral palsy, hearing loss, language delay, and related congenital abnormalities are also relatively common.

### Psychological development

Children with visual defects present from birth are often not identified immediately. Their parents only gradually realize that there is something wrong, perhaps because of delay in the smiling response, or the fact that the baby seems surprised when a bottle touches his mouth. In fact, in the first 6 or 8 weeks of life, the baby may indeed show peri-oral movements that resemble a smile but do not depend on visual stimulation and have no social function. It is only at around 2 months that the 'reflex' smile to visual stimuli and, later, the smile of recognition, become established. Professionals are often slow to accept that parental anxieties are soundly based.

Even in the absence of organic causes for slow development, children with visual deficits are often slow in their motor, language, and social development. This may be due to the fact that, ordinarily, vision-assisted learning takes place in achieving skills such as the location of sound and touch which do not obviously require an intact visual pathway for their performance (Sonksen 1985). It may also, at least in some cases, be related to the fact that visually impaired children receive inappropriate stimulation or too little stimulation from their parents. The infants' apparent lack of responsiveness may lead the mother and others around to avoid attempts at social interaction. The depression experienced by mothers after the diagnosis has been confirmed may also contribute to such stimulus deprivation. Mothers should be strongly encouraged to respond actively to their child's vocalizations, using touch as well as verbal stimulation. Initial expectations after diagnosis may be much too low, and the lack of visual cues may lead to delay in the development of attachment (Fraiberg 1977). Toys should be carefully selected for their tactile and auditory qualities. The development of object constancy (the capacity to retain an image in the mind when the stimulus is no longer present) is slow to develop in the visually impaired child, so that, once attachment has occurred, even brief separations from the mother can be more upsetting than they would be to a sighted child of a similar level of maturity (H. Hunt and Wills 1983). Hospitalizations therefore need especially careful planning to ensure that the child is not left unaccompanied by a familiar figure who can interpret what is happening to him. Assessment of intellectual level in blind or partially sighted children can be carried out using specially designed tests (e.g. Reynell and Zinkin 1975).

### Psychiatric disorders

The rate of behavioural and emotional disorders in visually handicapped children is higher than in the general population (R. D. Freeman 1977) with 45 per cent showing a moderate or severe disorder, but there is no reason to think that these are inevitable consequences. Factors relating to disturbance include the severity of the visual loss, the presence of multiple handicap, the frequency of early hospitalization, parental anxiety, and depression, as well as the quality of family relationships (R. D. Freeman *et al.* 1989).

Psychiatric problems characteristic of this group include feeding and sleeping problems as well as social isolation. Mannerisms, not necessarily related to other behaviour or emotional problems, are common. The most common are stereotypies, especially rocking, eye pressing, repetitive noises, and hand-flapping. The term 'blindisms' has been used for these mannerisms, but, in fact, with the exception of eye-pressing they also occur in other conditions such as autism. They are much more characteristic of blind than of partially sighted children. They appear related to the level of stimulation the child receives, and may represent a form of self-stimulation when unstimulated or a response to over-stimulation or tension.

Management of psychological and psychiatric problems in children with visual defects requires more than ordinary skills and experience of, for example, the ways in which the visually impaired child responds to frustration. Mannerisms may, however, respond well to formal behaviour modification of operant type (Section 6.2.2).

The special educational needs of a child with a visual defect require careful planning to reflect individual needs and especially the stage of development. This should be achievable with the child living at home but an older child or adolescent may benefit from the special facilities available in a residential placement.

Follow-up studies of blind children into adolescence and early adulthood suggest that a significant proportion are able to obtain employment and make rewarding social relationships, but this is less likely to occur in the multiply handicapped. It is certainly more difficult for the blind teenager to make and sustain friendships, but this may be made easier if the young person has the opportunity, at least part of the time, to mix with others with similar disabilities. The acquisition of sexual knowledge and appropriate sexual behaviour is problematic and requires special attention. Indeed a high proportion (nearly a quarter) of blind teenagers report having been sexually molested (R. D. Freeman *et al.* 1989).

# 2.12 Sphincter control

## 2.12.1 NORMAL BOWEL CONTROL

In the infant, distension of the rectum with faeces stimulates periodic automatic voiding by relaxation of the internal and external anal sphincters. As the child gets older, the external sphincter and levator ani muscles that consist of striated muscular tissue come under voluntary control. In order to prevent faecal expulsion, the child learns to tense these muscles when he experiences a sense of fullness of his rectum. The internal sphincter (which is not under voluntary control) is probably triggered into a long-term contraction by this voluntary activity. The internal sphincter then operates until further rectal distension requires the child once again to exercise voluntary control.

Most normal bowel 'training' is, in fact, a cooperative achievement by mother and child, the 'potting couple'. Around the second half of the second year of life the child (now physiologically ready to achieve continence and able to sit still for a longer period of time) usually indicates he is about to pass a motion by making a characteristic sound, adopting a characteristic posture, or both. He may be eager to imitate adults or other sibs in their mode of defecation. Alternatively, the child's bowel habit is so regular that the mother knows when he is likely to pass a motion. In any event, at a promising moment, she encourages him to sit on the pot and, if he performs, shows great pleasure. Appropriate defecation is reinforced in the context of a mutually rewarding relationship, and 'training' usually occurs within a few weeks. It is usually more appropriate to think of the child as having matured sufficiently to gain a new skill rather than to conceive of the process as one of 'training' by parents.

## 2.12.2 CONSTIPATION

Constipation involves difficulty or delay in the passage of faeces. Encopresis is the passage of faeces of normal or near normal consistency into socially inappropriate places (including clothing). Soiling is the frequent passage of liquid or semisolid faeces into clothing (Agnarsson and Clayden 1990).

## Clinical features

Constipation may present at any time from the first few weeks of life, but is an unusual presenting feature after 6 or 7 years. It usually, though not always, begins in infancy. The usual complaint is of failure to pass faeces except at infrequent or irregular intervals, and there may be associated pain, abdominal distension, and soiling. As constipation becomes more severe, it is likely to be accompanied by faecal soiling. On examination, faecal masses are often palpable on abdominal examination and hard faeces are palpable on rectal examination. In older children, constipation may occur in association with encopresis (see below), and findings on examination may vary over time, with faeces palpable on some occasions, but not on others.

## Aetiology

### Physical factors

Constipation present from birth may be due to anal stenosis, other minor anatomical abnormalities, and Hirschsprung's disease (a condition in which there is a failure of development of ganglion cells in the internal sphincter, and a segment of rectum, so that the neuro-muscular defecatory mechanism is deficient).

Infants may become constipated if they are not given enough to eat or drink, or because of overstrong, artificial feeds. In older children, dietary factors (e.g. a diet low in fibre) may also be a factor. Very occasionally, constipation may be a presenting feature of organic disease, such as hypothyroidism or idiopathic hypercalcaemia. Other local physical causes include an anal fissure or fistula. These may make defecation painful so that the child withholds. Constipation may also sometimes follow a prolonged febrile illness.

### Psychosocial factors

1. *Parental factors*. It is not at all uncommon for mothers to begin to 'train' their babies to be clean in the first year of life (Newson and Newson 1963) by holding their children over the pot after meals. This is a futile exercise but, in itself, probably does no harm unless the mother becomes angry and frustrated when the child fails to perform. The 'training' of older children may also be adversely affected by parental behaviour. Mothers under stress because of tense family relationships, financial hardship, or overcrowded housing will have less time and patience to attend to their child's cues. If they react to the child's failure to achieve bowel continence by punitiveness, this is likely to make the child more anxious, and to achieve the opposite result to that intended (Woodmansey 1967).

2. *Factors in the child*. The child may be poor at providing cues when he is ready to defecate. This may be because of global retardation, because he

is generally uninterested in pleasing his mother, or because he is using his capacity to control defecation in a battle for power with one or both parents. Younger children may have fantasies about their faeces and, once one is aware of these, the tendency to withhold faeces may become understandable. Thus, some children withhold faeces because they regard them as a necessary part of themselves they do not want to lose. They may therefore be worried about what happens to them after they are flushed away. Others regard faeces as a dirty part of themselves they do not want to have anything to do with, so they deny their rectal sensations and retain faeces.

3. *Constipation and sexual abuse.* This is a complex and controversial issue. Children traumatized by sexual abuse may withhold faeces as part of a generalized response to stress. It is reasonable to assume that this is more likely if the abuse has involved anal penetration. The clinician should therefore consider sexual abuse as one possible explanation for intractable constipation and consider whether there is other evidence (see Section 1.3.3). If, on examination of a constipated child, on parting the buttocks, the anus is patulous, this is, in itself, not supportive evidence of abuse. It occurs in about 15 per cent of severely constipated children (Clayden 1988). If the anus is initially closed, but then dilates (reflex anal dilatation), this is thought by some to be supportive evidence of abuse. It is, however, generally agreed not to be pathognomonic of abuse, and other evidence must be sought (see Section 1.3.3).

As a result of constipation, whatever its cause, a hard mass is formed in the rectum and lower colon. The internal sphincter becomes chronically stretched, and the child may have no sensation of fullness. Ballooning of the bowel above the mass can produce a functional megacolon. Liquid faecal matter accumulates, and oozes out around the sides of the constipated mass producing intermittent soiling. Partial intestinal obstruction may occur, especially in young children, with abdominal pain and vomiting.

### Management

An account should be taken of the development of the symptom. Constipation present from the first few days of birth suggests an organic cause. In other circumstances, the constipation is more likely to be functional. An account of early training practices should be obtained. The presence of associated behaviour problems is relevant, and the quality of parent–child interaction should be assessed.

Physical examination should include palpation of the abdomen for faecal masses, and examination of the perianal area for evidence of a fissure, as well as a rectal examination.

If constipation is present, it should be directly treated regardless of the level of psychological disturbance. Children cannot hope to achieve bowel continence if they have a large mass of faeces distending the rectum. Initially the mass should be cleared using a few doses of a simple but effective aperient such as liquid paraffin. If oral treatment is not effective, a brief course of enemas may be necessary, but prolonged courses of enemas or bowel wash-outs are unnecessary and undesirable. In any event, children should be carefully prepared for these potentially upsetting procedures (see Section 4.6).

Subsequently, to establish the bowel habit, it is usually necessary to continue with maintenance treatment for some months. This should consist of a diet high in roughage, fruit, vegetables, etc. and Senokot daily or every other day. A softening agent such as docusate (Dioctyl Paed) is helpful initially for a couple of weeks, followed by maintenance senna, or, in infants, an osmotic stool softener such as lactulose.

Associated with these physical forms of treatment, parents should be counselled to relax in their approach to the child's bowel problem. A positive approach with warm encouragement to the child to use the pot or toilet should be advised. The child should be praised initially just for sitting on the pot. Attention should be given to wider aspects of parent–child interaction with identification of more mutually enjoyable activities, so that there is less focus on the child's bowel problem. Children and families can be given 'bowel diaries' to record the frequency and consistency of bowel motions as well as where they are passed (potty, lavatory, or pants).

With this regime the great majority of children seen in primary care or in paediatric departments with constipation will improve considerably. Failure to improve should lead to a reconsideration of the possibility of a physical disorder with reference to a specialized paediatrician. If this is ruled out, more intensive psychiatric assessment and treatment along the lines described below in relation to encopresis are indicated. The possibility of sexual abuse should also be considered (see above).

**Outcome**

Constipation has a good prognosis. Given appropriate treatment, over 90 per cent of those treated will respond rapidly. A small proportion develop chronic constipation that persists into adulthood.

## 2.12.3 ENCOPRESIS

This involves the passage of faeces of normal or near-normal consistency in inappropriate places (including clothing). Encopresis may alternate with soiling—the passage of liquid or semisolid faeces into clothing.

By 3 years, 16 per cent of children are still showing signs of faecal incontinence once a week or more, but by 4 years only 3 per cent (N.

Richman *et al.* 1982). By 7 years only 1.5 per cent of children have still not achieved continence (Bellman 1966), and by 10–11 years this figure has decreased to 0.8 per cent. Most children identified in surveys are probably showing soiling secondary to constipation, and only a minority true encopresis. Boys are affected three to four times as commonly as girls. There are no consistent social class differences.

### Clinical features

Encopresis may be continuous from birth or may occur after a period of continence. It is unusual for it to begin after the age of 5 or 6 years. Mild constipation may be a constant feature, intermittent, or absent. Sometimes children with encopresis develop severe constipation with the clinical picture described above.

The soiling is often frequent, occurring several times a day. It may, however, not occur at school, but only when the child returns home. Parents will often interpret this as a hostile act on the part of the child, but, in fact, the soiling in the home setting may occur because the child feels more relaxed there.

The child's attitude to the soiling is variable. Some who soil their pants deny that it really occurs, or that they smell, and these have to be reminded to change clothing. They may have become habituated to the smell. Others are ashamed and embarrassed and go to great lengths to hide soiled pants. The faeces can be used to express aggressive feelings. Some children use them to smear walls and furniture, others deposit well-formed faeces in places likely to cause maximum distress—in glasses used for drinking, or in the parental bed. The soiling may represent one aspect of a child's attempt to retain the benefits of a younger child (regressive soiling). Despite normal intelligence, the child may be unable to cope with the increasing independence he is expected to show at school and elsewhere.

Associated psychological and psychiatric problems are almost invariably present, but, apart from the fact that the children are usually rejected by others because of their smell, are not of any very characteristic type. Children are usually of average or below average ability, though they may show significant educational retardation. They may be unduly anxious, show seriously aggressive behaviour, or neither of these. It is interesting that even 3-year-olds who have failed to gain bowel control show an unduly high rate of behavioural problems (N. Richman *et al.* 1982). About a third of children who soil are also enuretic at night, and a smaller proportion are wet by day.

### Aetiology

Encopresis, the passage of normal stools into inappropriate places, is never physically determined. A variety of factors in the parents, child, and in parent–child interaction have been identified as probably causative. None is invariably present, but there is a certain amount of consistency in the background factors usually in operation.

1. *Parental factors.* These may seem to be of little significance. Alternatively, some parents of children with encopresis are quite unusually aggressive and punitive both in their general attitude and in relation to the child in question. They may also show unusually high standards of cleanliness and have expectations of conformity for their children's behaviour that are quite unreasonable. As a result of these characteristics, early bowel training may have been coercive and resulted in the child feeling himself a failure.

2. *Factors in the child.* The child may be showing general developmental retardation with delay in achieving bowel continence as only one aspect of general immaturity. In children of normal ability, irrational fantasies about the faeces as described in relation to constipation (see above) may be present. These may result in a phobia of the pot or toilet. Encopresis is not uncommonly one manifestation of a generalized conduct or emotional disorder. Thus the child may be showing associated disobedience or aggressive behaviour, often linked with specific educational retardation. Children with associated aggressive problems are more likely to use their faeces for aggressive purposes, for example by smearing walls or windows. Less commonly the child may be generally depressed, fearful, or inhibited.

3. *Family factors.* Family relationships may be characterized by a good deal of tension and disharmony. Communication may be poor. Not infrequently, the child with encopresis is scapegoated. There may be a constant battle for control, with the child using his capacity to control his own defecation as a weapon in the family conflicts.

4. *Life stress.* Especially in children who have only recently gained bowel control, an unpleasant experience such as a frightening, acute illness, or an episode of bullying at school may precipitate the onset of encopresis.

## Management

A combined approach involving both psychological and physical measures is desirable. In hospital practice this may be best achieved by a joint paediatric–psychiatric team.

1. *General principles.* Whatever the specific approach used, certain general principles need to be followed. The child should be told he is not the only sufferer from this problem—it occurs in many other children. He should be given some idea, appropriate to his age level, of the mechanisms of defecation. The parents should be encouraged to ignore the child's soiling except at those times when they are involved in carrying out a behaviour modification programme (see below). In particular, parents should be discouraged from punishing the child for soiling. An attempt should be made to encourage activities which allow more positive parent–child relationships to be built up. While psychological treatments are in

progress, the child should be occasionally checked for the development of constipation. Appropriate physical treatment should be given if this occurs (see above).

2. *Behaviour modification.* As with other forms of behaviour modification, careful baseline monitoring, both of the undesired behaviour and of the reactions of the parents, is an essential first step (O'Brien *et al.* 1986). The most successful approach involves the shaping of an appropriate toileting habit. The child is encouraged to sit on the toilet for about 10 minutes after each meal. The behaviour is charted, and the child rewarded either materially or socially (using stars, stamps, hugs, praise, etc.) for appropriate behaviour—first for sitting on the toilet and then for passing a motion into it (see Section 6.2.2 for a more detailed description of behavioural programmes). Behavioural techniques may be best discussed at family interviews, as their administration may give rise to parental disputes (father, for example, thinking the child should be punished regardless). The presence of sibs is often helpful, but if the affected child is deeply embarrassed by his habit, he may not be willing to attend if they are present.

3. *Psychotherapy.* As indicated, the child's encopresis may have an aggressive component or (in regressive soiling) be linked to a wish to be treated like a younger child. Understanding the reasons for these feelings and taking measures to help is usually best achieved in family sessions. Aggressive tendencies may, for example, be linked to a parent's lack of ability to show affection and concern. Aggressive behaviour may be covertly encouraged by a father. Family therapy offers an opportunity to explore issues such as these. Individual psychotherapy may also occasionally be indicated, especially when the problem appears part of an emotional disorder, or irrational fantasies appear to be present.

4. *Medication* (e.g. imipramine) has been recommended as a treatment for encopresis, but there is no good evidence for its effectiveness.

5. *Hospitalization.* Encopresis is usually managed on an out-patient basis. Failure to respond to an out-patient treatment programme over several weeks is an indication for admission. Encopresis may respond to behavioural management on a paediatric ward, but, if the facility is available, admission to a child psychiatric in-patient unit, where more intensive psychologically orientated treatment can be carried out, will often be desirable. In both settings, the work carried out will need to involve parents; otherwise symptomatic improvement will occur in hospital, but not be maintained in the home setting.

### Outcome

Encopresis virtually always clears up by adolescence (Bellman 1966) and, although adult encopretics have been described, they are extremely rare. Treatment usually shortens the natural course of the condition, so that over 90 per cent may be expected to improve over a period of a year (Levine and

Bakow 1976). Outcome is better for those children who have a reasonably well-developed 'internal locus of control'—a sense of being in control of their fate. Although the encopresis disappears, associated symptoms, especially aggressive behaviour, may not, and here the outcome will depend on similar factors to those described when the prognosis of conduct disorders was discussed (Section 2.10.2).

## 2.12.4 NORMAL BLADDER CONTROL

Children are incontinent of urine at birth because they do not have the necessary physiological maturity, the understanding to know what is required of them, or the motivation to achieve continence. By the second and third year of life, they have usually achieved all three of these requirements. Continence will occur if:

1. The central nervous system has matured sufficiently. In babies reflex emptying occurs when the pressure inside the bladder reaches a certain point. Older children become able, as a result of maturation of the nervous system, to inhibit emptying by contraction of one or more of the bladder sphincters by day or at night when asleep. Maturation of nervous structures is partly genetically determined, and this explains why failure to gain bladder control at night often runs in families.

2. There is no defect of the nervous system supplying the bladder and no congenital malformation.

3. The parents can take a relaxed attitude to the achievement of dryness, and not make the child anxious about failure. Many mothers start toilet-training before the age of a year. Parents who begin at 18 months to 2 years and confine their 'training' to providing opportunities for voiding on the pot, especially when the child shows he is ready, and praising the child for a successful performance, are likely to see their children become dry earlier than those who punish for wetness or are upset by it (Brazelton 1962).

4. The family is going through a settled time when the child is 18 months to 3 years old. There is evidence that if the family is under stress when the child is of this age, he or she will be slower to become dry.

## 2.12.5 ENURESIS

### Definition

Enuresis can be defined as involuntary emptying of the bladder in the absence of an organic cause in a child over the age of 5 years. The term 'urinary incontinence' is usually employed if involuntary emptying occurs due to a physical cause. Enuresis may be nocturnal, diurnal (by day), or both.

**Background information**

At 3 years of age about 34 per cent of children are still wetting by night, and 26 per cent by day (N. Richman *et al.* 1982). By the age of 5 years, about 10 per cent of children will be wet by night and about 3 per cent by day (De Jonge 1973). By 8 years 4 per cent are wet by night and 2 per cent by day (N. Richman *et al.* 1982). At 14 years about 1 per cent of children are still wet by night. Boys are about twice as likely to wet as girls, and working-class children are slightly more likely to wet than middle-class.

Factors sometimes associated with wetting include:

(1) a family history of bed-wetting in parents or sibs;
(2) unsettling events in the preschool period, such as instability of family relationships, divorce, or separation (Jarvelin *et al.* 1990), or a history of unplanned separation of the child from his family;
(3) the presence of behavioural problems in the child;
(4) general developmental delay.

Enuresis may occur in any stage of the sleep cycle (Section 2.3.1).

**Causation**

Nocturnal enuresis is occasionally caused by physical disorders to be discussed below. Much more commonly it is produced by an inherited delay in maturation of the relevant nervous structures, or less commonly, by an interaction between delay in maturation and environmental circumstances which, for one reason or another, fail to promote the acquisition of bladder control. Maturation delay may be shown by a small functional bladder capacity, as children with enuresis appear unable to retain as much urine in their bladders as non-enuretics before micturating. However, the fact that children with small functional bladder capacity have an increased tendency to show psychiatric disorders suggests that there is an interaction between psychological and physiological factors, rather than distinct groups of children with either physical or emotional disorders (Shaffer *et al.* 1984).

Long-standing environmental circumstances of significance include:

(1) a rigid approach to toilet training in which the child is expected to conform to unreasonable parental standards of toileting, fails to comply, is punished for failure, and is then made anxious and thus even more liable to wet himself;
(2) negative or indifferent parental attitudes that result in the child failing to care whether he wets or not, because he has such low self-esteem.

In addition to long-standing adverse circumstances, precipitating events may be of causative importance, perhaps by altering sleep rhythm. They include:

1. Events that make the child anxious, such as sleeping in a strange house, seeing a disturbing TV film the previous evening, witnessing a parental argument, or having difficult schoolwork to do. Anxiety may also result in a child that normally wets having a dry night.

2. Events that result in the child feeling more relaxed, for example the school holidays, may also affect the sleep rhythm so that the child is either more or less likely to wet at these times.

Finally, although this is unusual, some children deliberately wet the bed as a hostile or angry gesture towards their parents.

## Clinical features

Most wetting is dealt with at the primary health-care level by general practitioners or health visitors. Children referred to paediatricians and child psychiatrists have usually failed to respond to symptomatic measures, or show signs of emotional disturbance in addition to their wetting. Wetting may have been present from birth (continuous) or it may occur after months or even years of continence (onset enuresis). It may be present by day or by night, or only by night. The wetting may be regular and frequent, or intermittent. Causative factors, other than a presumed inherited failure of maturation of nervous structures supplying the bladder, are often absent.

Other urinary symptoms are also usually absent. Pain on passing urine (dysuria), blood in the urine (haematuria), or dribbling of urine, are all indications for full investigation and usually indicate an organic cause for the wetting. Girls are more prone to urinary infections than are boys.

## Management

A careful history should be taken to elicit the onset, frequency of the wetting, and any precipitating circumstances. Enquiries should be made for possible signs of associated emotional or behavioural disturbance. Diurnal enuresis is more likely to be associated with emotional disturbance. Finally, the motivation of the child and parents to achieve change should be assessed, as this will affect management. For many children and parents, wetting is a trivial and unimportant symptom; for others it has come to dominate their lives. The possibility of nocturnal epilepsy producing enuresis can usually be excluded by the history. The presence of other urinary symptoms requires physical examination and further investigation. Otherwise, physical investigation should be limited to microscopic examination of a clean specimen of urine to exclude infection.

Management will depend considerably on the age of the child, the frequency of the symptoms, and the motivation of the child and family. However young the child or infrequent the symptom, if the parent is worried about the problem, the doctor or health visitor will need to take time to consider the problem seriously and provide helpful advice. The presence of associated emotional or behavioural problems is not a contraindication to a direct attempt to treat the enuresis. Indeed, sometimes such associated problems will improve if the enuresis is successfully treated. However, associated problems will occasionally need to be tackled initially. This arises if, for example, the child is too aggressive or inhibited to cooperate with a direct approach.

The first and often only important measure to take in monosymptomatic enuresis is the provision of reassurance about the nature of the problem, i.e. that it is not a symptom of disease, it is widely prevalent, and there is an overwhelming likelihood that it will clear up on its own, given some time. The child should not be regarded as lazy—he is doing his best. The information should be provided in a form that can be understood by the child as well as the mother. If the symptom is perceived as burdensome, then the doctor should sympathize with this and share the disappointment that the child is one of the minority to show the problem. Five-year-old children at school in classes of 30 can be told it is likely that at least two other children in the class have the same problem. The doctor should also sympathize with the mother's natural tendency to feel angry and irritated about the problem. In young children the doctor should suggest a return visit in a few months to check on progress.

In a child over the age of 5 or 6 years, it is reasonable to take a more active approach.

1. First, it is sensible to find out what the parents have already tried. Usually parents have tried rather unsystematic rewards and punishment, as well as restricting fluids and lifting at night before they themselves go to bed. The last of these is indeed sometimes successful, but is unlikely to be so if the child is upset by being awakened and refuses to use the toilet.

2. *Rewards.* Children may be helped by keeping a star chart or a chart in which they can stick in farm animals for 'dry nights'. Star charts emphasize the child's achievements rather than his failures, and they also ensure that a record of the child's progress is kept. This approach is more likely to be effective if the child and parents are motivated, and if the doctor or nurse show regular interest and enthusiasm.

3. *Bladder 'training'.* Daytime wetting may be reduced if the child is encouraged, with parental supervision, to pass urine at gradually lengthening intervals. The assumption behind this approach is that some children have bladders of small functional capacity, and these need gradual expansion. (The basis of the treatment is of doubtful scientific validity, but nevertheless it is sometimes effective.) During a school holiday, the child is encouraged to pass urine (micturate) at half-hour intervals on the first day, hourly intervals on the second day, one-and-a-half hourly intervals on the third day and so on, until there is an interval of three or four hours between

voiding. Improvement in daytime wetting may be accompanied by an improvement in nocturnal enuresis.

4. *Enuresis alarms.* This is the treatment of choice for children aged 6 years and over, living in good social circumstances, for whom reassurance and systematic rewards have proved ineffective (Dische *et al.* 1983). There are various types of alarm available. Usually, the child sleeps on two wire-mesh bed mats, connected to a bell or buzzer, which goes off when electrical contact is made by the child wetting. When the buzzer goes off the child is awakened, if not already awake, and taken to the toilet. The machine is then reset.

The procedures require a cooperative child usually sleeping in a room by himself. Parents need to be motivated, and a doctor or nurse experienced in the use of the technique and willing to be in contact personally or over the phone at weekly intervals (or even more frequently initially) to advise over the mechanical problems that may arise. Given these conditions, about two-thirds of children respond favourably to this method of treatment. Some children relapse, but many of these respond to a second course of treatment.

Treatment failures may be due to failure of the parents to understand and follow instructions, false alarms, and failure of the alarm to wake the child (putting the alarm in an empty biscuit tin enhances the noise). More recently, a mini-alarm system incorporating a tiny perineal wet sensor attached to a small alarm worn on the child's clothing has been devised. This is about as effective, and especially for girls, is more comfortable, but it is somewhat more likely to develop a mechanical fault (Fordham and Meadow 1988).

5. *Medication.* Tricyclic drugs used in adults for treating depression do produce improvement in a significant number of children. Imipramine or amitriptyline 25–50 mg *nocte* may be prescribed for a 4–6 week period, the dose depending on the age of the child. The mechanism of action of these drugs is unknown. They may work by altering sleep rhythms or by an adrenergic or anticholinergic mechanism, though anticholinergic drugs prescribed alone are ineffective. Relapse usually occurs when the drug is stopped. Bearing in mind the danger of accidental overdosage by toddlers in the family, and the short-lived benefit obtained, most clinicians are reluctant to use drugs for this relatively benign condition, although in one trial they were found to be as effective as an alarm system (Fournier *et al.* 1987). They can be useful when it is important for the child to be dry for a short period, for example when going away on a school journey.

6. *Psychotherapy.* Monosymptomatic enuresis is not an indication for interpretive psychotherapy, though certainly children benefit from the support and understanding of a sympathetic doctor or nurse. However, if the enuresis is associated with symptoms of emotional disturbance, psychotherapeutic measures for the disturbance may also improve the

wetting. Conversely, successful symptomatic treatment of enuresis as described above may sometimes reduce other signs of emotional disturbance, perhaps because the child gains generally in confidence and self-esteem.

### Outcome

The great majority of children with enuresis will stop wetting before or during adolescence. Ninety per cent of children wetting at 7 years will have stopped by 14 years. A very small proportion do continue to wet into adulthood, and there are no well-validated methods of identifying these. Even they are sometimes helped by symptomatic measures applied in adulthood.

# 2.13  Sexual development

## 2.13.1  NORMAL SEXUAL DEVELOPMENT

Differentiation between the sexes begins in the first 3 months of fetal life. The presence of a Y chromosome in the normal XY male leads to the development of a testis which secretes two types of fetal androgen, up to the time of birth. One androgen promotes masculinization, and the other inhibits feminization. In the chromosomally normal (XX) female, the absence of these androgens results in normal feminization (Money and Ehrhardt 1972).

During the first 7 or 8 years of life, there is little secretion of sex hormones, and little *further* differentiation between the sexes. At about 7–8 years, there is an increase in pituitary gonadotrophin secretion, leading to a rise in androgen (in both sexes) and oestrogen secretion (in girls).

Normal puberty begins in girls with breast development (between 8 and 13 years) and extends over about 4 years, with menarche occurring in Western countries on average at about 13 years. In boys, the first pubertal changes to occur are in the size of the testis, though this is occasionally preceded by growth of pubic hair. Pubertal changes begin between 10 and 15 years, and last over about 5 years before full sexual maturity is reached. The first ejaculation occurs, on average, round about the age of 13 years.

### Effects of early and late puberty

There is a very wide range of ages over which normal puberty occurs. Within the normal range, there are certain characteristics associated with early and late maturation. Early-maturing boys are likely, on average, to be slightly more intelligent and educationally advanced, and to retain these advantages after puberty (Douglas *et al.* 1968). They are also likely to be more confident and assertive (Clausen 1975). Early-maturing girls are also likely to have some intellectual advantages, but positive personality characteristics are more likely to be linked to average, rather than to early, physical maturation (Tobin-Richards *et al.* 1982). In most societies, girls do not particularly like to be different from their peers, and either very early or late maturation may result in social discomfort. Adolescent boys do not wish to be conspicuous either, but early male maturation provides physical advantages relating to muscular strength that are particularly valued by boys.

Psychosexual maturation involves the development of: (1) gender identity; (2) sex-typed forms of non-sexual behaviour (gender role); and (3) sexual behaviour.

## Gender identity

This is the degree to which an individual perceives himself or herself to be male or female. By 3 or 4 years most children can correctly identify their own sex and that of other children, and have a strong sense of their own essential maleness or femaleness. From then on, gender identity usually remains firmly established.

## Sex-typed forms of non-sexual behaviour (gender role)

These are types of behaviour in which the sexes differ. Totally sex-specific behaviour is unusual in our culture, though certain characteristically feminine bodily mannerisms are commonly seen in girls by the age of 5 years, but rarely seen in boys at any age. Doll play, sewing, and cooking activities are much commoner in girls, and rough-and-tumble games, and play with guns and cars more common in boys. But there is much overlap, and the differences are closely related to child-rearing practices. A more universal sex-typed characteristic is that, from about the age of 5 years, at least up to the onset of puberty and sometimes well beyond this time, boys prefer to mix and play with boys, and girls with girls. From an early age boys show more aggression and less prosocial behaviour than do girls (Maccoby and Jacklin 1980).

## Sexual behaviour

**Sexual drive,** the wish to achieve sexual pleasure from stimulation of the genitalia, is present from birth and persists through early and middle childhood. As a result of increased androgen secretion, sexual drive increases sharply in early and mid-adolescence to reach a peak in late adolescence and early adulthood. Throughout childhood and adolescence there is a wide individual variation in the degree to which sexual drive is experienced. This variation is probably partly physiologically determined, but the presence or absence of sexually stimulating experiences is also of major importance. Such experiences may affect not only the intensity of sexual drive, but also the choice of sexual object.

**Masturbation.** In infancy, occasional self-stimulation of the genitalia to produce pleasurable sensations (masturbation) is common in both boys and girls. It may occur from the first few months of life. Such self-stimulation can lead to orgasm-like phenomena in which the usually somewhat older child of 2–3 years can be observed to be in a state of acute excitement, with rhythmic pelvic movements, rapid respiration, and flushing of the face, followed by a sudden release and period of quiescence. Masturbation is somewhat more common, but is perhaps better concealed, in middle childhood. It is usually of no pathological significance whatsoever. Excessive masturbation may, however, occur and this is discussed below.

The great majority of adolescent boys masturbate at least occasionally, and probably most masturbate regularly. The habit is probable less common in girls, most of whom may masturbate only occasionally (Sorensen 1973), though accurate estimates of prevalence are not readily obtainable. There is no evidence that the presence or absence of masturbation in adolescence is of any pathological significance.

## 2.13.2 ANOMALIES OF GENDER ROLE

### Clinical features in boys

A distinction must be drawn between anomalies of gender role behaviour in which boys merely show effeminate characteristics, and anomalies of gender identity in which, in addition, boys would clearly prefer to be of the opposite sex and, in a minority, actually have powerful fantasies that this is the case.

Effeminate boys are likely to prefer to dress in girls' clothing, to play with dolls rather than toy cars, soldiers, and guns, and to prefer to play in groups of girls rather than boys. They may have effeminate mannerisms, but they do not usually wish to be girls.

Gender-identity disorder of childhood is present when, in addition, a pre-pubertal boy shows persistent and intense distress about being a boy, an intense desire to be a girl, and either preoccupation with female stereotyped activities, as shown, for example, by cross-dressing and participation in female games, or by persistent repudiation of his own male anatomical structures (penis or testes) (American Psychiatric Association 1987; World Health Organization 1990).

In the great majority of boys, there is no evidence of any physiological explanation. A number of familial factors have been identified as important in the development of effeminacy and gender-identity disorder (R. Green 1974; Coates and Zucker 1988). The pattern is probably usually set in the first 3 years. A high proportion of mothers have experienced severe trauma in the first year of the child's life, and this may have resulted in withdrawal from the child. There is often parental encouragement of feminine behaviour, including cross-dressing, in the first few years of life, and, as time goes on, an unusually close relationship with the mother, with associated absence of a male figure with whom the boy can identify. Lack of boys in the neighbourhood to play with, and a 'pretty' girlish appearance in the boy are probably additional factors that play a part in some cases. There is a variety of reasons why mothers and, less frequently, fathers may encourage feminine behaviour in their sons. A wish for a girl to have been born and fear of normal masculinity are probably the most prominent. Where a close male relative has a record of aggressive antisocial behaviour, parents, and mothers in particular, may react with relief and pleasure to a boy showing girlish tendencies. Some mothers of these children show severe overt or covert hostility towards men.

*Management*
Increasingly in our society, boys and girls are encouraged to share activities and interests from an early age. Despite this tendency, sex-role activities remain rather sharply distinguished and, in middle childhood, children still seem to prefer to mix in same-sex groups. The effeminate boy is therefore still seen as 'different' by his parents, teachers, and other children, and this may lead to considerable distress in the child and family. Further, a boy with marked feminine characteristics has a high risk of developing homosexuality (see below). The more tolerant attitude in society towards adult homosexuality fortunately means that this need no longer be seen as a disaster, but, for obvious reasons, most parents have a marked preference for their children to grow up to show a heterosexual orientation. Consequently, it seems ethically desirable to offer advice to parents who are unhappy about their son's effeminate behaviour, and sometimes treatment for the child and family.

In assessing the problem, particular attention should be given to the extent of effeminacy and gender-identity disorder, and to the presence of associated signs of emotional disturbance. The presence of emotional disturbance in the child, and distorted relationships in the family may be secondary to the effeminacy, but is also likely to be partly responsible for the effeminacy in the first place. Family interviews held as a preliminary to further treatment may clarify the attitude of the different members of the family to the boy's effeminacy. For example, if the mother is ambivalent, or even enthusiastic about her son's effeminacy, then this will need discussion before further treatment is attempted.

Family interviews can be followed by the application of a behavioural approach (Rekers 1977), probably best carried out in the home. Once identification of types of behaviour requiring change has occurred, the parents can be requested to make systematic observations of these behaviours, for say 10-minute periods, during the day. When a base line has been established, a reward system is developed to reduce the rate. Thus, for example, the child can be helped to become aware of his feminine mannerisms by losing a token every time one occurs. Masculine types of behaviour can be positively reinforced (see Section 6.2.2 for further discussion of behavioural approaches).

Alternatively, children and parents can be seen separately in psychotherapy, and those experienced in such work suggest that, for meaningful change to occur, such psychotherapy may need to be prolonged. There is evidence that psychotherapeutic work carried out with parents is of greater importance than that carried out with the child (Zucker 1985).

### Outcome

Effeminacy in boys and *a fortiori* anomalies of gender identity, especially if marked and accompanied by homosexual activity, often proceed to adult homosexuality or bisexuality. This may be accompanied by transvestism (cross-dressing), transsexualism (a desire to achieve membership of the opposite sex), and wider personality problems as occasional accompaniments. The personality problems are likely to be partly produced by society's rejecting attitudes to what is widely seen as sexual deviation. Many adult homosexuals recall a childhood in which they were unusually interested in dolls, liked to dress in women's clothing, preferred playing with girls, were regarded at school as 'sissies', and showed more sexual interest in boys than in girls (Whitam 1977).

Treatment of the effeminate boy, as described above, has been shown to produce short-term change, but longer-term alterations in adult life following treatment in childhood have not so far been demonstrated.

### Clinical features in girls

The significance of 'masculine' behaviour in girls is uncertain, though it is possible to identify groups of girls showing tomboyish characteristics to a marked degree (R. Green *et al.* 1982). Now that girls in our society are encouraged to take an interest in activities previously thought of as exclusively male, their significance is even more difficult to evaluate.

In the absence of evidence of undesirable outcome of tomboyishness, it seems inappropriate to do other than offer qualified reassurance to parents worried about their girl's preference for mixing with boys, or playing traditionally male sports or with toys usually regarded as reserved for boys' use.

However, a small minority of pre-pubertal girls show definite disorders of gender identity. This problem is much less common in clinical practice in girls than in boys. Such girls show persistent and intense distress about being a girl, desire to be a boy, and either have marked aversion to girls' clothing or persistently repudiate their female anatomical form.

Again physiological abnormalities are not usually present, though girls with the adrenogenital syndrome (Section 5.12.6) are more likely to show this pattern of disturbance (Money *et al.* 1984). In most cases it is thought that the pattern of child-rearing is of greatest importance, though aetiology is often more difficult to elucidate than is the case with boys with the disorder.

Similar family and behavioural treatment approaches to those used in boys may be employed, but there is a lack of follow-up studies to evaluate their effectiveness.

### 2.13.3 ANOMALIES OF SEXUAL BEHAVIOUR

**Excessive masturbation** of worrying degree in infancy and early childhood exists when it is occurring for hours in the day and interferes with the child's regular daily activity and sleep pattern. The child should be examined for the presence of local skin irritation on the penis or vulva, though the presence of sores is more likely to be secondary than primary. Occasionally, a primary lesion may be noted. In an infant, if a local lesion is present, the baby needs restraint to prevent further rubbing, while topical treatment is applied. In the absence of a primary cause, the presence of excessive masturbation should lead to the suspicion that the child is overly self-stimulating because it is generally understimulated. An alternative explanation that should be considered is that the child has been sexually abused (Section 1.3.3). If understimulation is suspected, enquiry into the amount the child is left alone to its own devices should clarify the issue and, if deprivation is present, the implications for management are obvious. In infants where there is no evidence for deprivation, a behavioural approach to reduce the frequency of the habit is indicated (Section 6.2.2). In older preadolescent children, excessive masturbation is very rare, and is usually accompanied by other signs of emotional disturbance. The possibility of sexual abuse should be considered. Assessment should explore wider aspects of the functioning of the child and family and, if sexual abuse is excluded, psychotherapeutic treatment for the child and family is likely to be indicated.

**Problems involving masturbation** in adolescence are of two main kinds. In some mentally handicapped teenagers, the habit is practised in public, often to the embarrassment and distress of parents. The teenager should be told to go to his own room when he starts to masturbate, and the parents can be encouraged to be very firm about this. Usually, but not always, this approach resolves the problem. Some teenagers of normal intelligence experience severe guilt feelings in relation to masturbation. Reassurance about the universality of the habit does *not* usually resolve the problem, as the teenager has usually been reassured by others many times before. The guilt may be associated with obsessional preoccupations about pollution, and accompanied by depressive feelings. Alternatively, the teenager may be in a state of acute concern about religious issues, the nature of existence, etc., and persistent, distressing masturbatory guilt may be linked to worries of a much wider nature. In either event, if the youngster is motivated to attend, assessment of the psychiatric state followed by the opportunity for a series of focused psychotherapeutic interviews may be helpful.

### Homosexuality in boys

The frequency with which preadolescent and adolescent boys are involved in occasional or regular homosexual activity (usually mutual masturbation) is unknown. It is, however, fairly common and surveys suggest it probably occurs in about a fifth of the male population. It is more common in boys at single-sex boarding schools and opportunity, or lack of it, for heterosexual activity is likely to play a considerable part. In that only about 3 per cent of the adult male population is homosexual, it is clear that only a small proportion of boys involved in homosexuality in childhood and early adolescence go on to show adult homosexuality.

### *Aetiology*

Chromosomal and endocrine studies have not identified biological causes of homosexuality. Concordance for homosexuality is higher in monozygotic than dizygotic twins, but the reasons for this are uncertain. Studies of children with intersex conditions strongly suggest that both gender-identity and sex-role activity are largely determined by child-rearing practices—the way in which the child is brought up.

### *Clinical features and management*

Although homosexual behaviour is not at all uncommon in adolescence, it is an unusual presenting problem in both psychiatric and paediatric practice. More commonly, depression in adolescence is accompanied by fears of homosexuality that may or may not be realistic. Alternatively, parents may have justified or unjustified fears that their son or daughter is showing homosexual behaviour. In both these circumstances, the main task is to help the teenager or parents express their thoughts and feelings about the matter, and to provide relevant information. A non-judgemental approach with communication about the frequency of the problem and the degree to which it will realistically interfere with the individual's life will often be found supportive and helpful.

Parents who are concerned that their prepubertal sons are involved in homosexual activity can usually be reassured about the significance of the behaviour for later sexuality. Mid- and late-adolescent boys who show preference for sexual activity with other boys or men, when opportunity for sexual activity with girls is available, are more likely to be homosexual later. In earlier childhood there is probably little significance in homosexual or bisexual activity, unless it is accompanied by effeminacy (see above).

## Homosexual activity in girls

Homosexual activity in girls is less common than it is in boys, probably occurring in about 10 per cent of preadolescent girls in our society. Again, as homosexuality is only present in 1–2 per cent of the adult female population, most girls involved in homosexual activity must grow up to be heterosexual or bisexual adults. A homosexual outcome may be more common in girls who show marked tomboyish tendencies as well as in those whose homosexual activity persists into mid- and late adolescence, but the evidence is lacking.

### 2.13.4 PREGNANCY IN SCHOOLGIRLS

The rate of pregnancy in girls under the age of 15 years increased sharply in the UK during the late 1960s and early 1970s, but subsequently levelled off in the 1980s at about 9 per 1000 per year in girls aged 13–15 years (Office of Population Censuses and Surveys 1986). Official statistics suggest that about two-thirds of such pregnancies are therapeutically aborted (Office of Population Censuses and Surveys 1983).

### Background information

Most pregnancies in young teenagers occur in girls who have had relationships with the fathers for several months or even years, and are not the result of a brief promiscuous relationship. A significant number of such pregnancies are not seen as unwelcome by the girls themselves. A common problem is that the couple are in dispute with their parents about their relationship, and one or both may feel rejected. They are likely to be relatively ignorant about birth control methods. In a significant minority of cases, the pregnancy appears to be an unconscious attempt to achieve status and love in a girl with low self-esteem who feels unwanted by her parents.

There are strong biological and social disadvantages to such early pregnancy (J. K. Russell 1974). Pregnancy complications, such as toxaemia, and birth complications are more frequent. The girl's schooling is usually interrupted and may well terminate at this point. Her social relationships are often disrupted, both within the family and with her friends. The psychological aftermath of a termination may be very distressing. If the pregnancy is continued and the child adopted, this too is very upsetting, while if the girl continues to look after her baby, it will often receive inadequate care. It should be emphasized that these aspects of outcome are distinctly less prominent in girls who become pregnant in their late teens, though even here some social disadvantage to mother and child is likely to occur.

### Management

The management of the physical aspects of the pregnancy is of special importance because of the increased risk of complications. This needs to be accompanied by counselling of the girl, her family, and the father of the child. Both individual and family sessions can be helpful in reducing the degree of family disruption and emotional rejection of the girl. A decision concerning abortion should be taken early, with the girl's wishes being of the greatest importance. There is an increasing tendency for girls who continue with pregnancy to be able to remain in school but, if this is not possible, special arrangements for continuing education should be made. Further unwanted pregnancies are less likely to occur if practical and emotional support for the girl is provided after the abortion or delivery (Osofsky *et al.* 1973).

## 2.14 Personality development

### 2.14.1 NORMAL PERSONALITY DEVELOPMENT

Like adults, children show wide variation in their styles of behaviour. We can all recognize, for example, that, while some children are cheerful and resilient, others have an enduring tendency to show anxiety or to dissolve into tears when facing stress. Different terms are used to describe such variations in everyday behaviour. The term 'personality' is used in a non-specific way to describe the totality of predisposing behavioural tendencies. 'Temperament' is used when there is an implication of genetic causation for such tendencies. 'Traits' refer to relatively circumscribed aspects of personality.

There are numerous theories of personality, each of which puts emphasis on different aspects of the way in which children develop and manifest their behavioural predispositions. Some of these theories are discussed elsewhere in the book, in relation, for example, to social and emotional development. Such theories vary especially in the degree to which personality characteristics are seen to develop as a result of interaction with the environment, to what degree they are innately determined, and to what extent unconscious processes are relevant. One fruitful approach in this field has been the study of children's temperamental characteristics. Defining temperament in terms of the style of everyday behaviour, a group of New York workers carried out studies based on maternal reports of child behaviour which suggested that children show continuities of behavioural style that may be partly constitutionally determined. In this New York Longitudinal Study (Thomas and Chess 1977), nine categories of temperament were delineated—activity level, rhythmicity, approach or withdrawal, adaptability, threshold of responsiveness, intensity of reaction, quality of mood, distractability, and attention span and persistence. In summary, children could be classified as 'easy', 'difficult', or 'slow to warm up' in their characteristics. 'Difficult' or active children, intense in their emotional reactions and irregular in their feeding and sleeping patterns, are predisposed to develop behavioural or conduct problems (P. Graham *et al.* 1973). 'Slow to warm up' children are more likely to develop withdrawal or anxious reactions. The 'easy' group are relatively protected against the development of behavioural or emotional difficulties. There is some evidence that these temperamental characteristics are partly genetically determined, but the influence of parental attitudes and child-rearing practices is probably also considerable.

More recently, Chess and Thomas (1990) have put emphasis on the importance of the 'goodness of fit' between temperament and environment. 'Goodness of fit' results when environmental expectations, demands, and opportunities are consonant with the individual's temperament and 'poorness of fit' when such expectations are excessive.

Another approach to the classification of temperament has used improved methods, and come up with somewhat different temperamental categories (Buss and Plomin 1984). Infants and young children can be categorized on dimensions of emotionality, activity, and sociability that are relatively stable over time and independent of each other. It has been pointed out (P. Graham and Stevenson 1987) that these categories relate fairly closely to the major child psychiatric diagnostic groups—emotional disorder, the hyperkinetic syndrome, and conduct disorder—and that it would not be surprising if children showing specific temperamental characteristics were at risk for specific disorders.

## 2.14.2 PERSONALITY DISORDERS

Both major classification schemes express some reservations about using the concept of 'personality disorder' in children. Thus the glossary to ICD-10 (World Health Organization 1990) states that it is unlikely 'that the diagnosis of personality disorder will be appropriate before the age of sixteen or seventeen years'. DSM III-R (American Psychiatric Association 1987) is less emphatic and, though it cautions against using the term 'antisocial personality disorder' in children who have conduct disorders, suggests that other personality disorder categories (e.g. paranoid, schizoid, schizotypal, borderline, histrionic, narcissistic, avoidant, dependent, obsessive–compulsive, and passive–aggressive) may be used 'in those unusual circumstances in which the particular maladaptive personality traits appear to be stable'.

In any event, because the diagnosis of personality disorder implies a chronic disability, clinicians are reluctant to apply what might be regarded as a damning label to a child or adolescent (Wolff 1984).

Although, therefore, some children do show early signs of enduring personality characteristics—obsessional, hysterical, antisocial, etc., it is not generally regarded as desirable to refer to such children as showing personality disorders. More commonly, they are described as showing unusual patterns of personality development without a firm diagnostic statement being made.

# 3

# Adult-type psychiatric disorders

The major psychiatric classification systems distinguish between (1) behaviour and emotional disorders with an onset usually occurring in childhood or adolescence; (2) developmental disorders; and (3) forms of disorder usually occurring for the first time in adulthood. These last may occur in childhood and indeed, as knowledge has advanced, it has become clear that many of them do begin earlier than had been thought to be the case. Further, even though these adult-type disorders may not begin in full-blown form in childhood, recent studies have demonstrated that the adults suffering them have shown specific behavioural traits in childhood. The distinction between childhood-onset disorders and adult-onset disorders is therefore more blurred than the chapter headings of this book might suggest.

## 3.1 Psychoses

### 3.1.1 INTRODUCTION

**Definition**

These are conditions in which there are major abnormalities of thinking (shown by the expression of incoherent and illogical thoughts), beliefs, and perception (shown by delusions and hallucinations). Such abnormalities are usually accompanied by gross behavioural changes.

In psychotic disorders there is a loss of contact with reality and a lack of insight on the part of the patient that there is anything wrong, whereas patients suffering from neurotic disorders, though they may be severely handicapped, are in touch with reality and realize they have problems. The distinction has been encapsulated in the aphorism 'neurotics build castles in Spain whereas psychotics live in them'. The distinction is not clear-cut, and is less frequently made than used to be the case, but it still has value, especially in the older child and adolescent.

Psychotic disorders appear less commonly in infancy and childhood than in adolescence and adulthood. This is probably because their actual prevalence is lower, but diagnostic difficulties may also be responsible. Especially in very young children, the identification of abnormalities of thinking and perception presents major problems.

**Classification**

Psychotic disorders can be divided into:

(1) schizophrenia;

(2) bipolar effective disorder;
(3) organic psychotic states;
    (a) unaccompanied by intellectual deterioration/dementia;
    (b) accompanied by intellectual deterioration/dementia;
(4) atypical psychoses.

Until recent times, childhood autism was regarded as a form of childhood psychosis, but it is now realized that this condition has more in common with disorders and delays of development such as developmental aphasia and mental retardation.

### Aetiology

Apart from organic psychotic states, the underlying pathophysiological bases of psychoses are not known. Information is, however, available on both genetic and environmental risk factors. The lack of a known organic causation for schizophrenia and manic-depressive psychosis has led to their being labelled as 'functional', i.e. caused by disturbance of function rather than physical structure. This term is not very helpful, and is likely to be less used as the aetiology of these conditions becomes clearer.

### 3.1.2 SCHIZOPHRENIA

### Definition

A condition of unknown aetiology but with both genetic and environmental risk factors. There are characteristic disorders of thought, perception, and posture.

### Clinical features

The condition most commonly presents for the first time in adolescence or early adulthood. It may, however, present before puberty, and, in one series of children, the mean age of onset was 7 years (A. T. Russell *et al.* 1989).

Characteristic features are:

1. *Thought disorder*. The child is likely to have difficulty expressing his thoughts, and be incoherent and apparently illogical. 'Loosening' of thought associations may result in the child moving from one word or idea to another in an apparently random manner.

2. *Delusions*. These are false beliefs impervious to reason. In older children they usually take a paranoid form, i.e. the child believes those around him are hostile and threatening. Primary delusions, which are characteristic of schizophrenia, are those that arise directly out of normal experience and are not secondary to hallucinations.

3. *Hallucinations*. These are false perceptions without sensory stimulation. In schizophrenia they are usually of auditory type—voices outside the

child's control, talking to him directly or referring to him in the third person.

4. *Disorders of mobility*. Catatonic behaviour in which the child takes up abnormal postures or enters into an unresponsive state or 'stupor' may occur.

In addition to these central features of schizophrenia, other less specific symptoms may also occur. These include disturbances of mood, acute excitement, depression, and anxiety, as well as mannerisms, stereotypies, and inappropriate social behaviour.

The condition may present in an acute or insidious manner. In the latter case, it may be many months before the child is realized to have a serious mental disorder, being regarded in the meantime either as lazy or as showing a less serious psychological problem.

Prior to onset the child may have been normal, or, more commonly, have shown some degree of developmental delay (Zeitlin 1986). Early language delay is common, and other developmental abnormalities such as clumsiness may have been present (Watkins *et al.* 1988). Other neurodevelopmental anomalies such as muscular hypotonia have been described. The child may have had difficulties with information processing, deficits of attention, signs of minor neurological abnormalities, autonomic hyperresponsiveness, social difficulties, and emotional lability (Neuchterlein 1986).

Following an acute phase, the child may return to normal or enter into a 'negative state' in which active symptoms such as delusions and hallucinations are absent, but the child is apathetic and lacking in motivation. Such a negative state may be followed by another acute attack, and recurring cycles of this nature may persist for years or throughout the individual's life.

### Aetiology and background features
The exact nature of the aetiology of schizophrenia is unknown.

*Genetic factors*
There is strong evidence that genetic factors are involved, as:

1. First-degree relatives have a substantially increased risk. The child of a parent or sib with schizophrenia has about a 12 per cent risk compared with 1 per cent in the general population. There is also increased risk in relatives of schizotypal disorder—a personality disorder with many of the features of schizophrenia, for example remoteness, inability to make friends, and oddities of speech, but without adequate evidence to make the full diagnosis.

2. Monozygotic twins show distinctly higher concordance rates than dizygotic. The exact difference in rates is uncertain, but that there is a difference seems clear-cut.

3. Fostered and adopted children of parents with schizophrenia have a similar risk of developing the condition when compared with those brought up by their schizophrenic parents (Heston 1966).

4. Children with schizophrenia have an increased rate of developmental abnormalities in their early lives. Indeed, some now regard schizophrenia as a developmental disorder (Murray and Lewis 1987). Considerable efforts have been made to identify neuroanatomical, neurophysiological, and biochemical abnormalities in schizophrenia. Ventricular size appears increased. The fact that symptoms of schizophrenia are specifically improved by drugs that block dopamine receptors suggests that abnormalities of these receptors may play an important part in aetiology.

*Other organic factors*
Many other organic factors may, in an individual case, play an important part. Perinatal trauma may have predisposed the child to develop a psychosis. Other trauma, encephalitic infections, neoplasms, and irradiation may also precipitate an apparently typical attack of schizophrenia with 'organic' features (see below).

*Environmental factors*
Various characteristic patterns of family interaction have been described as having aetiological significance. The 'double-bind' hypothesis suggested that schizophrenia may be precipitated by communication of inconsistent messages from parent to child. It is uncertain whether such communication is:

(1) secondary to the child's condition;
(2) a result of the parents showing a minor form of the child's condition; or
(3) of genuine aetiological significance.

In any case, the theory now appears discredited (Leff 1978). It does seem clear that if a teenager with schizophrenia is in contact with parents with high emotional involvement, this is likely to precipitate relapse. Such emotional involvement may sometimes precipitate the onset of the schizophrenic illness in the first place.

In older patients, the role of other non-specific 'life events' in precipitating relapse has also been demonstrated (Birley and Brown 1970), and these are probably also of importance in children and adolescents predisposed to the condition.

## Prevalence

Schizophrenia occurs in 0.5–1 per cent of the adult population, but is much rarer in the prepubertal period. The rate in late adolescence is about 3 per 10 000. In this younger age group, boys are more commonly affected than girls. There is a link with low social class, and this may be because the effects of the illness reduce social competence, and result in downward social drift.

## Assessment

In most cases of schizophrenia, diagnosis presents little difficulty. In young children, however, establishing the presence of thought disorder, delusions, and hallucinations may be difficult. The fact that a child mutters to himself and appears preoccupied should not, for example, be regarded as evidence that he is hallucinated, though certainly such behaviour should arouse this suspicion. Further, the presence of hallucinations need by no means always indicate a diagnosis of schizophrenia. Instead, other psychotic or neurotic disorders (M. E. Garralda 1984) or a drug-induced state may be responsible. It is important to exclude organic disease. The presence of disorientation or clouding of consciousness (again sometimes difficult to detect in young children) should raise the possibility of an organic condition. The presence of epilepsy should be carefully considered. Epileptic attacks may be followed by confusional states with psychotic features. Further, children and adolescents with long-standing temporal lobe epilepsy are at special risk for developing schizophrenia. A full neurological examination should be carried out, and positive findings will require further investigation.

In children who have abnormalities from the preschool period, childhood autism should be considered, though very occasionally autism may be followed by a schizophrenic state (Petty *et al.* 1984). In affective psychosis (manic-depressive disorder), delusions and hallucinations may also occur, but their content will be different (see below). Mixed pictures (schizoaffective states) with features of both types of psychosis, have been described in adulthood.

## Treatment

In general, this should follow lines used to treat schizophrenia in the adult patient, but the educational needs of the child and his family situations are likely to require special consideration.

### The acute phase

Psychotropic medication such as chlorpromazine or haloperidol is indicated (Section 6.2.6). The child or adolescent is likely to require admission to an in-patient unit. The family members require support to understand the nature of the illness and its likely outcome.

### Chronic phase

If the child or adolescent enters a state of low motivation with a continuing predisposition to future acute attacks, continuing close supervision is required. Maintenance phenothiazines, given orally or by depot injection will reduce the risk of relapse. If tardive dyskinesia, a serious complication of long-term phenothiazine therapy, is to be avoided, periodic withdrawal of the drugs every few months to test the continuing need for them is necessary. The use of electroconvulsive therapy should be limited to cases where catatonic stupor is a prominent feature. In these cases, excessively rare in childhood and adolescence, it is usually highly effective.

Counselling of family members to reduce emotional overinvolvement should be combined with supportive psychotherapy for the child. In children of school age, special educational facilities are likely to be required, possibly in a special day or residential setting. Genetic counselling should be made available, if appropriate, to the patient himself, as well as to the parents or sibs.

## Outcome

This will depend on the nature of the illness. The insidiously developing form with previous developmental and personality abnormalities has a poor outcome, with the risk of subsequent prolonged dependency. On the other hand, a single, acute attack in a previously normal child or adolescent may have no significant sequelae. The distinction between affective and schizophrenic psychosis in adolescence is not easy, and a teenager apparently suffering from schizophrenia may have later episodes more characteristic of manic-depressive psychosis (Zeitlin 1986).

### 3.1.3 AFFECTIVE DISORDERS

#### Classification

The psychiatric classification of disturbances of mood or affective disorders is complex, but two main groupings are described in DSM III-R (American Psychiatric Association 1987). The ICD-10 (World Health Organization 1990) classification is in most respects similar.

1. *Unipolar or bipolar disorders.* In these there are reasonably clear-cut episodes of hypomania and/or severe depression often with psychotic features.
2. *Depressive disorders.* These consist particularly of major depression and dysthymia, and are described elsewhere (see Section 2.9.5).

The distinction between uni- and bipolar disorders and depressive disorders is blurred, but they are described separately here because, unlike depressive disorders, bipolar disorders cannot easily be seen as arising as extremes of normal development. Consequently, in this book, while depressive disorders are described in Section 2.9.5, bipolar disorders are described as one of a number of adult-type psychiatric disorders.

In unipolar disorders there are single or recurrent episodes of either hypomania or depressive disorders (but not both). Children with bipolar disorders have recurrent episodes of both depression and hypomania.

## Clinical features

In **depressive episodes**, there is marked lowering of mood and loss of energy. Mood may be lowest in the morning. Appetite and weight may be lost. Sleep disturbance is usually absent in this age group, but may occur. There is often retardation of movement and speech, though occasionally agitation and pressure of talk are present. The teenager may express thoughts of unworthiness, guilt, hopelessness about the future, and desire for death. Suicidal behaviour may occur. Delusions and hallucinations are usually absent and, if present, their content appears to arise from the youngster's mood. Thus he may hear voices accusing him of crimes he imagines he has committed, or believe that his body is rotting.

In **hypomanic episodes**, there is elevation of mood with excitement and pressure of talk. Irritability is a prominent feature. The teenager will be unduly energetic and require less sleep than usual. Disinhibited behaviour may lead to financial extravagance or sexual misdemeanours. The youngster may have inflated ideas of his own capacities that may reach a delusional level. Hallucinations may occur, but are uncommon.

## Prevalence and background factors

These disorders are rare before puberty, but increase in frequency of presentation during adolescence. The condition occurs with equal frequency in boys and girls, though, at the upper end of this age group, females predominate.

### Genetic factors

Evidence in favour of a significant genetic contribution arises mainly from studies of older patients. About 12 per cent of first-degree relatives of patients with this condition suffer major affective disorders themselves, compared to 1–2 per cent in the general population. Studies of twins and adopted children support the genetic hypothesis, though no clear-cut pattern of inheritance has been determined.

### Other organic factors

Major depressive disorder sometimes follows head injury, viral infections such as influenza or infectious mononucleosis, and cerebral disease. Various metabolic abnormalities have been identified in this condition, particularly affecting monoamine neurotransmitters (especially 5-hydroxytryptamine), and cortisol metabolism, but no clear-cut physiological basis has emerged.

### Environmental factors

In older sufferers from this condition it has been demonstrated that predisposing or vulnerability factors include early parental loss and separations.

Precipitating factors include life events that could be construed as posing a threat to the well-being of the patient. The study of life events in childhood and adolescence is still sketchy, but it seems plausible to suggest that environmental factors operate similarly in this age group, producing low self-esteem, a sense of helplessness, and an inability to think other than pessimistically.

## Assessment

The severely depressed child or adolescent requires a full assessment with particular attention given to the family history and possible precipitating events. In assessing the mental state, it is important to consider carefully the likelihood of suicidal behaviour. Organic disease should be excluded by physical examination and, if necessary, further investigation. Differentiation from depressive or dysthymic disorder is often difficult but, in any case, the practical implications of making the distinction are often not clear. If psychotic symptoms do not appear to arise directly from the patient's mood, a diagnosis of schizophrenia should be considered.

## Treatment

### Acute depressive phase

1. *Counselling or supportive psychotherapy.* Although it may be difficult to conduct interviews with a very depressed and mute, or unduly excited, overactive teenager, every effort should be made to form a relationship with him or her. Relatively brief but frequent sessions for this purpose are indicated. This will allow an opportunity for examining the onset of the illness, with a view to preventing recurrences. The development of a trusting relationship also probably reduces the risk of suicide. The family as well as the teenager will benefit from a counselling approach.

2. *Securing the safety of the patient.* Youngsters with this condition may attempt to harm themselves, and admission to a psychiatric unit may be necessary for this reason (see Section 2.9.6 for assessment of suicidal risk).

3. *Medication.* In depressive disorders of this severity, antidepressant medication should be prescribed. In the retarded depressive child or teenager, imipramine in dosage beginning at 2 mg/kg per day up to 5 mg/kg per day in three divided doses can be given. At higher doses, routine electrocardiography should be performed weekly to detect evidence of heart block. In agitated, depressed patients, similar doses of amitriptyline should be prescribed.

4. *Electroconvulsive therapy* (ECT). If, after a period of about 4 weeks, antidepressant medication has had no effect, and the patient remains severely depressed, consideration should be given to the use of ECT (Section 6.2.7.3). The presence of depressive stupor and severe motor or verbal retardation are the main positive indications. In general, ECT seems less

effective in the younger age group than in older patients, but the indications for its use are similar (Bertagnoli and Borchardt 1990).

*Acute hypomanic phase*
The principles of counselling and securing the safety of the patient and others around him will again form a major component of the treatment plan.

Haloperidol in dosage up to 0.05 mg/kg body weight/day in three divided doses is likely to be effective. This may be combined with an antiparkinsonian agent such as orphenadrine 50 mg t.d.s.

*Chronic phase*
In teenagers subject to recurrent attacks of acute disorder, lithium carbonate is usually, though not always, an effective prophylactic (Section 6.2.6). The existence of a relative who has responded to this medication is a useful indication for its use (Delong and Aldersdorf 1987).

## Outcome

Recovery from the acute phase of the condition nearly always occurs in a few weeks. An acute onset with a precipitating factor and a complete return to normality usually indicates relapse will not occur. Absence of a precipitating factor, family history of recurrent illness, and a failure to return to normality indicate likely relapses. If psychiatric disorder recurs in adult life, depressive features are again likely to be prominent. A severe disorder of apparently affective type in childhood or adolescence may, however, recur later with more obviously schizophrenic features.

### 3.1.4 ORGANIC PSYCHOTIC STATES

## Delirium

Delirious states are not uncommon in young children with acute infections. Clouding of consciousness with diminished response to external stimuli, misperceptions, visual hallucinations, and disorientation can also occur after the ingestion of excess quantities of certain drugs such as ephedrine, atropine, and tricyclic antidepressants, as well as non-prescribed drugs such as solvents and amphetamines. Delirious states also, although rarely, occur as part of epileptic attacks in the ictal or postictal phase.

## Organic psychoses without dementia/intellectual deterioration

Non-progressive cerebral conditions are occasionally associated with psychotic reactions. These are usually of acute or subacute schizophrenic type (see above). Underlying conditions include cerebral infections, especially in encephalitic conditions, withdrawal states from certain drugs (especially alcohol and amphetamines) and epileptic attacks.

Suspicion of an organic cause for an acute schizophrenic illness should be raised if there is a history of epilepsy or drug ingestion, if the psychotic symptoms are accompanied by disorientation in time and space, and if hallucinations are of visual type.

## Dementia

This involves irreversible deterioration of intellectual skills, and is often accompanied by change in behaviour or personality.

### Clinical features

In infancy and childhood this is shown first by a loss of recently acquired skills. A young child may stop speaking and develop incontinence of urine and faeces. Autistic features may also occur, but the condition is differentiated from autism by the fact that there is continuing deterioration, sometimes leading to death. In older children, handwriting may deteriorate, and this may be followed by the child being unable to read at a level previously attained. There may also be a psychotic reaction, most commonly of schizophrenic type. Non-psychotic behavioural changes such as undue fearfulness or aggressive behaviour may precede the psychotic symptoms and intellectual deterioration (Corbett *et al.* 1977).

These psychotic features are usually accompanied by other evidence of organic disease. Such evidence depends on the nature of the condition, but is likely to include especially epileptic fits and involuntary movements. Sometimes, however, these obvious signs of organic disease are absent. In these circumstances, it is quite common for the condition to be treated as purely psychiatric, the evidence for intellectual deterioration having been overlooked (Rivinus *et al.* 1975). This situation may persist for several years.

### Aetiology

There are a large number of causes of dementia in childhood, all of which are rare. For a full description of these progressive conditions see Brett (1983). They include:

(1) infections, especially subacute sclerosing panencephalitis;
(2) neurometabolic disorders, for example the gangliosidoses, leucodystrophies, and mucopolysaccharidoses;
(3) Huntingdon's chorea;
(4) chronic poisoning, for example with lead;
(5) neoplastic conditions (especially from leukaemic infiltration of the meninges and as a complication of treatment of leukaemia, methotrexate–radiation leucoencephalopathy);
(6) other conditions of uncertain aetiology such as Rett's syndrome (Section 2.8.1).

*Management*

**Assessment.** The physical investigation of these conditions is outside the scope of this book. Psychological assessment of children with dementia is often helpful to establish a baseline for intellectual functioning, to monitor progress, and to advise on educational approaches.

**Treatment.** Most of the conditions described above are progressive, with death occurring after several months or years. The terminal phase is often extremely distressing, partly because of its protracted nature. Many of the conditions have genetic implications, sometimes clear-cut, often rather uncertain. This means that anxiety not infrequently hangs over younger children in the family for a number of years, and decisions over further pregnancies are difficult.

Such stressful circumstances mean that it is particularly important for strong emotional support to be provided by the paediatric and primary health care team involved (Nunn 1986). Support and practical advice from a medical social worker, and the use of hospital or hospice care to provide respite for the family can produce considerable relief (see Section 4.7 for discussion of terminal care).

## 3.2 Neurotic, somatoform, and stress-related disorders

In these conditions, the child develops a maladaptive pattern of behaviour together with emotional disturbance. The pattern of behaviour may be very clearly related to stressful events or may appear to arise independently. More commonly the events that occur in the child's life are relevant to some degree, but the child's personality and early experience mould his response in a fashion characteristic to him.

Although in the major classification systems anxiety and depressive states are categorized as neurotic disorders, in this book they are dealt with in Section 2.9 as deviations of development. This is because, in childhood, maladaptive forms of behaviour and disturbances of emotional life seem more logically considered in terms of the child's development than as conditions arising *de novo*.

### 3.2.1 OBSESSIONAL DISORDERS

**Definition**

Normal children often show rituals in their early years. Particularly between the ages of 4 and 8 years, children may take great care to avoid the cracks while walking on pavements, hate to be parted from favourite toys, or line up their possessions in a very particular way. Such behaviour persists into adult life in the form of superstitious beliefs and practices.

When irrational thoughts of this type become handicapping, they form part of an obsessive–compulsive disorder.

Obsessions can be defined as recurrent and persistent ideas that are experienced as senseless and which the individual tries to suppress. Compulsions are repetitive forms of purposeful behaviour performed in response to an obsession and aimed in some way at warding off the obsessive idea or some dreaded event (American Psychiatric Association 1987).

## Clinical features

The onset may be slow and the disorder only gradually come to notice or there may be a fairly rapid onset (Rapoport 1986). Children as young as 6 years may be affected, but the condition is commoner in mid-childhood and adolescence. Characteristically the child is noted to be spending a longer and longer time in the bathroom, repeatedly washing to prevent contamination. Homework may be repeatedly checked and takes an inordinate length of time. Bedtime rituals may be prolonged for hours while the child adjusts and readjusts the bedclothes or objects around him. At mealtimes the child may only be prepared to use certain utensils. There may be odd behaviour—the child, for example, nearly always needing to go through a particular series of repeated movements with his feet before going through a particular door.

Obsessional thoughts are less common and may only be elicited on direct questioning. They are likely to take the form of unwelcome aggressive or depressive ideas (e.g. that the child's parents or he himself will die, or that he himself will harm someone) that the child cannot get out of his mind.

Attempts to change the child's behaviour by parents are met by resistance. There may be a transient improvement followed by a relapse. In severe cases the child will try, often successfully, to insist that family members alter *their* behaviour to fit in with his obsession. Manipulative behaviour may consist, for example, in getting the family to sit at table in a particular grouping.

The child is likely to be generally quiet and shy, as well as perfectionist. In some cases there is great variation in behaviour, the child being excessively neat and tidy in some respects, but slapdash in others. Children with tics, Tourette syndrome, and anorexia nervosa have high rates of obsessional symptoms (Rapoport 1986).

Associated anxiety, specific fears, and depression are relatively common. Such children do not usually show aggressive behaviour, but occasionally episodes when the child's obsessional behaviour is less prominent may be accompanied by increased rudeness and disobedience.

On examination the child may deny or attempt to minimize the extent of his ritualistic behaviour. He may also deny that he is in any way upset. Sometimes, however, there is considerable sadness and misery accompanying the child's feelings of helplessness in the face of his own behaviour. The child may be able to explain how his rituals arise—perhaps on the basis of superstitious thoughts. Unless he carries them out, he feels some terrible event will occur. Feelings of 'internal resistance' (i.e. sensations of an internal struggle against obsessional thoughts or compulsive behaviour) which are often described by adults are not commonly expressed by children.

## Prevalence

Relatively transient checking behaviour and bedtime rituals are common in childhood. Obsessional symptoms also commonly occur as minor features in emotional and depressive disorders. However, fully developed obsessional conditions are much less common. They have been described as occurring in 1–2 per cent of American high school students with a significant number describing their symptoms as having begun before puberty (Flament *et al.* 1988).

## Aetiology

**Family influences.** There is often a strong family history with about two-thirds of parents showing obsessional tendencies and about 5 per cent obsessional disorders. Twin studies suggest a genetic influence in obsessional personality traits. There is also, however, an environmental component, with parents of obsessional children having higher educational expectations (Clark and Bolton 1985), and generally high standards of conduct.

   **Mechanisms.** Obsessional symptoms have been seen as an unconscious defence against internal anxiety or as learned avoidance responses to anxiety-provoking situations. They can also be seen as exaggerations of adaptive behaviour, for repeated checking against danger is a form of behaviour with survival value. In fact, none of the various explanations put forward has a great deal of explanatory value, and the causation of the disorder is usually uncertain.

## Assessment

Particular attention should be given to the extent and severity of the obsessional symptomatology and the circumstances in which it occurs. Differentiation from normal forms of ritualistic behaviour needs to be made. The child's manipulation of other family members and responses of the family to the behaviour may be of great significance. Enquiry should be made as to the presence of associated symptoms of emotional disorder and aggression. The nature of obsessional thoughts should be probed to exclude the possibility of delusional beliefs. Obsessional symptoms may occur in emotional disorders and schizophrenia as well as in obsessional disorder.

## Treatment

This can usually be carried out on an out-patient basis, though where the condition is well entrenched and severely handicapping, and the responses of family members are perpetuating the problem, admission to an in-patient psychiatric unit may be indicated. Admission often results in an immediate rapid reduction of symptoms, but these may recur when the parents visit or the child goes home for a weekend.

**Counselling and support for the family** are therefore indicated, either in family interviews or with parents seen separately. Parents should be encouraged to resist manipulation by the child. In established, severe cases this approach is unlikely to be effective in itself.

**Behavioural approaches** have been reported to have considerable success in adolescents (Bolton *et al.* 1983). The most effective technique in 're-sponse prevention' (Section 6.2.2). Following a baseline period, the child is encouraged to reduce compulsive behaviour by a self-imposed graded regime. Parental cooperation in cutting down the behaviour is also required. External constraint by nursing staff on an in-patient basis may be necessary in severe cases that fail to respond to out-patient treatment.

**Medication.** The use of the tricyclic drug clomipramine has been recommended as of specific value in obsessional disorders (Section 6.2.6). There is evidence for its effectiveness in adolescence and it is worth using as an adjuvant to other therapy.

**Individual psychotherapy.** Insight-orientated individual psychotherapy has been reported to produce good results (Hollingworth *et al.* 1980). Probably neither individual nor family therapy alone is as effective as behavioural approaches combined with counselling and medication, but they can be useful adjuvants.

## Outcome

Reports of treated series of cases suggest that the condition is often persist-ent over time. In one series of children followed for 2–7 years, three-quarters persisted, and those treated with clomipramine did no better than those untreated with this medication (Flament *et al.* 1990). Many severe and intractable cases of adult obsessional disorder have begun in adoles-cence, so clearly there are a significant number of early-onset cases that have a chronic course.

### 3.2.2 SOMATOFORM DISORDERS

## Definition

Psychogenic physical or 'somatoform' disorders are of two main types. In the first, of which hypochondriasis is the most common form, the child perceives a physical disorder to be present despite the absence of any evidence for illness and despite medical reassurance. This is therefore a specific disorder of perception or belief.

In the second type (conversion disorder), physical symptoms such as paralysis or mental symptoms usually signifying physical disease (such as loss of memory) are present in the absence of evidence of such physical disease. These two types of abnormality often coexist. The hypochondriacal child may also complain of physical symptoms such as pain or tiredness which do not have a physical cause. The child with a 'conversion' syndrome may be preoccupied with his disability to a hypochondriacal degree. However, the two may also occur independently, perhaps most classically in the child with a conversion disorder who shows remarkable blandness (*belle indifference*) towards his symptoms.

Hypochondriasis and other forms of disturbed self-perception of illness such as somatization disorder (in which the patient complains of multiple physical symptoms over many years), are rare in childhood, although they occur more frequently in adolescence. In contrast, conversion disorders are somewhat more common and will be discussed in greater detail.

Conversion disorders (which used to be called 'hysterical' disorders) are classified in ICD-10 (World Health Organization 1990) as a form of 'dissociative disorder', and in DSM III-R (American Psychiatric Association 1987) as a somatoform disorder. 'Conversion' and 'dissociation' are terms used to describe unconscious psychological mechanisms. 'Conversion' is said to occur when an emotional conflict is transformed into a physical disability. This may occur in a way that has obvious symbolic significance, for example weakness of the hand used for hitting in a conflict over aggression, or it may merely allow the child to avoid an unwanted situation (secondary gain). Dissociation occurs when one part of mental life is split off from another. In childhood these concepts are often difficult to apply because, not infrequently, conversion symptoms are present, yet convincing evidence for these psychological mechanisms is lacking.

Conversion disorder can usually be seen as a type of 'abnormal illness behaviour' occurring when a child develops a need to be a patient when he is not physically ill even when he has to suffer inconvenience and disability (Mechanic 1978). Being ill can be seen as a pattern of behaviour. A physically ill child behaves appropriately when he goes to the doctor, takes time off school, acknowledges sympathy, etc. The child can be seen to be enjoying the benefits of the 'sick role' (Parsons 1951). Many clinicians now find this framework of 'illness behaviour' the most helpful, but in a minority of cases, it still seems necessary to invoke unconscious mental processes as important in causation.

## Prevalence

Conversion disorders are unusual throughout childhood, and very rare indeed before the age of 7 or 8 years. Their precise prevalence is uncertain because of problems of definition, but is probably less than 1 per 1000 children. The rate increases in adolescence. The sex ratio in childhood is equal, but, by adolescence, more girls than boys suffer from the condition. There is no particular relationship with social class.

### Clinical features

Children or teenagers may suffer from 'active' symptoms such as pain in a limb, seizures, or other abnormal movements. Alternatively, they may show a deficit of function. Types of deficit include aphasia, aphonia, visual or hearing defects, and unilateral or bilateral limb weakness and stupor. The symptoms may have come on after a particular stressful situation or when the child is about to face a stressful event. It may also arise during or following a minor physical illness, such as a viral infection. The child may show other signs of emotional or behaviour disorder, especially depressive features, but more commonly this is not the case.

The family background may show obvious signs of disturbance, but this is not invariable. There is likely to be undue parental concern over physical complaints, and, if the hysterical disability is chronic, the parents may well be attached to the notion that there is an underlying physical disorder.

### *Related conditions*

**Tension headache and abdominal pain.** Headache and stomach-ache are common non-organic symptoms and are discussed elsewhere (Sections 5.7.4 and 5.13.2). In a minority of cases, the function of these symptoms (especially the benefits of the sick role) may mean that it is helpful to manage them along the same lines as hysterical reactions.

**Physical disease accompanied by excessive disability.** Some children are more disabled by a physical condition than one would expect on physical grounds alone. Occasionally a conversion or 'hysterical' reaction appears to follow on a genuine physical illness, symptoms persisting when the organic cause is no longer present.

**Symptoms referable to various psychophysical interactions.** Asthmatic attacks, diarrhoea related to anxiety, and palpitations are all examples of symptoms in which psychophysical mechanisms, sometimes related to autonomic nervous system activation, are implicated. It is usually not appropriate to regard these as hysterical in nature.

**Epidemic 'hysteria'.** A number of reports have been published of groups of children or teenagers (especially girls), who have, *en masse*, succumbed to symptoms such as faints, dizziness, and headaches. Characteristically the episode begins with symptoms such as these, thought to be due to some infectious disease or toxic agent, often occurring in a girl of high status in her group. Other girls, especially the more anxious in the group, then experience similar symptomatology. Isolation of the affected girls, and strong suggestion to the remainder that the epidemic has now passed its peak and will gradually subside, is usually sufficient to bring it to an end. Intensive investigation is likely to prolong the problem and is unnecessary, as the true nature of the problem is usually clear after the first two or three girls have been examined.

## Assessment and management

The history, together with observation of the child and physical examination, may reveal obvious inconsistencies in the child's disability, making it highly improbable or impossible that the child is suffering from a physically determined condition. It may be possible to demonstrate muscular power in some situations but not in others. Children with conversion blindness may show inconsistent performance on charting their visual fields. The pattern of signs elicited will not correspond to any known neurological lesion.

If a psychiatric diagnosis is obvious, then unnecessary further investigations should not be embarked upon, as these will only prolong the disorder. On the other hand, where real doubt exists, physical investigations should be thorough, as a significant number of childhood conditions diagnosed as hysterical ultimately turn out to have organic pathology (Caplan 1970). Conditions most commonly missed are epilepsy, maculoretinal cerebral degeneration, spinal tumours and abscesses, and (in the case of postural symptoms) dystonia musculorum deformans. Epilepsy is not infrequently missed as many children with pseudoseizures also suffer with epileptic fits (Section 7.7.2).

Further discussion may reveal the meaning and purpose of the physical symptoms. They may allow the child to avoid a painful situation by adopting the role of a sick person. Symptoms may have been inappropriately rewarded with undue attention—not only by parents and the family. Doctors who find the child's symptoms puzzling and challenging may also reward the child for his abnormal behaviour with the attention they provide. The symptoms may be necessary for family adaptation, allowing the parents to focus on their child's symptoms, and thus avoid talking about their marital problems. They may also reflect deep-seated conflicts of which the child is unaware. Finally, the symptoms may be a sign of another form of mental disorder, such as severe depression or schizophrenia.

## Treatment

Once the diagnosis of conversion syndrome has been made, the initial discussion with the parents and child needs careful planning (Schulman 1988). The clinician should first be sure he fully understands how the parents and child see the problem. Do they acknowledge the possibility of a psychological cause for the disability? If they do, the task of the clinician will be that much easier. Having established existing family attitudes, it should be explained to the family that there is no evidence for any current physical cause for the condition. The physical investigations that have been carried out should be gone over carefully with the family and the reasons for not making a physical diagnosis fully explained. Nevertheless, there is obviously something very wrong. It is important to explain the genuineness of the symptoms—the child is definitely not 'putting it on'. It can be explained too that, in some children, upsets of one sort or another sometimes produce physical symptoms, and this provides an opportunity for discussion of the role of stress in the child's life. This may or may not be apparent.

The principles of treatment, described below, can then be explained to the parents. In cases where parents are 'stuck' with the notion that the child's symptoms must be physically determined, then it may be best for the clinician to accept the parents' view that this may be the case. However, the clinician needs to go on to suggest that, as no physical diagnosis has been made and therefore no physical treatments can be applied, unless the child is to remain crippled, some form of 'rehabilitation' needs to be applied.

It is wise to give a good prognosis early on, and to suggest strongly that recovery will occur over a few weeks or months. The child can then be told to expect to gain greater control over his symptoms. How this is to be achieved will become apparent as treatment proceeds. It is important to arrange follow-up appointments to monitor progress and ensure the child is not taken elsewhere for yet more investigations. If a child is under the care of a family doctor or paediatrician, and the symptoms are not improved with this form of management within a few weeks, then referral to a child or adolescent psychiatrist for transfer of care should be arranged. A joint interview with the family involving both the paediatrician and the psychiatrist is helpful as an initial step.

A range of treatments may be applied to children with conversion disorders, some of which will require admission to an in-patient psychiatric unit. Some children benefit considerably from physiotherapy, enabling them, for example, gradually to regain power in paralysed limbs without losing face (Dubowitz and Hersov 1976). Family interviews may be required to improve communication between family members or to help family members to become more aware of maladaptive functioning. The child may benefit from individual psychotherapeutic interviews. Suggestible children may improve with hypnosis.

The details of treatment will vary depending on the nature of the problem, but the principles remain the same. An attempt is made to understand why the child has developed such abnormal illness behaviour so that he can be helped to achieve a more adaptive solution to his problems (Goodyer and Taylor 1985). During the treatment, previously unsuspected stresses at home or school may be revealed, but quite often improvement occurs without obtaining full understanding why the problem should have developed in the first place.

## Outcome

Most children with conversion disorders improve symptomatically with suggestion and active treatment. The subsequent personality outcome of these children is unknown, however, and may not be so favourable. Good prognostic signs include an acute onset and the presence of parents prepared to consider psychological factors seriously.

### 3.2.3 POST-TRAUMATIC STRESS DISORDERS

## Definition

Children who are exposed to unexpected, overwhelming stress in the form of a natural disaster or a major trauma such as being raped or seeing their parents murdered or forcibly taken away may react in a characteristic way. In both major classificatory systems, this is categorized as a post-traumatic stress disorder, and in both systems the reactions children show are not considered separately from those of adults.

## Clinical features

The characteristic reaction involves (American Psychiatric Association 1987; Eth and Pynoos 1985) re-experience of the traumatic event as a recurrent, distressing recollection in the waking or dreaming state, sudden acting or feeling as if the event was recurring in a 'flashback', and intense psychological distress when events (such as anniversaries) occur that symbolize the original trauma. The child persistently avoids stimuli associated with the trauma, for example by making efforts to avoid recollecting them, experiencing a feeling of detachment from other people, and may regress in feeding or toileting habits. There may be symptoms of increased arousal, including sleep disturbance, irritability, difficulty in concentrating, and hyper-vigilance.

In childhood, this pattern of symptomatology is probably most likely to occur after rape or other forms of sexual abuse, but a similar picture has been described after different forms of disaster, including bush fires (McFarlane 1988) and civil war.

Vulnerable children, already predisposed to psychiatric disorder by family disharmony, school failure, or other stresses, are those most likely to be affected, but previously normal children can also show the characteristic picture.

## Management

In natural disasters affecting large numbers of people, this needs to be considered at a community level, as well as at a family and individual level. Indeed there is some evidence that the length of time it takes the child to adjust is closely related to the time it takes the community in which the child lives to reorganize (Galante and Foa 1986). The effectiveness of relief workers in promoting self-help activity among victims and keeping family members together is therefore a major factor in outcome. In addition, children seem to benefit if they are encouraged (but not forced) to recount their experiences individually, with a group of peers, or with other family members. Play material, drawing, and painting may be useful adjuncts if the child is being seen individually.

The outcome of post-traumatic stress disorder is very variable and depends, among other factors, on the personality of the child, the presence of other stresses, the availability of the opportunity to talk about the experience, the reactions of other family members and, if relevant, the speed with which the affected community is able to reintegrate. It is notable however that McFarlane (1988) found the effects of severe stress still present 26 months after the bush fire experience of the children he studied.

## 3.3 Drug and alcohol use and abuse

### 3.3.1 DRUG USAGE

Drugs play an important part in the group culture of children and adolescents. In particular their use:

1. Promotes the individual's identify with the group. A child who takes up smoking or glue-sniffing may feel he is drawn closer to a group where this is normal behaviour.
2. Marks an individual child's entry into the adult world. A teenager who goes into a pub for the first time has undergone a rite of passage.
3. Reduces group anxiety. Alcohol at parties removes social inhibitions and thus promotes enjoyment.
4. Results in enhanced, shared experience. Small groups of teenagers smoking cannabis achieve an extension of normal experience which they regard as valuable and enjoyable.

Unfortunately, all drugs, except when taken in small quantities, damage health and impair social functioning. When taken in large quantities they all can and sometimes do lead to marked social or physical impairment and occasionally to death. Measures taken to reduce their use, whether directed towards individuals or society, need, however, to take into account the positive social functions already mentioned, for otherwise such measures are likely to be ineffective.

## 3.3.2 DRUG ABUSE

### Factors influencing the excessive use of drugs

*Social factors*

1. *Cost and availability*. Use is strongly related to expense. There is good evidence that, as the relative cost of alcohol has gone down in relation to the cost of living, its use has increased. Drugs only available in large cities, such as cocaine and heroin, will not be used in rural areas.

2. *Pattern of use by adults*. The use of certain drugs, such as solvents, has grown up relatively independently of use by adults. The use in childhood of more socially acceptable drugs, such as tobacco and alcohol, is however heavily influenced by the degree they are used by adults.

3. *Media effects*. Portrayal in the media of cigarette smoking and alcohol consumption as glamorous has an influential effect, especially among the disadvantaged. Lyrics of popular songs not infrequently sound the attraction of drugs, particularly marijuana.

4. *Legal status of use*. Banning a drug legally may enhance its attractiveness to anti-authority teenagers, but, in general, illegality discourages use. The example of cannabis usage (see below) makes it clear that this is not a universal rule.

5. *Availability of other sources of gratification*. Drug use is likely to be higher among demoralized groups of unemployed teenagers living in poor social circumstances. Counterbalancing this effect is the fact that drugs are expensive and thus less readily available to such groups.

6. *Health education measures*. There is, so far, little evidence that traditional health education measures (lectures and group discussions on harmful effects of drugs, etc.) have any effect on drug use. More specific measures, for example social skills training (Section 6.2.2), may be more effective.

*Personal/individual factors*

1. *Sex*. Boys are heavier users of all drugs than girls.

2. *Genetic factors*. With some drugs, for example alcohol, there is evidence that genetic factors may predispose to addiction once a pattern of regular use is established (Bohman 1978).

3. *Domestic circumstances*. Poor housing, overcrowding, and disharmonious family relationships are likely to encourage teenagers to spend more time out of the house and thus increase contact with other drug users.

4. *Pattern of drug use in the family*. Youngsters who see their parents and sibs smoking, drinking alcohol, etc. are likely to take up similar habits. Teenagers whose friends use drugs are likely to be even more heavily influenced.

5. *Personality characteristics*. The teenager who is extrovert in personality, impulsive, and more inclined to take risks than his contemporaries, is

also more likely to be a heavy drug user. Low self-esteem and depression are also related to drug use.

6. *Educational and occupational failure.* An individual denied these sources of self-esteem will seek other ways to boost his morale.

The importance of social and cultural factors overrides that of personal factors. Most drug users do not have significant adverse psychosocial factors in their background.

## Drug abuse, drug dependency, and patterns of usage

The effects of **drug abuse** can be classified into those that are physical, psychological, and social. All these effects are likely to be present once a state of drug dependency is reached.

**Drug dependency** has occurred when an individual suffers physically or psychologically if the drug is withdrawn over a short period of time. Adverse effects, and therefore drug abuse, may however occur long before a state of dependency is reached.

A useful review is provided by G. W. Bailey (1989). It is clinically important to distinguish between different patterns of abuse (Sbriglio *et al.* 1988). Usage may be recreational, experimental (mixing drugs), regular, or compulsive. Clearly, occasional recreational use has very different implications from regular or compulsive use.

Parents and teachers are usually quick to recognize behavioural changes in teenagers taking drugs other than cigarettes. Changes in behaviour include slowed movements, lethargy and lack of motivation, slurred speech, drowsiness, and irritability. There may be a decline in school performance due to absence from school and lack of concentration. Sudden, increased demands for pocket money may be indicative.

The fact that usage may be determined mainly by social or by personal factors means that, in appraising a youngster abusing drugs, the clinician needs to bear in mind the possibility that he is dealing with someone of normal personality under unusually strong social pressure, or with a seriously disturbed, perhaps very depressed child living in stressful circumstances.

## Glue-sniffing

### Pattern of use

Solvent misuse (Sourindrhin 1985) occurs most commonly in boys aged 11–17 years, with a peak age of 13–15 years. Adhesives, cleaning substances, petrol, and lighter refills are the most common sources. Substances are inhaled from paper bags, saturated rags, or direct from the containers themselves. Inhalation usually though not always occurs as a group activity, often in derelict houses in inner-city areas. Mostly glue-sniffing is a transient form of behaviour, engaged in as a risk activity by bored youngsters. Occasionally it fulfils a more important psychological need—relieving anxiety and depression. In this minority of cases, a progression to more dangerous drug usage is more likely.

## Clinical features and complications

Inhalation of solvents is usually followed by a short period of intoxication with light-headedness and slurring of speech. Physical symptoms following sporadic usage include nausea, vomiting, headache, abdominal pain, and tinnitus. Clouding of consciousness and visual hallucinations may also occur as a transient toxic psychosis (D. H. Skuse and Burrell 1982).

Chronic usage of these substances in higher concentration can produce serious effects. Chronic encephalopathy and fits have been reported in children (King *et al.* 1981), and older users have suffered kidney and liver damage. Death is not uncommon, with 140 deaths in the UK being recorded from 1971 to 1981, the rate increasing towards the end of this period (H. R. Anderson *et al.* 1982).

## Treatment

Boys with this habit are most unlikely to come to attention at health clinics merely because they are glue-sniffing. Most will never appear at health agencies. A minority will be referred to child guidance or child psychiatry units because of associated psychiatric disorders, or attend hospitals or accident and emergency departments because of physical complications.

Disturbed children and adolescents with glue-sniffing as one of their problems will require full psychiatric assessment and appropriate treatment depending on the nature of their psychiatric disorder. This may include consideration of ways in which they can be helped to find alternative ways of relieving anxiety and depression, or to find alternative group activities.

Physical complications such as fits and renal damage will also require appropriate treatment. Again, during and after such treatment, consideration needs to be given to ways in which the youngster can be rehabilitated and helped to avoid reinvolvement in the habit.

## Prevention

Legal measures to restrict the sale of solvents to children have been introduced in the UK. Early detection by parents and teachers can be achieved if they are aware of the signs. The finding of empty containers, the presence of children in an intoxicated state, and the appearance in a child of erythematous spots around the mouth and nose can all suggest that glue-sniffing is occurring.

A non-alarmist health education approach to parents and teachers, stressing the need to discourage the activity while finding alternative occupation for children involved, without making martyrs or heroes out of them, promises to be reasonably effective.

## Cigarette-smoking

### Pattern of use

In 1988, about 14 per cent of girls and 12 per cent of boys in UK secondary schools were regular or occasional smokers (Dunnell 1990). There is strong continuity between smoking in childhood and adulthood. About one in three of those who become regular smokers in adulthood had begun before the age of 9 years. Eighty per cent of all regularly smoking children in the UK remain regular smokers in adulthood.

Various studies have established that children are more likely to smoke if they are boys, come from working-class families, are educationally failing, are extrovert in personality, go to schools where teachers themselves smoke (Bewley 1979), and (of overriding importance) their parents and close friends smoke (Bewley *et al.* 1974).

### Clinical features and complications

Children who smoke are more likely to suffer from upper and lower respiratory infections, and indeed they are more likely to suffer from these conditions if their parents smoke, even before they begin to smoke themselves. (As they get older they are predisposed to chronic respiratory disease, lung cancer, and myocardial infarction.)

There are no established direct, psychological ill-effects of cigarette-smoking in childhood, but secondary problems sometimes occur. Thus teenagers involved in stealing money from their parents are not infrequently stealing to obtain money to buy cigarettes.

### Treatment and prevention

It is most uncommon for teenagers to request help to overcome a smoking habit. More frequently, boys with antisocial problems referred because of some aspect of their conduct disorder turn out to be heavy smokers. It is unusual for them to want help with the habit.

Over recent years, reasonably successful preventive measures have been developed. Lectures and small group discussions emphasizing the dangers of smoking are ineffective. By contrast, the use of role-playing techniques in which, for example, a youngster is given skills to refuse a cigarette at a party when offered one, are more successful (Botvin *et al.* 1980; Telch *et al.* 1982). Health professionals involved in dealing with children and teenagers suffering from conditions such as hypertension and cystic fibrosis, in which smoking has a particularly deleterious effect, should remember to provide advice about smoking even though they are dealing with a young group of patients. The advice may be more relevant than they may suspect.

## Cannabis

*Pattern of use*

Smoking an extract made from the flowers and leaves of the Indian hemp plant, *Cannabis sativa*, is a widespread practice throughout the world, though illegal in the UK. The prevalence is uncertain, but probably about a third of people in their late teens and early twenties in the UK are occasional smokers and a proportion of these smoke regularly. Many who begin the habit in adolescence continue it intermittently into adulthood.

*Clinical features and complications*

There is no evidence that intermittent, moderate usage is harmful, though users often report that the sense of relaxation and mild euphoria for which they smoke in the first place may persist over 1 or 2 days, producing a state of generally low motivation over this time. More prolonged usage may result in mild chronic apathy.

Heavier, intermittent or regular usage is sometimes associated with clear-cut adverse reactions. Acute intoxication with gastrointestinal symptoms may result. Episodes of perceptual distortion and occasionally frank though transient psychotic reactions may also occur (Naditch 1974). Psychological though not physical dependence has been reported. Regular users may withdraw from the challenges of the external world into a state of chronic low motivation. Such individuals then become at risk for entering into more dangerous drug-taking activities.

*Treatment and prevention*

Adverse psychological reactions other than reduction in drive and motivation are rare, and in any case do not usually come to medical attention. If they do, withdrawal from the drug is sufficient to produce improvement. An apparently prolonged reaction of psychotic type is unlikely to result from the effects of the drug alone, and is more probably a response to a constitutional predisposition to psychosis.

Preventive action is of doubtful validity in a drug that apparently has fewer adverse effects than tobacco-smoking, but teenagers and parents should be better informed of those adverse effects that do occur.

## Alcohol

*Pattern of use*

About 5 per cent of men and 2 per cent of women in the UK are problem drinkers. As alcohol consumption virtually always begins in childhood and adolescence, and as problem drinking in young people is increasing in frequency, there has been increased interest in and concern about the alcohol consumption of children and young people.

The average age at which UK children have their first alcoholic drink is

10–11 years. By 15–16 years (Plant *et al.* 1982) the average male weekly consumption is about 9 pints of beer (or equivalent in spirits) and about half this amount in females. At this age about a quarter of the teenage population have had a hangover in the previous six months and about one in 20 have had a drink in the morning to steady the nerves or get rid of a hangover. About 3 per cent have missed a day's schooling due to drinking.

Evidence is lacking, but it is probable that heavy drinking is associated with similar characteristics to those occurring with other heavy drug users (see above). In general, social and cultural factors outweigh personal factors in importance in defining those at greatest risk.

### Clinical features and complications
The picture of acute alcohol intoxication is well known and requires no description. Probably the main danger of acute intoxication lies in proneness to accidents (especially road traffic accidents) and predisposition to violent behaviour.

Chronically heavy alcohol consumption is, even in adolescence, likely to lead to problems that are:

1. *Social.* Financial difficulties, educational failure, problems in obtaining employment (due to unpunctuality, irritability, etc.).
2. *Psychological.* Confusion, memory lapses, irritability.
3. *Physical.* Especially chronic gastritis.

True alcohol dependency, with inability to cease alcohol intake without withdrawal symptoms and a tendency to relapse, is unlikely to occur until early adulthood, but does occasionally occur in teenagers.

### Treatment and prevention
A discussion of the treatment of alcoholism is beyond the scope of this book. Good accounts may be found elsewhere (see Edwards 1982). The prevention of alcoholism by conventional health education methods does not seem effective. It is likely that social skills training directed towards individuals in helping them to refuse drinks would have more success. Evidence does suggest that fiscal measures raising the relative cost of alcohol would also reduce the rate of alcohol consumption in the young population.

### Other drugs
Relatively small groups of teenagers, mainly those living in inner-city areas, are also at risk from other drugs. These include:

1. *Amphetamines.* Taken by teenagers because of its euphoric effect, this drug can produce a state of acute excitement with overtalkativeness, insomnia, raised pulse, and blood pressure. On withdrawal, a paranoid psychosis with delusions and visual hallucinations has been described.

2. *Barbiturates*. These may be taken orally or intravenously. Young users are usually on multiple drugs. Periodic drowsiness, slurred speech, and depressive mood are common.

3. *Cocaine*. This is usually taken by sniffing, but also by injection for its euphoric effect. Psychological dependence occurs early in its use. Withdrawal may be followed by a paranoid psychosis and characteristic 'formication'—a feeling of ants underneath the skin.

4. *Heroin (acetyl morphine)*. This is the most commonly abused narcotic drug of morphine type. It is usually taken by injection for its euphoriant effect, followed by relaxation. Again psychological dependency occurs early, and withdrawal is extremely unpleasant.

5. *Lysergic acid diethylamide (LSD)*. This is a synthetically manufactured hallucinogen, taken in tablet form because of its intensifying effect on sensory perception. Other effects include acute excitement, panic, and unpredictably aggressive behaviour. The effects generally last up to about 12 hours.

The treatment and prevention of abuse of these drugs is described elsewhere (see Schechter 1978). Addiction to cocaine and heroin is particularly dangerous. The expense of maintaining an adequate supply is likely to lead to stealing and other offences. Lack of money for food results in malnutrition with proneness to infection. Although natural recovery can occur, the risk of death from suicide or infection is considerable. Needle-sharing leading to HIV infection remains a common hazard in many large cities. Teenagers addicted to these drugs, especially if motivated to receive help, require urgent referral to specialist centres. Parents need support whether or not their children are motivated to receive treatment. Illicit drug use in adolescence puts the individual at risk for major social problems later in life (Kandel *et al.* 1986).

## 3.4 Child–adult continuities in psychiatric disorders

### Introduction

Knowledge about the continuity of psychiatric disorders into adult life is relevant to all child health professionals. Parents wish to know what the future holds for their disturbed offspring. Professionals may need such information in order to evaluate their own therapeutic and preventive efforts. After all, if all children with a particular condition, if left untreated, do badly when followed into adult life, then even a small degree of professional success would be worthwhile. On the other hand, adult outcome should not be regarded as the only criterion of success. Even if all children suffering a disorder were eventually to get better untreated, treatment in childhood might still be worthwhile if it reduced the length or severity of the problem in childhood.

There is now a considerable amount of information concerning both adult outcomes of childhood mental problems and childhood precursors of adult psychiatric illness (L. Robins and Rutter 1990). The interpretation of the information is complicated in a number of ways:

1. Continuity of a disorder may be simple (homotypic) or transformed (heterotypic). An individual may show the same problem in adulthood as in childhood (simple continuity), or a childhood diagnosis may be followed by a different, adult diagnosis (transformed continuity). Alternatively, of course, the disturbed child may grow into a mentally healthy adult (discontinuity).

2. The level of continuity may appear to be less than it really is if service contact is used as the criterion for disturbance. Thus few adult neurotic patients have attended child psychiatric departments as children, but this might be because such a small proportion of emotionally disturbed youngsters attend clinics. This issue highlights the need for long-term epidemiological studies examining total population samples.

3. Quite different levels of continuity may be found, depending on whether one starts with a disturbed childhood group showing a particular diagnosis and follows forwards, or starts with an adult group showing the same diagnosis and looks backwards. Most children with a particular diagnosis may improve, but if one looks at adults with the same diagnosis, nearly all of them may have shown the problem in childhood. This is in fact the case with conduct (antisocial) disorders: see below.

If diagnostic continuity, whether simple or transformed, is present, the processes underlying such continuity need to be understood if rational preventive action is to be taken. Here again, there are various issues to be considered:

1. Continuity may be present because the psychopathological process is internalized and impervious to external changes.

2. Continuity may be present but only because if an individual faces adverse life circumstances in childhood he may well be suffering the same adverse circumstances throughout life and into adulthood. Continuity of diagnosis may merely be a reflection of continuity of stress.

### Findings

Child-to-adult and adult-back-to-childhood continuities will be considered separately in this section. Information in this section can be supplemented by reference to the relevant sections of the book dealing with different diagnoses.

*Child-forward-to-adult continuity*

1. *Simple (homotypic) continuity*. Autism and schizophrenia generally persist relatively unchanged into adult life. Similarly, although they have a

better prognosis and many more cases improve than is the case with autism and schizophrenia, when anorexia nervosa, Tourette syndrome, and obsessional disorders persist into adult life, as they not infrequently do, there is strong continuity of diagnosis.

2. *Mixed simple and transformed continuity*. Most children with conduct disorders do not grow up to be adult criminals, but when they do have problems in adulthood, they are likely to show a mixture of conduct (aggressive personality) and affective disorders (Quinton *et al*. 1990).

3. *Transformed (heterotypic) continuity*. Children with the hyperkinetic syndrome or attention-deficit hyperactivity disorder (ADHD) often show conduct disorders in adult life (Gittelman-Klein and Mannuzza 1989).

4. *Discontinuity*. In general, children with emotional disorders do not have an increased risk of disorder in adult life (L. N. Robins 1966), though there is now evidence (Quinton *et al*. 1990) that children with emotional disorders whose parents have suffered mental illness themselves do have a raised rate of affective disorders in adult life.

The mechanisms underlying child–adult continuity probably vary considerably from diagnosis to diagnosis. For example, in schizophrenia and autism, the underlying psychopathology appears largely intractable to environmental change. This may also be true for the severe conduct disorders of childhood and, to a lesser extent, the hyperkinetic syndrome. In contrast, continuity shown by children with less severe conduct disorders and emotional disorders may be much more a result of the persistent nature of the family and wider social adversity they are likely to experience.

### Adult-back-to-childhood continuity

1. *Simple continuity*. Most but by no means all cases of adult autism, obsessional disorder, Tourette syndrome, agoraphobia, and anorexia nervosa begin in childhood or, much more commonly, in adolescence.

2. *Transformed continuity*. Only a small minority of cases of schizophrenia begin in childhood or adolescence. However, when information about the childhood of those adults who develop schizophrenia later is available, it suggests that they have tended to show other personality features. These pre-schizophrenic precursors include antisocial behaviour restricted to the family setting, and social isolation arising from rejection by peers, perhaps because of odd, incongruous behaviour (Zeitlin 1986).

3. *Discontinuity*. Although there is rather inconsistent evidence that they may have suffered traumatic experiences in childhood such as loss of a parent, most adults with either major or minor affective disorders have not shown significant emotional behaviour problems in childhood. This is, of course, not true for that minority of sufferers whose disorders have begun in childhood or adolescence. It is likely that as further information accrues, this view of the pattern of adult–child continuity in affective states will

need to be modified, for example by considering chronic disorders separately from episodic.

Finally, it should be emphasized that information in this section on continuity relates entirely to probabilities. For most diagnoses, there are many individuals who do not show the characteristic pattern of continuity and discontinuity described here.

## Further reading

Rapoport, J. (1990). *The boy who would not stop washing*. Collins, London.
Steinberg, D. (1975). Psychotic and other severe disorders in adolescence. In *Child and adolescent psychiatry: modern approaches* (ed. M. Rutter and L. Hersov), pp. 567–83. Blackwell Scientific Publications, Oxford.

# 4

# Psychosocial aspects of physical disorders: general

## 4.1 Introduction

Depending on the definition used, between 5 and 10 per cent of children of school age suffer from disability arising from a chronically handicapping physical condition (Rutter *et al.* 1970*a*; Pless and Douglas 1971). Of the physical conditions from which such children suffer, asthma is the most common, affecting 2 per cent of all children of school age, though some studies have found higher rates. Eczema occurs in about 1 per cent of children, so that illnesses with an allergic component account for a large proportion of chronic childhood physical disability. Neurological conditions are the next most common types of disorder—epilepsy occurs in about 8 per 1000 and cerebral palsy in about 2–3 per 1000. The remainder consist of a wide range of other physical disorders, of which the most common are sensory defects, congenital heart disease, diabetes mellitus, and orthopaedic disorders.

As medical treatment improves, and the period of survival for many conditions increases, although incidence (numbers of new cases) remains the same, the prevalence (numbers in the population) of disabling disorders rises (Newacheck *et al.* 1986). Further, the nature of an illness may be transformed by a new treatment. The five-year survival rate of children with leukaemia has improved over three decades from around 0 per cent to around 60 per cent. For parents, the stress of bereavement has been replaced by the stress of uncertainty. For children, a relatively brief and painless illness has been replaced by a prolonged series of hospital visits and treatments with many unpleasant aspects such as painful procedures and alopecia. Doubtless for most such children the new technology has had positive results, but for many it has not. Similarly, new forms of operative procedure, including cardiac transplantation, have transformed the picture in congenital cardiac disease, and new antibiotics and other forms of treatment have markedly prolonged survival in cystic fibrosis.

A broad account of prevalence gives a rather misleading picture of the pattern of disability, because it takes no account of the severity of disorders. Although asthma and eczema are common, and each is occasionally extremely disabling, most children who have asthma and eczema are able to lead normal lives. Some of the less common disorders such as cerebral palsy are usually much more disabling. Further, consideration of the pattern of physical disability in isolation fails to take account of the fact that most severe disability in children with physical disorders arises from associated handicaps, especially mental handicap and psychiatric disorders. Thus, cerebral palsy is frequently associated with mental retardation and specific learning disabilities which hamper the child's life and impair development to as great a degree as the physical condition. The allergic conditions are not, in general, accompanied by associated serious learning difficulties. In order to obtain a true picture of the pattern of physical disability, it is therefore necessary to consider the severity of different conditions and the other problems with which they are associated. Summarizing findings from the National Survey, Pless and Douglas (1971) found that just over half the chronic disorders were mild (preventing the child only from strenuous activity), a third were moderately severe (interfering with normal daily activities), and about a tenth were severe (requiring prolonged periods of immobilization, confinement to the house, or absence from school).

## Psychosocial causes of chronic physical disease

Psychosocial factors may be important in aetiology, either of the development of the physical condition in the first place, or of its maintenance and persistence. Thus, brain damage associated with cerebral palsy sometimes occurs as a result of injuries inflicted by parents on the child in infancy (Diamond and Jaudes 1983). Asthma is an example of a condition which is produced by an inborn lesion (hyperreactivity of the bronchi), but in which persistence often depends on the presence of psychological triggering mechanisms (Section 5.11.1). In other conditions, such as the various forms of childhood cancer, social and psychological factors are probably of little importance, either in primary aetiology or in maintenance.

Social factors are generally of less importance in the aetiology of chronic physical diseases than is the case with mild mental retardation and some behavioural disorders. Some chronic conditions, such as failure to thrive, obesity, and recurrent upper respiratory tract infections are, however, commoner in the more deprived sections of the population. Inappropriate diet, difficulty in maintaining hygienic conditions, and chronic financial stress leading to parental depression are probably the most important reasons for the association between these physical conditions and low socio-economic status.

In considering causation, it is usually unhelpful to think of some conditions occurring in children as 'psychosomatic' and others as 'non-psychosomatic'. In any particular condition (cerebral palsy and asthma are examples), psychological and social factors may be prominent in aetiology in some children but absent in others.

## 4.2 Impact on parents

The effect on parents of the discovery that their child has a chronically handicapping condition depends on a variety of factors including the nature and severity of the condition (especially the amount of physical care the child needs), the age at which the condition is diagnosed, the presence of associated handicaps, the personality and previous experience of the parents, the temperament of the child, the amount of help available from relatives and friends, and the quality of health, social welfare, and educational services available.

When a child is born with a handicap, or develops a serious condition after a period of good health, parental reactions tend to follow a characteristic course (Drotar *et al.* 1975). Perhaps surprisingly, these reactions, although they may be severe, do not appear to influence the attachment of the physically handicapped child to a significant degree (G. A. Wasserman *et al.* 1987). The reactions are akin to those following bereavement and indeed, although the child remains alive, the parents have suffered the loss of the normal child they expected to rear. Initially there is a stage of *shock* during which parental feelings are numbed, and a sense of unreality is experienced. The parents are in such a state of anxiety that they find it difficult to absorb information given to them even when it is presented repeatedly and in a very simple form. There may be then a stage of *denial* when the seriousness of the condition is questioned, and the parents may have fantasies that the child will be magically cured or take the child to a variety of different physicians. This is followed by a stage of sadness and anger when depressive feelings and sensations of guilt predominate, but parents may also rage against fate, each other, the child with the handicap, or the physicians who have been involved. Sometimes their anger may be justified—often it is not. Finally, the parents go through a phase of *adaptation* to the situation and reorganize their lives accordingly. They are able to see the child and his future in more realistic terms, and make plans for his care, education, and future that accord with his real potential.

Not all parents go through these various stages, and there is no reason to think that there is anything amiss if they reach the stage of adaptation by some other route. Some parents, for example, appear never to go through a stage of shock or denial. The absence of a period of sadness is, however, a cause for concern. There is evidence that women who are widowed who do not go through a period of mourning suffer more psychological and physical complaints later on (Raphael 1975), and, by analogy, one would expect parents who do not experience depressive feelings after the loss of their normal child to be similarly at risk.

The speed at which parents reach a stage of adaptation varies considerably. Usually this takes a period of several months, but it may occur more rapidly than this, and the process may also take several years. Some parents get stuck at the phase of denial or anger, and spend the whole of their lives either pretending to themselves and others that the child's handicap does not exist, or raging against the supposed (or actual) people responsible for producing the handicap in the first place. The doctors or midwife present at the delivery may be blamed, for example, for congenital defects.

As a disability becomes chronic, so parents' longer-term coping mechanisms come into play. Coping has been defined as 'the effort to master, reduce, or tolerate the demands arising from a stressful transaction'. Coping is made possible by physical resources (such as money and employment), social resources (such as friends and relatives), and psychological resources (such as beliefs, problem-solving skills, and personality). Various ways of coping have been defined. They include recourse to practical measures, wishful thinking, stoicism, the seeking of emotional support, and passive acceptance. Clearly, some of these are more likely to be successful than others. Coping strategies are discussed further below in relation to children's reactions.

### The parental relationship

Parents often do not pass through the period of initial adjustment at the same pace. Characteristically fathers are likely to spend more time in the phase of denial, perhaps protecting themselves against depressive feelings by preoccupation with the medical details of the handicap. They are likely to be particularly deeply affected by having a handicapped male child. Mothers are more likely to suffer protracted periods of guilt and sadness. The capacity of the parents to communicate and share their feelings affects the pace at which each of them will complete the 'mourning' process. Thus, if a mother is experiencing deep and incapacitating feelings of sadness, her husband may come to believe that he cannot afford to feel sad, or the care of the family would break down. In fact, if a father can admit to his depression and discuss it with his wife, this often leads to a reduction in intensity in her sad feelings. The opportunity to share sad feelings may enable a woman to experience a lightening of her mood. In non-communicating or poorly-communicating families, this is less likely to occur.

The rate of marital breakdown in parents of children with physical and mental handicaps is either the same as that in the general population or very slightly raised (Martin 1975). These findings are at first difficult to reconcile with clinical impressions that the presence of a physically handicapped child has a profound effect on the marital relationship. The reason for the inconsistency probably lies in the fact that while, in some marriages, the child's problem is something about which the parents take quite different attitudes and cannot communicate, thus producing a rift between them, in others the child's difficulties act as a focus of shared concern and bring the parents closer together than otherwise they might have been. Apsley

Cherry Garrard wrote 'The mutual conquest of difficulties . . . is the only lasting cement of marriage'. The presence of a child with a handicap will alter the functioning of a marriage in a variety of ways (Sabbeth and Leventhal 1984). The closeness of the couple, decision-making processes, and patterns of communication may all be affected. The effects of these changes may, however, be positive as well as negative.

### Impact on parental social life

The effects of having a handicapped child on social life have been well documented. For a minority of parents of children with serious handicaps, leisure activities are curtailed, mothers cannot go back to work when they want to, and there is a restriction in the number of additional children. Nevertheless, when comparisons are made with families without a handicapped child, there may be surprisingly few differences noted. Many families with handicapped children would have a rather restricted social life even if their child had been normal. The financial costs of handicap may include loss of the mother's income, visits to hospital, purchase of aids, and alterations to the home to meet the needs of the child. These may be considerable, and the financial effects, though mitigated by the availability of benefits, may still seriously affect the life of the family.

### Attitudes to the child

Most parents eventually develop a warm and loving relationship with their handicapped child, though inevitably they are sometimes irritated, perhaps by the child's slowness to learn, by behaviour problems, or by the extra demands upon them which the presence of the handicap may entail. However, a minority of parents show attitudes that are less helpful in the promotion of their child's development. The child may be unrealistically perceived as more vulnerable than he really is (M. Green and Solnit 1964), and this may result in the child being overprotected and infantilized. Alternatively, the child may be rejected and treated with indifference or neglect. For most parents, the healthy appearance and normal development of their children is a source of self-esteem. For some, especially those whose self-esteem is vulnerable for other reasons, the presence of a handicap is a severe blow to their pride, and marked negative reactions develop. Morbid reactions of parents to their handicapped children are not common, but when they do occur, they are often characterized by ambivalence with a mixture of overprotection and rejection.

## 4.3 Impact on the child

Like children without handicaps, those with chronic disability can be viewed as requiring to gain gradually increasing mastery of their environment (skill acquisition). To achieve this, as well as acquiring skills, they need to be able to enjoy mutually satisfactory relationships with others and a realistic view of themselves as worthwhile people (a positive self-concept). In considering the impact of handicaps on children's development, it is useful to think of these interrelated areas (skill acquisition, self-concept, and emotional development and relationship formation) separately.

## Skill acquisition

Some physical handicaps almost inevitably result in delay in the acquisition of particular skills. Motor handicaps of cerebral-palsied children and the language deficits of deaf children are examples. However, there is also the possibility that failure to acquire skills may arise indirectly, not as a result of the physical disability itself, but for a variety of other reasons—such as lack of stimulation from parents, inadequate schooling, repeated absences from school, or low self-expectations on the part of the child himself.

Most children with physical disabilities do not suffer from the indirect consequences of their handicap in this way. Language development, reading ability, and non-language skills usually develop normally. However, there are exceptions. In the Isle of Wight study (Rutter *et al.* 1970*a*) it was found that children with physical disabilities had two or three times the rate of specific reading retardation compared with children in the general population. Fourteen per cent of physically handicapped children were more than 28 months behind their expected reading ability, compared with 5 per cent of the general population. A variety of other studies have produced similar results (Schlieper 1985). Children with epilepsy and other brain disorders are particularly likely to be affected in this way (see below), but it is not just these who show educational disabilities. In children with other physical disorders, absence from school and low educational expectations, both from the children themselves and from their teachers, as well as other factors already mentioned, may be responsible. Repeated absence from school is likely to be especially important in those subjects, such as arithmetic, where new knowledge is often acquired in steps, each of which depends on a previous skill having been learned. In middle-class families parents are likely to ensure that a child has not missed out on a particular important step, but in working-class families where communication with the teacher may be less satisfactory, ground may not be made up in this way.

## Self-concept

There are three important aspects to self-concept in the child with a physical disability—the child's body image, his self-esteem, and his view of the cause of his disability.

### Body image

The handicapped child's view of himself, if it is to be realistic, needs to take account of whatever it is that is wrong with him. In some children this may be a visible physical deformity such as a skin lesion. In others there may be an invisible physical defect such as a leaky heart valve; in yet others, such as children with epilepsy, the child may be anatomically intact, but have a functional disability.

Children with cerebral dysfunction may have a specific inability to visualize their own bodies accurately as part of a deficit in visuospatial perception. In most handicapped children, however, the capacity to achieve a realistic body image is present. Distortions then are more likely to occur as a result of information obtained from parents, teachers, and other children. As in the case with non-handicapped children, the handicapped tend to see themselves as they believe others see them, so if they are led to think, by what a parent says, that a large birth mark on the face is hardly noticeable, or that a tiny blemish is grossly disfiguring, this will certainly have an impact. As children get older the effect of other children's comments on their disability may become as important as their parents' views.

## Self-esteem

There are two components to the degree to which an individual values himself—the cognitive and emotional. Thus a child can see himself as brighter or less bright, more or less attractive than he really is. Independently of the cognitive self-appraisal, a child can feel unduly sad or negative about himself and have a tendency to self-criticism, or he can view himself positively and with confidence. It is quite possible for a child to recognize his own perhaps quite severe limitations as a scholar, and yet feel he is a person who really matters in the world. Children with physical handicaps vary considerably in both these components of self-esteem. Most have a realistic cognitive self-appraisal and a positive view of themselves. However, some do show evidence of low self-esteem. In the early years, the problem is likely to arise from the depressed and negative feelings that their parents may have about them, but in later years, the views of other children grow in importance to the handicapped child.

Both perception of body image and self-esteem are likely to become of greater emotional importance in adolescence, when there is increased concern about identity, future career, and attraction to the other sex. Just as normal children who have not shown any particular worries about their appearance may become preoccupied with it soon after puberty, so children with physical handicaps may, at this stage of life, become acutely concerned about their disability for the first time in their lives. Despite this possibility, low-esteem is not a frequently occurring characteristic in adolescents with chronic illness.

## Perception of causes of disability

There is a characteristic development of the way in which children see the cause of illness (Brewster 1982). In the preschool years (aged 4–6) they are likely to see their disability as inflicted on them by others, perhaps as a punishment for misbehaviour. In middle childhood, the notion of contagion becomes more important, and they may see their problem as something they have 'caught' from another child or adult. Only by about the age of 10 years have they developed more mature concepts of personal vulnerability, with more or less appropriate ideas of physiological or anatomical causation.

### Emotional development and formation of relationships

*Prevalence of psychiatric problems*
Most children with physical problems develop a satisfactory relationship with their parents, sibs, and other children, are able to control their aggressive impulses as well as other children their age, and do not suffer from undue anxiety, depression, or other symptoms. However, most studies (e.g. Cadman *et al.* 1987, in a total population study carried out in Ontario, Canada), though not all, suggest that children with physical disabilities do have a somewhat higher rate of behavioural and emotional problems than do children in the general population. In the Isle of Wight survey (Rutter *et al.* 1970*b*) children with physical disorders had about double the rate (12 versus 6 per cent) of psychiatric problems than did normal children, and the rate was higher still in those with epilepsy (about 28 per cent) or other evidence of cerebral dysfunction (40 per cent). Breslau (1985) has confirmed that children with brain dysfunction are at increased risk for psychopathology, and that the quality of family environment does not modify this risk (Breslau 1990).

The long-term adjustment in adulthood of physically disabled children suggests that most regard the quality of their lives to be good (Query *et al.* 1990). However, in terms of educational achievement and behavioural difficulties, those who have been brought up in economically and socially disadvantaged circumstances do less well than others more favoured (Pless *et al.* 1989).

*Family factors*
As in children without physical disorders, the emotional development of the physically disabled child will depend very considerably on family factors. To a considerable degree the mechanisms whereby family factors produce a childhood disturbance are likely to be very similar in the handicapped as in the non-handicapped child. However, in the child with a physical problem, adverse parental attitudes may be particularly focused on the disability. Thus a parent who would have been somewhat overprotective anyway may become much more so with a physically vulnerable child. Children who are treated as fragile may come to see themselves as such and become unduly anxious, for example, about physical contact with other children. Alternatively, they may react against parental over-

protection and become unnecessarily reckless. A child who is rejected by his parents will often take over the low valuation his parents have of him. Thus, a handicapped 13-year-old child who eavesdropped and overheard his parents say that it would have been better if he had never survived, developed a depressed view of himself and made a suicidal attempt.

### Individual coping and defence mechanisms

A child with a physical disorder is subjected to a specific stress and an unusual cause for anxiety. The way in which individuals deal with anxiety-provoking stresses can best be viewed as a mixture of mainly conscious behavioural strategies (coping behaviour) and mainly unconscious psychological defence mechanisms. These interact with each other (successful coping can reduce the need for an unconscious defence), but it is easier to consider them separately.

**Coping behaviour** (Hamburg and Hamburg 1980). People who deal successfully with stress tend, among other strategies, to:

1. Ration the amount of stress they cope with at any one time. Very bad news is absorbed bit by bit. Thus a child may say to himself: 'I am just going to think about my operation tomorrow. I don't want to think further ahead than that.'

2. Try to obtain information about their problems from a number of different sources.

3. Rehearse to themselves behaviour that is going to be difficult for them. Thus a coping child with a scar who is frightened of teasing will rehearse what he is going to say before the teasing occurs.

4. Try a variety of ways of dealing with a problem, rather than stick to an unpromising approach.

5. Construct buffers against disappointment, for example a coping child having an operation would prepare himself beforehand for the failure of the operation as well as its success.

These various coping mechanisms are, of course, used by parents as well as children in dealing with the unpleasant facts of a physical disorder.

**Defence mechanisms.** A child faced with unacceptable aspects of a physical disorder whose coping behaviour is unequal to the task may use unconscious psychological mechanisms to deal with problems he is unable to accept at a conscious level. He may:

1. Deny the problem exists. 'I do get badly teased at school, and I don't know what to do about it.' = 'I don't get teased at school.'

2. Rationalize the problem. 'I feel miserable about how short I am.' = 'Everyone feels miserable from time to time.'

3. Project the emotion on to someone else. 'I feel angry with myself for not being able to stop having fits.' = 'My mum feels angry with me because of my epilepsy.'

4. Regress. 'If I were three years younger, it would be all right for me to want my mum around more, so I'll act younger.' = 'I just like my mum around a lot of the time.'

5. Repress the problem. 'I feel very sad because I think no girl will ever want to have sex with me because of the way I look.' = 'I don't have any sexual feelings.'

6. Displace the emotion. 'I feel guilty about my eczema and I don't know why.' = 'My room is in a mess and I feel guilty about not clearing it up.'

Commonly, children show a mixture of at least partially effective coping strategies and defence mechanisms. As coping improves, so the need for unconscious processes is diminished. The use of defence mechanisms may be healthily adaptive, and occur while coping mechanisms are being strengthened.

## 4.4  Impact on the sibs

The brothers and sisters of children with physical problems may themselves develop behavioural disturbances. For example, though most sibs of children with chronic epilepsy and Down's syndrome do not have problems, some studies have shown them to have significantly more disturbances than one would expect by chance (Breslau *et al.* 1981). The sib may be relatively neglected, feel unable to bring friends home because he is frightened of what they will think of his handicapped brother or sister, or show emotional reactions for a variety of other reasons. Some girls with a younger, handicapped sib may suffer severe limitations of their social lives because they are expected to spend a great deal of time looking after the handicapped child. As time goes on, sibs of chronically handicapped children have been shown to suffer an increase in depression and social isolation (Breslau and Prabucki 1987). There may also be benefit to sibs in having a handicapped brother or sister. These include an increased capacity to cope with stress, and an enhancement of their caring, altruistic tendencies.

## 4.5  Principles of psychosocial management of physically handicapped children and their families

### Communicating a depressing diagnosis

Bad news comes in various forms in paediatrics. In some cases, such as myelomeningocele, a diagnosis of serious import can be made immediately after birth. In others, such as diabetes mellitus, the diagnosis is usually made over a period of 1 or 2 days when the child is much older. Some diagnoses, such as cystic fibrosis, may be preceded by a long period of ill-health and uncertainty, but may also be made at or shortly after birth, especially if there is a previously affected sib. In yet others, such as asthma or epilepsy, the seriousness of a diagnosis may only be realized after routine treatment measures have failed to improve what is usually a relatively benign condition. The principles of imparting the diagnosis (D. C. Taylor 1982) will be similar in all these circumstances, although the details will vary considerably (see also Section 2.6.2 for discussion of communication of diagnosis of mental retardation). Principles include:

1. The doctor involved in imparting information of serious import should be the most senior and experienced in the team.

2. If at all possible both parents should be present, even if this means a delay of a few hours or a day or two and (if the mother is a single parent and agrees) even if the parents are not living together. The greatest support parents receive will be from each other, and they should be able to share the burden from the start.

3. It is useful to start by spending a moment or two finding out about the parents' current state of knowledge. 'I wonder what you yourselves have thought might be wrong with X.'

4. Aside from the above provisos, the news should be broken as early as possible, and the practitioner should be as open and honest as possible about it. It is usually helpful to begin by listing the symptoms and investigatory findings, and then give the diagnosis before pausing to give the parents the oppportunity to ask questions and express their feelings. The use of simple language to explain the problem without shirking technical labels, if this seems appropriate, is helpful.

5. After receiving the diagnosis, parents will often ask a large number of questions. This will usually enable the clinician to give information about causation, outcome, forms of treatment available, and genetic implications. As parents are likely to be in a state of shock, they will not take all this information in when it is first imparted. They are likely to need a number of opportunities to ask the same questions. If a team of people are involved, it is important that they decide beforehand who can take responsibility for communication of information (junior doctors, nurses, etc.), and ensure the same message is put over by all concerned. If parents, after a session or two, have not asked questions about a particularly important aspect— causation, outcome, etc.—then the subject should be raised with them to give them an opportunity to discuss it further. They may or may not be ready to do so.

6. The positive as well as the negative aspects of outcome should be presented. 'In the UK, all children are able to receive education. It should always be possible to control pain effectively', etc.

7. Emotional responses to the information should not be discouraged. If

a parent starts to cry, then one can say 'I know this is terribly upsetting. It is very natural you should show your feelings.' It does no harm for the doctor, nurse, or social worker involved to admit they themselves feel sad or worried about the situation.

8. When the initial interview with the doctor is over, it may be helpful for the parents to have someone (nurse, social worker) with them to continue to talk to. This person ought also to be sensitive to the parents' wishes to be alone at this point.

9. The child may be of an age to understand the diagnosis and its implications. In these circumstances, it is usually sensible to see the parents first and discuss with them how and in what form information is to be given to the child. There are many views on how much information should be given to children of different levels of understanding, and this is discussed further below.

### Continued counselling

The family with a physically handicapped child is likely to require continuing medical advice, but also continued counselling. The social and personal needs of the child and family will change with age. Inevitably the doctor with ongoing responsibility for medical care will find himself aware of parental attitudes, and discussing educational and other aspects of the child's life. For many families who would be reluctant to seek help for non-medical or psychiatric reasons, the medical consultation to review a child's needs provides the only opportunity for wider discussion of social and psychological factors. As the well-being of the child may depend at least as much on these as on his physical state, and the two are probably closely related anyway, it is important that doctors dealing with physically handicapped children on an ongoing basis have reasonably well-developed counselling skills.

The principles of ongoing counselling include:

1. An opportunity, built into the system, to discuss social and psychological development and related problems. Right from the start, families should feel this opportunity is not just an extra service available on demand, but something they are expected to require. For this to happen, either the doctor must make routine enquiries along these lines and have time to deal with issues raised, or there should be someone available, such as a social worker or psychologist, with whom he or she works closely, who can take this responsiblity. Counselling does not, however, merely involve giving advice. It includes being able to listen to anxieties and, on occasion, respond with sympathy without attempting to offer solutions.

2. As with the breaking of depressing news, continuing counselling is more likely to be effective if both parents are present.

3. It is important to make sure that the clinician is aware of new skills

the child has acquired since last seen. These are as important as any problems that may have developed.

4. When considering how to advise concerning problems that may have arisen, it is helpful to focus particularly on what the parents and child are getting right at the present time, rather than on what they are doing wrong. It may be easier to build on existing coping skills rather than to eradicate faulty practices. A mother may be generally rather overprotective, but allow her child to spend the night with relatives or friends. It may be more useful to spend time on working out with her how the night spent away from home enhances the child's confidence and independence than focusing on what she will not let the child do.

5. It is best not to give blanket advice, but to be as specific as possible. If a father is emotionally distant from his child, working out a particular activity the two might do together is more useful than just suggesting that father spends more time with the child.

6. Some doctors are discouraging to the expression of emotion in parents because they are embarrassed by it or fear that dealing with it will be too time-consuming. The expression of emotion and understanding its cause often clarifies issues and saves time in the long run. Once a doctor gains confidence in dealing with the expression of emotion, embarrassment is usually no longer a problem.

7. When parents differ about an issue, for example letting a child go to a special school, it is usually better not take sides. By contrast, working out with each parent why they feel the way they do will often result in the parent with the more irrational view ceding the point.

8. Families change over time. It is dangerous to stereotype a family or family members and forget that they may have undergone considerable development since last seen.

9. Communication to others working with the child and family is an important part of the job. It is very rare for people like schoolteachers to feel they have too much information and often teachers are surprisingly unaware of the implications even of relatively common conditions (Eiser and Town 1987). There may be difficulties in knowing whether to transmit information that is relevant, and yet might be shameful or embarrassing to the family. It is often sufficient in these circumstances to indicate that there are serious domestic problems affecting the child without being explicit about a father's alcoholism or a mother's depressive illness. Often teachers know about these matters from what even quite young children write in their essays. It is important to ask permission from parents before communicating with other agencies, and to explain what information it is intended to communicate. If in doubt whether to communicate a particular piece of information, then it is always sensible to check with parents what their views are on the matter. The most useful communication is two-way, and this is easier face to face or on the telephone than by letter. Teachers

and social workers often have information concerning, for example, compliance with medication, that is of great relevance to medical management.

10. Many children with chronic physical disorders have multiple medical problems and require a team of professionals to assist them. In these circumstances it is useful for one of the team to be nominated as coordinator for each family, and for other specialists not to attempt to take over this role. If the child's needs are complex, it is likely that this coordinator will be a member of the hospital team rather than the family doctor, but if this is the case, it is still vital that the family doctor, who is likely to be the first port of call in emergencies, is kept fully in the picture.

11. Health professionals are likely to be the counsellors that families see least of, although, because of their status and sometimes their expertise, their influence is likely to be disproportionate to the time they spend. However, many parents obtain great help from discussion with relatives outside the family, friends, teachers, and other non-health professionals. Of course, such people can sometimes be unhelpful, as can doctors, but, on balance, their supportive value far outweighs their negative effects.

Doctors may have a special part to play in introducing parents to others who may be especially helpful, including parents of children with similar problems. Regular parent groups organized by social workers (Bywater 1984) or other professionals, can be practically helpful and emotionally supportive. Parent organizations, linked to particular disabilities, provide a most helpful forum for information-sharing and mutual support, and often doctors or social workers can indicate how these can be contacted. The National Children's Bureau (1985) provides a useful list of addresses of organizations in the UK. A small minority of parents seem to manage to avoid facing their own problems by putting all their energies in these organizations, leaving nothing for themselves and their own families, but the overall positive contribution of these bodies is considerable.

12. It is often difficult for paediatricians to know when to refer problems to psychologists and psychiatrists (P. Graham 1984a). Obviously not all emotional, behavioural, relationship, and learning problems can be referred. Relevant factors in deciding on referral include the availability of interested mental health professionals, the severity and persistence of the psychological problems, and the motivation of the family. When paediatricians, psychiatrists, and psychologists meet regularly at ward liaison meetings or on psychosocial rounds (see below), there are usually few real problems over referral.

## 4.6 Hospitalization

### Introduction

Admission to hospital is a common experience for children. In the UK about 6 per cent of children are admitted each year and, by the age of 5 years, about one in four children have been admitted at least once. The mean duration of admission is about 6 days, and the majority of children have relatively short admissions of 3–4 days or less. In the UK, about one in seven children are admitted to adult wards rather than to special children's facilities. The commonest causes for admission are acute respiratory infections, but other infections, injuries, and ill-defined conditions without a definite diagnosis are also very frequent causes of admission (Department of Health and Social Security 1976).

While most admissions to hospital for illness are short-term, about 4 per cent of admissions are for more than a month, and some children remain resident in long-stay hospitals, on a more or less permanent basis, usually because of a combination of severe mental retardation, sensory impairment, and physical disability.

## Short-term effects of hospitalization

The reaction of young children admitted to hospital varies considerably, and will depend particularly on whether separation from parents occurs, and on how discrepant the hospital environment is from the child's everyday experience at home. Thus, if a child is not separated, retains familiar toys, is given food he likes to eat, is called by the same name as he is at home (perhaps Jamie and not James) and is able to continue to watch his favourite television programmes, the reaction to hospitalization may be minimal (B. Brown 1979). On the other hand, young children separated from their parents into new surroundings are likely to show distress that is much more marked. Initially they often show a wary approach to their new environment. At the point of separation from their parents, an angry protest with crying, sometimes leading to uncontrollable sobbing, is relatively common. Eventually, this active distress will give way to an expression of general misery, often with food refusal and sleep disturbance. Subsequently, the child will apparently become resigned to the situation. When the mother visits, he will, however, be likely to show angry feelings towards her for apparently abandoning him, and the cycle of protest, despair, and detachment (Bowlby 1975) will be repeated when she once again leaves him. This pattern of behaviour may last for several days or even weeks. Many young children, however, do become adjusted to ward life, make special relationships with particular nurses, and lose their tendency to distress when their mother leaves, with the realization that she will indeed return before too long.

Older separated children of school age are also likely to be wary on admission, but are distinctly less likely to show anger and despair to the same intense degree. They may, however, become preoccupied with reunion with their parents, counting the hours or minutes until the next visit. Both older and younger children are likely to develop fear reactions to specific nurses or technicians who have been involved in administering painful procedures. Again younger children are likely to show these re-actions more intensely.

After discharge from hospital, younger children in particular may show unusually difficult behaviour with their parents, apparently testing them out to see if they are still loved or likely to be sent away again. Feeding and sleeping problems are relatively common. There may, however, also be positive effects arising from hospitalization. Sick children, and those around them, may gain valuable experience and knowledge from a hospital admission in terms of a sense of generosity and concern for others (Parmelee 1986).

## Long-term effects

Children who have been repeatedly admitted to hospital in early and mid-childhood are slightly more likely to develop behavioural and emotional disturbances in adolescence (Quinton and Rutter 1976). The effects are particularly likely to be present if the admissions are for long periods and if the children come from disadvantaged homes. The reason for this is unclear, and it is possible that, with improvements in hospital practice, this adverse effect may be minimized or disappear altogether (Shannon *et al.* 1984).

## Factors modifying distress

The short-term effects of admission to hospital will depend on a variety of factors including especially:

1. The age of the child. Above the age of 1 year, the younger the child, the more severe the distress is likely to be.

2. The social circumstances of the family, for example the degree to which they are already financially stretched, and the relationship between the parents.

3. Adequacy of preparation for hospitalization (see below).

4. The condition for which the child is admitted. Children whose illness produces greater distress and discomfort will be more disturbed.

5. The necessity for frequent painful procedures (Saylor *et al.* 1987).

6. The presence of familiar figures, especially the parents, during the admission. Thus, if rooming-in facilities are available, this will markedly cut down short-term distress.

7. Previous experience of hospitalization.

8. The parent–child relationship, and especially the level of anxiety of parents and the degree to which their anxiety is communicated.

9. The temperament or personality of the child.

10. The coping style of the child in the face of stressful medical pro-cedures. This may be active (information-seeking) or avoidant (information-denying) (Peterson 1989).

11. The availability of adequate play and education facilities.

12. The attitudes of hospital staff.
13. The degree to which ward organization is child-centred.

## Prevention of adverse effects of hospitalization

Distressing reactions to hospitalization can be reduced by a humane and thoughtful policy towards child care (Vernon *et al.* 1965). If, in addition, extra resources are available for staff and equipment, distress can be reduced still further. The following measures have been demonstrated to have a beneficial effect:

1. *Preparation for the admission.* For planned admissions the hospital should provide a booklet describing what will happen when the child is admitted, ward routines, and procedures, etc. Parents of young children can be recommended suitable storybooks to read to their children. If at all possible, parents and children should visit the ward beforehand—perhaps immediately after the admission is decided upon in the out-patient department.

2. *Rooming-in facilities* should be available for parents of all children under the age of 5 years, and for older children who are very ill, or likely to be unusually upset by the admission.

3. *Visiting by parents and sibs* should be permitted at all times and should be actively encouraged. Parents should be particularly encouraged to visit on critical days, for example days when an operation is to be performed or the child is to have an important investigation. Parents should be made to feel welcome on the wards, and a sensitive, practical attitude taken to the fact that some may well be bored, angry, distressed, and manipulative at different times. The advantages of involving parents in the care of their children far outweigh the dangers of attempting to exclude them.

4. Children should be *adequately prepared* for all painful or unpleasant procedures by discussion and, with young children, suitable play material. No attempt should be made to disguise the fact that a procedure is unpleasant. Specialist wards will need special equipment for this purpose. It is particularly important to prepare children for postoperative discomfort and procedures.

5. Nursing and medical staff working on children's wards should receive *adequate instruction* in procedures necessary to reduce distress. In particular, an attempt should be made to increase their awareness of distress and its likely causes, so that they can take action accordingly. They should be aware of the developing child's changing concepts of illness.

6. *The presence of trained teachers* is necessary to provide appropriate education for children fit enough to receive it. Trained play staff should also, if at all possible, be available to assist in preparation for procedures, enable anxieties to be expressed and understood more generally, and to

occupy children, especially those young children whose parents cannot be available all the time.

7. *The ward atmosphere and organization* should be child-centred. Thus, as far as possible, mealtimes, waking and sleeping times, etc. should be geared to the needs of the children and families rather than those of the staff. Children should, if possible, have nurses specially allocated to them so that they have the opportunity to develop trusting relationships.

8. *Social work, psychological, and psychiatric back-up services* should be available to all ward staff. If at all possible there should be a regular psychosocial meeting once a week to discuss both psychosocial aspects of ward organization and procedures, and specific children causing concern. Such psychosocial meetings can have a therapeutic effect and also provide an educational experience for paediatric medical and nursing staff. Educational goals can include the encouragement of a developmental approach to emotional and psychosomatic reactions, and greater psychosocial understanding of illness in relation to children's cognitive level and coping techniques. The meetings can also assist in earlier recognition of emotional and management problems, especially those concerning severely ill children as well as anxious or manipulative parents.

9. In addition, the conduct of *joint paediatric–psychiatric work* with families of sick children in which physical or emotional problems interact will prevent tendencies to see children and family problems in either/or (physical/psychological) terms (Bingley *et al.* 1980).

10. Medical and nursing staff working with sick children and their families or in a special-care baby unit are often under severe stress themselves. They may benefit from a regular opportunity to discuss their own feelings in relation to their work, and provide emotional support for each other. The sharing of feelings of anger or depression that often arise in relation to paediatric care can benefit not only junior nurses and medical staff but also more experienced professionals. Such regular staff support groups may be initiated by, and involve, mental health professionals, or be organized entirely by the paediatric, nursing, and social staff themselves. Staff on wards where there is a relatively high level of mortality or chronic disability among patients are particularly likely to benefit from such group activities.

*Effectiveness of preventive measures*
There is evidence that the measures described above are indeed effective in reducing short-term distress due to hospitalization as well as emotional disturbance after discharge (Wolfer and Visintainer 1979). The long-term effects of hospitalization appear to have been reduced following the introduction of more liberal visiting and admission policies in the 1960s (Quinton and Rutter 1976). In well-run units, the level of distress is now a good deal less than it used to be (Shannon *et al.* 1984).

# 4.7 Care of the dying child

The death of a child is an uncommon event in the developed world. In the UK just under 1 per cent of pregnancies end in stillbirth, the psychosocial aspects of which are discussed in Section 5.1. After birth about a further 1 per cent die in the first 4 weeks of life—mainly of congenital anomalies. Subsequently, the death rate, still mainly due to congenital anomalies, is about 7 per 10 000 for the rest of the year, remains at about 7 per 10 000 for 1–5 years, and declines to about 3 per 10 000 from 5 to 15 years.

After the first year of life, accidents are the most common cause of death, accounting for about a quarter of the total mortality. Deaths from accidents are much commoner in lower socio-economic groups. After accidents, congenital anomalies and neoplasms (cancers) are responsible for most deaths from 1 to 5 years, and neoplasms remain the second commonest cause throughout the rest of childhood. The remaining deaths are caused by a wide variety of conditions, especially respiratory and cardiac disorders.

Although the experience of death is a quite frequently occurring event for those working in specialized centres dealing only with serious or complicated children's disease, for most other health professionals dealing with children it is fortunately not a common experience. For the family of a child with a fatal illness, the experience is made even more harrowing by the infrequency with which death nowadays occurs in the young. Because the event is so unusual, family members, friends, and people at work may feel uncertain how to behave or show sympathy, and leave family members isolated and unsupported.

## Phases of grief

### Parents
Parents of children with chronic illnesses with a frequently fatal outcome, such as leukaemia and cystic fibrosis, first experience grief when they are told the diagnosis and realize they have lost part of the future of their child. The term 'anticipatory mourning' is sometimes used to describe the grief reactions they experience at this time. Subsequently the parents experience a second phase of mourning, 'terminal grief' when the child dies. If the illness is acute, or the first phase of parental mourning is protracted, the second phase may supervene before initial mourning is completed.

The process of parental grief is similar to that described above in relation to reactions to the diagnosis of chronic physical handicap, but, of course, the emotions of parents faced with the death of their child are likely to be more intense (Burton 1974; Gyulay 1978). The period of shock when the parents hear the news may be accompanied by denial ('This can't be happening to me') and guilt ('What could I have done to make this happen?'). Somatic complaints such as dizziness, palpitations, weakness of the legs, and ringing in the ears are common. Subsequently, over the next few weeks, parents feel a variety of other emotions. Irrational fear and anger predominate. They may withdraw from relationships with others. Inability to accept the news may lead to internal 'bargaining' ('If I pray then this terrible thing won't turn out to be true'). Eventually the reality of the situation is more or less accepted and the parents are, for the first time, able to experience appropriate sadness.

Communication between parents may be difficult during these phases of grief, for mourning often proceeds at a different pace for the two of them. Pre-existing tensions may be worsened. A mild drinking problem may turn into a more serious one. Parents already on the verge of separation may stay together 'for the sake of the child', but their resentment towards each other and towards the child may increase. In contrast, some parents may genuinely be drawn closer together. These are likely to be the parents whose coping mechanisms in the face of stress (see above) are effective, and who find they cope better when they share the tasks placed on them by the child's illness.

### Children

Young children develop mature concepts of death much earlier than is generally appreciated, so that by the age of 5 years, about two-thirds of children in the general population have complete or almost complete understanding of the finality of death (Lansdown and Benjamin 1985). The concept of death may be more advanced if the child has experienced the death of a pet or of an elderly relative in the family. The development of a mature concept of death is a gradual process, that may be speeded up in children who spend a good deal of time in hospital because they are inevitably made more aware of its implications.

The fears of the fatally ill child are usually less to do with death than with fears of separation and painful procedures. Older children also experience these fears, but their reactions to the situation are usually much more complex. Although they need their parents for security, support, and information about their predicament, parent–child communication is often poor. Children see their parents to be anxious, upset, and perhaps angry. They may not wish to upset them further by expressing their own anxieties, and may indeed feel protective towards their parents. The child's own anger, perhaps stimulated by inability to lead anything like a normal life, may be directed towards the parents and inhibit communication further.

In the terminal phase, children may lose interest in their surroundings. With older children, it is often clear to those around them that they are aware of their predicament, even if there has been no discussion of the outcome. Fears of pain and discomfort become more prominent. In many cases the blurring of consciousness that precedes the final moments of death produces brief agitation before coma and death supervene.

## Management

The above account should make it clear that families respond in very different ways to the fatal illness of a child, and the main task of professionals involved with them is to ensure that family members feel supported, well informed, and understood whatever their particular reactions may be (Burton 1974). This is much more likely to occur if there is a 'key worker', usually a social worker, whose task it is to coordinate the psychosocial aspects of care provided by nursing, medical, and play staff, and mental health professionals. This 'key worker' may also take responsibility for linking the local community support network, including the primary health care team, school teachers, and local hospital staff with the specialist team.

During the initial grief phase, family members may wish to go over the details of the condition a number of times before they can absorb the information. Angry responses may elicit resentment in staff, but need to be seen as understandable. Parents who are involved in their child's care, and made to feel their contribution is important, will find it easier to cope.

Some parents, lost in their own grief reactions, may benefit from having the needs of the ill child and his or her sibs gently indicated to them. The needs of the sibs are often similar to those of the patient. Parents who feel discouraged and rejected by their children need particular support. Formal or informal groups of parents with children suffering from similar conditions may be helpful.

The problem of what to tell the child with a fatal condition is not easily resolved. Most of those experienced in the care of dying children now feel that openness concerning the outcome, linked to reassurance about the ability of nursing and medical staff to control pain and discomfort, is the policy of choice. However, parental views on this matter have to be paramount and, in a number of cases, the child will make it clear he is not ready to discuss the matter. A flexible approach is therefore desirable. If children are given information by professionals, it is important that parents are present at the time. Often parents will prefer to communicate information themselves.

Whether the final terminal phase is in hospital, at home, or as is very occasionally the case, in a hospice for dying children, strong support from professional staff is particularly important over this period. Requests for autopsy, procedures for allowing the parents privacy in which to show their feelings, permission for donation of organs for transplantation, explanation of the information on the death certificate—these matters need to be planned beforehand so that parents do not suddenly feel abandoned by familiar staff after the death of their child. Most hospitals dealing with children with fatal illnesses arrange a routine post-bereavement visit with paediatric staff a 4–6 weeks after the death, so that issues concerning the final stages of illness and the genetic implications of the disease can be further discussed. Some hospitals and children's units arrange post-bereavement counselling for parents who think they would find this helpful.

## Further reading

Eiser, C. (1985) *The Psychology of Childhood Illness.* Springer Verlag, New York.
Lansdown, R. and Goldman, A. (1988) The psychological care of children with malignant disease. *Journal of Child Psychology and Psychiatry* 29, 555–67.
Petrillo, M. and Sanger, S. (1972). *Emotional care of hospitalised children.* J. B. Lippincott, Philadelphia.

# 5

# Psychosocial aspects of specific physical conditions

## 5.1 Stillbirth and neonatal death

### Prevalence

In England and Wales stillbirth (late fetal mortality) occurs at the rate of about 8 per 1000 live births. Death *in utero* is usually associated with the presence of a condition of the placenta or umbilical cord, or a congenital abnormality in the fetus. However, a range of other factors may be responsible. Death in the first 4 weeks of life (neonatal mortality) occurs in about 9 per 1000 live births. Congenital abnormalities, the respiratory distress syndrome, infections, and complications of delivery are responsible for most such neonatal deaths.

### Psychological reactions

A comprehensive review of parental reactions to perinatal loss is provided by Zeanah (1989). Parental reactions to the unexpected death of the baby they have anticipated vary in severity. Severe emotional shock is the commonest immediate reaction, and this may be followed by feelings of numbness, preoccupation with thoughts of the dead baby, and difficulty in accepting that the baby is dead (Forrest *et al.* 1982). A mourning reaction is likely to persist for several months, and may normally last up to about a year. During this time recurrent feelings of sadness and thoughts of the dead baby are likely to recur. Significant factors affecting the severity and resolution of grief include the overall physical health of the mother, the gestational age at the time of the loss, the quality of the marital relationship, and pre-loss mental health symptomatology (Toedter *et al.* 1988). Occasionally more prolonged reactions occur with longer-lasting implications for the marriage and family life, but this is probably unusual, though there is a lack of systematic follow-up data.

### Psychosocial aspects of management

Arrangements made following stillbirth need to be sensitive to parental feelings (Forrest *et al.* 1981). The issue of a certificate of stillbirth or death certificate may be distressing. Arrangements for the funeral need to take into account parent's wishes. Mourning may be facilitated if the parents are encouraged to see, touch, and even perhaps hold their dead baby. A photograph of the baby may help parents to remember and come to terms with the death. Parents should be encouraged to attend the funeral and the baby's grave should be properly marked. Results of necropsy should be communicated to them by the paediatrician or pathologist with concern for their anxieties about future pregnancies. Parents appreciate open and honest discussion with the paediatrician and nursing staff after the loss, and are very sensitive to attempts staff may make to ignore or avoid talking to them. They do not appreciate advice from doctors to have another baby so as to 'get over' their loss, but value information about future risks to enable them to make their own decisions.

Brief counselling for bereaved parents to help them come to terms with their loss has been shown to abbreviate short-term adverse effects (Forrest *et al.* 1982).

## 5.2  Sudden infant death syndrome (cot death)

### Definition

This involves the sudden and unexpected death of an infant that is not linked to any clear-cut cause.

### Clinical features

Babies involved are almost always between 1 month and 1 year of age, with a peak incidence between 4 and 6 months. The baby is usually found dead by the parents either first thing in the morning or after a period, sometimes quite brief, of being left unattended. There is a story of mild respiratory symptoms in the previous 2 or 3 days in about a third of cases. 'Near-miss' cot deaths in which a baby is suddenly noted not to be breathing, and is resuscitated by shaking, also occur. There may be no serious after-effects, but in some cases anoxic brain damage sustained may result in severe mental retardation.

### Prevalence

The condition occurs in about 2–3 per 1000 live births. In the UK, it is the most common cause of death in the period from 1 month to 1 year, and its prevalence is rising. It occurs in all social classes, but is more common in babies living in deprived circumstances. It is distinctly more common in families where there are adverse social circumstances and the father is unemployed or absent (E. M. Taylor and Emery 1988). There is a slight male preponderance. There is an increased risk of the condition in babies where a sib has previously died of it. The risk in these children is about 1 per cent.

### Aetiology

Basically this is not understood, but it is thought that sudden respiratory arrest occurs as a result of a wide range of different infective, circulatory, biochemical, and immunological abnormalities. In a minority of cases (probably a small minority) deliberate and concealed suffocation by parents is the cause of respiratory arrest (Emery 1985), and this possibility should be considered more carefully where there is a recurrence.

Usually, affected babies have had a normal development, but in a minority there have been previous apnoeic episodes or other, less obviously related abnormalities of the pregnancy and neonatal period.

## Psychosocial aspects

Parents' reactions to the sudden and unexpected death of their young infant are, not unexpectedly, very severe. There is usually a period of disbelief, then shock, followed by an acute bereavement reaction. Mourning for the dead infant usually lasts several weeks or months, but may be further prolonged. Some parents, though by no means all, experience a sense of guilt for inadequate care and failure at not being available to their baby in his terminal moments. Parents may express anger towards each other, the dead child's sibs, or professionals who have previously seen the child. The grief experienced by older siblings may go unnoticed, but it is as often prolonged as that of their parents.

There are some reports of serious subsequent impact on parent relationships and attitudes to siblings, though such reports have generally been anecdotal and may have overemphasized the frequency of deleterious effects.

**Immediate management** may be complicated by police enquiries which may, even if sensitively carried out, add to the distress of parents. Inquest proceedings may also exacerbate guilt and anger.

## Counselling

Families where an infant has died in this way will be helped if strong, unequivocal emotional support is available from members of the primary health care team, especially the health visitor and family doctor. Additional help can be provided by the paediatrician, especially if he has already been involved in the care of the child. A full explanation about the nature of the condition, with some idea of the various possible causes, should be given. Parents should be reassured about their own responsibility, except in cases where this is definitely in question. Continued support, at least over several weeks, will be necessary and, if mourning is prolonged, referral to a psychiatric or counselling service is desirable. All UK parents should be given a leaflet available from the Foundation for the Study of Infant Deaths, London, which provides information about the condition and availability of counselling. A description of a children's hospital-based bereavement counselling service for parents who have lost a child in this way is provided by Woodward *et al.* (1985).

## Home monitoring

In a proportion of babies who have had a 'near-miss' cot death, who have been observed to have apnoeic attacks in hospital, or who have been born as sibs to babies who have died in this way, alarm systems to wake parents if the child develops respiratory arrest can be installed. False alarms, with enhancement of anxiety, are common with this apparatus. The impact on family life may be considerable (A. L. Wasserman 1984), with marital breakdown apparently a not infrequent occurrence. Other factors may have affected family relationships, but such alarm systems should not be instituted without due consideration of their benefit and possible adverse effects. There is, however, no evidence that adverse cognitive or emotional effects on the child persist, when children monitored in this way are followed up several years later (Kahn *et al.* 1989).

# 5.3  Chromosomal abnormalities

## Introduction

The most common defects involving chromosomes are:

1. *Numerical.* There are one or more additional chromosomes, or a chromosome is absent.
2. *Mosaic.* In a single individual, some cells have normal chromosome patterns, others abnormal.
3. *Deletions.* Part of a chromosome is missing.
4. *Translocations.* A chromosome normally present at one site has transferred to another site.

Abnormalities may affect either one of the two sex chromosomes or one or more of the 22 pairs of autosomes.

## Autosomal abnormalities

### Down's syndrome (trisomy 21)

This is the most common chromosomal abnormality and accounts for about one-third of all cases of mental retardation with IQ less than 50. In 19 out of 20 cases of the syndrome the cause is an additional free chromosome arising from an abnormality of cell division. In the remainder (translocation type) the extra chromosome at position 21 is displaced from a different site, so there is no overall increase in chromosome number. The trisomy (but not the translocation type) is associated with increased maternal age, with the condition occurring in about 1 per cent of babies born to mothers aged over 40 years, but only about 0.1 per cent in births overall. However, most babies with Down's syndrome are born to younger mothers.

**Physical features.** The condition is evident at birth. Children have upslanting eyes with epicanthic folds, a short nose with flat nasal bridge, white (Brushfield) spots in the iris of the eye, and a large tongue. The fingers are short and stubbed with an incurved little finger and a single palmar crease rather than the double crease normally found. The children are of short stature and show muscular hypotonia. There is a high rate of associated abnormalities, especially malformations of the heart (septal defects) and gut (atresias), some of which are incompatible with life. Upper respiratory tract and ear infections may produce chronic hearing difficulties.

**Intellectual development.** The very early development of children with this syndrome is often rather little below average (Carr 1970). However, by the age of 5 years the level of development is nearly always definitely retarded with a mean IQ of about 45 and a range from 25 to 70. Higher IQ levels have been recorded and a small number of Down's syndrome children are educable in ordinary schools. Early stimulation programmes help to promote development, and home-reared children achieve higher educational attainment in reading and arithmetic (though not higher IQ) than do non-home-reared children (Carr 1988), but there is no evidence that achievement of intelligence within the normal range is other than a most unusual outcome. The number now educable in ordinary schools has increased, partly as a result of more intensive early stimulation programmes, and partly because of new policies integrating mentally handicapped children into the mainstream of education. There are usually no specific deficits, although visuomotor skills and tactile discrimination may be particularly impaired. The IQ level tends to remain stable during middle childhood, though it is influenced, within limits, by the quality of care provided. Institutionalized children generally do worse, and those given extra stimulation programmes rather better than average.

In adulthood, individuals with Down's syndrome are particularly prone to develop presenile dementia of Alzheimer type.

**Impact on family life.** The presence of a child growing into adulthood but achieving only partial independence will alter the lives of both parents and siblings in a most significant way. Depression in mothers of affected children is more common than it is in the general population, and this is often related to the experience of loss of the normal child the mother expected to have. Guilty feelings are not usually a prominent feature, though they may be present. Rates of divorce are not higher in parents (Gath and Gumley 1984; Carr 1988) and a significant number of parents speak positively of the experience of bringing up a child with this condition. Behaviour problems in sibs, whose life is often markedly affected, are more often found if the child with Down's syndrome has disturbed behaviour (Gath and Gumley 1987). There is a higher rate of psychiatric problems in the older sisters in these families, possibly due to their carrying an undue burden of care. Academically and socially, siblings do as well as their peers (Gath and Gumley 1987).

**Behavioural development.** Children with Down's syndrome do not show specific behavioural features, although many are unusually good-natured, contented, and warm in their personalities, characteristics that have been confirmed in systematic comparisons with siblings. A minority are aggressive and difficult, and these are more likely to be of translocation type. The rate of significant behavioural disturbances is, if anything, slightly higher than that in the general population, but lower than that found with other mentally retarded children (Gath and Gumley 1984). Odd, eccentric behaviour with ritualistic habits, and mannerisms unrelated to autism are rather characteristic.

**Management.** The general assessment and management of these children should follow the lines described under the heading of general mental retardation (Section 2.6.2). Identification of the specific chromosome abnormality is important, as the translocation type is more likely to recur in subsequent children than the more common form.

*Other trisomies*
Trisomies at other chromosomal sites also cause multiple congenital abnormalities and mental retardation, but occur much less frequently. Trisomy 18 (Edwards' syndrome) is the most common. It is not usually compatible with life, although some babies with this condition do survive for several months.

*Other autosomal abnormalities*
Deletions and absences of chromosomes usually result in early spontaneous abortions, but there are occasional survivors. Babies with the *'cri du chat'* syndrome due to deletion of a short arm of chromosome 5 have severe developmental delay and microcephaly as well as the characteristic cry and an abnormal facial appearance.

**Sex chromosome abnormalities**

These are less common than autosomal abnormalities, and their effects are usually less obvious and more readily compatible with life. As with autosomal abnormalities, diagnosis is confirmed by microscopic examination of the chromosomes, using special techniques. Screening for sex chromosome abnormalities may be achieved by examining cells obtained from a buccal smear. Where more than one X chromosome is present, so-called Barr bodies, or clumps of chromatin, can be seen at the periphery of the nucleus. This technique is therefore often helpful when a male is suspected of having an extra X or a female an absent X chromosome.

*Turner's syndrome*
These girls have 45 chromosomes, one X chromosome being absent. The chromosome cell structure may be homogeneous, but about half the

patients show a mosaic chromosomal pattern with some cells containing XO and others XX chromosomal material.

**Clinical features.** The child is likely to be short and stocky. The neck may be webbed and the carrying angle at the elbow increased. External genitalia remain infantile, and the internal reproductive organs do not develop either. Other congenital abnormalities (especially coarctation of the aorta) are common. Menstruation does not occur at puberty, and the girls are infertile. The condition usually presents with growth delay, but, surprisingly, bearing in mind the short stature, the diagnosis is not infrequently delayed until puberty when the girl presents with amenorrhoea.

**Psychological aspects.** General intelligence is not usually affected, but there is often a specific defect of visuospatial ability, and mathematical skills are also frequently impaired. Moderate and severe intellectual retardation does not, however, appear in excess (Silbert *et al.* 1977).

These girls have to contend with unusual looks, although quite often their facial appearance is unremarkable. They are unusually small in childhood, and the mean average adult height is around 4 feet 6 inches to 4 feet 9 inches. They need opportunities to express their concern about their appearance and stature. This is all the more necessary as the diagnosis is so often delayed, and the girl is unprepared for the future in store for her. There may be anatomical problems with sexual intercourse, and infertility is certain. Menstruation can be induced with cyclical oestrogen therapy, and this may be desirable to help the girl feel more normal. Recent treatment with oestrogens, androgens, and growth hormones holds promise for improving growth and sexual development.

Social relationships are often impaired, and there may be other behaviour problems. This does not appear to be due to the short stature itself, as comparisons with children with other types of short stature suggest that girls with Turner's syndrome are specifically at risk for these difficulties (McCauley *et al.* 1986). Girls with the syndrome show a stronger feminine sex-role identity than do girls with constitutional short stature (Downey *et al.* 1987).

*Noonan's syndrome*
Individuals suffering from this condition, who are usually male, but may be female in appearance, have most of the features of Turner's syndrome, but the chromosome abnormality is absent. Presumably a closely related chromosomal defect is present, but this has not so far been detected.

*Klinefelter's syndrome*
This occurs in boys who have an XXY chromosomal structure. The appearance in childhood is not particularly abnormal, although patients tend to be rather tall. At puberty the testes remain small, and breast development occurs in about half the cases. The condition, which occurs in

about one in 600 male births, is usually not diagnosed until adulthood when patients present with infertility.

**Psychological aspects.** Retrospective information suggests that the children are likely to be anxious and dependent, and to have rather high rates of emotional disorder (Nielsen 1969). This has been confirmed by a study of a small total population sample (Bancroft *et al.* 1982). In adulthood the rate of homosexuality and schizophrenia may be slightly increased, but most men with this condition have a normal sexual orientation and do not have serious forms of mental illness. Intelligence of affected boys is usually within the normal range and compatible with high intelligence, though the mean IQ is somewhat below average (Ratcliffe *et al.* 1990) and both oral and written language may be affected, with implications for anticipatory guidance (J. M. Graham *et al.* 1988). Most people with Klinefelter's syndrome are within the average range of ability, but mild mental retardation is not uncommon.

### XYY syndrome

Children with this karyotype are often unusually tall, but show no other clearly characteristic physical features. The condition was once thought to be associated with criminal behaviour, but the evidence for this is weak. In fact children in the general population with the karyotype are definitely more likely to have early language delay and later reading (but not arithmetical) retardation (Ratcliffe *et al.* 1990). In childhood they do show high rates of temper outbursts and antisocial behaviour, and psychiatric referral occurs more frequently than in the general population. Psychological intervention should not, however, be influenced by the presence of the karyotype which will, in any case, usually be unsuspected (Ratcliffe and Field, 1982).

### Fragile X syndrome

In this condition, a break is detectable at q 27.3 on the long arm of one of the X chromosomes, provided that the cells are cultured in a folate-deficient medium. Laboratory diagnosis is sometimes problematic, as, even in fragile-site-positive individuals, only a relatively small proportion of cells show the defect. A fragile site in over 2 per cent of cells is now generally reckoned to confirm the diagnosis (Hagerman 1989). The number of cells demonstrating the site bears no relationship to the nature or degree of intellectual impairment.

Although the condition is sex-linked, the pattern of inheritance is not that of a simple sex-linked Mendelian disorder, and the precise mechanism of inheritance remains to be resolved. Only four-fifths of affected males show cognitive defects, and about one-third of 'carrier' females are themselves mentally retarded (Davies 1989).

Fragile X accounts for about 10 per cent of all boys with severe mental retardation and for about 6–10 per cent of otherwise unexplained mild mental retardation (Thake *et al.* 1987). It is therefore the second most common cause of mental retardation after trisomy 21.

Affected children show characteristic physical features, especially a high forehead, prominent supra-orbital ridges, 'simple' low-set protruding ears, a prominent jaw and post-pubertal macro-orchidism.

Mental retardation may be associated with autism or autistic features, but autism probably does not occur to a greater extent in fragile-X-associated mental retardation than in other forms of mental retardation (Payton *et al.* 1989), although many males with fragile X do show autistic-like features (Hagerman 1989), especially hand-flapping, hand-biting, and poor eye contact.

Most pre-pubertal males are at least mildly retarded. Reading and spelling skills are relatively advanced, in contrast to poor arithmetic performance. Speech is frequently 'cluttered' with frequent dysfluent, rapid, tangential remarks and poor topic maintenance. The simultaneous presence of visuospatial difficulties, impaired constructional ability and shortened digit span may be indicative of non-dominant hemisphere dysfunction. As the pattern of inheritance is so poorly worked out and there are, in addition, questions about the reliability and validity of laboratory diagnosis, genetic counselling is problematic. Bearing in mind that about 6–10 per cent of children with unexplained mild mental retardation (IQ 50–70) show fragile sites (Thake *et al.* 1987), it is advisable to request laboratory investigation of chromosomes of all individuals with IQs below 70, specifically requesting examination for fragile sites. A positive result from an experienced laboratory requires explanation to the parents, who should be advised that further male children have about a one in two chance of mental retardation and female children a roughly one in six chance of being affected. (These figures are based on knowledge that about 10 per cent of males and 70 per cent of females with a fragile site are of normal ability.)

*Prader Willi syndrome*
This condition is produced by an autosomal anomaly—a deletion on the short arm of chromosome 15. Mental retardation with severe early speech and language delay is accompanied by a voracious appetite leading to obesity, hypotonia, and hypogonadism.

The special psychological feature of this condition is the management problem posed by the severe obesity (Laurance 1985). Results are improved if strict food restriction is undertaken from an early age. Behaviour modification is the most effective means of reducing weight, but, as with other forms of obesity, it may be possible to obtain weight reduction, though difficult to sustain it. Appetite suppressants are of little or no value.

Other associated psychiatric problems include anxiety reactions and obsessive compulsive symptoms, though aggressive behaviour is unusual

(Whitman and Accardo 1987). These problems may be a reaction to the stress of experiencing food restriction in the presence of a voracious appetite. Parents need a great deal of support and in many developed countries there is a parent support association specifically for this condition.

*Other sex chromosome abnormalities*
There is a large number of such abnormalities, all of which are extremely rare. Most of them are incompatible with survival of the fetus to term.

# 5.4  Malformations

## 5.4.1  CLEFT LIP AND PALATE

### Clinical features

The baby is born with a defect of the lip and/or palate, resulting from failure of the normal process of fusion of embryonic tissues during fetal development. If the defect involves the lip, it will be obvious at birth. Isolated defects of the palate may only be identified when the palate is examined during routine physical examination of the newborn, or when the baby begins to feed. Occasionally an isolated defect of the palate is covered with mucous membrane (submucous cleft), so that the defect is not identified until the child's speech is noted to have a nasal quality. Cleft lip and palate are associated with other congenital abnormalities, especially congenital heart disease, in about 10 per cent of cases.

The condition is surgically correctable, and, in skilled hands, the cosmetic results are excellent with little obvious deformity. Depending on the nature of the defect and the practice of the surgeon, operative procedures may be carried out in one or two stages. Surgical repair of the lip is now carried out at 1–2 months of age and of the palate between 12 and 18 months. Consequently, if the palate is affected, the baby will require a dental plate and/or a special teat for feeding for the first few months of life. Babies whose cleft is confined to the lip can usually be breast-fed.

Apart from the minor cosmetic defect, the outcome may be complicated by recurrent ear infections leading to hearing loss. Articulation defects may also arise either as a result of the structural abnormality or the hearing loss.

### Background factors

Cleft lip and palate syndromes occur in about 1.4 per 1000 births, cleft lip alone being rather more common than cleft palate. The condition is mainly genetically determined with a risk of 1 in 25 for children born to an affected parent. The concordance rate is 40 per cent in monozygotic twins, but 5 per cent in dizygotic twins and sibs. The administration of certain drugs, including lithium and epanutin, to the pregnant mother also increases the risk.

## Psychosocial factors

This is one of the commonest congenital abnormalities, and there is extensive information on psychosocial aspects (Lansdown 1981; Tobiasen 1984). Parents naturally respond with severe shock to the first sight of their deformed babies, but this usually dissipates in a few days. Feeding difficulties are common, not just for the obvious practical reasons, but because many mothers are made anxious by the child's defect, and this anxiety is transmitted to the baby.

The later psychological adjustment of children with repaired clefts is usually good, assuming that the surgical repair has been effective. However, even minor physical defects attract teasing at school, and this is especially the case with deformities of the mouth and teeth (Lansdown and Polak 1975). The great majority of children have intelligence and educational attainment within the average range, although the mean level of ability is somewhat lower than average. However, teachers tend to underestimate the intelligence of children with unsatisfactorily repaired clefts (L. C. Richman 1978), and the presence of undetected hearing problems may also lead to a misdiagnosis of mental retardation.

## Psychosocial aspects of management

Following a warning about the baby's appearance with an explanation that the defect, although apparently seriously deforming, is surgically correctable, the baby should be shown to both parents, preferably together, as soon as possible. The parents will need continuing emotional support and explanation, especially over the first few days. Before-and-after photographs of children with repaired clefts are usually very reassuring. It is, of course, important that other congenital defects that may be associated are identified, so that the implications of these can also be discussed with parents.

Procedures and advice concerning feeding should take into account the fact that mothers may often find the situation upsetting as well as practically difficult. However, the great majority are able to cope and become very deft in helping their children to feed. It is important that parents are given full details of the nature of the operation with information about how the baby is likely to look immediately postoperatively and during the subsequent weeks.

As children get older, the combined advice of an audiologist and speech therapist will become more important in the detection of hearing impairment and remediation of speech defects. These professionals are likely to be able to give good advice on how to talk to the child about the nature and cause of his defect. It is usually possible for parents themselves to help children cope with teasing at school through discussion with the child's teachers. Referral to a psychologist or psychiatrist will be necessary only if psychosocial problems are persistent, and, in this case, factors other than the cleft deformity are likely to be of significance. Psychiatric problems may centre around feelings of inferiority and an exaggerated sense of feeling rejected, with low self-esteem. Counselling with behavioural advice to improve social skills will often be effective, but a minority of affected children, usually those who have experienced additional family problems, may require more formal psychotherapy. As children reach their teens, it is important for parents to explain to them the genetic nature of the defect and the risk of inheritance. Prenatal diagnosis is not possible at the present time.

## 5.4.2 HYPOSPADIAS

This is a congenital deformity of the penis found in about 1 in 350 male babies. The cause is unknown, but there is a familial, probably genetically determined transmission. The external meatus (orifice) is on the under-surface of the penis rather than at the tip, and this is often associated with bowing of the penis (chordee). Treatment is surgical, but surgical practice is variable. There is now a tendency to operate in single-stage operations within the first two years of life, often before one year. Results are usually good for appearance and function, and depend mainly on the severity of the deformity. Postoperative complications requiring reoperation are not at all uncommon.

### Psychosocial aspects

Parents are often particularly anxious about this condition and the operation necessary to correct it, for a number of reasons. They are likely to be concerned (with some reason—see below) for the later sexual competence and fertility of their sons. The condition is one that many parents feel embarrassed about, and secrecy can mean it is more difficult for them to receive emotional support from family and friends. Parents of children attending playgroups, etc. before they are operated upon are also often concerned that the child will be discomfited when asked by others why he has to urinate sitting down.

Operative procedures need careful explanation to both parents and children well before surgery. The child is often in considerable discomfort following operation, and the appearance of bandages and tubes is anxiety-provoking, especially if unexpected. The postoperative course is sometimes accompanied by an excess of aggression or withdrawn behaviour in the child.

Despite apparently normal endocrine function, a usually relatively normal appearance of the penis, and a normal capacity to achieve erection, there is an increased risk of sexual and psychological maladjustment of the adult who has had a hypospadias repair in childhood. Most will not have such problems, but follow-up studies of variable quality summarized by Schultz (1983) suggest that, as a group, they experience intercourse later and less frequently than controls (R. Berg *et al.* 1981). They are less likely to marry and have children. They may suffer from feelings of inadequacy and avoid competitive activities (Schultz 1983).

**Psychosocial implications**

Clearly, particular care needs to be taken in the preparation of both the boy and his parents before operation. The timing of operation has been a matter of controversy. A good surgical result may be easier to obtain if the operations are delayed to between 3 and 5 years, although some claim a satisfactory outcome with operation in the first 2 years of life. The fact that the period of 6 months to 3 years is a time when separation anxiety and possibly fears of mutilation are at their height has led some to suggest that operation should not be performed at this age (American Academy of Pediatrics 1975). However, others believe that from a psychological point of view, operation is preferable before 1 year, and that, in competent hands, full correction is feasible before 2 years (Schultz 1983), and this is now common practice.

The apparently unsatisfactory sexual and personality outcome of a number of affected children suggests that counselling both in childhood and adolescence might have a beneficial preventive effect. It is important that youngsters who might well believe themselves still to be deformed or inadequate should have an opportunity to express their feelings and thoughts on this matter, and that adults who have required the operation in childhood should have access to sexual counselling if they are experiencing difficulties in this sphere.

# 5.5 Injuries to children

## Introduction

In developed countries injuries are the major causes of death in the 1–14 year age group. The term 'accident' with its implication that random forces outside human control are responsible, is inappropriate to describe the events producing injuries. In fact, much is known about the factors involved in injury occurrence, and the behavioural characteristics of injury victims and their families have been found to be of significance.

## Prevalence and background factors

In England and Wales, injuries cause over a tenth of deaths in the 1–14 year age group and about half the deaths in the 15–19 age group. Road traffic accidents are responsible for about half these deaths. Drowning and burns also produce significant mortality. Injuries producing less serious results are very common. About one child in every six presents to a hospital casualty department with an injury at least once *each year* (Child Accident Prevention Trust 1989). Boys are injured overall about twice as commonly as girls, and in the 15–19 year age group they have a much greater vulnerability to road traffic injuries. There is a strong social class link with fatal injuries, in that children of manual and unskilled workers are much more likely to sustain fatal injuries and burns at all ages. Children in social class 5 are seven times as likely to sustain a fatal road injury than those in social class 1 (Department of Health and Social Security 1980). Social class trends for less serious injuries are much less marked, and some studies have failed to find any social class differences at all.

*Personality and behaviour*
Boys who are aggressive, competitive, impulsive, and overactive are prone to injury, and temperamentally 'difficult' children are especially at risk. Children who have had repeated injuries in the preschool period are several times more likely to continue to suffer repeated injuries during school than those without such an early history (Bijur *et al.* 1988), but persistent adverse social circumstances may be as important in explaining this finding as childhood personality.

*Family type and functioning*
Children living in single-parent or reconstituted families, and children of less well-educated mothers are at extra risk (Matheny 1986). In addition, maternal depression is associated with the occurrence of injuries to children (G. W. Brown and Davidson 1978). Children are more liable to injury at times of family crisis and change, as when another family member is suffering a physical illness or the family is involved in a house move. All these factors are likely to exert their effect by reducing the level of supervision.

**Psychosocial aspects of management**

When a child sustains an injury, physical assessment and appropriate medical treatment must take immediate priority. However, where resources in terms of time and available staff permit, psychosocial considerations can contribute significantly to management.

*Assessment*
Psychosocial factors need consideration because of:

1. Relevance to medical management. A child of distressed and panic-stricken parents may need to be admitted because supervision of care at home will be inadequate.
2. The possibility that injuries have been non-accidentally caused. The nature of the injuries and mode of presentation may provide clues (Section 1.3.2).

3. The circumstances of the injury, which may give rise to serious parental guilt and self-blame. The quality of parental supervision at the time of the accident may need attention if recurrences are to be avoided.

4. The possibility that behavioural disturbance in the child or emotional problems in other family members may have made an important contribution to the occurrence of the injury.

### Management
Again, emergency physical treatment must take precedence. In deciding whether further management measures need to be taken in the light of the injury, it is necessary to:

1. Establish whether the injuries could have been caused non-accidentally. If this is the case, further management is described in Section 1.3.2.

2. Enable parents to express feelings of guilt and anger with themselves, their child, or someone else thought responsible for the accident.

3. Provide treatment for children with behavioural and emotional problems which have come to notice as a result of the injury. Similarly, occurrence of an injury may, even if the injury is 'accidentally' caused, indicate that the quality of parental supervision or level of family conflict and non-communication require further attention.

## Psychological and psychiatric outcome following injury

### General
The majority of injured children do not develop any significant behavioural or emotional sequelae. In a minority, however, an injury can act as a trigger precipitating an emotional or behavioural disorder. In a small group, injury to a limb can be followed by a hysterical reaction (Section 3.2) in which the disability is much greater than the extent of the physical injuries would warrant. Other forms of neurotic disability following accidents appear to be uncommon in childhood (Lishman 1987). The nature as well as the severity of the injury is likely to be of significance in determining seriousness of behavioural and emotional outcome. Thus head injuries (see below) may have a direct effect on subsequent cognitive and behavioural development, owing to interference with brain function. Burns (see below) may have specifically deleterious effects for other reasons, for example the necessity for repeated painful procedures and the prolonged hospitalization required to achieve rehabilitation. Severe, multi-system injuries to children are likely to have profound psychological and social effects on the survivors and their families, lasting for years after the injuries occurred (B. H. Harris *et al.* 1989).

### Head injury
**Cognitive outcome.** The best indication of the severity of a head injury lies in the duration of post-traumatic amnesia (PTA)—the period of time from

the occurrence of the injury to the point when memories again become continuous. The relationship between the severity of injury and psychological sequelae has been thoroughly investigated by Klonoff *et al.* (1977) and by Rutter *et al.* (1983).

In children with PTA lasting less than 2 weeks, cognitive impairment does not appear to occur. The child's intellectual status following the injury can be expected to remain the same as before it. In children with PTA lasting 2–3 weeks, transient or persistent cognitive impairment occurs in virtually all children, and persistent impairment is the commonest outcome if PTA is longer than 3 weeks. Visuospatial and visuomotor skills are impaired to a greater degree than verbal skills. The pattern of cognitive functioning following injury does not provide a clear guide to the site of the damage, although academic attainment may be more affected following left-hemisphere lesions. In children with severe injuries most cognitive improvement takes place in the first year, though minor improvements can continue into the second year, or even longer than this (Klonoff *et al.* 1977). Rather surprisingly, the age of the child is not a good predictor of outcome (Oddy 1984).

**Psychiatric outcome.** The rate of subsequent behavioural and emotional problems does not appear raised in children with a PTA of less than 1 week. Thus, if a child with an injury of this level of severity shows a behavioural or emotional problem after the injury, it is improbable that this is a direct result of the brain injury.

In children with PTA lasting longer than a week, the rate of psychiatric disorder rises sharply, whether or not cognitive impairment has occurred. Such psychiatric disorder is present in excess even in those children without obvious neurological sequelae. In general, the behavioural sequelae that occur are rather non-specific in type, except that marked social disinhibition (undue cheekiness and proneness to make embarrassing remarks) is especially common. The type of symptom is unrelated to the site of the lesion, except that depression appears more common when the lesion is in the right frontal or left posterior region.

Although age and sex appear little related to psychiatric outcome, the intensity of family reactions and the number of social problems in the family do appear to have a modifying effect. Children whose parents react least intensively and those from families living in advantaged circumstances appear to have the best outcome.

### Accidental poisoning

**Epidemiological findings.** In the UK about nine in every 1000 children under the age of 5 years require attention in accident and emergency departments each year because they might have ingested a harmful substance. However, of these a significant number (perhaps two-thirds) turn out not to have swallowed anything harmful (Calnan *et al.* 1976). Over 30 child deaths a year occur in the UK as a result of accidental poisoning. The peak age is between 2 and 3, with boys somewhat more at risk than girls. Drugs and household products together account for over 95 per cent of substances ingested (Wiseman *et al.* 1987). Analgesics, anxiolytics, bleaches, and detergents are most frequently involved.

**Social and family factors** are often prominent in the background of accidentally poisoned children (Shaw 1977). The children tend to come from overcrowded homes in which there have been a number of changes of accommodation. Their families are characterized by high rates of marital disharmony and other types of family conflict (Sobel 1970). Rates of mental ill-health in parents are high. Surprisingly, there is rather little relationship between the degree of care over storage of drugs in a safe place and the risk of the child being poisoned. The personal characteristics of the child, however, are relevant. Poisoned children are more likely to show high levels of overactivity.

Principles of *management* are similar to those described for children suffering injuries (see above). The *outcome* of children who have been accidentally poisoned does not suggest that the poisoning itself is likely to have long-term adverse effects. Intellectual status and rates of emotional disturbance have not been found to be significantly different than in matched groups of children. However, the adverse psychosocial factors so often found in children who have been poisoned are themselves likely to have long-term consequences on behaviour.

**Non-accidental (deliberate) poisoning** of children is discussed in Section 1.3.2.

### Prevention of injuries and accidental poisoning—psychosocial aspects

There is increasing evidence that, with an active injury prevention policy, a decline in the rate of injuries can be achieved. An excellent, practical approach to prevention has been published in the UK by the Child Accident Prevention Trust (1989). Preventive measures that appear effective include:

1. Road safety precautions. Compulsory use of seatbelts in cars and compulsory use of helmets for motorcycle riders.

2. Domestic equipment design. Careful design can make paraffin heaters, ovens, saucepans, etc. more childproof.

3. Protective barriers for windows in high-rise council or municipal property.

4. Teaching children to swim reduces drowning accidents.

5. Design of playgrounds, for example use of impact-absorbing surfaces.

6. Child-resistant containers for drugs, both prescribed and non-prescribed.

7. The use of smoke alarms reduces the risk of domestic fires and burns.

8. Encouraging family doctors to provide antidepressants, etc. in small quantities and to give warnings to mothers of their dangers to children.

In theory, all these measures require back-up by an active health-education programme with instruction to parents on how to help their children avoid injuries from domestic equipment, drugs, and other dangerous or poisonous substances. In practice, evaluation of such health education appears to have disappointing results, while the practical environmental measures listed above are more successful (Gatherer *et al.* 1979).

Although there are various correlations between adverse background psychosocial factors and occurrence of injuries, there have so far been no clear-cut implications for prevention arising from these findings. For example, it would probably not be cost-effective to direct health-education programmes to those at high risk for psychological reasons. Nevertheless, attention to psychosocial factors in the management of these children found to be subject to repeated injury (see above) might well have an impact on the risk of subsequent re-injury.

## Burns

This type of injury presents specific psychosocial problems. Children living in socially deprived, overcrowded circumstances where accident control is more difficult are more likely to be involved. Children with certain physical conditions and behavioural states, such as epilepsy and the hyperkinetic syndrome, are particularly predisposed. Burns may be inflicted non-accidentally, for example by lighted cigarettes.

In the acute phase, after the burn has been sustained, there will be acute pain and, if the burn is extensive, intensive care for life-saving procedures. Subsequently the child may require numerous grafting operations with prolonged hospitalization.

Parents are particularly likely to experience guilt after their child has been burned. The extreme suffering of the child, and the fact that the injury may indeed well have been preventable, will exacerbate guilt feelings. Parents may find it very difficult to accept the scarred appearance of their child and respond by rejection or overprotection. Parents (preferably both fathers and mothers together) will benefit from counselling sessions to ventilate their feelings and to ensure that they are understood and emotionally supported. Group sessions of parents of burned children are also helpful, especially if attended by nursing staff who can provide practical advice, as well as emotional support.

The level of development and maturity of the burned child will influence the type of reaction that occurs (Sawyer *et al.* 1982). Many children regress in their behaviour and become more clinging. There are often intense, fearful reactions to the painful procedures involved in changing dressings, and phobic, over-anxious behaviour may persist after recovery from the burn (Stoddard *et al.* 1989). Low self-esteem and antisocial behaviour are relatively common in older children and teenagers. Burned children will benefit from careful preparation and explanation in relation to painful procedures. Allowing the child an element of control in the procedure will also reduce distress. Hypnotic procedures (Section 6.2.7) may be helpful in the older child and teenager. Children will also benefit from the opportunity for discussion of their feelings about their appearance. Before return to school, it is useful to arrange a visit by a nurse or psychologist to the school, so that teachers can be helped to prepare other children for the changed appearance of the burned child as well as the child's likely sensitivity to teasing.

## 5.6 Infectious diseases

The psychological and psychosocial impact of infectious disease is particularly affected by whether it occurs in the intrauterine period or after birth.

### Fetal infection

#### Congenital rubella

Congenital rubella occurs in the UK in about 1 per 50 000 live births, but more frequently when epidemics occur. The fetuses of women who contract the condition in the first two months after conception are most at risk (95 per cent), but the risk has diminished very substantially by the end of the fourth month of fetal life.

Commonly occurring sequelae are ophthalmic conditions (especially cataract), a wide range of neurological disorders, microcephaly, heart defects, sensorineural deafness, and enlargement of the liver and spleen. Mental retardation occurs in about 50 per cent of cases.

Among mentally retarded children affected by rubella, autism (Section 2.8.1) is a particularly common complication (Chess and Fernandez 1980). Autism is usually of no-onset type, but may occur after a period of apparently normal development, perhaps because the live virus may continue to survive after the birth of the child.

**Identification** of rubella as a cause of mental retardation or autism in an older child depends on the presence of the characteristic associated defects. Antibody tests are of value in excluding the diagnosis but, if they are positive, this is not necessarily significant, as the child may well have contracted a postnatal infection. The great majority of children with rubella-induced autism have been identified as suffering from rubella at birth.

**Prevention** of congenital rubella is possible with existing vaccines, and, in many countries, the routine immunization of young children with measles, mumps, and rubella vaccine (MMR) is achieving increasing success, though coverage in many areas of the UK is still far from the minimum desirable level of 90 per cent. Termination should be offered, after explanation of the risk of fetal infection, to women contracting the condition early in pregnancy.

*Cytomegalovirus (CMV)*
This also is a mild condition in adults, but may produce serious defects if transmitted from pregnant mother to fetus. In the UK about one in 200 babies is born showing evidence of active infection (excreting virus in urine), and about 10 per cent of these show subsequent evidence of damage. The child may be born with low birthweight, enlargement of liver and spleen, purpura, and eye defects, but most are normal at birth.

Older children who have suffered from the congenital infection may show mental retardation and specific learning difficulties. Among clinic groups of affected children, very high rates of expressive language delay have been identified (Williamson *et al.* 1982), but it is possible that rates are lower in infected children in the general population. There is, however, no good evidence that CMV causes mental retardation or learning difficulties in the absence of other characteristic clinical features.

*Toxoplasmosis*
This condition is produced by a protozoon parasite, toxoplasma gondii, which is widely prevalent in the population. Damage may occur if the mother is infected at any stage of the pregnancy, though the effects are worse the earlier in pregnancy the infection occurs. In late pregnancy, transplacental transmission may occur. The affected fetus may be immediately aborted or may be stillborn at term. Alternatively, it may born with signs of the disease, rapidly developing jaundice and a purpuric rash. Most infected children are, in fact, asymptomatic.

Brain damage produced by the parasite may result in severe mental handicap and/or epilepsy. There do not appear to be any specific aspects to the mental retardation or associated psychiatric states. Both autism and the hyperkinetic syndrome may occur. The condition will be suspected if radiological investigation of the skull shows intracranial calcification, and it is confirmed by serology, the toxoplasmosis dye test. Treatment is available, but its influence on the course of the condition is uncertain.

*Human immunodeficiency virus*
The human immunodeficiency virus (HIV), a slow virus belonging to the subgroup of retroviruses, produces loss of T4 lymphocytes with consequent loss of resistance to infection. Acquired immune deficiency syndrome (AIDS) is the end-stage of the condition, in which chronic degenerative neurologic disorder with recurrent opportunistic infections progresses to a fatal outcome. Children usually succumb to opportunistic infections without going through a stage of chronic neurologic disorder. Groups at particular risk are children with haemophilia (infected before HIV screening of blood was possible), male homosexuals and their partners (including women in the case of bisexual men), and drug abusers using contaminated

needles. Vertical transmission may occur to the fetus via the placenta in infected, pregnant women.

*Aetiology: psychosocial aspects*

Psychosocial factors are of major importance in understanding the development of HIV infection. The quality of early family life and relationships is probably of significance in the development of a homosexual pattern of behaviour (see Section 2.13), as well as in drug abuse. This generalization does not apply to infected haemophiliacs, and there are other exceptions. Apart from infected haemophiliac individuals, HIV sufferers tend to come from groups of young people alienated from society and living in socially deprived circumstances.

HIV infection may impinge on cognitive, emotional, and behavioural development. In the young child infected by vertical transmission, HIV may be suspected because of developmental delay, though it is unusual for this to occur without physical symptomatology. In teenagers, neurologic involvement may first become obvious because of memory impairment or the unexpected onset of learning difficulties.

Awareness of the infection and its probably fetal outcome may produce severe anxiety reactions and depressive states. Rage and subsequent aggressive behaviour may also occur. The effect on family relationships, often already seriously disturbed, may lead to family breakdown. Finally, even in the uninfected, fear of HIV infection is not infrequently a major focus of anxiety in teenagers with anxiety states, obsessive–compulsive disorders and (more realistically) those who realize they have a homosexual tendency or are already involved in homosexual activity or drug abuse.

There are several psychosocial components in the management of HIV infection (Krener and Miller 1989). An infant with a congenital infection, vertically transmitted, will often be identified at birth because of the known maternal infection. Infants of infected mothers will be HIV-positive (i.e. have IgG antibody) at birth and this may persist for up to 18 months. However only between a fifth and a quarter of these children are actually infected. Not surprisingly, this uncertainty has great implications for their relationship with their carers. For reasons of child protection, such an infant may require substitute care from birth if social circumstances are inadequate. Placement of the baby suffering from HIV infections will produce a need for support for the new carers. If a baby stays with the mother, as will usually be the case, the mother will herself need help in dealing with guilt reactions as well as reactions relating to her own fatal or potentially fatal disease. The child will in any cse require careful follow-up. The child may remain asymptomatic, at least until the early school years, although some degree of development delay and a rapid terminal course is more common (Ultmann *et al.* 1985).

A baby or young child with an unexplained infection or developmental delay may be suffering from an HIV-induced condition. In these circumstances, testing for HIV will require permission from parents, for whom the implications of a positive result will be considerable. Parents may well need the opportunity to ventilate their feelings about the test before the result is obtained and, in the event of a positive finding, further counselling will certainly be necessary. Parents should be given the opportunity to be seen separately as well as together.

Teenagers at risk of infection and those already infected will often require counselling and the opportunity for peer group support. Psychiatrists and paediatricians may be involved in the development of outreach programmes and other social measures for drug abusers and young people alienated from their families. Parents of these teenagers may themselves require counselling to cope with their own grief at the predicament in which their children find themselves.

Finally, teenagers who have been recipients of infected blood (a group to which fortunately no new cases are now being added) may benefit from counselling to help them deal with their anger and feelings of helplessness. Such teenagers and their families will already have had to cope with the need to accept the presence of a chronically handicapping condition, and this second blow will often produce very considerable adjustment problems.

### Other infections
Other infections, such as herpes simplex and syphilis, may also impair fetal development and, if they do not result in fetal death and abortion, may be followed by severe developmental retardation.

## Postnatal infections
Acute infectious disorders are the most common forms of illness in childhood. Acute respiratory infections alone account for about half the general practice consultations provided for children, and acute gastroenteritis is also widely prevalent. Most respiratory and gastrointestinal infections are caused by viruses, but bacteria are responsible for a significant proportion. Specific viruses are also responsible for the infectious 'fevers', such as measles, German measles (rubella), mumps, chickenpox, and whooping cough (pertussis).

### Aetiology: psychosocial aspects
The likelihood of a child developing an infection will depend on factors related to exposure to the infective organism (especially the intensity of exposure and the virulence of the organism) and factors related to the child (especially the level of immunity and specific host response). Both broad social and specific environmental factors have been demonstrated in population studies to be of importance in susceptibility and probably exposure to infection.

**Social factors.** There is a definite tendency for children from lower social class families to suffer more from respiratory infections and gastroenteritis. The reasons for higher rates of bronchitis in UK working-class children probably lie in the higher rates of parental smoking and large family size in the lower socio-economic groups (Colley and Reid 1970). A large number of siblings will usually mean overcrowded accommodation and greater likelihood of exposure to infection. Parents in lower socio-economic groups are also more likely to suffer from lung disease themselves, and their children are more likely to be exposed to environmental pollution, especially smoke from domestic fuel and industrial waste.

Working-class children are less likely to be immunized against infections such as whooping cough and measles, both of which have significant postinfective morbidity. They are more likely to face stressful circumstances in their lives (see below).

**Stress.** Stressful circumstances may predispose to infection. For example, beta-haemolytic streptococcal infection has been found to occur more often after a period of family stress (Meyer and Haggerty 1962). The level of stress in a family is also a predictor of whether antistreptolysin O titre increases following the acquisition of streptococci in the throat. Exposure to stressful circumstances probably influences resistance and speed of recovery in other infectious illnesses.

The precise mechanism whereby stress affects susceptibility to infection is unknown. Probably relevant are the effects of stress on corticosteroid metabolism and on neuroimmunological pathways mediated via the limbic–hypothalamic–pituitary system.

## Mental states associated with infection

### Acute and subacute confusion (delirious states)
Confusional states may occur as a result of toxic or metabolic processes without invasion of brain tissue during any infectious illness, but occur most frequently in young children and when transient or persistent encephalopathy or meningitis are present. During the course of the infection the child may first become irrationally angry, irritable, or miserable. This is usually rapidly followed by episodes of drowsiness or reversal of sleep rhythm. Hallucinations, especially of visual type, are quite common. Generalized convulsions, sometimes prolonged, may occur. A child may misperceive other people to be objects, and vice versa, and be disorientated in time and place. Sudden irrational fearfulness or depression may occur. A marked feature of delirium is its variability and transience. Very occasionally, depending on the underlying condition, the cerebral involvement may result in coma, which can be irreversible, but usually marked and rapid improvement in the child's mental state occurs as the child recovers from the underlying condition.

The differential diagnosis of delirium includes especially drug-induced states following accidental or non-accidental ingestion of toxic substances, or other forms of intoxication. A careful history and the absence of other signs of infection, such as fever, should clarify the diagnosis. Epilepsy may also present with acute or subacute confusion. Again the history, together with careful observation of the child to exclude non-convulsive status (Section 5.7.2), and investigations, especially an EEG, will be helpful here.

### Post-infective psychiatric and psychological sequelae

**Depressive states.** Post-infective feelings of apathy, mild malaise, and depression are not uncommon, and usually last no more than a few days. Much more rarely, prolonged depressive reactions of considerable severity and sometimes associated with school refusal may occur (Section 2.9.3). These usually follow influenzal-type illnesses and infectious mononucleosis, but may occasionally follow other infectious conditions. The reason why some individuals are affected in this way is unclear, but it has been suggested that, in infectious mononucleosis, psychological reactions are associated with specific T-cell phenomena (Hamblin *et al.* 1983). Affected children are often living in adverse psychosocial circumstances, so that an interaction between this immunological phenomenon and environmental factors is probable.

Management of these severe postinfective depressive states will depend on the presence of associated family and school problems and the personality of the child. As with other depressive states, psychological treatment of family or individual type should be instituted (see Chapter 6), and antidepressant medication may be indicated.

**Personality changes.** Since the First World War, when physical recovery from episodes of encephalitis was apparently followed by the development of overactivity, antisocial behaviour, and learning difficulties, it has been suggested that encephalitic infections may produce long-lasting impairment of behaviour and learning (Sabatino and Cramblett 1968; Greenwood *et al.* 1983). The evidence for this view is rather thin, and it is more likely that subsequent behavioural disturbances are due to family disruption, separation, and deprivation at the time of infection.

**Dementia.** Progressive intellectual deterioration following infection is extremely rare, but occurs particularly in sub-acute sclerosing panencephalitis. Children with this condition usually present between 3 and 18 years, with a mean of 10 years. It is four times as common in boys as in girls. The onset is often characterized by a fall in school performance, thought to be due to laziness or some environmental stress. Deterioration in speech and handwriting is common. The child may become irritable and moody. The presence of an organic rather than a psychologically determined condition should be suspected because there is actual loss of skills rather than merely failure to make progress.

The condition is thought to be due to the measles virus, and follows several months or years after an apparently typical measles infection. A characteristic form of myoclonic jerk occurs in the more advanced stage of the condition. The diagnosis is made by a sustained rise in measles antibody titre in serum and cerebrospinal fluid. There is also a pathognomonic EEG pattern with bursts of high-voltage spike waves occurring on a background of slow activity. The disease is almost always fatal, though there may be a very prolonged period of chronic illness with slowly progressive dementia.

### Learning deficits

Infections not involving brain substance or meninges do not have subsequent direct effects on learning. There is some evidence suggesting that children suffering from meningitis (e.g. of tuberculous type) and encephalitis (e.g. mumps, measles, chickenpox, etc.) may show long-term general and specific deficits. Specific deficits of visuospatial and visuomotor abilities have been reported in a group of children as long as 6–8 years after an attack of haemophilus influenzae meningitis (H. G. Taylor *et al.* 1984). However, other studies (Tejani *et al.* 1982) suggest that low IQ may not be a feature of children who have previously suffered from haemophilus infections. There is some evidence that children contracting meningitis or other CNS viral infections before the age of 1 year are at risk for later learning deficits (Lawson *et al.* 1965; Chamberlain *et al.* 1983). In one follow-up study of children suffering from bacterial meningitis in the first year of life and then followed up 6–16 years later, about 10 per cent had some degree of mental retardation, and about 15 per cent had language delay, often linked to hearing loss (Sell 1987). However, the children studied may have been unusually severely affected, and the figures quoted may over-represent the risk of subsequent handicap. In isolated cases, severe brain damage, with associated mental handicap, may follow upon a meningitic or encephalitic illness.

### Parental reactions

A severe infection, particularly in a baby or young child, is an extremely frightening experience for parents. Even parents who are professionally qualified as nurses or doctors find the experience of coping with a somnolent child with high fever very stressful. Parents whose children suffer febrile convulsions often believe that their child is about to die (Baumer *et al.* 1981).

Reassurance of parents whose children are suffering from acute infection is emotionally supportive and reduces unnecessary anxiety. In those uncommon cases where infections lead to serious sequelae, there is often a long period of uncertainty before it is clear whether the child has suffered permanent brain damage. In these circumstances, sharing of uncertainty rather than unrealistic reassurance is likely to be more helpful. Specific psychological intervention may be indicated in some families, but long-term benefit of such intervention has not been established (Kupst *et al.* 1983).

# 5.7 Disorders of the central nervous system and muscle

## 5.7.1 CEREBRAL PALSY

### Definition

In this group of conditions there is a permanent, non-progressive disability of movement arising from damage or dysfunction of the immature brain. Motor impairment affects muscular power and tone as well as posture, and there are often involuntary movements. The brain damage present is often responsible for associated sensory, language, and visuospatial disabilities, and sometimes epilepsy. If the damage is widespread, there may be global mental retardation. Paralysis may involve all four limbs (quadriplegia), one side of the body (hemiplegia), or the lower limbs (diplegia). The main disability may be postural involving balance (ataxic cerebral palsy), or there may be involuntary athetoid or choreoathetoid movements (dyskinesia).

### Aetiology

The likely cause of the condition varies with the type of cerebral palsy (Hagberg and Hagberg 1984). Most cases of hemiplegia have a pre-natal, third trimester cause. Spastic diplegia and quadriplegia are frequently related to prematurity, often extreme prematurity. Dyskinesias are likely to be due to severe birth asphyxia. Simple ataxia, especially if linked to mental retardation, is often inherited.

### Prevalence

The overall rate is between 1.5 and 3 per 1000. Boys are slightly more affected than girls, and there is no particular association with social class.

### Clinical features

These vary considerably depending on the nature of the condition and its severity. A child with cerebral palsy may be handicapped by no more than slight weakness and incoordination in an upper limb. Alternatively, the child may be quadriplegic, blind, and grossly mentally retarded. Commonly there is a stormy birth, but the child appears to have made a good recovery after receiving intensive care. There may be serious feeding difficulties. Subsequently, later in the first year of life, the child is noted either to be unusually stiff or floppy. Motor milestones are delayed, but the child may appear alert and responsive. Disorders of articulation are common. A diagnosis is made on the basis of the history and clinical examination. It is often difficult to predict outcome in terms of disability in the first 3 or 4 years of life, but by about 5 years, the degree of likely future disability is usually reasonably clear.

## Psychological development

The cognitive development of the child with cerebral palsy is often hampered for two main reasons. First, the brain damage or dysfunction may have affected parts of the brain necessary for adequate intellectual function to occur. Such damage may have an effect on both general and specific cognitive functions. Second, the fact that the child has cerebral palsy may seriously limit his experiences and thus his opportunities for learning. A child who because of his disability is confined to one room in his home may well have difficulty learning about spatial relationships.

## General intelligence

Survey evidence suggests that about 50 per cent of children with cerebral palsy have IQs below 70 (the proportion is about 2.5 per cent in the general population) (Rutter *et al.* 1970*b*). Cerebral palsy is, however, quite compatible with normal or even superior intelligence. The type of cerebral palsy is relevant. Children with hemiplegia are less likely and children with diplegia and quadriplegia more likely to be mentally retarded. As a group, children with choreoathetosis have the best-preserved intelligence.

The degree of physical disability is related to IQ level, so that the more physically handicapped the child, the more likely he is to be retarded. However, physical disability can be a very misleading guide to intelligence. Some mildly physically handicapped and apparently alert children may, on testing, turn out to be profoundly retarded. Many cases also occur in which gross physical disability, especially of dyskinesic type, is accompanied by high ability. This means that systematic assessment of cognitive skills is necessary.

## Specific abilities

It is very common for the child with cerebral palsy to have a wide scatter of skills. In general, verbal skills are more highly developed than non-verbal, but there is wide variation.

1. *Language disability.* Articulation problems may be severe because the orolaryngeal musculature is affected by spasticity. Such articulation problems may be accompanied by difficulties in the comprehension or expression of language at a higher level. Aphasia is not common, but does occur.

2. *Perceptual abnormalities.* Even in the absence of damage to the eye, ear, or other related structures, the child may have a problem in perceptual discrimination. Children with auditory perceptual deficits may be unable to distinguish like-sounding consonants such as p and b, t and d. Even if able

to perceive differences, they may be unable to perceive the *meaning* of different sounds. Thus a child may be able to perceive that a doorbell makes a different noise from a running tap, but in the absence of visual cues, be unable to recognize that the one sound means someone at the door and the other that someone in the kitchen is washing up or getting a drink.

Perceptual problems may be visual, auditory, or tactile. Visual perceptual problems can be divided into those that are visuospatial, and those that are visuomotor. *Visuospatial difficulties* may show themselves, with inability to distinguish right from left or under from above. Distortion of body image may also result from a visuospatial deficit. *Visuomotor difficulties* occur when the child fails to perceive the changing visual stimuli produced by his own movements (for example, in drawing and writing) to help him coordinate further motor activity

3. *Abnormalities of motor coordination.* Disability in the execution and coordination of movement (clumsiness or motor dyspraxia) may be present to a degree unexplained by muscular weakness or abnormality of muscle tone. This may give rise to particular difficulties in using a spoon and fork, drawing, writing, and dressing. Again these problems may occur as a result of brain dysfunction or inadequate opportunity for learning, or as a result of both of these.

These motor and perceptual abnormalities may result in the child showing educational retardation in relation to his measured intelligence. Specific reading retardation (Section 2.5.5) is, for example, common in children with cerebral palsy.

### Psychiatric aspects

**Prevalence.** Children with cerebral palsy have a higher-than-expected rate of psychiatric disorders. In the Isle of Wight Survey, about 40 per cent of children with cerebral palsy showed significant emotional or behavioural problems, the rate being slightly lower in children without associated epilepsy and higher if epilepsy was also present (Rutter *et al.* 1970*b*). This figure compares with 6–7 per cent in the general population.

No specific types of psychiatric problems have been shown to be present, although physically handicapped children are especially likely to show poor relationships with other children, and to be more frequently the object of teasing and bullying. Cerebral-palsied children with associated mental retardation are likely to show high rates of the hyperkinetic syndrome and autistic syndromes.

Reasons for the high rates of psychiatric disorder in the cerebral-palsied group include:

1. *The direct effect of brain damage and dysfunction.* Children with physical disabilities not involving the brain do not show such high rates of psychiatric disorder even when their disabilities are severe, so some direct

effect of brain damage on behaviour in the cerebral-palsied child seems probable. There is no difference in the rates of behaviour disturbance between children with lesions in the dominant and non-dominant hemisphere, but dominant lesions are more often associated with 'externalizing' behaviour (e.g. aggression, disobedience) and non-dominant lesions with 'internalizing' problems (e.g. anxiety, fearfulness) (Sollee and Kindlon 1987).

2. *Rejection by peers and siblings.* Children with physical disabilities are less frequently chosen as friends by other children. It is not clear whether cerebral-palsied children are particular likely to be rejected (Richardson and Royce 1968), but certainly they often have serious social problems. When young, they are likely to be subjected to teasing and ridicule, and as they reach adolescence, although teasing decreases, they are less able to participate in the increasingly mobile social lives of their contemporaries.

3. *Poor self-concept and low self-esteem.* From the age of 4 years onwards (Teplin *et al.* 1981), cerebral-palsied children see themselves as different from others, and, as time goes on, they are increasingly likely to perceive themselves to be less valuable than others.

4. *Parental attitude.* Disturbed cerebral-palsied children may have parents who have unrealistic ideas about their disability, either denying or exaggerating its extent (Minde 1978). Some parents also withdraw warmth and emotional commitment from their physically handicapped children as they get older.

5. *Family relationships.* The rate of divorce and separation is not raised in parents of cerebral-palsied children. However, when arguments and tension do occur they often focus over problems in the management of the handicapped child, and this predisposes the child to become disturbed.

6. *Educational failure.* Children who, in addition to their physical disability, are failing at school in relation to their apparent ability, are more likely to show disturbance.

*Implications for management*
Various measures taken by health professionals are likely to reduce the rates of psychiatric disorders in cerebral-palsied children (R. D. Freeman 1970).

1. *Early identification* with involvement of both parents in the assessment process is likely to help parents to value their handicapped children appropriately. If parents are given practical advice how to help their children gain mobility and communicate, this also will be helpful.

2. *Comprehensive assessment*, especially by a physiotherapist, speech therapist, and psychologist, will ensure that associated handicaps are identified. Systematic assessment for the presence of perceptual and perceptuomotor skills and disabilities will assist teachers and parents in

promoting development. Continuing emphasis on what the child *is* able to do, and on developmental progress rather than on deficits, is also supportive.

3. *Promotion of social development* can be achieved by ensuring that the child is given every opportunity to improve his communication skills. The use of an alternative communication system involving signing, such as Makaton, may be helpful here. The child needs opportunities for non-rejecting interaction with others of his own age that will result in avoidance of social isolation. Whether the child is placed in an ordinary or a special school may be less important than whether parents and teachers are aware of the child's social needs and go out of their way to encourage friendship. In the UK, PHAB (Physically Handicapped/Able Bodied) clubs for mixed groups of handicapped and non-handicapped adolescents have established themselves as useful in this respect.

4. Providing opportunities for the physically handicapped child to express anxieties and worries should ensure that communication about the disability is adequate. Health professionals can provide older cerebral-palsied children and adolescents with such opportunities, listen to their concerns, and, with the agreement of the child or teenager, feed back information to those who might be able to help. Sexual counselling is particularly important for adolescents who, because of reduced contact with other youngsters, may experience guilt concerning sexual feelings and masturbation.

5. The management of established psychiatric disorders is no different from that required in the non-handicapped population. Family approaches, as well as individual and behavioural psychotherapy, may be indicated. It is important to establish how the disturbed child and parent perceive the disability, and to what degree misperceptions and poor communication about the handicap are responsible for emotional disturbance. Children with such physical disorders are often disturbed for reasons that have little or nothing to do with their disability.

### 5.7.2 EPILEPSY

**Definition**

Epilepsy is present when there is a repeated paroxysmal discharge in the brain producing sudden episodic involuntary alterations of movement or sensory experience. Epilepsy is the most common but not the only cause of seizures (sudden episodes of loss of muscle power). Seizures may be produced by other organic as well as non-organic conditions, such as hysterical or 'sick role' disorders.

**Classification**

Epileptic attacks can be classified in a number of different ways according to their clinical and EEG features, their duration, their origin in the brain, their cause, and their association with fever.

*Clinical and EEG features of the attack*
### Classification of epilepsy
(Modified international classification)
I   Generalized seizures

Tonic, clonic, and tonic–clonic
Absence
Myoclonic
Infantile spasms

II   Partial seizures

Simple (motor, sensory)
Complex

III   Unclassified

Seizures may be **generalized** or affect the patient only partially. Major generalized seizures (tonic–clonic) usually involve a brief warning sensation (the aura) followed by loss of consciousness with spasmodic twitching of the limbs, clenching and unclenching of the jaws, and sometimes incontinence of urine. Afterwards there is usually a period of disorientation and sleep, often followed by headache. In generalized absence attacks there is usually a brief (10–15 seconds) episode of impaired consciousness with staring of the eyes, blinking, and rarely facial twitching. Such attacks often occur in clusters. In myoclonic seizures, there are sudden jerky movements and loss of muscle power often producing a brief fall to the ground or a fall back in the cot or bed. Infantile spasms, which nearly always occur in the first year of life, involve attacks of sudden flexion of the head, knees, and arms, associated with gross EEG abnormality of characteristic type.

Partial seizures may be *simple* or *complex*. A simple partial seizure may consist of an isolated disturbance of sensation or movement, for example twitching of a limb. A complex partial seizure is likely to involve some impairment of consciousness, the child, for example, being able to stand, but not respond to questions. Such partial loss of consciousness may be followed or accompanied by a variety of movements and sensation including, for example, brief flicking of the eyelids, sucking movements, spasmodic jerking of one limb and/or one side of the face. In seizures arising from epileptic activity in the temporal lobes, the child may experience a variety of symptoms including a constriction or feeling of fear in the lower chest, strange sensations of taste or smell, other hallucinations, giddiness, mood changes, and automatic behaviour.

## Duration of the attack

Most epileptic attacks last a few seconds to a few minutes. Prolonged seizures, referred to as status epilepticus for generalized seizures, or, for partial epilepsy, epilepsia partialis continua, call for immediate medical treatment.

## Site of the epileptic activity in the brain

The paroxysmal electrical activity may arise from the deeper central brain structures (centrencephalon) and spread in a generalized fashion over the whole substance of the brain. This will produce a generalized seizure described above. Alternatively, the epileptic attack may arise from a cir-cumscribed part of the brain such as the frontal or temporal lobe (see above).

## Cause of the attack

Most epilepsy is not symptomatic of an underlying condition. It probably occurs because of an inherited disposition to attacks. All individuals have a disposition to epilepsy and, under certain conditions (e.g. if given an injection of insulin and made hypoglycaemic), would suffer an attack. The reason why some people with non-sympatomatic epilepsy have an unusually strong disposition is usually unknown, and it is often unclear why attacks occur at particular times, though sometimes triggering factors, including psychological stress and exposure to flickering light, can be identified.

Some epilepsy, however, arises from an identifiable cause, for example from infection, intoxication, neoplasm, trauma, or vascular disorder. One common cause is the presence of a scar that acts as a focus or trigger for electrical activity. Such a scar may occur if a part of the brain dies as a result of anoxia occurring in an earlier prolonged seizure. Temporal lobe epilepsy is believed sometimes to occur in this manner following prolonged febrile convulsions (see below).

## Association with fever

Febrile convulsions are attacks that commonly occur in infants and young children aged from about 6 months to 4 years that are accompanied by fever. Non-febrile convulsions may occur at any age. About 3 per cent of children with febrile convulsions go on to develop epilepsy—the remainder are subsequently asymptomatic.

## Background factors: psychosocial aspects

Epilepsy is one of the most common physical disorders occurring in child-hood. Apart from febrile convulsions (3 per cent of children) the rate is about 7–8 per 1000 in school-aged children (Rutter *et al.* 1970*b*). Both sexes are equally affected, and there is a slight tendency for working-class children to be more at risk.

There are various important ways in which psychological and social factors interact with the presence of epilepsy.

*Behavioural changes*

1. For a day or two preceding a seizure, a child may become more excitable, moody, and irritable. This change in mood may be terminated by the seizure. Observation of the benign effect of an epileptic seizure on difficult behaviour in mentally ill adults led to the development of electro-convulsive therapy as a treatment for depressive psychosis.

2. Epileptic seizure (ictal) activity in the brain may be reflected in an alteration of mood and behaviour that can be mistaken for a non-epileptic psychiatric disorder. Children who are experiencing absence attacks with brief lapses of concentration may be mistakenly thought to be inattentive for other reasons. Surprisingly, this most commonly occurs in children who are known to have epilepsy because they are also having major attacks, but it does also occur in children not suspected to be suffering from epilepsy. In psychomotor attacks, usually arising from the temporal lobes or other parts of the limbic system, the child may experience a sudden acute sensation of constriction and fear in the lower chest and retrosternal area, and may run or cling to the nearest person. This may be mistaken for a non-organic anxiety attack. Hallucinatory experiences may be thought to arise from a psychotic disorder. Automatic behaviour may be thought to reflect a dissociative or hysterical state. In each of these cases, diagnosis will be assisted by the presence of the acute onset, by the apparently inexplicable nature of the symptoms, and by corroborating EEG changes. The increasing availability of telemetric and ambulatory EEG apparatus allowing EEG recording to be obtained during an attack is a significant advance in investigative technique (Stores 1985).

3. Seizures may be followed by behavioural change. In the period of neuronal quiescence after an epileptic attack, the child may wander around in an apparently purposeless manner, still mildly disorientated. This behaviour may also be mistaken for a psychiatric disorder of hysterical type, especially as the seizure disturbance itself may not have involved a major attack.

*Precipitation of seizures by psychological stress*

Probably about a quarter of children suffering from epileptic seizures have attacks brought on by stressful changes in the environment. There is no specific type of stress likely to be involved, though in schoolchildren, examinations are often an important factor. Family arguments are also occasional triggers. It has been suggested that epilepsy is more likely to persist when parental attitudes to the child are both intrusive and rejecting.

The significance of stress and the precipitation of attacks is often difficult to confirm because of the possibility that stress is coincidental. However, experimental EEG controlled studies have made clear that neurophysiological changes of an epileptic type can indeed be stress-induced, even though strong clinical evidence for the importance of stress in an individual child may be hard to come by (Stevens 1959).

Children are sometimes able to control the frequency of their attacks to some degree. Some children can inhibit attacks by controlling their thoughts in a particular way. It is always worth asking children if they can stop an attack if they want to, and useful to find out why they will sometimes do this and sometimes not. Others are able to induce fits. The most common method of fit induction is by finger or hand flicking while looking at the sun or other bright light. This behaviour occurs most commonly in mentally retarded children, but is not unknown in children of normal ability. Clearly, in these cases the occurrence of a fit is a rewarding experience for the child. Further evidence in favour of the possibility of at least a small self-control element in epilepsy comes from studies demonstrating that, in selected cases, behavioural intervention can reduce seizure frequency (Dahl *et al.* 1988).

### Associated behavioural and emotional disorders

The rate of psychological problems in children with epilepsy has been found to be over four times that in the general population (Rutter *et al.* 1970*b*). If the epilepsy is associated with evidence of an anatomical brain lesion (e.g. if the child also has cerebral palsy), then the rate of psychological disorders is even higher. These rates of psychological problems are much higher than those occurring in children with other physical disorders not involving the brain, such as diabetes mellitus and asthma. Psychological problems in children with epilepsy therefore probably arise partly because of the specific effects of brain dysfunction, rather than merely as a reaction to the presence of physical disability. Further evidence to support this view is provided by the fact that children with epilepsy have more psychological problems than children with other physical conditions at the time of onset of the fits (Hoare 1984*a*), and that subclinical epileptiform discharges do seem to affect psychological function (Siebelink *et al.* 1988).

There is no specific type of psychological problem linked to epilepsy. Such children are just as likely to show emotional problems with withdrawal and anxiety as aggressive behaviour disorders. At the same time, children showing some relatively uncommon psychiatric disorders, such as autism (Section 2.8.1), do have a high risk of the development of epilepsy.

Children with epilepsy are more likely to show psychological problems if their attacks are of psychomotor type, especially if they are of low intelligence (Harbord and Manson 1987). There are a number of possible reasons for this. Children who suffer such seizures often have experiences, such as unexplained anxiety and other mood changes, and strange sensations that are near enough to reality to be confused with it. It may be easier for children with generalized attacks which do not involve such strange but near-normal experiences to see their disorder in an objective, detached manner. Second, most psychomotor attacks arise from the limbic system of the brain, deep to the temporal lobes. The main nuclei of this system are the hippocampal formation, the amygdala, the hypothalamus, and the septal area. The brain neurotransmitters, especially dopamine, noradrenaline, and serotonin, that have been shown to have specific effects on mood and behaviour, exist in considerable concentration in these nuclei. This may explain why dysfunction in these areas of the brain may be especially likely to produce psychological problems. A proportion of teenagers with psychomotor epilepsy develop confusional psychotic states, sometimes following a run of seizures. Further, children with chronic temporal lobe seizures are at an increased risk for the development of paranoid schizophrenic disorders in later life. Such psychotic states occur, on average, 7–8 years after the onset of seizures (Slater *et al.* 1963). Finally, a very high proportion of children with infantile spasms (see above) later show developmental retardation with associated autistic features and/or hyperkinesis (Riikonen and Amnell 1981).

Most children with epilepsy have normal intelligence, and the mean IQ of children with this condition is only a little below average, provided that children with evidence of associated brain damage (e.g. cerebral palsy), are excluded from consideration (Rutter *et al.* 1970*b*). However, children with epilepsy have a high rate of specific reading disability and those with focal spike discharges (especially those with left temporal spikes) are particularly likely to show such a deficit (Stores 1978). The rate of learning and behavioural problems of schoolchildren who have previously suffered a febrile convulsion is slightly higher than that in the general population (Wallace 1984). The presence of neurodevelopmental abnormalities before the convulsions and the occurrence of frequent febrile fits increases the risk of such later problems.

Other background factors that make children with epilepsy more prone to psychological problems include low socio-economic status and depression or anxiety in the parents. This suggests that children with epilepsy have heightened vulnerability to the same sort of factors that produce psychological problems in children without epilepsy.

*Anticonvulsant drugs, behaviour, and learning*
Anticonvulsants are given with the intention of altering brain function in a specific manner—by raising the threshold for the development of epileptic attacks. It is not surprising that their effects, when administered, are less specific than this, and spread over into other areas of brain functioning, affecting behaviour and learning.

Phenobarbitone, sometimes used in the prophylaxis of febrile convulsions (see above), often produces irritability and overactivity, especially in the young child, though long-term effects on cognitive development and behaviour do not persist when the drug is discontinued (Wallace 1984). Phenytoin overdosage is occasionally accompanied by a confusional state, and long-term administration of phenytoin may produce chronic intellectual deterioration, which is not necessarily reversible. These effects may be produced by their influence on folate metabolism. Other anticonvulsants including those others most commonly used—carbamazepine, ethosuximide, primidone, sulthiame, and sodium valproate have all been described as having adverse effects on behaviour and learning in a minority of children to whom they are administered (Corbett *et al.* 1985). Inattention, lack of concentration, and minor disorientation and irritability are the most common effects observed. There is no anticonvulsant drug of which one can say with confidence that mental side-effects do not occur, but side-effects affecting intelligence and behaviour are most common with phenobarbitone and phenytoin, and least common with carbamazepine (Trimble 1987).

*Surgery for epilepsy*
In children with a clear-cut, unilateral temporal focus who do not also suffer from generalized seizures, temporal lobectomy may be indicated. Such children often show serious psychiatric disorders before operation, but, given careful rehabilitation, the majority can be expected to be free of psychiatric symptomatology subsequently if the operation reduces seizure frequency (Lindsay *et al.* 1984). Hemispherectomy can similarly be helpful in those unusual cases of hemiplegic epilepsy when unilateral paralysis is severe and fits are poorly controlled. Again, surgery that is successful in fit control will also relieve severe psychiatric disorders. It is likely that both temporal lobectomy and hemispherectomy have been underutilized as surgical procedures in the past.

*Impact on the child and family*
Epileptic seizures are frightening to witness, and even doctors who are not frequently involved in coping with them would agree they feel anxious when they see one. The affected individual appears in the grip of an unseen force, and the onlooker, especially in the presence of a grand mal fit, feels helpless. In many developing countries now, and in Western industrialized societies not so long ago, epilepsy indicated possession by an evil influence, causing children with fits to be kept from going to school and mixing with other children. In 1949 a representative survey of American adults revealed that 43 per cent of them would object to their child playing with an epileptic child, though by 1974 this proportion had dropped to 14 per cent (Caveness *et al.* 1974). In 1949, 55 per cent believed that people with

epilepsy should not be employed (19 per cent in 1974). Community prejudice against epilepsy is therefore widespread, although decreasing.

The impact of seizures on the child cannot readily be gauged by examining rates of emotional and behavioural disorder, as some children are probably disturbed not as a psychological reaction to the fits but as a more direct result of brain dysfunction. Nevertheless, there is ample clinical evidence to suggest that the impact is considerable (Ferrari *et al.* 1983). Children with generalized seizures are frequently made anxious by the fact that they do not know what goes on when they have a fit. They may fear that they will lose control, betray secrets, or make fools of themselves. Adolescent girls may be frightened that they will be sexually assaulted. It is common for youngsters to be worried that they will not be able to obtain a job or get married. Sometimes there are realistic fears of being teased or bullied at school if the fits are witnessed.

Other members of the family are sometimes profoundly affected. Many parents of infants and young children having their first febrile convulsion think that their child will die in the fit and some never forget the experience (Baumer *et al.* 1981). Parents often become depressed, especially if the fits prove difficult to control or affect the child's capacity to lead a normal life. There is an increased rate of psychological problems in both the parents and sibs of children with epilepsy (Hoare 1984*b*).

This account should not obscure the fact that most children with epilepsy do not show psychological problems, and that the majority of parents develop realistic ideas about the nature of their children's disorder and cope remarkably well with the problems it produces.

### Management

**Assessment** of the child with seizures begins with an appraisal of the nature of the attack, a physical examination, and laboratory investigations including an EEG. The possibility that the epilepsy is a manifestation of an underlying disease should always be considered, as should the possible role of psychological factors as precipitants.

In **counselling** parents and children in cases of so-called 'idiopathic' epilepsy, or epilepsy of unknown cause, various specific points need to be considered (Voeller and Rothenburg 1973).

The nature of the condition and its usually benign prognosis should be repeatedly explained to parents and children in appropriate language. The family should be encouraged to use a term like 'fits', 'attacks', or 'convulsions' when describing the seizures to friends, relatives, and teachers. Some explanation along the lines of a temporary excess of electricity in a particular part of the brain is often helpful. Parents should be encouraged to express their own views of the nature and causes of the fits, and this may reveal inappropriate fantasies concerning brain damage and the hereditary nature of the condition that require further discussion. The purpose and limited value of an EEG, and what happens when the child has an EEG need explanation. Some children think the EEG puts electricity into the brain rather than recording it. Parents and teachers will be helped by being given precise instructions about what to do if a fit occurs, and should be reassured that they are coping appropriately if this is indeed the case. Limitations of exercise, for example swimming, need to be regularly reviewed, and it is useful to check on each attendance what the child is allowed to do. Parents and teachers may be over-restrictive or, much less frequently, may fail to take sensible precautions to safeguard a child who is having frequent attacks. Monitoring of anticonvulsants should include parents or the doctor checking with teachers whether the child is unduly drowsy or inattentive. As indicated earlier, *any* anticonvulsant may cause side-effects in particular children.

Children with epilepsy whose primary care has been well handled will be at reduced risk for the development of psychological disorders, but inevitably a significant number will show signs of disturbance, mainly emotional and behavioural disorders. In general, the management of such problems will follow similar lines as in cases without epilepsy. Indeed the psychiatric problems of many children with epilepsy will be unrelated to their epilepsy and more closely linked to factors such as disturbances of family relationships. However, particular attention will need to be given to the possibility that behavioural disturbance is a manifestation of epileptic activity, that the child is being overprotected or rejected because of his attacks, that medication is producing adverse behavioural or learning side-effects, or that either the child or other family members are being made anxious by inaccurate fantasies about the nature of the condition.

The differential diagnosis of seizures and hysterical or 'pseudoseizures' sometimes presents problems, in particular because pseudoseizures not infrequently occur in children with epileptic attacks (M. Gross 1979). They are likely to be atypical in form and the child will probably show obviously purposeful movements. There may be inconsistent reactions to stimuli during an attack. EEG recording is likely to be helpful only if the child actually has an attack during the procedure, but this will only rarely occur unless telemetric ambulatory EEG is available.

In adults it has been reported that there is a rise in serum prolactin after an epileptic attack (not seen in pseudoseizures), and this may also be a useful diagnostic aid in children.

Assessment and management of pseudoseizures should follow the same lines as that described under the heading of hysterical or sick-role disorders (Section 3.2).

### 5.7.3 CEREBRAL TUMOURS

Brain tumours are a relatively common form of malignancy in childhood, second only in frequency to cancers of the blood-forming tissues. Types of cerebral tumour seen are quite different from those found in adults. About two-thirds occur in the subtentorial part of the brain, and of these, most are medulloblastomas and cerebellar astrocytomas. The remainder consist of supratentorial tumours and tumours of the brainstem and adjacent structures (Till 1975). Only 3 per cent of childhood tumours are metastic from other parts of the body.

## Clinical features

These will depend on the site of the tumour and the rapidity of its growth. Most subtentorial tumours present with headache, vomiting, and unsteadiness of gait. Psychological symptoms such as irritability and aggressiveness may be present, but almost always the physical symptoms are so prominent that there is little possibility of diagnostic confusion.

The presentation of supratentorial tumour will largely depend on the site where it occurs. Again psychological symptoms are not prominent, and epileptic fits or unilateral weakness of the limbs are much more common presenting features.

By contrast, children with brainstem tumours, in which cranial nerve involvement is usually an early sign, may show more marked behavioural symptomatology with lethargy, irritability, inability to concentrate, enuresis, and sleep disturbance sometimes present (Panitch and Berg 1970). Pontine gliomata may show even more striking behavioural changes and these may precede physical symptoms and lead to misdiagnosis. There is characteristically a period of withdrawal, apathy, and lethargy, followed by aggression, overactivity, and temper tantrums (Lassman and Arjona 1967).

## Treatment

Surgical removal may be possible depending on the site and nature of the tumour. Radiotherapy is used both as an adjuvant to surgery and, where surgery is impracticable, as in most tumours of the brainstem, as the main mode of treatment. The prognosis varies according to the type of tumour, the age of the patient, and treatment employed. The survival rate for children with medulloblastomas, the most common form of tumour, is between 40 and 75 per cent at 2 years and between 30 and 50 per cent at 10 years after diagnosis.

## Psychosocial aspects of outcome

Among survivors, some degree of physical disability is common and, in addition, the rate of behavioural and emotional problems is high, with nearly half the children showing such difficulties. These vary in type and include both aggressive and emotional disorders, with occasional occurrence of psychosis (Bamford *et al.* 1976; Kun *et al.* 1983).

Intellectual function may also be impaired, and there may be specific learning disabilities. About a third of the survivors require education in special schools. The intelligence of younger children is most seriously at risk, and this may be because intellectual function is most affected not by the tumour or surgical treatment, but by the subsequent irradiation (Silverman *et al.* 1984). There is strong evidence from follow-up studies of children with leukaemia that young children are particularly susceptible to post-irradiation effects on intelligence. The longer-term outcome of survivors of paediatric brain tumours is also often problematic (Lannering *et al.* 1990), with moderate or severe disability occurring in about one-third, when studied 5–16 years after diagnosis.

The postoperative course and psychosocial outcome of the rare craniopharyngiomata are somewhat different from other tumours. Postoperative hormone replacement therapy is always required in these cases. Intellectual deterioration is less common, but non-specific psychiatric problems (neurotic as well as aggressive in type) are more severe (Galatzer *et al.* 1981).

### 5.7.4 HEADACHE

#### Classification

Pains and aches in the head occur commonly in childhood, and are of three main types—organic, migrainous, and non-organic. Organic headaches are those with a recognizable organic cause and disappear when this cause is removed. Migraine headaches have certain characteristic features—their aetiology is usually multicausal, partly organic and partly non-organic. Non-organic headaches, which may be only poorly distinguishable from migraine, can be divided into those where the cause is psychological and those where it is unknown.

#### Clinical features

Headache of **organic origin** may arise for a large variety of reasons: infection of the teeth and sinuses; cerebral causes such as tumours and abcesses; refraction errors, especially shortsightedness; and vascular causes such as cerebral haemangiomata and hypertension. In general, pain due to such causes is reasonably well localized and, because few of the causes are benign, increases in intensity day by day. It is usually worse in the evenings and frequently wakes the child at night. There is often no family history of recurrent headache. Children under the age of 5 years rarely suffer from non-organic headaches or migraine, so headache in this age group is particularly likely to be organic. Any change in behaviour with this type of headache is usually fairly obvious, and secondary to the pain rather than just associated with it.

**Migrainous** headaches do not usually start until 8 or 9 years of age, but may occur earlier. In childhood auras are infrequent, but occasionally headaches are preceded by a scotoma (blind spot), blurring of vision, or simple hallucinatory phenomena. The headache itself usually lasts less than half an hour, but occasionally goes on for hours or even days on end. It is pulsatile in quality, unilateral or bilateral, and may be accompanied by nausea and occasionally vomiting. Photophobia is often present. There is often a family history of migraine and, in many cases, a parent or sib has typical attacks.

Such migrainous headaches may be precipitated by a variety of factors including psychological stresses as well as the ingestion of certain foods, especially chocolate, cheese, nuts, and food containing glutamates. Fatigue and strenuous physical exercise may also be triggers. Occasionally, migrainous headaches occurring over a period of some months may be precipitated by trauma to the head.

**Non-organic headaches** are usually of daily occurrence and do not vary much in intensity throughout the day. They are usually experienced as tightness or pressure around the head, and may be accompanied by mild throbbing. They do not wake the child up at night. A family history is common. When the cause is psychogenic, there is likely to be a relationship between circumstances and the occurrence of the headache. For example, the headache may be a presenting sign of school refusal, occurring only on school days and not at weekends or in the holidays (Section 2.9.3). Psychogenic non-organic headaches are usually accompanied by other behavioural and emotional changes, especially anxiety, depression, and other psychosomatic symptoms such as abdominal pain. Such headaches may also form part of a hysterical reaction (Section 3.2). In these circumstances, the child will be seen to benefit from the 'sick role' he has adopted. The complaints of pain will seem greater than the appearance of discomfort.

Differentiation between migraine and non-organic headaches is problematic. Often, for example, children have only one or two of the features of migraine and there are associated symptoms suggesting a psychogenic cause. It is also likely that the pathophysiological basis of some 'non-organic' headaches is similar to that of migraine.

### Prevalence

Headaches of non-organic or migrainous origin are very common in children. The exact prevalence is unknown, but Sillanpaa (1983) found that over one-third of 7-year-olds and over two-thirds of 14-year-olds had suffered from headaches. Rutter *et al.* (1970*a*) found that 11 per cent of 10- and 11-year-olds had suffered headaches at least once a month in the previous year. In one study about 3–5 per cent of 7–15-year-olds had headaches reasonably characteristic of migraine with at least two of the following four criteria: one-sided pain; nausea; visual aura; family history in a first-degree relative (Bille 1962).

## Mechanisms

Organic headaches are produced by a variety of mechanisms depending on the cause. Migraine headaches are thought to arise as a result of vasoconstriction and then vasodilatation of the branches of the external carotid, internal carotid, and, rarely, basilar arteries. The reason why these arterioles undergo such changes is uncertain. Histochemical alterations occurring as a result of immunological reactions, themselves triggered by stress, ingestion of certain foods, etc., seem one likely explanation. Non-organic headaches may occur as a result of muscular tension, or undue sensitivity to normal pain receptor stimulation. The physiological processes involved in the development of such commonly experienced discomfort are poorly understood.

## Assessment

In obtaining an account of the symptoms, focus should be on the frequency and nature of the headache, the presence of associated physical symptoms, especially nausea, vomiting, and abdominal pain, the relationship of the pain to psychological stress and ingestion of certain foods, and the presence of emotional or behavioural problems. The presence of headache in other family members should be noted.

In all cases of headache, physical examination should involve screening for neurological abnormality, including refractive errors, and observation of the fundi. Local causes of pain should also be excluded. An EEG is usually non-contributory, but skull X-ray and CAT scan may be indicated by any suggestion of possible cerebral pathology.

In the great majority of cases, physical examination will be negative and either migraine or non-organic headache diagnosed.

## Treatment of migraine and non-organic headaches

As already indicated, these two types of headache are poorly distinguished from each other, and can be considered together in relation to treatment. Attention to possible stress factors at home, at school, or in the neighbourhood is always worthwhile. The child is sometimes perfectionist in nature, and parents are occasionally unnecessarily pressurizing. Counselling both parents and child to find less-than-perfect performance acceptable is sometimes helpful.

Attention should be drawn to the possibility that certain foods (e.g. chocolate, cheese, nuts, etc.) precipitate the headache. Avoidance of these can be useful. With disabling migraine, a severely restrictive diet with gradual reintroduction of possibly offending foods may produce useful results.

Symptomatic relief may also be obtained by analgesic medication. Paracetamol often alleviates the pain, especially if taken early in the attack.

If this is ineffective, ergotamine tartrate may be indicated if classical migraine symptoms are present.

When headaches are but one manifestation of an anxiety or depressive state, the treatment should be directed towards the underlying condition (Section 2.9). Where it seems possible that the pain has arisen from muscular tension, relaxation techniques (Section 6.2.2) may be a useful addition to family, individual, or behavioural psychotherapy, although the evidence for this is inconclusive (McGrath *et al.* 1988). If the headache is the presenting feature of school refusal or a hysterical reaction, the treatment should follow the lines described elsewhere.

5.7.5 SPINA BIFIDA

### Definition

This is a condition present at birth, arising from a failure of fusion of the arches of the vertebral column. In most cases (spina bifida occulta), the failure of fusion is minor and of no pathological significance. In a minority, between 0.5 and 4 per 1000 births, there is herniation of the meninges and spinal cord as well. A meningocele or myelomeningocele, depending on the contents of the herniated tissue, is thus formed. Primary neurological defects comprise motor and sensory deficits (depending on the level of the lesion), partial or complete paralysis of the bladder and bowel, and sexual dysfunction. Hydrocephalus is usually associated with spina bifida, and there may be secondary neurological problems relating to this.

Surgical treatment at birth is now generally regarded as desirable only in those infants with the prospect of a reasonable quality of life (Lorber 1971). Assessment of the child born with spina bifida to determine its likely prognosis is a skilled technique that is carried out in specialized centres. The great majority (over 90 per cent) of children thought unsuitable for surgery die within the first year of life.

### Psychosocial aspects

The **early management** of children born with this condition requires sensitive handling with involvement of both parents in any decisions that are taken. Most parents will want to leave the responsibility for decision-making with the paediatrician and surgeon, but discussion with parents of the reasons for recommendations is likely to improve the chances of later psychosocial adjustment.

The terminal nursing of infants rejected for surgery usually takes place in hospital, but a few parents prefer to take their children home. In either event, emotional support for parents from the paediatrician and primary health care team should be provided. The situation is also distressing for nurses caring for the rejected child in hospital, and these too will need emotional support. A small minority of such children survive, and are subjected to surgery later. These have a poorer intellectual outcome than those operated on earlier. The uncertainty associated with these children's future is particularly distressing for parents.

### Developmental progress and intelligence

About two-thirds of children with surgically operated spina bifida show retarded development at 3 years of age (Spain 1974). However, only a small minority of surgically operated cases develop severe mental retardation, and the majority are within the dull average, mildly, or moderately retarded range. A smaller proportion are of average or above-average ability. Spina bifida is not associated with any specific cognitive deficits, but older children with the condition have been reported to show 'hyperverbalism' or the 'cocktail party syndrome' (Swisher and Pinsker 1971). These sometimes falsely convey greater verbal facility than they possess.

Poor intellectual development is associated with the presence of central nervous system infection and with the level of the lesion. The lower the lesion, the better is the intellectual outlook (G. M. Hunt and Holmes 1976). There is a reasonably high correlation between early development in the first two years of life and later IQ (Fishman and Palkes 1974).

### Psychiatric disorder

Although not all investigators have found children with spina bifida to show high rates of psychiatric disorder (Spaulding and Morgan 1986), most have obtained positive findings. Thus, Connell and McConnel (1981) found prepubertal children with hydrocephalus to have high rates of behaviour and emotional problems, probably partly related to the organic brain dysfunction and partly to environmental, especially family, influences. Adolescents with spina bifida have high rates of depressive disorder. Dorner (1976) found 85 per cent of 13–19-year-olds with this condition reported feeling miserable and unhappy to the point of being tearful and wanting to 'get away from it all', and a quarter of the girls had suicidal ideas. Depression in adolescents is related to problems of mobility and feelings of isolation. A low self-concept is common (Kazak and Clarke 1986). Children with bowel problems frequently report social embarrassment and feelings of shame. Concerns about sexual function and fertility are common in this age group, and these are not made easier to manage by the fact that medical and surgical opinion as to what the future holds for the individual in these respects is often unclear. Many teenagers with spina bifida are ignorant of the sexual facts of life (E. Anderson *et al.* 1982*a*).

### Education

Most children with spina bifida are able to attend normal school, but a minority require special schooling, either because of general or specific learning difficulties, or because it has not proved possible to modify school buildings, for example to take wheelchairs. As integration of handicapped children into normal schools continues, a smaller proportion will be excluded for this reason.

## Implications

In the early years, parents need emotional support as well as a good deal of practical advice to deal with problems of mobility and incontinence. The genetic implications of the condition require careful explanation. The risk of recurrence, together with the possibility of detection of spina bifida *in utero* with subsequent abortion if the child is affected, should be explained before further pregnancies occur. The possible preventive role of vitamin supplementation in the periconceptual period also needs to be understood and appropriate dietary advice given.

In middle childhood and adolescence, many of the issues discussed in Chapter 4 concerning the psychosocial problems of handicapped children in general will be relevant to this group. Specific problems, especially concerning children and adolescents with spina bifida (Thomas *et al.* 1985), include concerns for the social embarrassment arising from continuing incontinence. Practical advice over menstruation and contraception will be necessary, especially if there are sensory deficits in the pelvic area. Sexual information and counselling is currently lacking, but the Association for Spina Bifida and Hydrocephalus (ASBAH) has produced a helpful booklet on these subjects for affected individuals, their parents, and professionals (ASBAH 1983). Health professionals should check that advice concerning further education and employment is available to the teenager. As the child gets older, it will often be helpful to arrange that he or she has the opportunity to talk about feelings concerning the handicap and the future with someone outside the family. Mental health professionals who may be providing this service need to have a good knowledge of the practical difficulties experienced by those suffering from the condition.

## 5.7.6 MUSCULAR DYSTROPHY (DUCHENNE TYPE)

### Definition

This is the most common type of progressive muscle disease in childhood, occurring in about 1 per 3500 boys. It is often inherited as a sex-linked recessive, so that if one boy in a family is affected, the chances of another boy having the condition may be as high as one in two. However, the gene has now been cloned, and accurate antenatal diagnosis is possible. It usually presents with muscle weakness, especially of the legs, between 1 and 5 years. By 12 years, the majority of boys are wheelchair-bound, and death usually occurs by the age of 20. Other forms of muscular dystrophy, usually with a more benign prognosis, may occur in children of both sexes.

## Psychosocial aspects

**Intellectual development.** There is a tendency for boys with this condition to be of less than average intelligence, and there is a higher than expected rate of severe mental retardation. Many, however, are of normal ability. If cognitive impairment does occur, it is particularly likely to affect verbal ability with defective auditory processing (Dorman *et al.* 1988). Indeed, the condition may present with language delay. Most boys can be educated in ordinary school until their physical condition prevents it. Subsequently they should as far as possible be educated in special schools rather than have home tuition and lose contact with other children.

**Diagnosis.** Because the condition is rare (and there are many other causes of boys tending to be slow to walk or to fall frequently once having learnt to walk) the diagnosis is often delayed. Inevitably this leads to parental distress. When parents are told that their son is suffering from this devastating condition, they need to be told together; unfortunately this does not often occur (Firth *et al.* 1983).

**Communication.** Parents often complain they are not given enough information about the condition. In fact they need repeated opportunities to express their own ideas and fears as well as to receive factual information. The genetic implications are not easy to grasp, and again many parents will need more than one explanatory session.

Principles of communicating with parents and children where the child has a fatal condition are discussed elsewhere (Section 4.7). Parents will wish to communicate between themselves and with their child and his siblings about the muscular dystrophy in their own way, but it may be helpful for professionals to suggest early on that the more open the communication about the illness and its implications, the easier the family atmosphere is likely to be. Psychological disorders in sufferers and their sibs are more common where there is lack of open communication.

Boredom, apathy, depression, and angry reactions are common psychological problems shown (Fitzpatrick *et al.* 1986). These are less likely to occur if a positive attitude to learning, education, hobbies, and contact with friends is maintained by the child's parents and teachers; much ingenuity is required to retain a positive approach when the child becomes very weak.

# 5.8 Metabolic disorders

There is a very large number of such disorders occurring in infancy and childhood. Most are very rare and occur as a result of inborn errors affecting amino acids, carbohydrates, protein, lipid, purine, and other aspects of metabolism. Psychosocial aspects of a small number of the less rare disorders with particular psychiatric or developmental significance, are described here.

## 5.8.1 PHENYLKETONURIA

### Definition

This is a disorder of amino acid metabolism in which there is a deficiency in phenylalanine hydroxylase, the enzyme responsible for converting dietary phenylalanine to tyrosine.

### Aetiology

The condition, which occurs in about 1 in 15 000 births, is an autosomal recessive disorder. The enzymatic defect results in excessive production of phenylpyruvic acid, phenylacetic acid, and phenylacetylamine. There is consequent interference with maturation of grey matter, defective myelination, and cystic degeneration of white matter in the central nervous system.

### Clinical features and treatment

If untreated, affected children are usually markedly developmentally delayed and become severely or moderately mentally retarded. Normal intelligence has, however, occasionally been reported in untreated cases. Epileptic fits, eczema, and behavioural disturbances (especially hyperkinesis and autism) are common acompaniments.

The condition is *treatable* if diagnosed within the first few days of life. Universal screening from the fifth to the eighth day of life now ensures that virtually all children born with the condition are identified. Treatment involves a low-phenylalanine diet, with close monitoring of blood phenylalanine levels throughout the first 7 or 8 years of life. Subsequently, a relaxed diet with some restriction of phenylalanine intake is usually advised until mid-adolescence (but see below).

Dietary treatment should be instituted in a specialized centre. Specially manufactured low-phenylalanine milk substitutes provide the main source of nutrition in the early years, and other foods have to be severely restricted. This means that the child is unable to lead a normal social life. In the preschool years this is not usually problematic, but in older children conflict over the diet may occur, especially if family relationships are tense for other reasons, or if the child is temperamentally difficult (Kazak *et al.* 1988). Counselling can be helpful in both preventing and managing these kinds of difficulties.

Treated children are virtually always within the normal range of ability, though their mean IQ is slightly lower than that of their siblings, and they have an increased likelihood of minor neuropsychological anomalies. Relaxation of diet in mid-childhood results in a small but definite decrement in intelligence (Holtzman *et al.* 1986). Rates of behavioural deviance are raised in treated children with phenylketonuria (I. Smith *et al.* 1988) when

compared with classroom controls. Affected children show more mannerisms, fidgetiness, restlessness, and poor attention, as well as being more
anxious, solitary, and miserable. Behavioural deviance is inversely related
both to IQ and to the quality of biochemical control in the early years.
Residual abnormality of phenylalanine metabolism, or the effect of having
to take a rather unpalatable diet, may be responsible for high rates of
behavioural deviance. Genetic counselling is an important component of
family management.

**Outcome**

The outcome of early treated cases is usually good and, in adulthood, a
normally independent life is to be expected. Affected women are, however,
likely to give birth to children with congenital abnormalities and mental
retardation as a result of the exposure of the fetus to high levels of maternal
phenylalanine, unless they return to a diet before conception (Lenke and
Levy 1980). Phenylketonuric women planning to have a baby should
therefore go on a carefully monitored low-phenylalanine diet, which
should be monitored from before conception to delivery.

### 5.8.2 GALACTOSAEMIA

This is a rare disorder of carbohydrate metabolism in which there is a
deficiency of the enzyme galactose 1-phosphate uridyl transferase. Insulin
production is stimulated so that the brain is exposed to persistent hypoglycaemia and other metabolic abnormalities, producing brain damage and
dysfunction. Damage to the liver, spleen, and eye (with cataract formation)
also occurs. It is transmitted as an autosomal recessive condition. Brain
damage can be largely prevented by the administration of a galactose-free
or low-galactose diet, but this is highly restrictive and is necessary for the
whole of the individual's life.

**Intellectual development** in treated cases is satisfactory, although compared with their normal sibs, affected children have a slight decrement in
IQ with specific visuospatial deficits (Komrower and Lee 1970).

Studies of **emotional and behavioural development** have found that
treated children have high rates of problems (Fishler *et al.* 1966). Younger
children tend to be fearful and anxious, and older children are more
aggressive and anti-authority in their attitudes. It is likely that these difficulties arise partly as a result of residual brain dysfunction, and partly as a
reaction to social problems arising both in the family and outside from the
need to persist with an unusual diet.

### 5.8.3 MUCOPOLYSACCHARIDOSES

This is a group of disorders affecting storage of carbohydrate. They are
classified according to the precise type of enzyme deficiency that is present.

The terms Hunter and Hurler syndrome and, more descriptively, gargoylism were used to describe varieties of the condition before the metabolic defects were identified.

The conditions are normally identifiable by the characteristic facial appearance at or shortly after birth. In all the mucopolysaccharidoses there is a normal early development for about 6–12 months, followed by a failure to develop further and subsequent mental deterioration. There are associated skeletal and other organ defects. No particular emotional or behavioural difficulties are associated with the condition. Sufferers may die in childhood or survive into early adulthood. A precise metabolic diagnosis is desirable, as different variants have different genetic implications—the conditions may be inherited in an autosomal or sex-linked recessive form.

### 5.8.4 LESCH—NYHAN SYNDROME

In this very rare disorder of purine metabolism, there is a defect of the enzyme hypoxanthine–guanine phosphoribosyl transferase. It is of particular psychiatric significance because of the association with mental retardation and severe self-abusive behaviour. It is a sex-linked condition, and all reported cases have been boys.

Affected children show choreoathetosis and moderately severe mental retardation as well as self-destructive behaviour. Self-mutilation is of a different order from that occurring in other forms of mental retardation and autism. Biting of the child's own lips and cheeks produces loss of tissue with sinus formation. Severe head-banging can produce further brain damage.

Treatment of the self-mutilation can be extremely difficult. External restraint is nearly always necessary. Behavioural modification with removal of attention from the undesirable behaviour has produced short-term improvement (McGreevy and Arthur 1987). The use of carbidopa and tryptophan has also produced short-term beneficial change (Nyhan 1976).

## 5.9 Endocrine disorders

### 5.9.1 THYROID DYSFUNCTION

#### Hypothyroidism

Deficiency or abnormality of thyroid hormone secretion occurs for a large number of reasons, of which by far the most common is incomplete development of the thyroid gland. If the gland is absent or virtually absent at birth, severe hypothyroidism (cretinism) ensues. If some function in the gland remains, mild hypothyroidism occurs. In some children, deficiency of thyroid secretion only becomes apparent in later childhood (juvenile myxoedema).

The **clinical features** of untreated hypothyroidism (coarse facial features, neonatal jaundice, early feeding difficulties, and constipation) are associated with mental retardation. The degree of mental retardation depends on the severity of the hormonal condition and may therefore range from mild to severe.

*Prevalence*
Congenital hypothyroidism is relatively common, occurring in about 1 per 4000 births. Screening for the condition has now been widely introduced, and is commonly carried out 4–6 days after birth. Once the diagnosis is confirmed, treatment consists of a daily dose of thyroxine. This should not impose a significant burden on children and families.

Studies of children treated with thyroxine following identification shortly after birth suggest that their mean intelligence is within the normal range, but somewhat below that of matched controls (Murphy *et al.* 1986). Minor neurological problems may be present, especially motor incoordination and other motor disorders. There is a raised rate of psychiatric deviance, and this tends to be of neurotic type, although there may also be raised levels of activity and attentional deficits. A poor intellectual outcome is usually associated with the presence of an additional physical condition or poor compliance with treatment due to family problems (Murphy *et al.* 1986).

## Thyrotoxicosis

This condition may occur in the neonatal period or in later childhood (juvenile thyrotoxicosis), but is very rare. As in adults, psychological changes may be among the presenting features. In particular, there may be anxiety symptoms, lack of concentration at school, and restlessness. The associated loss of weight and goitre (swelling in the neck) make it unlikely that the physical condition will be missed, but sometimes the child is thought initially to show a primary psychiatric disorder.

## 5.9.2 DIABETES MELLITUS

### Definition

Diabetes mellitus is a condition in which there is a failure of production of insulin, the hormone responsible for clearing glucose and ketones from the blood. There are two types of diabetes mellitus. In juvenile-onset diabetes, the patient is dependent on insulin for survival. The onset of this condition is usually between 3 and 15 years of life, but it may occur later. In maturity-onset diabetes, which usually comes on much later in life, a different pathological mechanism, independent of insulin production is involved. In this section, only juvenile-onset diabetes will be considered.

## Background features

The condition occurs in about 2 per 1000 schoolchildren in the UK and North America, though its frequency varies in different parts of the world. The cause of the condition lies in a failure of the islet cells in the pancreas to produce insulin, but the reasons for this failure are unknown. Nevertheless, there are now some strong indications. An immunological defect is probably involved, and specific HLA antigens have been identified. Chromosome studies have provided further evidence of a genetic cause. Animal experiments suggest that a virus infection may, at least sometimes, be responsible. In addition, twin studies and the familial incidence of the condition make it likely that there is an inherited predisposition or vulnerability to environmental agents such as viruses. The more remote possibility that emotional stress may be a precipitant is discussed in more detail below.

## Clinical features

Children who suffer from this condition are likely to present with abdominal pain, excessive thirst, and polyuria (excretion of excessive quantities of urine). The development of nocturnal enuresis may be a very early symptom. Sufferers tire easily and their appetite, though initially increased, is later reduced. Symptoms develop from onset to a state of serious metabolic disturbance over a period lasting a week to a month. As the disease progresses, severe vomiting occurs and the child gradually enters a comatose state. Without rapid and competent medical treatment, the condition is fatal. Death is, however, now extremely rare in developed countries.

Treatment of the acute diabetic condition or ketoacidosis is followed by a period in hospital in which the child is treated with an appropriate diet and regular insulin injections. After initial adjustments, it is usually possible to achieve very satisfactory stability of the condition with one or two daily injections administered by the parents, or, in older children over the age of 10–12 years, by themselves. Glucose levels are monitored by urine- or blood-testing, carried out by the parent or child. Some diabetics, however, remain unstable, and psychological factors, discussed below, are of definite significance in some, perhaps most, of these cases. In unstable diabetes the danger of coma from low blood sugar due to an excess of insulin is greater than the risk of coma from excess glucose in the blood. Insulin coma is often entered rapidly, but the child or parent will usually have sufficient warning from feelings of dizziness to prevent it occurring with a sweet drink or lump of sugar. Dietary management is also important.

## Links with psychological factors
### Onset

There is no good evidence for the existence of a specific diabetic type of personality—a suggestion made in the 1950s. More recently there have been suggestions that external stress factors, either of a non-specific nature or specifically related to loss of a parent, might precipitate the illness. It does seem established that there is an excess of life events in the few months preceding the onset of the condition, particularly in older children. It is, however, possible that the events that occur may not be independent of the early symptoms. If, for example, the child's resistance to infection or tolerance of stress is reduced in the early stages of diabetes, then the child might be admitted to hospital or put down a class in school for this reason. An investigation limited to the study of events preceding the diagnosis of diabetes might wrongly imply that the hospitalization or school difficulty was the cause rather than the result of the condition.

If external stressful events are of significance in precipitating the onset of the condition, they might exert their effect either through their influence on pituitary secretion and catecholamine production, or by reducing resistance to virus infection. In the present state of knowledge, it seems improbable that they play an important contributory part in the onset of the condition, although, as seen below, they may well exert an important influence on its course.

### Diabetic control

While the majority of children with diabetes achieve very satisfactory control with minimum inconvenience to themselves and their families, a minority present control problems and, in a smaller proportion still, these difficulties produce a major disruption in the child's life, requiring frequent out-patient visits, hospitalizations, and absence from school. Such 'brittle' or 'labile' cases of diabetes may be due to specific metabolic or endocrine factors, but they are more likely to be due to the direct or indirect influence of stressful factors operating on the child.

1. *Direct factors.* Laboratory studies have demonstrated that the production of pituitary hormones and catecholamines induced by stress can lead to a decrease in insulin production and an increase in free fatty acids in the blood. A stressful interview has been shown, for example, to produce an increase in urinary epinephrine. A beta-adrenergic blocking substance administered before the stress interview prevents these physiological changes. It has therefore been suggested that the effect of stress on diabetic control is mediated by endogenous catecholamines acting on adrenergic beta-type receptors. However, as exposure to stressful situations does not result in changes in circulating levels of glucose, ketones, or free fatty acids (Kemmer *et al.* 1986), it seems unlikely that stress has a direct physiological effect producing hyperglycaemia.

2. *Indirect factors.* The success of diabetic control depends on the compliance of the patient and family with treatment—diet, injections, and urine-testing. Compliance can be affected by the quality of organization of

care in the home. It is likely to be poor in those families where, for reasons of financial hardship or personality conflicts between the parents, or a combination of the two, the home care is relatively disorganized (White *et al.* 1984). The confidence and self-esteem of the parents is an important key factor (Grey *et al.* 1980). Some parents react to their own anxiety about the diabetes by becoming more overcontrolling and rigid, or alternatively rejecting and neglectful.

The control of the diabetes also provides a ready-made focus for parent–child conflicts. Diabetic children are expected gradually to take over their own care as they move into adolescence. If the child is having difficulty in achieving separation from his parents smoothly, he may, consciously or unconsciously, use the threat of refusal to comply with treatment as a means of manipulation. Failure to comply may also occur in depressed children who have lost interest in their own survival. Such low self-esteem may occur because of neglectful behaviour on the part of parents pre-occupied with their own problems.

Studies of poorly controlled diabetic children show high divorce rates in the parents, frequent chronic family conflict, and generally inadequate parental care (White *et al.* 1984; Marteau *et al.* 1987). The children themselves are often depressed, angry, low in self-esteem, and show learning and disciplinary problems in school. Major psychiatric problems may be associated with suicidal attempts. Family relationships have been found, albeit in uncontrolled studies, to be characterized by overprotection of the children, lack of flexibility, and poor resolution of conflict (Minuchin 1974).

### Learning problems
Children with diabetes show an increased rate of learning problems. This probably occurs as a result of mild, chronic metabolic disturbance, as cognitive impairment in younger children is related to time since diagnosis. Children diagnosed before 5 years are more likely to show specific visuo-spatial problems. Later-onset diabetes is more likely to have an effect on verbal, left hemisphere performance.

### Behaviour and emotional problems
Rates of psychiatric disturbance in children with diabetes have been extensively studied, and most investigators find increased rates (Johnson 1988). A wide range of problems is described, and there is no particular behavioural profile. High rates of anxiety, depression, low self-esteem, anger, obsessive–compulsive symptoms, hyperactivity, and schizoid traits have all been found by different workers. Although in poorly controlled diabetics there are high rates of individual and family disturbance (see above), those who have

looked at general population groups of children find that, if anything, it is the more rigidly controlled who show higher rates of problems (Fonagy *et al.* 1987), though the reverse has also been found.

## Psychosocial aspects of management

Most children with diabetes and their families will cope well with the social and psychological stresses imposed by the illness. Coping behaviour in these well-functioning families can be enhanced by early and repeated full explanation of the illness and its implications. Practical advice over diets, perhaps with the suggestion that the whole family should modify its nutritional intake to some degree, will be helpful.

The implications of the condition for the child and family will alter as the child grows into adolescence (Cerreto and Travis 1984). Advice on insulin management should include discussion of ways in which children can gradually become responsible for their own injections and urine- and blood-testing, certainly as they move into their early teens and usually earlier. Both children and parents often find it useful to experience a deliberately induced minor hypoglycaemic attack in hospital so that they can learn to recognize the signs and gain confidence in how to deal with it. Parents and children should be encouraged to think that diabetes is quite compatible with a normal, active life, and there are a number of helpful booklets available describing how this can be achieved. Adequate knowledge of the condition predicts a better psychosocial adjustment (Sullivan 1979).

The **psychiatric appraisal** of poorly controlled diabetes requiring repeated hospitalization for recurrent ketoacidosis can be a crucial aspect of management of these difficult cases. It has been estimated that physical factors such as intercurrent infection are responsible for only a minority of these problems (White *et al.* 1984), though it remains possible that there are metabolic reasons why some children rather than others go out of control so rapidly when faced with stressful situations.

**Assessment,** which may be carried out by the paediatrician, social worker, psychologist, or psychiatrist, should involve, at least at some point, a whole family interview, an interview with the parents, and an interview with the child or teenager. The focus should mainly be on aspects of family life other than the diabetes and its control, for it is here that the key to the problem usually lies. A home visit, especially in families where cooperation is poor, will also often be helpful.

Two psychological approaches to **treatment** in cases of poor diabetic control have been found helpful. Good results have been claimed for family therapy. The focus in family sessions is likely to be on the way in which the parents find it difficult to allow the child autonomy, and on achieving better ways of reaching resolution of conflicts. Behavioural methods of treatment, with the setting of short-term goals for the child and family, have also been found to be useful. Principles of behavioural management are described elsewhere (Section 6.2.2). A combination of family and behavioural approaches is likely to be most effective, but, at the present time, there is a lack of controlled intervention studies from which clear-cut implications for interventions can be drawn.

# 5.10 Blood disorders

## 5.10.1 HAEMOPHILIA

### Definition

This is a condition in which excessive bleeding occurs because of deficiency of one of the factors (factor VIII or IX) necessary for blood clotting to occur. The type of haemophilia depends on which factor is deficient.

### Clinical features

In mild cases, in which the level of the deficient factor is between 5 and 50 per cent of normal, unusual bleeding only occurs when the child is exposed to serious trauma, as, for example, with a serious accident or surgical operation such as tonsillectomy. In moderate cases (1–5 per cent of the deficient factor present) bleeding occurs with mild trauma or dental extraction, and in severe cases (deficient factor less than 1 per cent of normal) spontaneous bleeding occurs.

In the severe form, abnormal bleeding may occur from superficial skin cuts. Bleeding into soft tissues and muscle may produce painful bruising, and such bruises (haematomata) may occasionally prove dangerous if they put pressure on vital structures. Nosebleeding and haematuria (blood in the urine) may occur. Bleeding into joints may result in swelling and pain requiring immobilization. If this occurs frequently, or is inadequately treated, the immobilization may result in atrophy of muscles because of disuse.

Treatment consists of stemming the bleeding and replacement of the deficient factor with a plasma concentrate. Such treatment can now be carried out on an out-patient basis or in the home.

### Aetiology

Haemophilia is a sex-linked recessive disorder and the defective gene is carried on an X chromosome. Women who are carriers are clinically unaffected, but 50 per cent of their sons develop the condition. All the daughters of affected males are carriers. Sons of affected males are unaffected except in the highly improbable circumstance that their mothers are carriers. The condition occurs in about 1 per 15 000 males. There is no particular link with social class, and psychosocial factors are of no importance in primary aetiology, though they may influence the course of the condition (see below).

## Psychosocial aspects

Haemophilia of mild or even moderate severity usually has rather little impact on the psychosocial life of the child and family. Occasionally, where there are pre-existing factors contributing to vulnerability, the presence of a disorder, even with relatively minor implications, will have a significant effect on the child's personality development and family life, but this is unusual. Severe haemophilia more frequently has a major impact (Markova *et al.* 1980*a*).

**Parents** of severely affected children are generally well-adjusted in the degree to which they protect their children from hazards such as sharp objects. Some, however, are markedly overprotective, and take unnecessarily strict precautions. Such parents are also those most likely to show guilt, as well as bitterness and resentment over the child's development of the condition (Mattson and Gross 1966). Many parents, however, successfully cope with their children's vulnerability by denial, apparently experiencing little anxiety about the possibility of injury or excessive bleeding. Fathers are generally quite highly participant in child care, but if complications of the condition become frequent or severe, this may put strain on the parental marriage, and the couple may find themselves in frequent disagreement over how the child should be managed.

Severely affected children are usually outgoing in personality, experiencing only relatively brief periods of sadness and withdrawal after a bleed. Although unable to participate in contact sports, they are often fascinated by such activities and may become preoccupied with them. Young affected children may be unusually clumsy in their use of sharp objects such as knives and scissors because they have been prevented from using them. School performance is usually within the average range, even though the children almost inevitably miss an unusual amount of school because of their condition. With modern forms of treatment, the great majority, even of severely affected children, are able to attend normal school (Markova *et al.* 1980*b*). In the mid-1980s a new, stressful development occurred in the lives of some children with haemophilia. As a result of transfusions, a number of affected children became infected with HIV (human immunodeficiency virus). Although the risk of transmission of this condition to other children is very low, parents of unaffected children may express serious anxiety about the risk of infection and this can lead to ostracism of the haemophiliac children, and a threat of exclusion from ordinary school. At this time, satisfactory policies to deal with this situation have been worked out, and it seems likely that, if parents of unaffected children are kept well-informed, their anxieties will be alleviated.

**Emotional factors** may have an influence on the likelihood of bleeding. Usually bleeding is precipitated by minor trauma, but apparently 'spontaneous' bleeding sometimes occurs. In a minority of children this may occur just before some exciting event such as an outing or party. The mechanism may involve an autonomic response to emotional stress altering the integrity of the capillary wall (Mattson and Gross 1966).

**Differential diagnosis** of the condition may include non-accidental injury (Section 1.3.2). The occurrence of spontaneous bruising may mean that parents of mildly or moderately affected children have been wrongly suspected of injuring their children. The crisis that this may have produced has occasionally had long-lasting effects on family functioning.

*Specific psychosocial effects of management*

1. *Genetic counselling.* The genetic aspects of the condition are not easy to grasp, and both parents, and, as they get older, affected children and their sibs, need frequent opportunities to discuss this issue. Decisions about further children must, of course, be left to parents following provision of full information. In the early 1990s antenatal detection of this condition is becoming increasingly possible.

2. Good communication between family and paediatrician is vital. Parents and children should have a feeling of confidence that they know precisely what to do if bleeding occurs, and have clear advice on the level of protection the child really needs.

3. That minority of parents who suffer adverse emotional reactions need continuing counselling. Their children are likely to be overprotected. Professionals should encourage as normal activity as possible. The supervised use of sharp tools such as knives and scissors when the child is still quite young (4–5 years) is helpful in preventing clumsiness (Markova *et al.* 1984).

4. Teachers in the normal schools these children now mainly attend need adequate information about the nature of the condition, necessary protective measures, and what to do if bleeding occurs. The fact that many boys prevented from participating in sporting activities nevertheless enjoy play-fighting with their friends often exasperates teachers, who need opportunity for discussion.

## 5.10.2 LEUKAEMIA

This is a cancer or neoplastic condition of the white cells of the blood. It is the most common form of malignancy in childhood, accounting for about half the total number of cancers in this age group. About 5 per 100 000 children are affected. Acute lymphoblastic leukaemia is the most common form of the condition.

### Clinical features

The condition can occur at any time during childhood, but it is most frequent between 2 and 4 years of age, and is unusual after puberty. There may be an acute or chronic onset. Usually there is a period of several weeks of ill-health with susceptibility to infection and general malaise. The child will often be pale and look unwell. Then an episode of bleeding or some other traumatic event may make it clear something serious is amiss. Clinical examination may reveal bruising, an enlarged liver and spleen, and bone tenderness. The diagnosis is made by examination of a blood film and sample of bone marrow.

### Aetiology

The cause of the condition is unknown, although in a minority of cases genetic factors or exposure to irradiation may be relevant. The incidence is raised in Down's syndrome. A viral cause has been suspected, but not confirmed.

### Treatment and outcome

This is usually carried out in specialist centres. After initial acute treatment, it consists of chemotherapeutic drugs to induce a remission of the neoplastic process, and cranial irradiation to prevent the spread of the condition to the central nervous system. With this treatment a considerable majority of the patients go into a remission, although a proportion relapse. Children who relapse may be treated with bone marrow transplantation. If persistent response to treatment occurs, currently the chemotherapy is stopped after about 3 years. Because the treatment suppresses body immunity, even the common childhood complaints such as chickenpox and measles can produce a fatal outcome, so if the child is in contact with these and other conditions, immune globulin is given. Over the past two decades, treatment has gradually improved the prognosis. The condition used to be universally fatal, but now the 5-year survival rate is in the region of 60 per cent and, although inevitably some uncertainty remains, it is anticipated that many children have been permanently cured.

### Psychosocial aspects

*Initial diagnosis*
Because the initial presenting features are often so non-specific, the diagnosis is quite frequently missed in the first few weeks by the family doctor. Once suspected, however, the diagnosis can be rapidly confirmed. Consequently, the parental reaction is likely to be one of severe shock, coupled sometimes with anger at the family doctor. Parents are likely to be unable to absorb complex medical information at the first interview with the physician. This is particularly the case with this condition because of the complexity of medical treatment required and the wide range of uncertainty about prognosis, with a real possibility both of death and of complete cure. Parents are not unnaturally bewildered, and require repeated interviews for clarification. They sometimes believe that the illness has been precipitated

by some stressful event. The evidence in favour of a precipitating role for stress in the condition is poor.

*Active treatment phase*
During the initial active phase of treatment, frequent visits to hospital are required. Bone marrow investigations are painful and may be the aspect of the condition and its treatment most feared by the child. Adequate preparation linked to the availability of play facilities in the out-patient department to divert the child before the procedure will reduce distress. Short-term side-effects of treatment, including nausea and vomiting, mood changes, and constipation, are relatively common. Hair loss is a complication of radiotherapy which makes the child conspicuous at school and almost inevitably leads to embarrassment and sometimes teasing from other children. Older children can be offered a wig or prefer to wear a cap to cover their baldness.

The susceptibility of the child to infection causes a great deal of parental anxiety. The danger is a real one, yet the quality of the child's life can be greatly impaired if exaggerated precautions are taken to keep the child out of contact with others who just might be infected. Direct communication of parent with headteacher and classteacher to explain the situation and request information where infectious diseases are reported in classmates can be backed up if necessary by direct doctor–school contact by telephone. Active concern by all professionals involved with psychological and social problems is usually found helpful at this stage.

*Later phase*
While children are in remission on maintenance treatment, anxiety is usually, but not always, reduced. Parents usually maintain a more optimistic attitude than the physicians concerned, and children are generally more optimistic still. Many families cope well with the chronic stress and anxiety (Kupst and Schulman 1988). However, rates of depression, anxiety, and sexual problems in parents have been reported to be high, and many parents are reluctant to disclose such problems, so that their professional advisers may be unaware of them (Maguire 1983). Factors associated with good coping include harmonious family relationships, a supportive network of family and friends, lack of other stresses, open communication, and an attitude of living in the present (Kupst and Schulman 1988). Successful coping is achieved with a range of processes and what suits one family may not suit another. Some mothers find talking about the problem with other people is helpful, while with others this is most unhelpful. Some may function well with full acceptance of the problem, others by the strong use of denial. Fathers more commonly cope by not talking about the condition. In general, parents tend to hold more optimistic beliefs about the outcome than the physicians involved (Mulhern *et al.* 1981). Some

families, perhaps most, are drawn more closely together by the stress; some find it distances them. Most help is obtained from family members and friends.

School problems are relatively common. Probably as a result of the irradiation, children who are diagnosed and receive treatment before the age of about 5 years may develop specific cognitive deficits, especially loss of short-term memory and poor concentration (Cousens *et al.* 1988). These problems do not seem to affect children irradiated at an older age (Jannoun 1983). Younger children, however, tend to find difficult those school subjects such as arithmetic which involve sequential learning. It is currently uncertain to what degree special remedial help can overcome these deficits; nor is it known for how long they persist, but extra educational help is often indicated.

Not unnaturally, teachers are inclined to allow children with leukaemia extra leeway, and avoid putting even the normal amount of pressure on them to do homework and conform in the classroom. Good physician–school communication can ensure that teachers are aware that children with leukaemia tend to thrive best when they are treated like others and normal pressures are applied. School refusal (Section 2.9.3) has been reported in children with this condition, and this is likely to be related to parental anxiety.

Relapses terminating in a fatal outcome still do unfortunately occur in about 40–50 per cent of children with leukaemia, even with the best available treatment. Psychosocial aspects of terminal care are discussed in Section 4.7.

### 5.10.3 THALASSAEMIA

This is an inherited disease caused by a failure of synthesis of normal adult haemoglobin. It is commonest among people living in the Eastern Mediterranean, but also occurs in other countries. It is inherited as an autosomal recessive with the homozygote suffering from the major form of the disease. The heterozygote carrier shows minor haematological abnormalities, but is usually clinically normal.

It usually presents in the first 6 months of life with pallor due to anaemia, and failure to thrive. The facial appearance is often unusual, because of skull deformities. Treatment is symptomatic and involves frequent blood transfusions, but eventually organ failure from chronic iron overload results in premature death.

The psychosocial aspects of this condition have been little studied, but are of considerable significance. Blood transfusions can now be carried out on a day-patient basis, so that repeated hospitalizations are not required, but, nevertheless, young children especially may develop phobic reactions to the medical procedures involved. Short stature and failure to enter puberty are not infrequently associated with low self-esteem and with-drawal reactions. Children are likely to be especially vulnerable when a sib or other close relative has died of the condition.

Discussions in groups of parents of children with this condition have revealed (Tsiantis *et al.* 1982) that communication about the disease within the family is often poor, with parents showing maladaptive denial and expressing a good deal of guilt about their responsibility for the condition. It is uncertain how far this is a general phenomenon, but certainly parent groups would seem to offer an opportunity for the provision of mutual support and the improvement of family communication. Genetic counsel-ling is of major importance, as identification of the heterozygous state and antenatal detection of the homozygous condition, either by amniocentesis or, in the first trimester, by chorionic villus sampling, with the possibility of therapeutic abortion has now become possible.

# 5.11 Respiratory disorders

## 5.11.1 ASTHMA

### Definition

This is a condition in which there is intermittent, reversible obstruction of the respiratory airways due to hyper-reactivity of the airway lining to a variety of stimuli. The characteristic wheeze is produced by forced exhala-tion through narrowed bronchi.

### Background factors

The condition occurs in between 5 and 10 per cent of the population in childhood, and is the commonest chronic physical disorder in childhood (Gergen *et al.* 1988). It often begins in the first 2 years of life, when distinction from 'wheezy bronchitis' is difficult. Occasionally, first attacks occur later in childhood. There is no clear-cut link with social class. In young children, in whom infection is prominent as a causative factor (see below), working-class children are probably more commonly affected, whereas in the older age group it is more frequent in middle-class children.

### Causation

A family history of allergic disorder is common, and it is likely that a genetic predisposition to asthma is inherited as part of a general tendency to allergic illness. Children who develop asthma have often previously suffered from infantile eczema, and may show other atopic manifestations.

Attacks are precipitated in vulnerable children by a variety of environ-mental agents or events. Allergens may be inhaled or ingested. They include pollens, housedust (in which the house mite may be the principal causative agent), feathers, and animal fur. A variety of foods, including eggs, shellfish, and cow's milk, may induce attacks. Attacks may also be precipitated by upper respiratory tract infections of a bacterial or viral nature, so that often asthma occurs when the child gets an ordinary cold.

In addition to these allergic precipitants, attacks are also produced by non-allergic factors. Of these, psychological stresses, to be discussed in more detail below, are amongst the most common, but physical exercise and cold weather also fall into this category.

## Links with psychological factors

Psychological factors are often of major importance in the precipitation and maintenance of asthmatic attacks, but earlier-held views that the condition is primarily of psychological nature were largely incorrect. There is no evidence that major life events play a part in aetiology, and no specific personality profile characteristic of asthma. However, the condition does have some specific effects on family relationships.

### Direct psychological precipitants

In vulnerable children any event likely to produce an emotional reaction may also precipitate an attack. Events causing anxiety, anger, and anticipatory excitement are most commonly responsible, and, in these circumstances, the attack often occurs as the emotion is experienced. Situations about which the child may reasonably be worried, such as a separation from one or both parents, or school tests and examinations, those which might induce anger, such as any sort of frustration, and the onset of pleasurable activities such as Christmas or a birthday, may all therefore be responsible.

The mechanisms whereby emotional states produce asthmatic attacks are not well understood, though there is no doubt that intense emotion can reduce pulmonary flow rate. It is likely that the hypothalamic–limbic–midbrain circuits are activated following appreciation and interpretation of a psychologically meaningful stimulus at cerebral level. The activation of these circuits is known to influence immune systems. Presumably there is a final common pathway, involving the release of chemical mediators immediately preceding the asthmatic attack, and this may be reached either by direct influence of neuroregulatory pathways from the brain in arousal states produced by emotion, or as a result of hypersensitivity to allergic substances. It is also possible that the various forms of respiratory behaviour associated with emotional expression—laughing, crying, etc.—may sometimes act as specific triggers of attacks.

### Indirect psychological contributory factors

Indirect factors include those that affect the child's likelihood of experiencing emotional states productive of attacks as well as those that affect the child's vulnerability to allergic stimuli, for example by non-compliance with medication.

**Family factors.** These are probably the most common and important psychosocial influences. Mrazek *et al.* (1987) have shown that insecure attachment is more common in severely asthmatic preschool children, in comparison with healthy controls, and J. H. Kashani *et al.* (1988) that parents of asthmatic children are indeed usually anxious. Asthma in its acute form is a very frightening condition for anyone to observe, and for parents it is often terrifying, because death may seem so near. It is therefore only natural that between attacks parents should show a great deal of anxious and protective behaviour towards their asthmatic children. Such anxiety is readily communicated, and the child is then more likely to experience an attack. Family studies of children with asthma, in comparison with healthy and diabetic children, have demonstrated that the families of asthmatic children tend to rigidity and over-involvement or enmeshment in their relationships (Gustafsson *et al.* 1987).

The occurrence of the attack provides a further reason for parental anxiety, and the cycle is complete. The circular chain of events may be temporarily broken by medical treatment, and admission to the safe environment of a hospital ward is often a considerable reassurance in these circumstances. However, if every time a minor attack occurs the child is admitted to hospital, a pattern of unnecessarily frequent admissions may be established. The way in which contact with parents can elicit asthma has been well demonstrated in an experiment in which parents of selected asthmatic children were encouraged to spend 2 weeks away from home while their children were looked after by substitute parents. The design of the experiment meant that the children were still exposed to the same allergens in the domestic environment. In a sizeable number of the children, there was a marked reduction in asthmatic attacks, followed by an increase when their parents returned (Purcell and Weiss 1970). In contrast, a small minority of parents can also show marked denial of the presence of the condition in their child and fail to ensure adequate medical care. Such rejecting behaviour may result in the child becoming aggressive, and also at risk of unnecessary invalidism or even death as a result of medical neglect.

### Psychological factors in the child

1. There is no specific type of *personality* associated with asthma. Nevertheless, a proportion (about 10 per cent) of asthmatic children show significant emotional and behavioural problems of different types, and asthmatic preschool as well as school-age children do show higher rates of disturbance than healthy controls (Mrazek *et al.* 1987). Although such disturbances may be caused by factors unrelated to the asthma, for example family disturbances, school stresses, and adverse temperamental characteristics, the asthma or some aspect of treatment may be used by the child as a means of expressing anger or frustration, or of gaining attention. Although

the evidence is not consistent, it appears that severely asthmatic children show the greatest levels of emotional disturbance, aggression, and anxiety (McNichol *et al.* 1973).

2. *Illness-related abnormal behaviour.* The occurrence of an attack may allow a child to avoid a stress he would otherwise have to face. Some children develop the capacity to induce attacks deliberately with this in mind. Children may not take medical measures they know will abort an attack, if the occurrence of an attack will bring them advantages. Non-compliance with medical treatment may also occur in children unable to face the fact that they do have an illness. By 'forgetting' or refusing to take their prophylactic medication, they are able to deny the fact that there is anything wrong with them. Some parents who cannot cope with the idea that they have an ill child may also behave in this way. Asthmatic children, depressed and discouraged for a variety of reasons, including parental neglect and rejection, may fail to abort attacks because they have such a low self-image that they do not see themselves as worth troubling about. Parental neglect has in these cases encouraged self-neglect. There is some evidence that children who are low in allergic potential are generally more likely to have asthmatic attacks in which psychological mechanisms are prominent.

### Assessment

Assessment first involves the establishment of the diagnosis and of precipit-ating factors leading to attacks. A broad approach to assessment from the beginning is desirable. Parents and children should be asked if they have observed the effects of the various substances breathed in or eaten that are known to induce attacks. In addition, they should be asked about the possible effects of other factors such as exercise and psychological stresses in precipitating and maintaining attacks (Matus 1981).

It should be possible to assess the attitudes of the child and family to the asthma. Is the degree of worry and concern commensurate with the severity of the disorder? Is there overconcern about a relatively minor problem or denial of a serious disorder?

Where psychological factors appear prominent, either at the onset of the condition or once it is well established, an account of the quality of family relationships and the child's experiences at school should be obtained. The possibility that factors unrelated to the asthma are affecting the way in which the illness is experienced or reacted to should be borne in mind.

### Treatment

Treatment of asthma in childhood first involves competent medical man-agement, both in acute episodes and between attacks. As well as appropri-ate medication, this should include advice on the avoidance of allergens. The attitude of the child and other family members to the asthma should be monitored. It is often helpful to encourage the child and parents to express anxieties they may have about the recurrence of acute attacks and the possibility that the child may die in one of them. In fact, the mortality from asthma is low, and the physician can be reassuring in this respect. Some forms of medication such as theophylline may have an adverse effect on behaviour and learning (Rachelefsky *et al.* 1986), and the use of such medication should be accompanied by monitoring of school performance. Steroids, however, do not appear to affect psychomotor adaptation as has been suspected (Bender *et al.* 1987).

The child should be encouraged to lead as normal a life as possible and, in most cases, no restrictions at all other than those related to diet and avoidance of allergens need be imposed. Some children have exercise-induced asthma, however, and these should be discouraged from pro-voking attacks by exercising to the degree that produces one. Normal school attendance should occur as far as possible. There is evidence that asthmatic children are, as a group, of average or a little above average ability, but may develop educational retardation probably due to pro-longed school absence (Hill *et al.* 1989).

In cases where there is non-compliance with medication or allergen avoidance, or the child seems to be deliberately inducing attacks, there needs to be a more intensive investigation of the child and family. It needs to be understood, for example, why the child prefers to have an attack rather than to go to school, or why it is so important to a child and family that he denies the fact that he has an illness to such a degree that he forgets to take medication. The physician, however, should not forget that parental non-compliance with medication may not always be against the child's best interests. Surprisingly, one study has shown there is little relationship between treatment compliance and medical outcome in this condition (Deaton 1985).

The management of psychological factors leading to exacerbation of asthma will depend on the formulation of the problem. When a child is clearly gaining more from having attacks than he would if he did not have them, an operant behavioural approach is often helpful (Section 6.2.2) Where there is a problem within the family in the communication of feelings and perceptions of the asthma, family therapy may be effective. There is evidence that brief family therapy designed to improve family communication produces a modest but definite improvement in lung func-tion (Lask and Matthew 1979). When the child's pattern of behaviour suggests internal conflicts resulting in self-destructive tendencies or ill-understood outwardly directed aggression, brief, focused individual psychotherapy is indicated. If, however, the child is generally anxious, but there is no good evidence that deep-seated conflicts are responsible, a course of relaxation may be helpful. It is in these circumstances that hypnosis is likely to have an effect, though there are no controlled trials demonstrating its efficacy in asthmatic children. In a small minority of children where asthma is severe and protracted, and there is evidence from the effects of repeated hospitalization that the child fares better away from home, residential schooling will be indicated. With more effective medication, the need for such special schooling has decreased.

### Outcome

Most (about three-quarters) children with asthma improve considerably at or before early adolescence (Blair 1977). A minority of those who remit do, however, relapse in late adolescence and early adulthood. There is no evidence that the prominence of psychological factors has either adverse or benign influence on the course of the physical condition.

Psychological aspects of the condition may, however, have a long-lasting impact both on the personality of the child and on the life of the family. In particular, because of the fact that emotional arousal provokes attacks, some children become emotionally constricted and frightened of the experience of powerful feelings. This may lead to the development of a rather introverted and inhibited personality. Most children with asthma will, however, develop normally, and there is no evidence that they are especially prone to psychological problems later in life.

### 5.11.2 CYSTIC FIBROSIS (MUCOVISCIDOSIS)

### Definition

This is an inherited condition which affects the function of all exocrine glands—those glands that discharge their secretions directly into the body. The secretions of these glands are abnormally viscid and pathological effects occur due both to deficiency of the secretions and to obstruction of the various ducts. The most seriously affected organs are the lungs and pancreas, but other organs such as the liver, sweat glands, and intestines are also involved.

### Clinical features

Depending on the severity of the condition in the individual, it may present at any time of life, but most commonly the first signs occur in infancy or early childhood. In general, the earlier the symptoms begin, the more serious the condition.

At birth the intestine may already be obstructed by secretions—a condition known as meconium ileus. This condition occasionally requires early surgical intervention if the infant is to survive. In early and mid-childhood, chronic respiratory infections and failure to thrive are the most common presenting features. However, cystic fibrosis may also present in other ways, including diarrhoea due to intestinal malabsorption, or prolapse of the rectum. The diagnosis is confirmed by testing the patient's sweat for salt concentration.

Usually the most serious and prominent symptoms are respiratory. The child becomes chronically infected with both common and uncommon (especially pseudomonas) organisms. Eventually these become more and more difficult to control, and death occurs due to respiratory failure. The patient may, however, survive for many years following diagnosis, and many treated with antibiotics, regular physiotherapy, and vitamin supplements now live on into adulthood. The prognosis depends on the severity of the condition, the availability of treatment, and treatment compliance.

### Causation and incidence

The condition is genetically determined as an autosomal recessive, so that the risk for future births after the birth of an affected child is one in four. It occurs in about 1 in 2000 births, and there is a roughly equal sex ratio. Recent rapid progress has been made in identifying the locus of the affected gene and, on the basis of prenatal investigation of gene markers, it is now possible to give increasingly accurate information to parents about the likelihood of their child being affected.

### Psychosocial aspects

There are a number of specific psychosocial features of this condition which complicate its management (Burton 1975). The **outlook** for patients is very variable so that it is difficult for both parents and children to develop firm expectations for the future. Often, before a decline sets in, the child will remain relatively well for several years, requiring only intermittent treatment for respiratory infections. With other children the progress is much more rapid. The terminal phase of the condition is also variable in length, with some children experiencing a rapid deterioration leading to death, and others suffering from apparently overwhelming infection, but then recovering and returning to a life of good quality for several more months or even years. The recent introduction of heart–lung transplantation in children and adolescents nearing the terminal phase of their illness may improve the prognosis significantly and will introduce a new set of psychosocial issues to be considered (see Section 5.14).

**The demands of treatment** made upon the child and parents are very considerable. Thus parents, especially mothers, are likely to be involved in physiotherapy (postural treatment to drain the lungs) 2 or 3 times a day, to a total of several hours a week. The need for replacement enzymes, antibiotics, and vitamins adds further to the burden of treatment. The daily active involvement of the parents with treatment of the condition means that the risk of child and parent becoming emotionally overinvolved with each other and totally preoccupied with the illness is considerable.

**Parental understanding** of the nature of the condition is often limited, and this impedes communication with the sick child and sibs. A significant proportion of parents remain unaware of the genetic implications and the risk of further affected children, perhaps partly because they are not adequately informed, and partly because they may have difficulty absorbing the information.

**Family relationships.** Although family functioning in families with a child with cystic fibrosis has been found to be normal in terms of adaptability and cohesiveness (Cowen *et al.* 1986), and such families do not perceive themselves to be under particular stress (Walker *et al.* 1987), there is no doubt that family life is greatly affected by the birth of such a child. Parents often treat the child with cystic fibrosis more indulgently than their other children (Burton 1975). They set fewer demands for independence and conformity. This is particularly likely to be the case if they have previously lost a child with cystic fibrosis. Not infrequently this leads to jealousy between the affected child and non-affected brothers and sisters. Other behavioural problems, such as temper tantrums, disobedience, and problems with other children, may also arise for similar reasons.

Despite these various special stresses, the evidence from controlled studies (Gayton *et al.* 1977; Bywater 1981; Drotar *et al.* 1981) suggests that the rate of behavioural and emotional disorders in children with cystic fibrosis is not higher than that of children with other non-cerebral physical disorders. Some investigators (e.g. Cowen *et al.* 1986), however, have found rates of disturbance to be high in uncontrolled studies and have pointed particularly to rates of aggressive behaviour in young children. Mothers are more likely to be depressed the more severe they perceive their child's illness to be (Walker *et al.* 1987). The level of disturbance in affected children is related, to some degree, to the amount of social support the family receives (Frydman 1981). Although the children do miss school because of their condition, and sometimes have less educational pressure placed upon them than they might otherwise, their educational standards are usually not retarded.

The sibs of children with cystic fibrosis are also exposed to especially stressful circumstances (Cowen *et al.* 1986). Parents often do not explain the nature of the affected child's condition to brothers and sisters, who may consequently feel confused and uncertain how to react. Older children tend to be solicitous towards and make allowances for the younger affected child, but where an unaffected child is younger than the affected, this occurs less often and negative reactions are much more common (Burton 1975). Reactions of the affected child to his or her own illness are various. Most, despite being inadequately informed about the nature of their condition, come to terms with their disability and cope with the symptoms of their illness and demands of treatment with great fortitude. Many affected children particularly enjoy sporting and other outdoor activities when they are well. A small proportion react to their disability with denial and refusal to accept their treatment—physiotherapy or tablets. Usually denial in the child is linked to a similar problem in parents, who are likely to have difficulty in telling the child about the nature of the illness.

**Management of psychosocial problems**

Ensuring that both parents (fathers as well as mothers) of children diagnosed as suffering from cystic fibrosis are well informed about the nature of the condition and its genetic implications will prevent much misunderstanding both between parents, and between parent and child. Parents should also be encouraged to have the same level of expectations for behaviour and self-help for their affected child as for their other children.

There is evidence that children are better adjusted if parents can communicate with children about their illness and its implications (Burton 1975). The same holds true for non-affected sibs. This does not mean that parents should emphasize the fatal outcome of the condition from the outset, but it does mean that children should know they have a chronic condition, and it is not their fault or that of their parents. They should know that they will always be likely to have symptoms, and that these are likely to become more severe as time goes on. They should have the opportunity of sharing their feelings of disappointment, grief, and anger with their parents, who will themselves adjust better if they (the parents) can share their own feelings with others, relatives, friends, and professionals. Management of the psychosocial aspects of terminal stages of illness are discussed in Section 4.7.

# 5.12 Genitourinary disorders

By far the most common urinary tract disorder of psychological significance is enuresis (Section 2.12.5). However, a number of other urinary tract disorders also raise some specific psychosocial issues, and these will also be briefly discussed in this section.

## 5.12.1 CONGENITAL MALFORMATIONS

Severe congenital malformations of the urinary tract are relatively uncommon and, when they occur, are not infrequently incompatible with survival. However, relatively recent surgical advances have improved the prognosis for some conditions. The least uncommon is the presence of posterior urethral valves, occurring in boys, usually presenting in the first year of life. The valves obstruct the outflow of urine and this leads to distension of the bladder, ureters, and kidneys. Surgical treatment is often fully successful, but partial success may require further hospitalization, for example for implantation of the ureters into the colon. Such procedures often understandably raise anxieties in parents concerning matters such as the subsequent sexual function of their children.

## 5.12.2 URINARY TRACT INFECTIONS

These are relatively common throughout childhood, and occur more frequently in boys during the neonatal period but subsequently more commonly in girls. In the first year of life, failure to thrive with irritability and poor feeding may be caused by chronic or recurrent urinary infections. In babies from socially deprived backgrounds, it may be unclear to what degree failure to thrive is due to an infection or to relative neglect. Young children presenting with failure to thrive and identified as having urinary infections which are then successfully treated should therefore be carefully followed up to assess whether other factors might be responsible for growth failure.

In girls of school age, recurrent urinary infection will require physical investigation to determine whether there is a structural or physiological abnormality such as reflux. In the absence of such abnormalities, aetiology of psychosocial significance should be considered, though this will not frequently be present. Poor social circumstances linked to inadequate hygiene may be relevant. Aspects of the history or examination of the external genitalia may raise a suspicion that the girl is subject to sexual abuse (Section 1.3.3).

## 5.12.3 NEUROPATHIC BLADDER

The presence in a child of a distended bladder with a poor stream of urine is almost always due to a defect in the bladder nerve supply. Most commonly this is associated with spina bifida. In a small minority of children, however, a neuropathic bladder occurs in the absence of neurological signs. This raises the possibility of a hysterical conversion syndrome (Section 3.2) and psychiatric investigation is indicated. It is uncommon, however, for a clear-cut psychiatric diagnosis to be made in these circumstances, and the cause of the condition usually remains uncertain. Treatment, of which psychosocial support should be one component, is symptomatic.

## 5.12.4 CHRONIC RENAL FAILURE

Failure of kidney function is usually caused by chronic glomerulonephritis, pyelonephritis, or congenital abnormalities, but sometimes by a variety of other conditions. Progressive renal failure results in uraemia and death unless modern treatments, particularly haemodialysis and renal transplantation, are available. The general indications for haemodialysis and transplantation are outside the scope of this book. However, some paediatricians regard instability of the family and the presence of psychological disturbance in the child as relative contraindications to these procedures. Psychosocial appraisal has not reached a stage where successful prediction of compliance is possible, though there is some evidence for the importance of particular adverse factors.

Technical improvements in haemodialysis, which mean that home-based treatment is now possible, have reduced the stress on children and families. Nevertheless, psychological problems occur more frequently than normal in children with chronic renal failure on dialysis (Garralda *et al.* 1988). The more severe the physical problem, the more likely is the child to show emotional difficulties, particularly anxiety, loneliness, and over-dependency on parents. There is a reduction in levels of behavioural disturbance after dialysis has been established, probably arising from alleviation of parental anxiety. Nevertheless, the stress on family life remains considerable. Treatment takes 12–18 hours a week, diet has to be restricted, and complications are not uncommon. Nevertheless, children are usually able to attend normal school on a reasonably regular basis.

Renal transplantation is a preferable form of treatment, but is not always available. About two-thirds of transplants are successful. Organ rejection is therefore fairly common and is associated with high rates of depression and anxiety in the children and adolescents concerned. Aggressive and hostile reactions may also occur in this age group. Psychiatric problems in children with renal transplants occur at about twice the rate of those in healthy children (S. D. Klein *et al.* 1984). The children are preoccupied with their short stature—an almost inevitable complication of the renal disease and steroids used to suppress organ rejection. Girls are particularly upset by the bloated facial appearance produced by steroids, and this and other factors not uncommonly lead to non-compliance with medication. Good preparation for the various procedures and for operation, with warning of the side-effects, as well as involvement of the adolescent patient in the decision to transplant, may reduce subsequent non-compliance, so that transplantation may be expected to result ultimately in overall psychological improvement. Mental health professionals may have a useful part to play in this aspect of management.

## 5.12.5 UNDESCENDED TESTIS

This is a fairly common condition occurring in about 1 in 50 boys and is often diagnosed either at birth or on school entry medical examination. The testis may have failed to descend or may have descended to an abnormal position. In either event, if natural descent does not occur in the first year, surgical treatment is necessary for various reasons (risk of malignancy, preservation of fertility, cosmetic reasons, etc.) and is usually carried out before the age of two years. Some surgeons give a prior trial of hormonal treatment. Surgery is generally successful.

Because of anxieties that both parents and professionals experience when talking to children about their genitalia, preparation is often inadequate. Adequate preparation of the child for the procedure is, however, particularly important because boys often have fears about being permanently mutilated in the genital area. Anxiety about castration, which used to be thought a common fantasy, is probably less common than fears of mutilation of the penis. There is no good evidence that boys undergoing this operation, if adequately prepared, need suffer long-term adverse psychological consequences, nor is there evidence that the timing of operation has long-term psychological significance (Cytryn *et al.* 1967).

## 5.12.6 INTERSEX DISORDERS

These are conditions in which sexual appearance is ambiguous or in which there is inconsistency in features of sexual differentiation. The criteria for sexual differentiation include chromosomal structure, internal morphology, external morphology (the appearance of the external genitalia), the pattern of sex hormone secretion, gender identity (the degree to which individuals perceive themselves as male or female), gender role (pattern of sex-typed behaviour), and choice of sex object. Although chromosomal or gonadal sex are sometimes suggested to be the indicators of 'true' sex, there is no logical justification for this, and indeed gender identity and role might just as reasonably be regarded as the best indicators.

In so-called 'true' hermaphroditism the gonads contain both ovarian and testicular tissue. Conditions of this type are extremely rare. More common are male and female pseudo-hermaphroditism in which chromosomes and gonads are clearly differentiated, but the external genitalia are ambiguous. Thus in male pseudo-hermaphroditism the chromosome structure is XY, and testes are present, though there is cryptorchidism and the external genitalia have a female appearance.

*Congenital adrenal cortical hyperplasia* (CAH), when it occurs in chromosomal females, causes female pseudo-hermaphroditism and is by far the most common intersex condition. It results from an inherited (autosomal recessive) enzyme defect, usually of 21-hydroxylase. This affects steroid production and there is excessive androgen excretion *in utero* and subsequently. In boys this produces precocious puberty. Girls present at birth with ambiguous genitalia. A salt-losing syndrome is often associated. The condition is diagnosable after birth and treatable with steroids.

Decision as to which sex the child should be reared should be taken as soon as possible after birth when the condition has been investigated and the diagnosis made. Decision-making should involve parents, paediatricians, and the paediatric surgeon, and a psychological or psychiatric input is often helpful. At this stage, the most relevant consideration is the likely external appearance after plastic surgery.

Issues of reassignment of sex after a period in which the child has been brought up in one sex-role require very careful consideration with strong psychological or psychiatric input. In general, reassignment of sex after two or three years of age presents very considerable problems and involves a major family crisis, as gender identity and parental perceptions are relatively fixed by this time (Money and Ehrhardt 1972). Reassignment has been described at puberty with five α-reductase-deficiency males brought up pubertally as females, but this occurred in a closed community where the condition is relatively common and accepted by members of the community.

Although the physical outcome of CAH is usually satisfactory, given early diagnosis, psychosocial outcome may be much less so (Hochberg *et al.* 1987). Chromosomal girls raised as females tend to be tomboyish in manner and, post-pubertally, have a higher than expected rate of homosexuality (Money *et al.* 1984). Chromosomal girls raised as males usually have a small penis even after surgical reconstruction, and show poor peer relationships and later low sexual potency. If early reassignment of CAH chromosomal females raised as boys is possible, this should be discussed with parents.

Children with intersex conditions and their parents are likely to benefit from continued psychological and psychiatric counselling (Money 1975). Practical advice concerning how to deal with situations in which, for example, the child's genitalia may be seen by other children will be helpful. Information about the future can be presented in a positive manner. For example, parents and, as they get older, children, can be told marriage will be possible as well as child-rearing, though this may need to be by adoption. Self-esteem is often closely linked to a clearly defined sexual identity, and children with intersex conditions may need special encouragement for their performance in school or in recreational activities so that they can maintain a good self-image. Counselling for children and families should not be discontinued after the pre-pubertal years, but should continue well into adolescence, when new difficulties may need to be faced.

# 5.13 Disorders of the alimentary tract

## 5.13.1 ORGANIC GASTROINTESTINAL DISORDERS

### Peptic ulcer

Duodenal ulcers are not common, but occur occasionally in childhood. As in adults, the pain is likely to be epigastric, but there may be little relationship with mealtimes. Episodes of pain are periodic, and the pain may wake the child at night. The condition often presents with an attack of gastrointestinal bleeding. There is often a positive family history.

Psychological factors are usually not prominent, but sometimes the pain may be precipitated by stress. Peptic ulceration has been reported as occurring in association with school refusal. Children with the condition have also been reported to be shyer and more tense than controls, with an increase in school difficulties (Christodolou *et al.* 1977). Early separations from parents and parental loss have been reported to occur in excess.

Attention to stress factors and family tensions should form part of management, which will otherwise consist mainly of dietary advice, antacids, and possibly other specific medication. Many ulcers occurring in childhood do become chronic, and it is possible that attention to stress factors improves the prognosis.

## Crohn's disease

This is a chronic, inflammatory condition, most commonly occurring in the ileum, but also appearing at times in other parts of the alimentary tract. The condition runs a chronic course and, when it occurs in childhood, often produces significant social disability and growth failure.

Psychological factors are not thought to be of importance in aetiology. Children suffering from the condition have a high rate of associated behavioural and emotional problems (B. Wood *et al.* 1989), with high rates of psychiatric disorder linked to greater disease activity. Stress may precipitate relapses. Paediatric management of the condition is often highly complex. It is therefore important for the clinician and others involved with the child and family to be aware of the family circumstances and school progress to a degree which allows monitoring of these factors and, occasionally, intervention to reduce stress. Psychiatric referral may occasionally be indicated if associated emotional problems become prominent.

## Ulcerative colitis

This involves chronic inflammation of the colon and rectum. It is uncommon in childhood, but can occur in children as young as 7 or 8 years. The symptoms, especially pain and the passage of mucus and blood in the stool, usually continue over several years with frequent remissions and relapses. Partly because of the risk of cancer in parts of the bowel affected over very long periods, surgical treatment with creation of an ileostomy is now more readily undertaken. Most, but not all, children adjust well following stoma surgery, but careful preparation is required, giving the child and family the opportunity beforehand to meet others who have had the operation (Lask 1986). Adverse psychological factors should not delay operation if the medical condition of the child suggests this is necessary.

Psychological factors are thought to be of little importance in aetiology (Feldman *et al.* 1967), but they often play a part in the maintenance of the condition and in the precipitation of relapse. The rate of psychological disturbance among childhood sufferers is high (Raymer *et al.* 1984) and is not of any particular type, though obsessive–compulsive symptoms occur more commonly than in other physical conditions (Burke *et al.* 1989). The presence of disturbance is not related to the severity of the condition or to specific physical features other than growth failure.

Psychotherapy has been recommended as an adjuvant to surgical and medical treatment (McDermott and Finch 1967), but attention to psychological factors is probably best provided by the clinician responsible for medical management. His long-standing acquaintance with the family will enable him to be sensitive to likely stresses operating on the child. Psychiatric referral should be limited to those cases where there are associated persistent behavioural and emotional problems or non-compliance with treatment.

## 5.13.2 RECURRENT NON-ORGANIC ABDOMINAL PAIN

### Clinical features

The condition usually presents in children aged between 5 and 12 years. Pain is usually diffusely experienced in the abdomen and may be accompanied by headaches and pains in other parts of the body, especially the limbs and joints. It may be accompanied by nausea and vomiting. Sometimes the pain may come on in direct relationship to stress, such as at the beginning of a new term at school, but more often there is no such obvious relationship. The child may, however, be generally tense, anxious, or depressed. Often the child has had a rather tense personality since early childhood.

Physical examination reveals a lack of definite, localized tenderness or rebound tenderness. If physical examinations are carried out, they prove negative. The pain tends to last a few hours and then remit, but it may be more persistent than this and relapses are common. In a small proportion of children with this condition, abdominal pain occurs in regular association with vomiting, headache, and low-grade fever. This constitutes the so-called '**periodic syndrome**', a poorly defined condition possibly related to migraine (Section 5.7.4) or to emotional disturbance (Apley and MacKeith 1968). If vomiting is present, this condition is usually referred to as '**cyclical vomiting**' and, in these circumstances, headache and fever are absent. Usually attacks of cyclical vomiting last only a few hours and the child is unwell for a day or so. Very occasionally the vomiting is persistent and produces life-threatening dehydration.

### Background features

Recurrent abdominal pain is a common feature of growing children and occurs in 5–10 per cent of the population. Rutter *et al.* (1970*a*) found that 5 per cent of 10- and 11-year-olds had suffered from stomach-aches at least once a month over the previous year. Boys and girls are roughly equally affected, and there is no specific link with social class.

The proportion of children whose recurrent abdominal pain is due to psychological factors is uncertain. Some clinicians regard the link as almost invariable, but there is in fact a rather weak relationship between the presence of abdominal pain and psychiatric disorder (McGrath *et al.* 1983), though migrainous headaches, by contrast, are associated with emotional disorder, as well as abdominal pain and vomiting (Kurtz *et al.* 1983). Children with this type of pain may be predisposed to somatic symptoms by family attitudes to illness. Their symptoms may be an out-come of interaction between difficult temperament and environmental stresses (Davison *et al.* 1986). On the other hand, extensive medical investigations are nearly always negative in their findings, and occasional suggestions of organic aetiology, for example that the syndrome is pro-duced by lactose intolerance, are rarely confirmed.

The mechanism of pain production is usually unknown. Pain may be due to muscular tension, undue sensitivity to normal abdominal sensations, or vasoconstriction and vasodilatation of migrainous type.

### Assessment

Chronic or recurrent pain in the abdomen in childhood can occur for organic and non-organic reasons. Among the non-organic group are those where the aetiology is psychogenic, and others where the cause is unknown.

Identifiable organic causes account for only about 10 per cent of cases of children with recurrent abdominal pain, the remainder being either of psychogenic or unknown cause. Renal tract infections, mesenteric lymph-adenitis, chronic pancreatitis, peptic ulcer, Crohn's disease, and ulcerative colitis are the most common organic conditions found, but none of these occurs frequently. A host of other conditions, both intra- and extra-abdominal, are very occasionally responsible.

*Clinical features distinguishing organic and non-organic pain*
(Table 8)
Pain of organic cause is likely to be well-localized and restricted to the abdomen. Non-organic pain is more likely to be diffuse and accompanied by pain elsewhere in the body, especially the head and limbs. Severity of the pain and the presence of associated vomiting are not good guides to its aetiology. In non-organic pain there are often signs of associated emotional disturbance—especially anxiety, depression, and a tense personality. Organic pain is more likely to be linked to tiredness and pallor.

Physical examination is the best single indicator, with localized tender-ness and rebound tenderness indicating an organic cause. Even severe non-organic pain is usually unaccompanied by tenderness on palpation. Physical investigations, especially examination of the blood, urine, and radiological tests, may be useful in excluding organic causes.

The quality and time relationships of the pain therefore, together with the presence of nausea or vomiting and other abdominal symptoms, should be identified. Are there associated aches and pains in other parts of the body? Has there been any loss of weight? What is the association of the

**Table 8**

|  | Abdominal pain | |
|---|---|---|
|  | Organic | Non-organic |
| Frequency (per cent) | 10 | 90 |
| Pain | Localized | Diffuse |
| Pain wakes at night | Often | Rarely |
| Pain elsewhere in body | Unusual | Common |
| Vomiting | May be present | May be present |
| Emotional state | Usually normal | Usually anxious, tense, depressed |
| Abnormalities on examination and investigation | Present | Absent |

pain to stressful events and, in general, how anxious and tense has the child's emotional state been? What potentially upsetting events have there been at home or at school at the time the pain started? Do other family members suffer from stomach-aches or headaches?

Physical examination should be carried out with particular concern for the degree of abdominal tenderness. Physical investigations may be indicated, but in most cases of recurrent abdominal pain these will not need to extend beyond examination of the urine.

### Treatment of non-organic abdominal pain

In those rare cases where there is vomiting producing dehydration, medical management may require intravenous therapy and anti-emetic medication. Stemetil, diazepam, or chlorpromazine may be helpful. Subsequent management should, however, proceed along the lines described below for non-organic abdominal pain.

Assuming organic causes have been virtually excluded, it is helpful to explain to parents and child that no serious physical condition is present and that the child has a relatively common problem that can be expected to get better over the next few weeks or months. It can then be explained that often, though not always, the type of pain and other symptoms the child is experiencing can be due to tension because of emotional upset. Have the parents or child thought this possible in their case?

The response to this question will enable parents and older children to express their present understanding of the problem in psychological terms. They may have already come to this conclusion and be ready to discuss psychological management. Alternatively, they may be indignant at the suggestion.

Once this information is available, it is likely that the clinician will be able to conclude that the non-organic pain experienced by the child falls into one of three groups:

(1) somatic symptoms part of a wider emotional disturbance with anxiety, depression, etc.;
(2) monosymptomatic response to stress in the child's circumstances;
(3) pain of uncertain aetiology in which psychological factors may or may not be playing a part.

When abdominal pain forms part of a generalized emotional disorder characterized by anxiety, treatment should proceed along the lines described in Section 2.9.2. If the pain is monosymptomatic, but due to tension, then consideration should be given to relieving known stress factors at home and at school, and teaching the child relaxation techniques (Section 6.2.2). In cases of uncertain aetiology, the child and family should be reassured that, although the cause is unknown, the pain is very likely to disappear. Follow-up should be arranged to check on progress and, if the pain continues and the family members are motivated, to consider what stress factors might now be operating to maintain it. It is important in these cases to ensure that usual activities are continued as far as possible, and invalidism avoided.

### Outcome

With the type of management described above, most paediatricians describe an excellent short-term outcome in 60–80 per cent of their cases. No controlled trials of psychological management have been reported, so it is impossible to know whether this makes a difference to outcome.

Follow-up studies of children with this syndrome suggest that between a third and a half continue to suffer abdominal pain in adulthood (Apley and Hale 1973). Children of parents who continue themselves to suffer abdominal pain are especially likely to show pain persistence (M. F. Christensen and Mortensen 1975). This suggests either that the persistent form of the symptom has strong genetic loading, or that family patterns of responding to stress by somatic symptoms are learnt within certain family systems.

## 5.14 Congenital heart disease

### Introduction

About 1 in 150 children is born with a congenital heart defect, and this is therefore the commonest type of congenital malformation. The defect may vary from the trivial (e.g. a small ventricular septal defect) to a defect incompatible with life even if skilled surgery is available. About 1 in 10 children with congenital heart defects has an associated defect elsewhere in the body.

The condition is usually polygenically determined. However, children with some chromosomal abnormalities (e.g. trisomy 21, Down's syndrome) have high rates of heart defects, and some defects arise because of abnormalities in the pregnancy (e.g. maternal rubella).

Detection of the condition may occur at a routine postnatal examination or during a medical examination for some other reason in an older child.

Occasionally an abnormality may be suspected antenatally following ultrasound scan. In the majority of children who present in the neonatal period or early infancy, there is cyanosis, slow feeding, breathlessness, failure to thrive, or other symptoms.

Many cardiac lesions can be confirmed by non-invasive echocardiography but a significant proportion of children will proceed to cardiac catheterization, a procedure not without risk. Subsequently, many affected babies will receive surgery, but a few will have conditions too severe for successful operation. These receive symptomatic treatment but have a fatal outcome.

Surgery is often remarkably successful, but a number of children require re-operation, either as a planned procedure or because the initial operation was only partially satisfactory.

## Psychosocial aspects

### The trivial heart condition

In a number of children with 'innocent' murmurs, follow-up reveals that the detection of the murmur and subsequent parental anxiety have led to overprotection and invalidism. The same outcome occasionally occurs with minor heart conditions of no functional significance. In these children, great anxiety may occur every time the child has a minor procedure or operation such as a dental extraction, because of the need for antibiotic cover. The danger of overprotection is unrelated to the seriousness of the cardiac condition (Garson *et al.* 1978). Once it has been decided that a murmur is of no pathological significance, review appointments are likely to raise unnecessary anxiety. Complete reassurance with discharge from the out-patient clinic is more likely to avoid overprotection and invalidism.

### The inoperable defect

Some children are likely to have a relatively brief life. Parents will require considerable support over this period and after the death. They may be particularly upset because they will often be attending clinics where the prevailing mood is one of optimism, with most children doing well. For a discussion of psychosocial aspects of terminal care, see Section 4.7.

### Operable lesions

With advances in paediatric cardiac surgery, these now form the largest group. All such children and their parents have to face cardiac catheterization, a major operation, and a period of postoperative uncertainty. A proportion have to face further operations and, in a minority, operative procedures fail to correct the lesion adequately and a fatal outcome ensues. Specific psychosocial aspects that need to be considered in the care of these children include:

1. Awareness of the fact that the heart is often regarded by parents as the most vital organ the child possesses. A defect of the heart is particularly frightening even if, to the clinician, the lesion seems rather minor.

2. *The anxiety-provoking nature of the procedures.* Parents need full explanation of the procedures before they occur. The nature of the catheterization should be fully explained with a prior visit to the catheterization room. Parents need to be given details of the cardiac lesion, the likely site and extent of the incision, and the likely appearance of the child after the operation. They should have the opportunity of a prior visit to the intensive care unit where the child will be nursed postoperatively and should be given a description of the likely rate of progress subsequently. They should be given an honest account of the risks of the operation and the rates of the various complications.

The recent introduction of heart and heart–lung transplantation requires special consideration. Psychological aspects of these procedures when carried out in childhood have so far been little researched, but parental anxieties are naturally intense. Specific fears concerning the nature of the donor, the viability of the transplant and its capacity to develop in a way that continues to meet the needs of the growing child may or may not be expressed, but parents should always be given the opportunity to discuss them. The greatest anxieties of the surgical team probably centre around the immunological status of the child postoperatively rather than around the operation itself and, although it is unlikely that parents will realize the significance of these concerns for themselves, they (and the child if old enough to understand) will benefit from discussion of them as early as possible.

*The nature of the preparation of the child* will depend on his age. Much surgery is now carried out with neonates and children under the age of 18 months, for whom no verbal preparation is possible. Good preparation of parents will ensure that the anxiety reactions of such young children are minimized (Glaser and Bentovin 1987).

Older children require explanations appropriate to their cognitive level (Section 4.6). It has been shown that good preparation of young children can reduce anxiety and increase cooperation during cardiac catheterization (Cassell and Paul 1967). Children over the age of about 2 years need some verbal explanation and the opportunity to visit the place in which they will be nursed after operation. From a slightly older age, children need explanations of the nature of the defect, the site and extent of the incision, as well as the likely degree of postoperative pain and discomfort and the steps that will be taken to alleviate these. They need to know for how long they will be in hospital and what their condition is likely to be on discharge.

3. *Uncertainty over outcome*. Many operative cardiac procedures are so well established that the risks of operation and uncertainty concerning outcome are very low. With other procedures this is not the case. Parents and children need extra support, advice, and explanation in these circumstances.

4. *Follow-up appointments*. Children with congenital heart diseases have about twice the expected rate of behavioural and emotional problems compared with children in the general population. Because of the extra risk of unnecessary invalidism in cardiac patients, it is important at follow-up to check carefully on school attendance and the degree to which the child is allowed independence outside school hours. Where overprotection is a problem and fails to improve after reassurance and advice, referral to a social worker or psychiatric facility will be desirable. The presence of psychiatric disorder may be unrelated to the cardiac lesion, and a wider assessment of the family may be helpful.

### Behaviour and emotional problems

The rates of behaviour and emotional problems in operated children depend to some degree on operative success (Kramer *et al.* 1989). Asymptomatic children do not, as a group, have higher rates of problems than healthy controls. On the other hand, children with physical conditions following surgery do appear to have higher levels of anxiety and impulsiveness as well as a greater sense of inferiority when compared with physically fit children and healthy controls (Kramer *et al.* 1989). Parents may not be aware of underlying concerns that children experience. The increasing numbers of survivors into adolescence will almost certainly present additional problems as, for many types of lesion, the prognosis remains uncertain. The need for re-operation and the occurrence of sudden, unexpected death present major stresses, and this situation has not yet been adequately researched.

### Cognitive development

Although very occasionally an operative disaster can lead to permanent brain damage, the intelligence of the great majority of children with operated congenital heart lesions is within the average range, though their mean IQ is somewhat lower than their sibs (Linde *et al.* 1967). Cyanotic patients function somewhat less well than acyanotic, but again, most are within the average range. There has been some concern that the profound hypothermia and cardiac arrest procedures commonly used in corrective cardiac surgery may affect intellectual outcome. Follow-up studies have, however, been reassuring in that mental retardation and neurological sequelae have been found to be very uncommon. The mean IQ of children subject to such procedures in infancy is about 10 points below average, but the great majority have abilities within the normal range.

## 5.15  Skin disorders

### 5.15.1 ECZEMA (ATOPIC DERMATITIS)

**Introduction**

This is a common skin condition in which there is dryness and scaling of the skin with associated itching and scratching. It usually begins in the first 18 months of life in children in whom there is a familial disposition. The condition is probably caused by an inherited immunological dysfunction, and most patients have elevated IgE levels and T-cell abnormalities. It may start at the time of introduction of artificial milk or mixed feeding, but other potential allergens may also be important. Treatment involves mainly the use of creams and ointments with antihistamines at night for sedation and to reduce scratching. The condition is nearly always benign, clearing up in the first few years of life. It is, however, often very distressing while it is present and, in a minority of cases, it is both severe and persistent at least into the middle school years.

**Psychosocial aspects**

In the early years, the presence of severe, widespread eczema may lead to problems in the development of attachment (Rauch and Jellinek 1988), as one or both parents may be repelled by the condition and have problems in cuddling or even holding the child. Later, particularly in children in whom the condition runs a chronic course, psychological problems often become prominent and may be involved in the maintenance of the condition. In some children, stressful situations appear to exacerbate the condition. In others, the need for very frequent bathing and bandaging, tasks often performed by the mother, leads to resentment on the part of the child, and this may distort the mother–child relationship (Wittkower and Hunt 1958). The mother may come to feel that she alone knows how to care for her child and be unwilling to share the tasks involved. This may make her feel hostile and resentful towards the child. Conversely, the child may become extremely dependent on his mother and angry and irritable with her. The parents may well feel the child could do more to stop himself scratching, and this too can lead to mutual irritation and arguments between the parents.

Although in adults there is work to suggest that individual psychotherapy can be helpful in eczema (D. G. Brown 1972), this may not be the case with children. Here it seems more probable that, in those cases where psychological factors are prominent, family sessions designed to improve communication between family members and allow more sharing of the child's physical care, would be valuable. An educational approach to parental counselling directed towards more effective limit-setting and self-understanding of ambivalent feelings may be helpful (Koblenzer and

Koblenzer 1988). Associated sleep problems, which are sometimes prominent and exhausting to both children and parents, may respond to behavioural management programmes (Section 2.3.2).

## 5.15.2 NEURODERMATOSES

These are conditions in which both the skin and nervous system are affected.

### Tuberous sclerosis

This is a condition in which there is abnormal proliferation of nerve cells giving rise to benign and occasionally malignant tumours in the brain. There may also be tumours of the heart, liver, and other organs. A characteristic skin lesion, producing a butterfly-shaped rash over the nose and cheek, is usually present. The condition is often inherited as an autosomal dominant.

The condition may present with epilepsy (infantile spasms, partial or generalized seizures) or developmental delay, and is usually diagnosed in the first 3 or 4 years of life. As well as mental retardation, there may be intellectual deterioration caused by repeated, uncontrolled epilepsy or the development of single or multiple cerebral neoplasms. There is a very high rate of associated psychopathology in this condition and, in one series, over 50 per cent of children showed psychotic features and 60 per cent hyperkinesis (Hunt and Dennis 1987).

Treatment is symptomatic and involves anticonvulsant medication to control the attacks and often special education. Behavioural abnormalities may respond to behaviour modification programmes. It is important to provide genetic counselling, although the existence of formes frustes (partial syndromes in which only one or two features of the condition are present) means that some uncertainty concerning the genetic risk is inevitable.

The outcome is variable, and depends on the spread of lesions and whether malignant changes occur. A chronic rather than a rapidly deteriorating course is usual.

### Neurofibromatosis (Von Recklinghausen Disease)

This condition is caused by a spread of cells from nerve sheaths to other parts of the body, where they may form tumours, usually benign, but with a risk of malignant change. One form of the condition occurs in infancy and early childhood. It is inherited as an autosomal dominant.

The clinical features of the condition depend on the site and spread of the lesions. Developmental delay is an unusual presenting feature, but epilepsy is relatively common. Sufferers from the condition nearly always have a number (usually more than five) of 'café-au-lait' skin lesions. Lesser numbers of such spots are commonly found in unaffected individuals.

Treatment is symptomatic, and the outcome is variable depending on the spread of the lesions and the development of malignant changes.

### Sturge–Weber syndrome

In this condition there is an angiomatous malformation of the skin in an area corresponding to the distribution of one or more branches of the trigeminal nerve. A similar angioma is present on the interior of the skull, usually overlying the occipital lobe on the same side, but sometimes occupying a more widespread area. The cerebral ischaemia produced by the internal lesion may give rise to epilepsy and hemiplegia as well as other neurological symptoms. There may also be some degree of mental retardation. Anticonvulsant medication may be required for the epilepsy.

### Management of neurodermatoses

These neurodermatoses may all sometimes be accompanied by severe psychiatric disturbances arising from brain dysfunction or as a psychological reaction to the condition. Involvement of the temporal lobes is particularly likely to result in behavioural and emotional disturbances. There are no characteristic psychiatric features, and aggressive behaviour as well as emotional disturbances are frequent. Frank psychotic states may also occur. Treatment is symptomatic and behavioural methods are likely to be especially useful. Emotional support for the families, with appropriate genetic counselling, should be a major part of management.

# 5.16  Immunodeficiency disorders

Bodily response to infection can be impaired by a deficiency in the immune system—the complex process involving especially immunoglobulins, proteins produced in plasma cells and circulating in body fluids (humoral immunity), and T-cells, lymphocytes produced in the thymus (cellular immunity). Deficiency in this system is sometimes responsible for an infant or young child being unusually susceptible to infection. The condition is occasionally inherited, but may be acquired as a complication of both infectious and non-infectious disease or its treatment. Some immunodeficiency is mild, transient, and of little significance to the child. Other forms, especially those produced by inherited conditions or drug therapy, may be very serious or even fatal. Treatment will depend on the nature and severity of the condition.

### Psychosocial implications

Children known to have unusually low resistance to infection often have to lead very protected lives. As in mild and moderate forms of immunodeficiency it is often unclear even to the physicians involved how necessary this protection might be, it is not surprising that parents are themselves confused and uncertain. This may result in unnecessary overprotection, or the child may be exposed to inappropriate hazards. If a child does contract or even succumb to an overwhelming infection, parental guilt may be considerable.

The condition may, if mild, require no specific treatment, but in some circumstances prophylactic intramuscular injections of immunoglobulin are indicated.

Bone marrow transplantation has been successful in some cases of immunodeficiency, as well as in severe aplastic anaemia, and is also used as a form of treatment in some types of leukaemia. The marrow is usually transplanted from a parent or healthy sib of the affected child. Studies of the psychological effects of bone marrow transplantation on the patient, the sib, and the rest of the family have revealed the serious degree to which family members are under stress (Atkins and Patenaude 1987). A detailed study of the psychosocial impact of bone marrow transplantation is provided by Pot-Mees (1989).

The patient is likely to require transfer from a local hospital to a specialist centre. The family may feel rejected and abandoned by their local medical and nursing staff, and travelling difficulties may lead to a loss of contact between the child and family members.

The patients themselves may develop serious anxiety and depression related to the isolated circumstances in which they are nursed, and fear of death (a realistic but often unexpressed concern) is frequently present. The provision of age-appropriate direct emotional support for the child is of primary importance. The degree of pressure put on a sib to cooperate may present ethical problems. Sibs whose brother or sister subsequently dies (and a fatal outcome does occur in a third to a half of cases) may have fantasies that they have been responsible for the death.

These psychological problems can probably be reduced significantly if communication between local and specialist centre, between nursing and medical staff at the specialist centre, between family members, and between staff and family is open and honest, and if account is taken of the immature understanding and apparently irrational fears and fantasies of the patients and sibs so that appropriate explanations of procedures can be given. The relative unpredictability of extremely unpleasant complications such as graft versus host disease, high fatality rate following transplantation (varying between 30 and 80 per cent), and intense involvement of staff with patients and families mean that psychological support for staff is particularly necessary.

In those conditions that are genetically determined, genetic counselling with the opportunity to discuss implications for future pregnancies is necessary.

## 5.17 Juvenile chronic arthritis

### Definition

This is a condition in which a number of joints are involved in an inflammatory process, often resulting in chronic limitation of movement, deformity, and pain.

### Clinical features

The illness commonly starts in the first 5 years of life, when fever, enlargement of the spleen, and lymph nodes are common. Arthritis is less prominent in this form of the condition, which is known as Still's disease. In older children developing the condition, in whom prognosis is rather better, polyarthritis is a more salient feature. Although some children suffer only a single attack of acute arthritis and remain well subsequently, others go on to develop a chronic illness with remissions and relapses. It is these who suffer chronic disability and pain.

### Psychosocial features

It has been suggested that the onset of the condition can be precipitated by stressful events, but the evidence for this is poor, as the events taken into account might have been caused by the illness itself. Pathological immunological processes are certainly of causative importance, and it is possible that, in some children, stressful events can produce exacerbation by their effect on the immune system.

Despite earlier reports to the contrary, children with this condition do not appear to have higher-than-expected rates of behaviour and emotional problems (Billings *et al.* 1987, Wallander *et al.* 1988). However, in adolescence, sexual anxieties are common, and girls are particularly prone to depression (V. A. Wilkinson 1981). The use of steroids in treatment not uncommonly leads to stunting of growth with consequent adverse effects on self-image and self-esteem. There is some conflict of evidence as to whether this is due to the immobility and social isolation produced by the condition (V. A. Wilkinson 1981), as others have found that it is the least-handicapped children who are the most disturbed (McAnarney *et al.* 1974). The latter found that children with minimal disability had high rates of psychiatric problems, perhaps because their disease was not taken seriously by parents and teachers.

Young children with chronic arthritis appear to experience less pain than older children and adults with the same condition (Beales *et al.* 1983). It is suggested that this is because the young child, unlike older patients, does not attach such unpleasant meaning (disability, chronicity, etc.) to his pain every time he experiences it.

These studies confirm that clinicians should be aware of the special psychosocial problems relating to this condition (Ungerer *et al.* 1988). Great efforts should be made to ensure that immobility does not lead to social isolation. Counselling of adolescents may need to focus on sexual concerns, and teenagers need help to form realistic plans for future employment. Counselling directed towards what the adolescent *will* be able to do in later life may reduce the level of hopelessness and counter the tendency to experience pain as a reminder of chronic disability. On the other hand, the adolescent may also need to experience appropriate depression in relation to the reduction in quality of life.

# Further reading

Apley, J. and Ounsted, C. (1982). *One child*. Heineman/Spastics International Medical Publications, London.

Blum, R. (ed.) (1984) *Chronic illness and disabilities in childhood and adolescence*. Grune and Stratton, London.

Eiser, C. (1985). *The psychology of childhood illness*. Springer Verlag, New York.

Hall, D. M. B. (1984). *The child with a handicap*. Blackwell Scientific Publications, Oxford.

# 6

# Prevention and treatment

## 6.1 Preventive approaches

### Introduction

Preventive activity in the field of the psychosocial disorders of childhood can, like other forms of medical prevention, be classified as follows:

1. *Primary prevention.* Action taken to prevent the development of the disorder in the first place—usually by removing the cause.

2. *Secondary prevention.* Action taken to identify the disorder at its onset, so as to prevent extension.

3. *Tertiary prevention.* Action taken to limit disability arising from an established condition.

In this section, preventive activity will be suggested that arises from our knowledge of the influences on psychosocial disorders (D. R. Offord 1987). In a sense, all positive approaches to health, education, and welfare are potentially preventive in their effects. In this section there will be an emphasis on those measures directly related to psychosocial well-being, though clearly, preventive action taken, for example, to ensure that congenital dislocation of the hip or undescended testes are detected without delay after birth, will have indirect psychosocial benefits.

The onus for prevention in this field lies largely, though not entirely, with those working in primary health care, such as general practitioners, community health doctors, and health visitors, and those working in education, particularly teachers. Emphasis in this section will therefore be given to activities these professionals might undertake. Those involved in secondary care, such as paediatricians, psychologists, and psychiatrists, can, however, also undertake activities of major preventive significance, particularly if there is a focus on high-risk groups (Boyle and Offord 1990). Further, there are, in addition, certain more general, social, and other public policy measures of potential preventive significance, and these will be discussed first.

### Public policy aspects of prevention

1. *Alleviation of poverty.* The presence of mild mental retardation is strongly linked to material disadvantage, and some conduct disorders also occur more frequently in disadvantaged groups. Although poverty in itself is only weakly linked to the presence of conduct disorders, its presence predisposes to a variety of family difficulties, such as family breakdown,

that are much more strongly associated (Rutter and Madge 1976). Measures taken to reduce gross social inequality, and the sense of injustice that this evokes, may be expected to reduce rates of these problems.

2. *Neighbourhood cohesiveness.* The development of self-help groups, family centres, and other neighbourhood activities can provide helpful emotional support, especially for isolated mothers of young children (Haggerty 1980). The legal and social discouragement of racism can reduce tension in families belonging to ethnic minority groups. It should be remembered that many children in indigenous families also suffer psychologically as a result of racial tension.

3. *Housing.* Families with children need adequate space, both for everyday activities and for play. Avoidance of high-rise and other forms of cramped housing for families with young children could contribute to the prevention of behaviour disorders. Children living in homeless families' accommodation are at major risk for health, including behaviour problems (Bassuk and Rosenberg 1990). Adolescents also need a certain amount of personal space in their own homes.

4. *Employment.* Availability of adequate work opportunities for parents, and good prospects of employment for teenagers, would almost certainly reduce the rate of maternal depression and minor affective disorders in teenagers (Banks and Jackson 1982). It would reduce anxiety especially in teenagers with learning difficulties—a group that experiences most serious problems in obtaining employment.

5. *Child protection.* Legislation and social policy to prevent child abuse and neglect is described in Section 7.3.3. Policies that encourage early adoption or long-term fostering in situations where parents are clearly not going to be able to look after their biological children are likely to reduce rates of antisocial disorder and mild mental retardation.

Social policy directed towards protecting children from exposure to illicit drugs and advertisements for cigarette smoking would reduce the numbers of children and teenagers whose health is affected by these substances. Protection of children from exposure to violence, and more especially perverted forms of violence shown on television and in videos, can also be seen as socially desirable and, at least possibly preventive in its effects.

6. *Accident prevention.* Although mainly an issue concerning prevention of physical handicap, this is also relevant to psychosocial prevention, for disturbed, overactive children are most likely to be affected. Health education, the development of safety devices in the home, and childproof tablet containers are all useful measures (see Section 5.5).

7. *Education.* Preschool education plays a significant part in the prevention of mild mental retardation. Parent participation on the acquisition of reading (see Section 2.5.5) can prevent reading disability. Promoting an academic ethos in secondary schools might reduce rates of truancy, as well as other forms of behavioural deviance (see Section 7.4.1).

8. *Marital and family disharmony* are major influences in the development of behaviour and emotional disorders. The ready availability of marital counselling from professionals who are sensitive to the needs of children might reduce rates of disturbance in youngsters inevitably affected by the marital problems of their parents (Stolberg and Garrison 1985).

## Specific preventive measures by health and other professionals

Preventive activity by professionals can be classified according to the stage in the life cycle in the individual towards whom it is directed.

### Preparation for parenthood

This begins in the childhood of the parent-to-be. An emotionally satisfying childhood is in many ways the best preparation for later parenthood. More specifically, sympathetic instruction and the opportunity for discussion in adolescence about the physical and emotional aspects of pregnancy, childbirth, and child-rearing will provide a basis for later learning from experience. Health-professional input into secondary school curricula on these subjects can be invaluable. Ready availability of contraceptive advice to adolescents can prevent unwanted pregnancies with their distressing sequelae. There is no evidence that the availability of such advice increases promiscuity.

### Antenatal care

Pregnancy provides many opportunities for preventive activity (Newton 1988). Good antenatal care with regular check-ups reduces the rate of birth complications and subsequent brain damage and learning difficulties. Practical advice concerning avoidance of smoking and excessive alcohol intake should also produce a reduction in later learning difficulties in the offspring. Providing parents-to-be with the opportunity for discussion of fears and fantasies about, for example, possible birth difficulties and congenital abnormalities will result in a more relaxed pregnancy and better outcome. Such discussion is particularly necessary when the pregnant woman suffers complications requiring complex technical investigation and treatment.

### Birth and the neonatal period

Technically competent and sympathetic management with good communication between the mother and the birth attendant about the progress of the delivery will promote an initial positive attitude to the baby. The presence of other social supports during parturition has been shown to have a similarly positive effect (M. H. Klaus *et al.* 1986). Allowing the mother early contact with her child and the availability of rooming-in facilities provide a basis for later positive interaction. Early detection of maternal depression and abnormalities of mother–child interaction may require immediate intervention or follow-up that is more intensive than usual. Parents whose babies have to be admitted to special-care baby units need extra support and as much opportunity as possible to continue to be involved in the care of their babies. Early screening for phenylketonuria and hypothyroidism, with provision of appropriate treatment, reduces the rate of later mental retardation.

## Preschool period

There are frequent opportunities for preventive activities during this age period. A critical evaluation of surveillance procedures has suggested that regular examinations of young children should take place at 6 weeks, 8 months, 21 months, 39 months, and 5 years (Hall 1989), but that the assessments carried out should be highly selective. These examinations allow checking of height and weight, a physical examination, identification of defects of hearing or vision, and an assessment of the child's developmental status. Further discussion of hearing, vision, language, and motor impairment is discussed in other sections of this book. The frequent attendance of young children at primary care consultations for minor infections and other health problems allows further opportunities for developmental surveillance, and discussion concerning child-rearing. Screening examinations of young children in primary care and attendance for health problems in primary or secondary care settings also provide an opportunity for prevention and early identification of emotional and behaviour problems. There is evidence that regular discussion of a child-centred approach to child-rearing reduces rates of later behavioural problems and enuresis (Brazelton 1962; Cullen 1976; Gutelius *et al.* 1977). Such discussion can be provided by psychologists attached to health centres as well as by family practitioners and primary-care paediatricians themselves (Kannoy and Schroeder 1985).

Observation of the parent–child interaction at clinic attendances may reveal causes for concern in a variety of ways. A mother may be noted to be slow and lethargic or unduly irritable in the way she handles her young child. Such behaviour may be symptomatic of maternal depression, perhaps secondary to difficulties she is having with her child or related to other life problems. A baby or young child may be noted to relate poorly to his mother, to be unduly active, to be markedly attention-seeking, or to show unusual behaviour in other ways.

When noting this type of behaviour in parent or child, or in the interaction between them, it is important not to jump to unwarranted conclusions. The observed behaviour may be highly situation-specific, the mother being under particular stress because of the clinic attendance, or the child showing behaviour that is atypical. Nevertheless, it is also important not to dismiss such observations as irrelevant, for they may be the first indicators of serious problems perhaps amenable to psychosocial intervention. Screening examinations should therefore include systematic questioning about the child concerning possible behavioural difficulties (eating, sleeping, compliance), relationships with other children (if age appropriate), and the presence of undue fearfulness. It may be useful to request the mother to complete a behaviour check-list before she is seen, so that this can be used as a basis for discussion, and N. Richman (1977*b*) has provided such a check-list suitable for children aged 2½–4½ years. There should also be routine questioning about the mother's physical and mental health, and the parental relationship. Questions can be framed in an unthreatening manner in relation to the child's needs. 'How are you feeling in yourself? Do you feel you are able to cope with all the extra work X is giving you?' 'What about your husband/boyfriend? Do you find he is helping as much as he can?'

As with screening in other aspects of child care, identification of problems is pointless, and indeed potentially harmful, unless appropriate advice and intervention is available if problems are identified (Baird and Hall 1985). As indicated in other sections of this book, the limited but definite effectiveness, especially of some informal counselling procedures and behavioural intervention, does justify problem identification in a wide range of childhood behavioural and emotional disorders, provided that the practitioner can apply them himself or refer to an appropriate agency.

Preschool programmes for disadvantaged children with a focus or promotion of development can result in an improved mental health outcome. Adjustment at age 19 years was found to be better in those young adults who had the experience and educational programme at 3–4 years when compared with controls who had not had this experience (Berrueta-Clement *et al.* 1984).

The preschool period also provides the best opportunities for the prevention of distress and subsequent psychiatric disorder in children who are hospitalized, although older children may also be affected. Preventive activity therefore needs to be age-appropriate. This issue is discussed in more detail in Section 4.6.

*Middle childhood*

At this age, preventive approaches naturally focus on prevention of learning failure. Parental involvement in teaching at the initial stages of reading, early identification by screening procedures of children failing to learn to read so that they can be given extra help before they become discouraged, and good parent–teacher communication are all likely to be effective preventive procedures.

In a significant number of children with health problems, it is of importance that teachers are aware of associated difficulties that may arise in school, either directly as a result of the physical condition or as a side-effect of treatment. Good family doctor or hospital communication with the child's teachers, either directly or through a school doctor or nurse, can prevent misunderstanding and unnecessary neglect or ignorance of the child's disabilities.

Various behavioural methods have been used in preventive programmes directed towards children of this age. The aims of such programmes have been, for example, to train children not to respond to the approaches of strangers in public places, to resist peer pressure to smoke and drink alcohol, to observe home-safety rules, and to cope with stressful situations (Lee and Mash 1989). In addition, cognitive, problem-solving techniques have been used to help children relate to their peers and to adults in more satisfying ways (Spivack *et al.* 1976), but the benefit to emotional adjustment is uncertain.

Finally in this age group, school-based, multi-level preventive programmes directed towards parents, teachers, and children themselves have been shown to enhance academic achievement in disadvantaged children (Comer 1985) and to reduce rates of bullying in schools (Olweus 1989).

## Adolescence

The life cycle of preventive activity is complete at this age, with the evident need for preparation for parenthood. Teenagers can benefit from the opportunity for group discussions of social (including sexual) relationships. Often such discussions appear to be most useful when they arise in relation to other subjects being studied, such as literature or human biology. The degree to which younger teenagers should be provided with information about sex and contraception is a controversial subject. Many health professionals take the view that the provision of such information generally leads to a more, rather than a less, responsible attitude to sexual activity.

Counselling in schools or neighbourhood centres for teenagers, so that they can discuss life problems before these lead to adverse psychiatric reactions can also be seen to have an important preventive role, though such facilities are little available in the UK in the early 1990s. Counsellors can also be helpful in advising youngsters at risk of becoming involved in drug-taking of techniques they can employ to resist encouragement from drug pushers.

In mid- and late adolescence the prevalence of suicidal activity has simulated interest in programmes directed towards reducing rates of self-destructive behaviour. A large number of school-based programmes exist in the USA, but their effectiveness is uncertain, and their exclusive reliance on a stress-related explanation of suicidal activity may actually be counter-productive (Garland *et al.* 1989).

Finally in this age group, programmes directed towards the promotion of non-academic skills such as sports and manual crafts in less academically able youngsters have been shown to have modest but promising results in the reduction of antisocial behaviour (D. R. Offord 1987).

**Organization of child health services: preventive aspects**

In all developed countries, a complex system exists to meet the educational, health, and social needs of children. This system is readily understood by parents whose children have only minor problems. If their children have a minor illness or injury, they go to the family doctor or accident and emergency department. However, as soon as children have special needs, parents enter into less familiar and often confusing territory. Confusion is not limited to parents. There are many teachers and doctors who are uncertain of the difference, for example, between a psychologist and psychiatrist, and many mental health professionals whose understanding of the education and social welfare system is extremely sketchy. In the organization and delivery of services for the child with a psychosocial problem, in order to reduce such confusion, certain principles can be borne in mind.

*Concern for the whole child and family*
Especially in young children, any attempt to separate physical and mental health needs of children is doomed to failure. As stated in the previous section, about a quarter of child consultations in general practice have a significant psychological component. Paediatricians estimate that about a third of their referrals are for 'psychosomatic' problems. Psychologists and psychiatrists need to be aware of physical conditions that may underlie the mental health problems with which they are dealing.

*Knowledge about and respect for associated services*
Professionals, however broad and comprehensive their own approach, need to be aware of what other groups of professionals have to offer, and how such help can be obtained. Thus general practitioners and clinical medical officers as well as psychiatrists need to have full awareness of the availability of special educational and social services in their locality.

*Communication between professionals*
Parents are sometimes confused by conflicting advice given to them by professionals who are unaware of previous opinions offered. The service provided, especially by the primary health care service (family doctors, health visitors, school nurses, etc.) can do much to prevent such confusion, but only if others who are in contact with families ensure that their participation is known to the primary health care-givers.

*Availability of mental health consultation*
Results of community surveys of disturbed children in the population make it clear that there will never be enough specialized psychologists and psychiatrists to deal with them. Even if there were, it is unlikely that it would be cost-effective to separate delivery of mental health services from

other health services at primary care level. Further, many parents do not wish their children to be referred to psychiatrists and, although prejudice against mental health professionals is doubtless less prevalent than it used to be, it is improbable that community attitudes will ever change to a degree that makes referral universally acceptable. For all these reasons, non-mental health professionals will, for the foreseeable future, be involved in assessing and treating disturbed children and their families. The availability of advice from more highly trained professionals in this field would be one way of ensuring that problem children can be identified and treated early on. Many staff working in child and family psychiatric clinics now spend as much as a third of their time in such consultation services.

Consultation most appropriately takes the form of discussion of policy issues and of individual children who are causing concern. A comprehensive child mental health consultation service in a health district might involve:

(1) psychosocial meetings on the paediatric ward in the district general hospital;
(2) consultation with social workers on difficult child care cases;
(3) attendance at meeting of teachers in local special and ordinary schools;
(4) discussion with residential child care workers concerning disturbed children in residential care;
(5) ready availability for consultation to primary health care workers such as health visitors and general practitioners.

Only a well-staffed child mental health service is likely to be able to meet all the demands for consultation made upon it, but staff working in such services will themselves benefit from greater awareness of community needs and resources if they devote a significant amount of time to such consultation work. There is evidence (Mannino and Shore 1975) that those in receipt of consultation services do derive benefit from such activity.

*Emphasis on parent involvement in decision-making*
Although, when highly complex technical decisions need to be taken, parents will appropriately wish professionals to take full responsibility, in many other cases (the need for a special psychological assessment, treatment of a sleep disorder, etc.) greater success is likely to be achieved if decisions are taken *with* parents rather than for them. In particular, referrals for psychiatric consultation are only likely to be useful if parents are motivated. If they are not, and a serious problem exists, then it is those who still have to care for the child who will be in need of advice and counselling, and this too should be readily available.

*Encouragement of voluntary activity*

Professionals are sometimes wary of voluntary organizations and suspicious of the motives of those involved in their activities. Nevertheless, many parents of children with chronic physical or mental disorders have found great comfort and support in sharing their problems with others who have children with similar problems. The voluntary organizations dedicated to serve the needs of families where a child has a specific type of disability are often particularly helpful, both in providing information and in putting parents in touch with others in a similar predicament. In the UK, the National Council for Handicapped Children has produced a useful booklet containing the names and addresses of relevant organizations (National Children's Bureau 1985).

## Conclusions

If the mental health needs of children are met, there is at least a reasonable hope that the adults into which children grow will themselves be more caring and more sensitive to the needs of their own children. But the justification for a positive mental health approach to children does not just lie in making better adults. As Jack Tizard once wrote, 'Childhood is not just a preparation for life, it is a part of life'. Respect and concern for children who, because of immaturity, need special care and protection, is one of the criteria by which a society can judge its own worth.

# 6.2 Treatment

## 6.2.1 INTRODUCTION

Treatment methods in child and family psychiatry are less well scientifically established than in most other branches of medicine. In particular, knowledge concerning the effectiveness of different types of treatment for particular disorders is often uncertain. Consequently, it is not surprising that treatment practice varies quite widely from centre to centre. Nevertheless, much useful knowledge is available (Kolvin *et al.* 1981; Rutter 1982*a*).

In the following section a number of methods of treatment are described, though it must be appreciated that many practitioners often use different forms of therapy in combination. In each case an attempt has been made to describe the form the treatment takes, the principles underlying it, indications and contraindications for its use, and evidence for its effectiveness. Some examples are given of its application in different types of facility—primary care, paediatric, and psychiatric. For the sake of brevity, the case examples do not contain all relevant information. Names of children described, and some identifying data, have been altered to preserve anonymity.

Clearly, clinicians will vary greatly in their skills and opportunities to carry out these forms of treatment. The more highly specialized the clinician dealing with child psychiatric disorders, the more skilled he or she is likely to be in therapy, but there is no reason why all professionals concerned with children should not be capable of applying some of the principles described here. As with other forms of treatment, initial experience is likely to be more useful if it is supervised, and supervision by direct observation with the use of video and one-way screens, if such equipment is available, has many advantages.

One notable form of variation in treatment in specialist centres concerns the degree of co-therapy or joint therapy that is thought desirable. In some centres, virtually all treatment is carried out conjointly, whereas in others this way of working is unusual. One agreed principle of therapy in this field is that it is desirable for all doctors working with disturbed children to have ready access to colleagues with skills in psychology and social work as well as to have good contact with teachers when dealing with children of school age.

### 6.2.2 BEHAVIOUR THERAPY

**Definition**

Behavioural methods of treatment are those that are directed towards the removal of symptoms or problems by methods based on learning theory. The focus is on observable behaviour rather than underlying thoughts or feelings, though later developments of behaviour modification such as cognitive therapy do deal with thought processes.

An essential and central feature of these methods, however, is that they involve systematic analysis of the symptomatic behaviour as part of *assessment*, and that the *treatment* is accompanied by equally systematic monitoring of its effects, with modification if the treatment is not producing the desired result.

The features of behaviour modification in children are similar to those in adults, except that with children, parents and teachers are frequently used as agents of behavioural change.

The following account of treatment is necessarily limited in scope. A more detailed review of behavioural approaches is provided by Werry and Wollersheim (1989), and a useful, highly practical account of the application of behavioural methods by Herbert (1981).

**General principles of assessment for behaviour therapy**

The form of assessment will depend on the time available, the nature and severity of the problem, and the skills and experience of the person making the assessment. Clearly a health visitor or family doctor using behaviour modification principles to deal with a sleep problem in a 2-year-old child will not carry out as intensive an assessment as a psychologist referred a child with a severe and persistent obsessional disorder. Nevertheless, the principles of assessment are likely to be similar.

### Defining a problem to be treated

Obtaining a description of the actual form of behaviour complained of by parents is usually an essential first step. Where, as is often the case, the child is showing more than one form of problem behaviour or deficit, it is necessary to establish which is to be the first target for change. Parents may have different views from professionals on this matter, and it is important to respect parents' views, as high parental motivation is often required to achieve success.

### Defining the circumstances of the problem behaviour

Before treatment begins, it is necessary to obtain a reasonably clear idea, not only of the frequency and severity of the problem, but also of the circumstances that precede and follow it when it occurs. A basic assumption is that whether a form of behaviour changes will depend on whether there is a change in what happens before the behaviour or after it.

$$\text{Antecedents} \rightarrow \text{Behaviour} \rightarrow \text{Consequences}$$

A 'functional analysis' is therefore necessary so that the clinician will understand what precipitates and maintains the problem behaviour—a necessary precursor to treatment. So, for example, if temper tantrums are the target behaviour, one would need to know how often they occur, who is present when they occur, in what setting they occur, what starts them off, how long they last, and what those present do when they occur (do nothing, give the child what he wants, smack the child, etc.).

### Baseline recording

Once it has been established which problem or problems are to be treated, it will be necessary to obtain a baseline recording. This is usually achieved by the child, parent, or teacher keeping a chart of the frequency with which the problem occurs. If the clinician is especially interested in treating a particular condition such as bedwetting, then he or she will probably develop a special chart for this purpose. It is useful, however, to have an all-purpose chart that can be used for a whole variety of problems. Figure 4 shows a chart that could be used to record a range of difficulties. Periods of time represented by each rectangle could involve a whole day (in which case the chart could be used for five weeks) or for a segment of the day.

What needs to be recorded on the chart will depend on the nature of the problem and the frequency of its occurrence. For example, if soiling occurring about three times a week is being treated, it will be necessary to chart all episodes that occur, or all periods during which the child is clean. If aggressive outbursts occurring several times a day are being treated, it may

| Time | Nature of problem | | | | | | |
|------|--------|---------|-----------|----------|--------|----------|--------|
|      | Monday | Tuesday | Wednesday | Thursday | Friday | Saturday | Sunday |
|      |        |         |           |          |        |          |        |
|      |        |         |           |          |        |          |        |
|      |        |         |           |          |        |          |        |
|      |        |         |           |          |        |          |        |
|      |        |         |           |          |        |          |        |

**Fig. 4.**

be more appropriate only to record what happens in one hour between 5 and 6 p.m.

The process of keeping a baseline recording is sometimes, though not often, sufficient to achieve significant change in parental attitudes or child behaviour. Once parents see a graphic and precise illustration of the frequency and severity of the problem for the first time, they may decide that the problem is more trivial than they had thought and not worth troubling about. Alternatively they may, once they see the severity of the problem, alter their own behaviour consequent on the child's behaviour, with the result that the child improves even before the clinician has made any active intervention. Finally, problems parents have in keeping the chart may be useful diagnostically. Low motivation for change, sharp differences between the parents over what constitutes a problem, and poor capacity for change may come to light merely as a result of keeping a chart.

## Operant methods

### Principles

The experimental work underlying this technique, pioneered especially by Skinner, indicated that the likelihood of a behaviour recurring depends on the significance of the event immediately following it to the person showing the behaviour. If the event following the behaviour is positively reinforcing or rewarding, then it will recur. If it is not reinforced or is punished, then it is less likely to recur and eventually will stop completely—a process known as 'extinction'. An alternative, but related approach is 'stimulus control'— changing the events preceding the behaviour, thus altering the likelihood that the same behaviour will recur. Figure 5 summarizes all this.

In childhood the most common form of *positive reinforcement* is social.

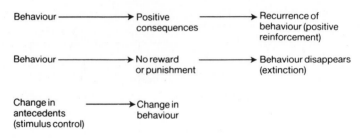

Fig. 5.

Children are likely to repeat behaviour which gives pleasure to those of whom they are fond. Usually, but not necessarily, their parents and teachers are the most important positively reinforcing figures, but, as they get older, other children increasingly take on this role. If a teacher pays gratifying attention to bad behaviour (even if the attention takes the form of shouting at the child), then bad behaviour will recur, especially if attention is something of which the child is generally deprived. Material rewards, such as money, sweets, chocolates, other favourite foods, and watching television are usually positive material reinforcers. Further reward comes from the sense of satisfaction a child achieves at having succeeded in a task.

**Extinction** generally occurs most rapidly following withdrawal of those things that are positive reinforcers. Thus, the withdrawal of love or attention from people of whom the child is fond is often the most powerful way of achieving extinction of the undesirable behaviour that children show. In other children, the withdrawal of material goods, such as pocket money, special food or drink, and opportunity to watch television is more important.

**Punishment** is a further technique used to remove undesirable behaviour. Any unpleasant consequence of behaviour which makes that behaviour less likely to occur can be seen as punishing. Physical punishment by parents is the most frequently used, but many children do not respond to it by a reduction in their undesirable behaviour. Probably the extra attention they get when they are punished has a positive reinforcing or rewarding effect, and this result overrides negative experiences of physical pain. The experience of negative emotional states—anxiety, depression, and a sense of failure—is, by contrast, strongly punishing to many children. If, as a result of an unpleasant experience (for example being bullied by a group of boys on a particular corner of the street), a child experiences severe anxiety when he returns to that particular location, then he is less likely to return on a subsequent occasion. Of course, if he does return and the unpleasant experience is not repeated, then he will have less anxiety about making subsequent visits.

**Nature of reinforcers.** Whether something is positively reinforcing or punishing depends on the effect it has on behaviour. Virtually nothing is intrinsically reinforcing. What may be positively reinforcing to one child will not be so for another. For example, usually food will be positively reinforcing, but to an anorexic girl who hates the sight of food it may be punishing. Pain is usually punishing, but to a child preoccupied with guilt or with masochistic tendencies, it will be positively reinforcing or rewarding. Further, the strength and direction of reinforcement will depend to some degree on the child's relationship with the person administering or involved in it. A game of football is likely to be more positively reinforcing for a boy if it involves his father than his mother. A star chart for bedwetting worked out in cooperation with a mother with whom a 6-year-old has a good relationship is likely to be more effective than if the mother and child are in serious conflict.

*Indications*
Operant methods may be indicated with:

1. Conduct problems such as aggressive behaviour, bullying, and stealing. Such behaviour may be reduced by stimulus control (e.g. by avoiding situations that produce conflict) or by systematic positive reinforcement (reward) for the non-occurrence of the antisocial behaviour. Introducing rewards for alternative, non-antisocial forms of behaviour is also a promising approach.

2. Problems related to anxiety such as social withdrawal and difficulties in making friends. A combination of social skills training (see below) making it easier for the child to meet potential friends in a safe situation at home (stimulus control) and reward for even small steps in increasing social contacts may be effective.

3. Deficits of behaviour such as mutism, aphasia, and language delay occurring in both otherwise normal and autistic children. Good results have been obtained using positive reinforcement methods, rewarding the child for utterances just a little in advance of those he is currently making.

4. Feeding difficulties both in infants and in adolescents. In early feeding problems and failure to thrive (see Section 2.2.3), removing the tension which may have built up around the feeding situation from the mother–child relationship by desensitization methods (see below) can be combined with social rewards, for example an increase in hugging and kissing if a child enjoys this, for more appropriate feeding behaviour. In the early treatment of anorexia nervosa (see Section 2.2.6), enforced bedrest with the reward of increased physical activity for weight gain can be used as a form of positive reinforcement.

5. Hyperactivity where it is hoped to build up the child's ability to sit still and concentrate. Avoiding over-stimulating situations (e.g. by

ensuring that toys and games are produced one by one rather than all at once) is a form of stimulus control. Rewards can be instituted for longer and longer periods of concentration. Tasks requiring concentration can be broken down into separate steps, and the child rewarded for progress in each of them. Thus, when faced with a task, the child may be rewarded for learning to stop and consider what action he is going to take, but not for an immediate, thoughtless reaction.

6. Habit disorders such as enuresis and encopresis: providing an immediate reward on the morning after a dry bed (see Section 2.12.5 for this and other behavioural approaches to enuresis) is a form of positive reinforcement. The 'shaping' of appropriate behaviour involving the use of the toilet by positive reinforcement methods (see below) is often a useful adjunct in the treatment of encopresis. Habits such as thumb-sucking may also be tackled in this way (A. P. Christensen and Sanders 1987). Techniques such as habit reversal may be used. The child is encouraged to identify the circumstances in which the habit occurs and then develop a competing response, for example enclosing the thumb inside the fingers in a closed fist for 20 seconds or so when these circumstances arise. Alternatively, such habits can be tackled by 'differential reinforcement'—rewarding with a token the absence of the behaviour for each 3 minute period over half an hour.

The indications, therefore, cover a very wide range of problems. Operant methods are particularly likely to be helpful where the problem is well circumscribed, and the family members (including the child) are motivated to achieve change. They are less likely to be effective when the problems are widespread, seem to emerge from deep-seated personality difficulties, or where motivation is low.

### Practical applications

The mode of application will depend very much on the setting where the problem behaviour occurs, or where the deficit is most in evidence. Operant methods can be used effectively in the home (when applied by parents), at school, or in an in-patient setting. Probably its widest application for the health professional consists in its out-patient use with parents acting as 'co-therapists' with the clinician in attempting to alter or add to the child's behaviour, and it is this approach which will be described here.

Following assessment involving a functional analysis and baseline charting (see above), the parents and child then identify potential reinforcers or rewards. The child is then told that appropriate behaviour, i.e. not carrying out some undesirable activity such as a tantrum or carrying out a desirable activity such as helping with the dishes, will be followed by reward.

When choosing a behavioural target, change must be clearly within the child's capacity. Often it is necessary to 'shape' behaviour in the desired direction, and not attempt to produce a radical change very rapidly. Thus, for example, a child with encopresis might be reinforced first for sitting on the lavatory, then for attempting to pass a motion, and only finally for passing a motion into the lavatory pan. A child with a speech delay might be rewarded for making more and more appropriate sounds before he is rewarded for articulating words.

Reinforcers or rewards should, as far as possible, be social rather than material. A hug, kiss, or word of praise are more likely to produce lasting behavioural change than sweets or small-value coins. Nevertheless, for some children, material rewards are necessary, especially in the first stages of a behavioural programme. Whatever they may be, rewards should be given as soon as possible after the desired behaviour has occurred. Appropriate timing is essential. Promising a bicycle at Christmas if an encopretic boy is clean in September will probably have less effect than an immediate word of praise.

As well as positively reinforcing desirable behaviour, it is often useful to concentrate on removing rewards attached to undesirable activities ('extinction'). Parental attention to bad behaviour may be rewarding, and parents should be encouraged to ignore, rather than take notice of, undesirable behaviour. It should be recognized that this is often easier said than done. It is often very difficult to help parents to work out ways of ignoring difficult behaviour, such as a child wrecking furniture or breaking crockery. Clearly in some circumstances, children do have to be given attention and stopped from doing things. Parents of children with such extreme behaviour are often best helped by the devising of techniques to help them avoid the development of conflicts such as this by the earlier application of positive reinforcement. An additional technique involves the use of the 'time-out' procedure, the parent removing the child from the place where the undesirable behaviour is gaining attention, to a separate room or corner of the room where such behaviour can be quietly and calmly ignored until it stops.

Assuming behavioural change is achieved in the desired direction, parents should be encouraged to 'fade' the rewards, so that the child is achieving well without them. This may need to be done slowly if the improvement is to be maintained.

### Desensitization

#### Principles
The development of normal and abnormal fear and anxiety reactions is described in Section 2.9. Thinking of fear as an emotion that can be both learned and unlearned has proved a useful concept in treatment. Normal coping mechanisms for dealing with a feared situation include a cautious but sustained approach towards the source of fear until the situation has been mastered and the fear conquered. At each stage the person involved must feel confident and assured before proceeding further. 'Desensitization'

involves a systematic application of this principle of a graded approach to overcoming anxiety.

## Indications

Phobias, especially those confined to one or two feared objects or situations, are the main indications for this form of treatment. In particular, fear of the dark, of dogs and other animals, and of separation from familiar figures may be treated in this way, but any reasonably well-defined phobia is likely to respond.

## Practical application

**Establishing a hierarchy of feared situations.** The first step in treatment is to develop a 'fear hierarchy' of situations, linked to the phobia. In a dog phobia, the least-feared situation might involve seeing a small dog on television while in the company of the father, and the most-feared might be meeting a large dog coming towards the child on the same side of the road while the child was unaccompanied. In severe separation anxiety, the least-feared situation might involve mother going into another room while the child is in the company of the father, and the most-feared might involve both mother and father going away for the weekend, leaving the child in the company of a relative. In each case, there might well be as many as a dozen intermediate feared situations.

**Graded exposure.** Parents should be encouraged to expose the child to the least-feared situation until this has been mastered before proceeding to the next stage. At each stage anxiety can be reduced by reassuring and praising the child, thus achieving greater relaxation. The procedure is usually best carried out in the home by parents, but occasionally the nature of the fear may mean that treatment needs to be carried out in school or a clinic. In children, exposure to the feared situation in reality is usually necessary, but in adolescents and adults, phobias may be treated in this way by getting the individual to imagine feared situations while gradually achieving greater relaxation and freedom from anxiety.

**Relaxation.** In adolescents and sometimes in younger children with phobias and anxiety states, teaching *relaxation* may also be of value. This involves a conscious attempt to achieve reduction of muscular tension by concentrating on one muscle group after another. The subject may first be encouraged to concentrate on ensuring his respiratory movements are easy and relaxed, then his stomach muscles, neck muscles, limbs, etc., until he has the impression that his entire body is in a pleasant relaxed state. The child or adolescent may be encouraged to practise relaxation on his own two or three times a day. Relaxation may be used on its own, or in association with desensitization techniques described above.

Parents of young children may be able to grasp the principles and apply desensitization methods so sensitively that very little help beyond an initial explanatory session is required. On the other hand, in the unusual case where the phobia is severe or parents have difficulties in applying the method, a great deal of regular supervision may be necessary.

## Other behavioural techniques

A number of other behavioural techniques have been described as useful in children, but appear to have more limited application.

### Classical conditioning

This method derives originally from the model of classical conditioning developed by Pavlov. Working with dogs, he paired an unconditioned stimulus (food) with another stimulus (a bell) until the latter itself became effective and 'conditioned' to stimulate salivary secretion. Classical conditioning methods are little used in child psychiatry today, although the bell and pad method of treatment of enuresis (Section 2.12.5) can be seen as one example of its use. One explanation of the effectiveness of the bell and pad lies in the view that the bell becomes a conditioned stimulus, paired with the unconditioned stimulus (bladder distension) and produces sphincter contraction. However, the treatment can also be considered as a form of operant conditioning, or as a combination of operant and classical conditioning.

### 'Flooding'

This method has been used in phobias, and involves sudden massive exposure to the object of which the child is afraid. It is not more effective than desensitization, but might have a place where desensitization has failed. Parents and child must agree that it is worth undergoing an extremely unpleasant experience so that the phobia can be relieved. One naturally occurring example of 'flooding' occurs when a child with no previous experience of hospital and with a fear of hospitals and doctors has to be admitted as an emergency. Such an experience may produce a sharp increase in anxiety, from which it may take the child months to recover. However, the 'flooding' of the feared situation can also result in the child losing his original fear. Rapid return to school in a child with school phobia is another example of 'flooding'.

### 'Aversion therapy'

This involves the use of the same principle as the operant methods described above. Although, of course, many parents do use physical punishment to reduce the likelihood of repetition of undesirable behaviour as a routine, the technique has very little application for professional use. It has, however, been employed on an in-patient basis in mentally retarded children who show very severe self-destructive behaviour. Mild electric shocks have been used as the aversive stimulus. Even in these cases, very careful consideration to ethical issues must be given.

## 'Cognitive methods'

Cognitive psychology, a field that expanded enormously in the 1980s, involves the study of the way information is processed. This work has considerable potential for therapy with children (Meyers and Craighead 1984). For example, the self-denigrating thoughts of a depressed adolescent can be seen as resulting from distortions of the incoming information from other people. The impulsivity of children with attention-deficit disorder can be seen as resulting from over-rapid processing of information. Pathological shyness in a child can be seen to arise from lack of ability to deal appropriately with information in social settings. Cognitive behaviour therapy involves attempting to alter the child's or adolescent's behaviour by changing the way he thinks about himself or approaches a task. Thus, a depressed adolescent may be encouraged to express his thoughts about himself as an unworthy person, less deserving than others of his age. He is then asked to examine the basis of this belief by considering his achievements and observing how other people in fact treat him. As he realizes that his depressive beliefs about himself are not confirmed by these observations, the depression may remit. Cognitive methods have also been used in inattentive children who cannot focus on a task. They may be trained to approach tasks in a more reflective manner by reinforcement of appropriate thinking processes—first stopping to examine the task, then proceeding to carry it out carefully step by step, and finally monitoring the result.

## 'Response prevention'

This is a technique used in children and, more commonly, adolescents who are compelled to carry out rituals as part of obsessional disorders. The conduct of the rituals relieves anxiety, and the cycle of behaviour is therefore self-reinforcing. By encouraging the patient to desist from the rituals for longer and longer periods, the cycle can be broken. This treatment may need to be carried out on an in-patient basis with the assistance of nursing staff (Bolton *et al.* 1983), if the symptom is resistant to out-patient treatment.

## 'Social skills training'

This technique, either in group or in individual sessions, may be used in children who have difficulty in their relationships with adults or others of their own age (Spence 1983). There may be difficulties in making or keeping friends, or the child may be a constant target of bullying. Such difficulties may have arisen because of constitutional factors, because children have been exposed to inappropriate models, or because the child has, for other reasons, not acquired the skills needed to develop relationships successfully. In any event, a series of social skills training techniques, especially those involving role-playing, have been devised to counter such problems. Parents may, for example, be encouraged to practise with their children the problematic situation that they have in relating to peers, so as to improve their skills in this direction. Alternatively the professional involved can undertake such role play with the child himself. Such techniques can be used to help children with obesity or physical deformities to practise how they are going to behave when they are teased or asked embarrassing questions by other children.

Social skills training may also incorporate cognitive methods described above in a form of treatment known as interpersonal cognitive problem-solving (Pellegrini and Urbain 1985). Children may, for example, be asked to reflect systematically on conflicts in which they have been involved, and then work out alternative strategies they could adopt in the future when similar situations arise, in order to avoid such conflicts.

*Massed practice*
This technique derives from experimental work showing that frequent repetition of a behaviour results in the inhibition of its recurrence. The method is used mainly with ticquers, who are encouraged to reproduce their tics consciously as frequently as possible for short periods each day. Long-lasting benefit has not been recorded.

## Evidence for effectiveness

There is now a large literature involving single case studies and studies of small groups of patients demonstrating that behavioural methods can be effective in a wide range of childhood disorders (Yule 1985). However, there is a lack of information on the degree to which these methods are generally applicable, on long-term outcome, and on the degree of expertise required to apply them. With few exceptions, beliefs that removing one symptom would merely result in the development of others (symptom substitution) have not been confirmed.

Finally, it should be emphasized that, as with other forms of psychological treatment, the principles of behavioural therapies can be applied by relatively untrained professionals, and in mild problems one would often expect a successful outcome. Resistant and severe problems will need referral to more highly skilled professionals, who will often apply rather similar methods, but more systematically, and with more attention to detail.

## Case examples

### Informal behaviour modification—general practitioner
Tracy was a two-year-old girl brought to her general practitioner by her mother because of her sleep problem. Her mother requested sleep medicine for Tracy, or for herself. Tracy would not settle in her bed. She kept coming downstairs crying if her mother left her in her bedroom. Eventually she would settle by about 11.00 p.m. but then she would wake around 2.00–3.00 a.m., and come into her parents' bedroom. They would take her back

to her room, give her sweet drinks, and play with her a little to try to settle her, but she would continue to cry and scream. Eventually after much coming and going they would take her into their bed. This was happening virtually every night. Tracy's young parents, who did not have a particularly good relationship between themselves, argued about how to overcome the problem, father thinking Tracy should be left at night to 'cry it out', while mother took a softer line. Mother was mildly depressed, tearful, and irritable. She said that she was at the end of her tether.

The general practitioner suggested that mother kept a chart to record what actually happened each night for a fortnight—what time Tracy went to bed, what she did, the times she woke in the night, and what the parents did when she woke. He arranged to see Tracy again in ten days and said that it was important that father also came on that occasion. He said that if mother felt she might lose control of her temper with Tracy, she was to contact him immediately beforehand.

When the family came, it became clear from the chart mother had kept that Tracy was not waking each night but only about three times a week. However, on one to two nights a week she seemed to be up a good deal of the night. The general practitioner asked the parents which behaviour they would most like to change and the parents chose her night waking as the most stressful. The general practitioner discussed with the parents how they would feel about just letting Tracy 'cry it out' when she woke in the night, but mother was not happy about this, and even father admitted the neighbours would probably complain.

The parents agreed a programme with the general practitioner whereby when Tracy woke and came into their bedroom in the night, father would take her back into her own room. Tracy preferred her mother to look after her in the night, and it would be less rewarding for her if father did it. She would not be given a drink and on no account would her father play with her. Father was asked if this would not be too exhausting for him, but he said he was having a quiet time at his work and could cope. Mother was asked to continue to keep a chart. Tracy was extremely difficult for the first few nights and the parents, especially father, were on the verge of giving in to her and taking her into their bed. However, on the fifth night she appeared to accept the inevitable—woke up only once at about 3.00 a.m., whimpered but did not call out and went back to sleep. Subsequently her bad nights decreased to about once a fortnight. The parents were still having a problem getting her to stay in her bedroom when she was put to bed in the evening, but decided they did not want any more help because they could cope with this problem themselves.

*Behaviour modification—paediatrician (with advice from psychologist)*
Jason was a seven-year-old boy who presented in the paediatric out-patient department with encopresis. He had been clean and dry by 3 years, but had

started soiling when he began school just before 5 years. He had found his reception class teacher unsympathetic and had been frightened of her. The soiling occurred every day—weekends and holidays as well as schooldays. He said he did not know when he was passing a motion. The motions were said to be normal in colour and consistency. There were no other significant behaviour problems and, although his parents seemed rather irritable with each other, there were no obvious family or other stresses. Physical examination was negative, and it was decided no physical investigators were indicated.

The paediatrician decided to institute a behaviour modification programme along the lines he had tried previously in consultation with a psychologist. He explained the nature of the problem to the parents and to Jason in terms of a habit that had become established. He asked the mother to chart the times each day when the soiling occurred, together with what happened before and after this event.

The results of this behavioural analysis made it clear that the soiling did indeed occur at any time of the day, and that after the soiling mother would become irritable with Jason and behave in an inconsistent way. Sometimes she would make him wash his pants, sometimes not. Sometimes she would try to make him go to the toilet after he had passed a motion, though she was not clear why she did this.

After discussion concerning Jason's particular likes and dislikes, a behavioural programme was instituted in which Jason was given a book in which to stick stars—one star for sitting on the toilet and two if he tried to pass a motion. At this stage no reward was given for passing a motion into the toilet, but only for trying. He was collecting football stickers, and six stars earned him one of these. His mother was encouraged not to comment on his soiling, but just to change his pants when she noticed he had messed himself. It was suggested that Jason, with his mother's help, rang the paediatrician's secretary (to whom he was introduced) to report progress each week, and a further appointment was given in three weeks' time. At this appointment mother reported that he had been to the toilet regularly for a few days, but then lost interest in the stickers and had started to refuse to go when she reminded him. A further discussion revealed that Jason would very much like to go to a football match with his father, and it was agreed that 25 stars, earned as previously, would win him this outing. He became dramatically cooperative and the paediatrician wondered whether his original formulation was accurate, or whether the boy was low in self-esteem partly because he had previously been rather ignored by his father.

Jason was in fact rather scared by the crowd at the football match he went to, so rewards were changed to small sums of money with which he bought sweets or football stickers. His father was encouraged to take him out more regularly. He began to pass motions into the toilet and his soiling improved, though he continued to mess himself about once a week, mainly at home.

*Behaviour modification and medication—psychiatrist*

Victor was a 12-year-old boy with a history of multiple tics going back four years. The tics were unusually severe and involved his head, neck, trunk, and arms. Over the previous nine months he had started to make involuntary screeching noises. He had never shown any other serious behaviour or learning problems, though he was occasionally in trouble at school for disruptive activities and was somewhat below average in his educational attainments. He was the youngest of three boys. His oldest brother had been in quite serious trouble for various offences, but appeared now, at 20 years, to have settled down to a regular job. The parents were warm and caring, but they had great difficulty in accepting that the movements were beyond Victor's control.

Victor's Tourette syndrome (Section 2.4.4) was initially treated with a combination of counselling and medication. He was put on Haloperidol in doses up to 0.5 mg three times a day. On this dose he had a severe dystonic reaction. The drugs were stopped for three days, and it was found that he could tolerate 0.5 mg b.d. On this dose his tics were definitely reduced, but still present.

It was decided to embark on a behaviour modification programme. A behavioural analysis, carried out by getting the parents to chart the tics (with Victor's knowledge and cooperation) showed that the tics were worst in the evenings when both parents were present. They were especially marked when his parents told him to stop the movements and also appeared worse on evenings when he had PE the following day (because of his movements he found PE very embarrassing).

With a knowledge of the current frequency of tics, Victor was encouraged to try the effect of massed practice (see above). He practised making his noise as often and as loudly as he could for ten minutes each day, morning and evening. This programme had no effect on the frequency of tics at other times, and Victor complained it gave him a sore throat. A further contact was made with the school, which agreed he could be excused PE.

His parents then agreed that Victor should chart *their* behaviour in telling him to stop making his noise. The mere fact that Victor was charting their behaviour produced a dramatic effect, and they almost completely stopped telling him off overnight. This also produced a further slight improvement in tic frequency.

Over the next three years, Victor was maintained on Haloperidol. Every three or four months an attempt was made to stop the medication, but it was clear each time that he was benefiting from it. At about the time he left school, at the age of 16 years, he was able to stop the drug and his tics gradually improved, though they did not cease completely.

## 6.2.3 INDIVIDUAL PSYCHOTHERAPY

### Definition

Psychotherapeutic techniques involve the use of understanding to help a child or adolescent become aware of the meaning of symptoms to himself and to other people so that he can, if he wishes, change his behaviour.

Psychotherapy can be **formal** or **informal**. If formal, it involves the practitioner setting aside a particular time (usually for 50 minutes once or more times a week) during which no interruptions are permitted. Formal psychotherapy is usually carried out by practitioners who have had training in a particular school of psychotherapy, and who would usually expect to see their patients or 'clients' at a regular time at least over several weeks, and sometimes over several months.

Informal psychotherapy is, as its name implies, less highly structured. Appointments may be made on a more *ad hoc* basis, though an agreement for a determined number of sessions might be made at the outset. The child might be seen for varying periods of time. Interruptions, though unwelcome, might be tolerated, if urgent. The practitioner is likely to be less highly trained in psychotherapy, but will nevertheless apply many of the same principles. Indeed, some highly trained psychotherapists practise informal as well as formal psychotherapy. Informal psychotherapy is practised by many general practitioners, social workers, psychologists, psychiatrists, and paediatricians. Elements of psychotherapy enter into many encounters between children, parents, and professionals. For example, a doctor working out with a teenager why he keeps on forgetting to take his anticonvulsant medication might well use his knowledge of the unconscious mechanism of denial during a medical consultation. Further, elements of psychotherapy are often involved in the application of other methods of treatment, both psychological (such as behaviour modification) and non-psychological (e.g. medication). But psychotherapy, formal or informal, can really only be said to take place when time is set aside specifically for the purpose of achieving greater understanding of feelings and behaviour.

### Principles of individual psychotherapy

Although there are many schools of psychotherapy, those who practise it share a number of principles (Reisman 1973) that can be summarized as follows:

1. The child should be allowed the time he needs to express his feelings and thoughts.

2. Practitioner and child should, early on in their meetings, try and focus on some particular goals. These goals might involve the eradication of particular symptoms, but they are more likely to involve the achievement

of understanding in some area of the patient's life, for example why he feels so angry with his parents, or why he has so much difficulty making friends. Sometimes, before this goal can be reached, the child may need to attain a preliminary goal—the ability to recognize his own real feelings and put a name to them.

3. The practitioner communicates his respect for the child's own feelings and thoughts, no matter how unacceptable these might be to the family, teachers, and others around. This does not mean that the practitioner communicates approval, only his acceptance, emotional empathy, and wish to help the child understand himself, so that the latter can change if he wants to.

4. While the practitioner expresses his willingness to help the child recognize and understand his feelings, he communicates at the same time that responsibility for trying to change rests with the child.

### Techniques

The following techniques have been found helpful in applying the above principles:

1. *Voluntary attendance.* At the initial interview it is important to establish with the child that he does indeed wish to come. Some children attend only because they are virtually forced to by parents or teachers. This may be acceptable as far as assessment is concerned, but is not compatible with ongoing psychotherapy, formal or informal. Therefore, if a child or adolescent does not wish to attend, this should usually be respected, though in some highly specialist units dealing with, for example, very severely disturbed mute or anorexic children, it is sometimes thought desirable for attendance to be insisted upon by parents and professionals acting together. Ambivalence can be dealt with by suggesting that the child might like to begin therapy and see how it goes. In accepting a refusal of therapy, one can make clear that the child is welcome to attend in the future if he changes his mind.

2. *Establishing the pattern of attendance.* At the beginning of the sessions, it is as well to make clear how many there are likely to be, and how they are going to be arranged, formally or informally. The practitioner should make clear how he is going to deal with possible interruptions.

3. *Clarifying the reason for attendance.* The practitioner should also make explicit at the beginning the purpose and nature of the sessions. This might be put along the following lines. 'You told me that you get angry with your parents/often feel very worried/get panicky about going to school/can't make friends/find yourself messing about with your insulin, etc. We are going to meet to try and work out why this happens so that you can change if you want to. If you tell me about yourself, your feelings, about what's happening in your life and how you feel about things, together we may be able to make more sense of what's going on.'

4. *Defence mechanisms*. In trying to make sense of the child's world, the practitioner will probably make particular use of the concept of unconscious defence mechanisms—means by which the child protects himself against emotionally unbearable thoughts (Freud 1966). These are described, in relation to ways in which parents deal with the unacceptable sides of having a child with a chronic physical handicap, in Section 4.2. If these defence mechanisms (denial, projection, repression, regression, etc.) are very strong, it is less likely they will be altered by psychotherapy, and indeed, in some circumstances, it may be unwise to attempt to do so. However, some children are able to become aware of their unacceptable aggressive, sexual, or other feelings, and, if the practitioner conveys *his* acceptance of these feelings, the child may have a reduced need to defend himself against them in a maladaptive way. For example, a child who is in trouble for fighting, and who blames other children for all the episodes that occur, may be using projection as a defence mechanism. 'I feel very angry with other people.' = 'Other people feel angry with me.' If he can be helped by the therapist to understand his aggressive feelings and then to accept them in himself, his need to use the defence of projection may be reduced and his behaviour may change.

5. *Making connections*. Another technique for helping the child make sense of himself is to point out possible connections of which the child may previously have been unaware. Connections may be made, for example, between:

(a) the child's behaviour in one current situation and another, for example in everyday life and in the session with the therapist;
(b) the child's behaviour in the past and in the present, for example the first time the child was separated from his parents and a more recent event;
(c) feelings and particular situations, for example the child may feel angry whenever he thinks he is going to be forced to do something;
(d) a symptom and a particular situation, for example a row with mother and suicidal thoughts.

6. *Making clarifying suggestions*. In making suggestions that might help the child to make sense of himself, it is preferable to put ideas forward tentatively. The child is the only arbiter of the helpfulness of a suggestion. However, sometimes it is fairly obvious that a child is rejecting an idea that would make some sense of his predicament. In these circumstances it may be worth going on to question how upset the child would be if the suggestion were valid.

7. *Observing the child in therapy*. In the therapeutic session, the practitioner should constantly be aware of clues the child is providing to the way he is thinking or feeling. The way the child reports events of the previous week is likely to be at least as important as what actually happened. In this way, the practitioner will be in a better position to pick up inconsistencies

in the child's behaviour that it may be helpful for the child to be aware of, for example if a child describes in positive tones a time when he played a game with his sister when he has previously described only arguments with her. The meaning of non-verbal messages should also be considered.

8. *Respect for the child.* The therapist should attempt to enhance the self-esteem of the child by taking him seriously, and approving of his coping behaviour. The child's self-esteem will not be enhanced by the therapist pretending to approve of undesirable behaviour, but attempting to examine why undesirable behaviour occurs without condemning it is likely to have a more positive effect.

9. *The silent or unresponsive child.* Many children, especially younger children, find it hard to talk about their feelings. They may require play and drawings to express feelings which have not yet been given a verbal form in their minds. The practitioner should have available a supply of toys (especially families of dolls), coloured pencils, and felt-tip pens or paints for this purpose. For formal psychotherapy, it is useful to have a small box of play material reserved for the child. Obviously, for the therapist, the technical quality of what is produced is irrelevant—the drawings, etc. are useful in so far as they provide a starting point to understand and talk about feelings, etc. Some children will not play, draw, or talk, yet they continue to attend in mute silence. The practitioner can probably best help the child by tolerating and accepting the silences for minutes at a time, but occasionally making suggestions why it is so hard for the child to communicate. The use of a projective test such as the Make-a-Picture story test (Shneidman 1947) may be a useful way of tapping the fantasies of an unresponsive child. A child's drawings or play can also provide clues to thoughts and feelings in a way that can be used in therapy. A drawing with a violent theme can be employed to discuss a child's fears of being harmed. Play with a family of dolls or the drawing of a family can help the clinician (and thus the child) understand the child's wishes concerning her own family.

10. *Therapist style.* Practitioners should use styles with which they feel comfortable. Some are more comforting than others: some frequently use humour, others only rarely. Some therapists find it helpful to use 'paradoxical' interventions—'I can't see why you should ever think of going back to school. You seem to have life very well organized as things are now.' Others would be troubled by using such techniques. A practitioner who can vary his style flexibly to meet the demands of different children and families is at a particular advantage. All therapists should frequently examine their own feelings and thoughts about a particular child and his problems. Indeed the therapist can use his own feelings in the session with a child as a guide to what the child is feeling but is, as yet, unable to express in words.

11. *Stopping treatment.* If, at the beginning of therapy, the practitioner and child have agreed on a particular number of sessions, there will be no

problem about deciding when to stop. In other circumstances, termination should ideally occur when the problem is resolved, but may well be necessary when no further progress is being made or when the child no longer wants to come. If termination occurs in a planned way, then there should be an opportunity at the end for the child to discuss how he feels about stopping. Children may well go through a sequence of feelings about treatment terminating (particularly if it has been prolonged). Denial may be followed by bargaining, anger, sadness, and only finally acceptance. At the very end there should be a firm arrangement whereby the child should either be seen at a follow-up appointment or know he can contact the therapist again if the need arises.

### Indications for individual psychotherapy with children

1. Emotional disorders: although the rather rarely occurring specific phobias are better helped by behaviour modification techniques, and behavioural management advice is also helpful in a wide range of conditions in which anxiety is prominent, psychotherapy is probably the treatment of choice when anxiety and depression are diffuse and chronic.

2. Physical conditions in which psychological factors are prominent in aetiology: usually as a supplement to family therapy, individual treatment can be helpful for children in this category with, for example, abdominal pain, chronic headaches, and asthma.

3. Less-well-established criteria include situations in which a child with a physical condition is failing to comply with treatment, social disorders in which the most prominent features are withdrawal and inability to relate to other children, and some special symptoms such as encopresis. In many of these conditions, psychotherapy can be supplementary to other forms of treatment, such as family therapy.

If these conditions are present, positive indications for psychotherapy would include situations in which the child and family are well motivated for psychotherapy, the condition is not obviously entirely reactive to factors within the family, and the child has already shown some capacity for making connections, i.e. has some degree of psychological mindedness for his symptoms. In specialist settings, it is often helpful for the parents to receive counselling at the same time as the child is undergoing therapy.

The age of the child is relevant. Children under the age of 6 or 7 years will find it difficult to communicate verbally about their problems. Although in theory, and to some degree in practice, children under this age can communicate their feelings through play, there is little evidence that connections made in therapy in this way generalize to everyday life. Other methods, including family therapy and behaviour modification, are more likely to be applicable. In contrast, the older the child the more likely he is to benefit from psychotherapy. If should, however, be added that some highly trained analytic child therapists do claim effectiveness for this form of treatment with very young children. Such claims are unconfirmed by systematic evidence.

## Contraindications

There is no evidence that the type of therapy described above, or indeed intensive psychotherapy, is helpful for children with autism (pervasive developmental disorder) or hyperkinetic syndrome, or for conduct disorders unless symptoms of emotional disturbance (anxiety or depression) are prominent. Psychotherapy is contraindicated in these conditions because the child is likely to show poor capacity for symbolic thought, short attention span, and difficulties in forming interpersonal relationships—all of which make interpretive forms of psychotherapy difficult or impossible to apply.

## Evidence for effectiveness

There is evidence for the effectiveness of child therapy in the conditions in which it is described above (Eisenberg *et al.* 1965; Rosenthal and Levine 1971; Gurman and Kniskern 1979). Most of the conditions for which psychotherapy is effective have a reasonably good prognosis untreated, for the main benefits of the treatment lie in its capacity to shorten the duration of disturbance. From work examining therapist effectiveness in student counselling (Truax and Carkhuff 1967) one can reasonably assume that improvement is related not only to the nature of the child's condition, but also to therapist qualities, especially capacity for warmth, accurate empathy, and genuineness. Also by generalizing from work with adults (Reid and Shyne 1969) one may assume that brief, focused therapy, with a finite number of sessions arranged at the outset is, in general, more effective than therapy conducted without clear goals.

## Case examples

*Informal individual psychotherapy—general practitioner*
An overweight 11-year-old girl, Jennifer, was brought by her very slim and attractive mother to the general practitioner. There were no complaints about her, apart from her obesity and tendency to overeat. She was indeed over the 97th percentile in weight, but only of average height. The GP knew there were considerable problems in the family. Father was now living at home, but had previously left home on three or four occasions for months at a time—unable to make up his mind whether to live with another younger woman or with his family. Jennifer was the second of four children and the oldest girl. The GP suspected that the mother was having an affair herself because he noted she came for birth control checks when her husband was away. Previous attempts by the GP to treat the obesity with a restricted diet had been unsuccessful.

He tried to talk to the girl and found her unresponsive. With her mother present she looked at him stonily, chewed gum, answered his questions monosyllabically. Yes, she would like to be thinner. No, there was nothing she was worried about. He asked to see the girl by herself and made little further progress. She denied she was worried her father might leave home again—they did not need their father. If he wanted to go, that was his business. They had managed without him before.

The GP said he would like to see mother and daughter again in a fortnight's time and put aside a further 25 minutes to talk to the girl. Ideally, he would also like to have seen the father, but it was clear he would not attend. On this second occasion the girl was initially somewhat more forthcoming, then began to cry. Her father was threatening to leave home again. She hated the idea, not because she missed her father, but because when he left, her mother was irritable with all the children, made her do more housework, so that she could not go out with friends. They were much more hard up. Her younger sister was taken into her mother's bed and Jennifer admitted she was a bit jealous. Her younger sister was always messing up her games with her friends. The GP said he really understood why she had to eat so much. She must be very unhappy. He said he did not think she could go on a diet in the circumstances. The girl said, on the contrary, she thought she could. He asked if she could talk to her friends about her problems and she said she could not—she was too ashamed. The GP suggested there might be other girls in the same boat. He said at least she should not feel a failure if she found she could not stop eating. He suggested her way of responding to feeling sad was to eat more, while her mother just got cross and irritable when she was sad. The girl agreed and added that she thought it would be better for the family if her mother ate more rather than being cross and irritable when she was sad. The GP offered further appointments, but the girl said she would come and see him if she wanted another talk. She did not lose weight, but at least managed to avoid putting on any more. The next time she came was for birth control advice when she was 15 years old. She was slimmer and more forthcoming. The GP was able to talk with her about her need for affection in a way she obviously found helpful in deciding whether to sleep with her boyfriend.

*Informal individual psychotherapy—paediatrician*
Paul was a 14-year-old boy with cystic fibrosis that had been diagnosed in infancy. An older brother had died of the same condition 2 years previously, aged 16 years. There was an unaffected 20-year-old sister. The family members had always had difficulty in expressing their feelings to each other, and to the social worker who had known them for the previous 5 years. The paediatrician had been responsible for Paul's hospital care since just after the diagnosis had been made, but he also found difficulty in knowing what was going on in the family. Paul had been rather little affected by his condition until about the age of 10 years, since then he had

had three hospital admissions with infections that had been difficult to treat.

Over the last 6 months, Paul had changed from being a rather cheerful, active boy, keen on games and with many friends, to being morose, sullen, and isolated. He was rather uncooperative when undergoing physiotherapy both at home and during the present admission. The nurses on the ward were worried about him, and the social worker was away.

The paediatrician assumed he was depressed because of increasing awareness of his prognosis. He decided to see the boy on his own and in fact saw him three times before his discharge. He and Paul talked about Paul's interests, how the football team Paul supported was doing. He then asked what sort of things Paul was worried might happen over the next few months. Paul said he supposed he would just go 'jogging along'. There was nothing more. The paediatrician asked if he was worried about his mother, and immediately Paul's eyes filled with tears. He said he thought his mother was going to crack up. She had been very upset by the death of Gerald (the older brother), but no one else but he, Paul knew about this. The paediatrician asked how Paul remembered his brother, and Paul immediately said that it had not been too bad for Gerald because his mother was always with him. It was obvious that Paul was angry with Gerald, and also that he was frightened of being left alone when he was very ill. The paediatrician commented that it was very natural for Paul to feel angry with Gerald, but he knew sometimes people felt bad about being angry with someone who had died.

He asked Paul later on what worries he had about himself and Paul again denied having any. However, as the paediatrician was about to leave the cubicle on one occasion, Paul asked him if the thought they might discover a cure for cystic fibrosis. The paediatrician said that was not impossible— there was research going on all the time, but he did not want to raise Paul's hopes.

After he left Paul, the nurses noticed the boy crying, and there was a certain amount of anger amongst the junior nurses towards the paediatrician who, it was felt, had upset Paul unnecessarily. This led to a useful discussion at the next ward psychosocial meeting about whether it was, in fact, better for teenagers to be encouraged to deny the implications of their condition, or to encourage the expression of sad feelings. No conclusion was reached, but both medical and nursing staff were left with a better idea of the complexity of the issue, and the need to treat each individual case on its merits. The paediatrician continued to see Paul in out-patients after his discharge, and felt he was now much more open in expressing his feelings.

### Individual psychotherapy—psychiatrist
Sally was a 10-year-old girl with widespread fears and anxieties. Her parents both complained that she took hours to settle in her bedroom

because she kept coming downstairs. She frequently woke in the night with nightmares. She was too fearful to stay in another's girl's home if she went out to play, so either girls came to her to play or she was unable to see her friends. She would not go to the shops by herself. School attendance was reasonably regular, but she often tried to get out of school by complaining of physical symptoms.

The family background was disturbed and provided some but probably not a full explanation why Sally was so anxious. Two family interviews were held, but it became clear that, for various reasons, father was not prepared to attend any more. The widespread nature of the fears made it improbable that behavioural methods would be effective, and it was decided to embark on an 8- to 10-session course of brief psychotherapy. The focus was to be on the understanding of the source of Sally's anxieties, so that she could obtain better control over them.

The first session was largely taken up with setting Sally at ease and discussing why she felt so anxious with the psychiatrist away from her mother. She was clearly uncomfortable in the situation, but managed to stay for 45 minutes. In the second and third sessions she started to draw rather stereotyped pictures of a house and garden. One house had three girls and their mother living in it (Sally was an only child). Why no boys? No, boys were rough. There were a lot of naughty boys at school. If they had any sense, girls did not want boys around. The psychiatrist wondered how far aggression was a problem for Sally. On the fourth session the psychiatrist was unavoidably 20 minutes late, and Sally was obviously fed up. The psychiatrist commented on this, but Sally denied it, saying she did not mind. She had been quite happy in the waiting room. The psychiatrist suggested that it was hard for Sally to admit she was angry. He suggested it might be hard for her to admit she was angry at home too. The next session was taken up with a similar theme. Prior to the sixth session the social worker, who was seeing the mother, told the psychiatrist that Sally appeared slightly less anxious, that her mother was finding her more disobedient and aggressive to her. The final three sessions were taken up with the psychiatrist preparing Sally for termination of the sessions, while Sally tested the psychiatrist out. She tried to mark the table and then the wall with felt pens, swore at the psychiatrist when he firmly prevented behaviour he had told her would not be allowed, and then sulked. The psychiatrist made connections concerning her mixed feelings about him and her feelings about her parents and teachers. He suggested it was very hard for her to be a little girl who had to do what adults told her to—both with him and with her mother. Sally agreed with this, and said this was because grown-ups were stupid. Towards the end of the final session she became tearful. The psychiatrist wondered about offering further sessions, but decided against this, partly as a result of pressure on his own time.

The girl's anxieties outside the session decreased during and after the therapy. Over the next three years, during which she was seen for review appointments, her general anxiety also decreased, perhaps partly as a result of the natural history of the condition and partly as a result of the therapy. She became a rather isolated, shy teenager, but her sleep problems completely waned and she was able to stay the night with friends without difficulty.

## 6.2.4 FAMILY THERAPY

### Definition

This involves the treatment of family members as a group, rather than as individuals (Lask 1987). Families vary in size and in the ways they function. The group treated might just consist (in an isolated one-parent family) of mother and child. It might, however, involve ten or more people with grandparents included as well as parents and their children. Those who conduct formal family therapy usually try to include all those who are in daily or more frequent contact within the same physical household. Sometimes, however, less formal therapy is conducted involving only some of the family members.

### Principles

Most therapists use more than one approach, or borrow from more than one theory in their practice.

1. *Behavioural.* Behavioural approaches are described in some detail in Section 6.2.2. As discussed in that section, these approaches are linked by an emphasis on changing behaviour without consideration of underlying unconscious mechanisms. They can be applied directly to the individual child, or a parent can be involved in their application. However, some behaviour therapists find it effective to use their technique with the whole family present. It may be easier, for example, to ensure consistency of reinforcement (reward), or to deal with problems that may arise if one child (the identified patient) receives a reward and another does not. Behavioural family therapy allows such issues to be tackled more directly.

2. *Psychoanalytic.* Many family therapists are influenced in their work by psychoanalytic concepts (Section 1.1.1) concerning the importance of unconscious mental processes acting on behaviour. These provide the most common background for individual psychotherapy (Section 6.2.3), but they may also provide the theoretical framework for family therapy. Thus, for example, a therapist may characterize a whole family as prone to the use of *denial* or *projection* as a means of dealing with anxiety. Alternatively, the therapist may observe that one family member is demonstrating the consistent use of an unconscious process in his relationship with another family member, and try to deal with this in the family context, i.e. with the

whole family present. Thus, for example, every time a father is angry with his wife, he may displace his anger on to the child identified as the problem in the family.

Family therapists have adopted or adapted concepts from behavioural and psychoanalytic schools. They have, however, developed other theories described in Section 1.1.1, when systems, communications, and structural theories are outlined.

## Assessment

When a child is referred for a behaviour or emotional problem, or is seen for a physical complaint which turns out to have a strong functional component, assessment of family relationships with a view to family therapy may be informal or formal.

A formal family assessment involves seeing together as many family members living under the same roof as possible, to observe their characteristics, mode of communication, and interaction. A specific period of time, normally between 1 and 2 hours, is set aside for this purpose, and during the end of this time, the therapist will usually feed back to the family the observations he is making and assess the impact this has.

### Informal family therapy or family counselling

Paediatricians and general practitioners are often involved in counselling small groups of family members. Usually this involves listening and talking to mothers or both parents and a child identified as having a problem.

For example, a general practitioner may note that the mother of a boy with asthma is unnecessarily overprotective. He may see the parents together with the child on a couple of occasions, and facilitate the father's feelings that his son could be allowed more independence. The family doctor might sympathize with the mother's predicament but support the father in his wish to help mother and son to separate. A paediatrician faced with a teenage diabetic girl failing to comply with treatment might arrange to see parents and child together to discuss the situation. He might discover that the girl's refusal was part of a wider rebellion. He might give the parents, on the one hand, and the girl, on the other, the opportunity to discuss their feelings about independence and autonomy. Then he might sympathize with the parents in their worries about their daughter's behaviour, but support the daughter's wish to achieve her identity in her own way. He might succeed in persuading the girl in this way to fight her battles in areas that do not involve her health.

The distinction between this form of family counselling and formal family therapy is not clear-cut. The following account of family assessment and treatment is probably most likely to be used by those working in a psychiatric department, but many of the principles and techniques can be applied in other settings and with less time available than is usually the case in the child and family psychiatric clinic.

### Formal family therapy

It is sensible to invite the whole family (i.e. all those living in the same physical household) to attend if a family approach is to be used. It is particularly important for both parents and the identified child to be present, but the effectiveness of the family approach is enhanced by the attendance of brothers and sisters.

To assess family functioning it is important to elicit family interaction. This can usually be successfully achieved by asking the family to focus on various topics. One can begin with the nature of the problem behaviour of the identified child, the way in which the referral came about, and the attitude of all the family members to the attendances. While discussing this particular topic, it is especially important to elicit the feelings of the identified child, and to ascertain whether or not he thinks there are other, perhaps bigger problems in the family. If the sibs do not speak spontaneously, they should also be asked their views. Members of the family can be asked to talk directly to each other.

Other topics which can be used to elicit family interaction include finding out about the way in which family tasks, such as shopping, keeping the house clean, and looking after the children, are shared; the out-of-home activities undertaken by family members, separately and together; the development of the family, including the way the parents met, the circumstances of the children's births, the times when the family moved house, children started school, etc.; and the impact on the different family members of recent stressful events.

Family assessments can and often should be combined with individual assessment of the child to explore the mental state or feelings in more detail. It will often also be sensible to offer parents separate sessions without the children. During family assessment the therapist will find it helpful to use certain techniques. He should be aware of the ways in which the more silent family members show their feelings and make a particular effort to comment on their form of non-verbal expressions and to get them to talk. He should refrain from taking sides, but make sure that everyone's opinions and feelings are as fully expressed as possible. He should make tentative observations of the family's functioning during the sessions and check out with family members whether they perceive these observations to be accurate. Such observations can be considered under the following headings:

1. *Family structure*. Who sits next to whom, speaks for whom? Are there any special alliances or coalitions in the family? Are family members overinvolved ('enmeshed') or too detached from each other? Are the inter-

generational barriers respected? Do the parents behave like children or are the children taking on parental roles?

2. *Communication.* Are feelings about important family matters readily expressed? Are messages about what the parents and children would like of each other clearly communicated? Do the children understand and accept the sanctions imposed by the parents?

3. *Family atmosphere.* Is this tense or relaxed, depressed or cheerful? Do the family use humour to release tension or heighten it? Does the atmosphere change, for example, when the child identified as a problem is being discussed or the parents start to talk about problems they have together?

4. *Potential for change.* How rigid and inflexible does the family seem when family members discuss and argue among themselves? Can they tolerate differences between themselves? How well do family members take up the tentative suggestions made by the therapist?

## Treatment

After the preliminary assessment has been completed, usually in one but perhaps in two sessions, the therapist should decide whether treatment is indicated and if it is, draw up a treatment plan with the family.

### Format

Treatment might involve family sessions alone, or family sessions combined with marital or individual sessions. The main advantage of not having separate individual or marital sessions is that inevitably those not present at such sessions develop fantasies about what is being said about them behind their backs. On the other hand it is inappropriate to discuss parental sexual problems with children present, and other family matters may also be regarded as secret from the children. When a member of the family requests a separate session, it is reasonable to check out with the family present whether this concerns something all family members do not already know about. If the therapist accedes to the request for a separate session and it turns out to concern a matter it *is* reasonable for the rest of the family to know about, then it should be suggested that the whole family be involved in further discussions. The length and frequency of treatment should be discussed with the family. It would be reasonable, for example, to suggest that there be 6–10 sessions held at fortnightly or three-weekly intervals over 3–6 months.

### Techniques

It is useful to develop a *focus* for the sessions with the family fairly early on. This might involve achieving improvement in communication, helping the family to understand why a particular behaviour problem in a child is occurring and how they might cope with it better, how a family member, for example the father, might be more integrated with family life, etc.

The therapist should comment on **positive** as well as negative aspects of family functioning. When examples of good coping behaviour, clear expressions of feelings, etc. come up, then this should be positively commented upon by the therapist. The therapist should also positively comment on the adaptive functions of behavioural problems. Thus, he can say of a child who is having temper tantrums that he is really very good at getting his own way. The question is whether he could not succeed in getting his own way in another more acceptable fashion. A child soiling his pants could be congratulated on finding such an effective way of showing how angry he is with his parents, etc. Of course, the therapist should make clear he does not approve of this behaviour, but establish that he sees its function and is intent on helping a child find other modes of expression. The therapist should facilitate **communication** of feelings as well as information. He can ask how one family member thinks another feels about a particular problem and then ask them to check this out in the session. He should promote the **establishment** of appropriate **boundaries** between the generations, communicating, for example, how confusing it must be for the children if parents behave like them at one point in time, and then expect to be obeyed at another.

The more experienced therapist will use a variety of other techniques (Barker 1981) to promote more adaptive family functioning. An example of a more advanced technique is the paradoxical challenge (Cade 1984). The therapist may suggest that it would be better for the family if a particular form of maladaptive behaviour did not change. Thus, for example, in a child not attending school the therapist might suggest that, if the child did attend school, the mother, unable to cope with the separation, would have a breakdown, and that the father would have to stay at home to look after her. Perhaps it would be better for the child to remain off school. Some families respond to such a challenge when they have failed to respond to more conventional therapy by taking positive moves to achieve better functioning, but clearly this is not the sort of technique to be used lightly by a less experienced therapist.

### Indications

Most child and adolescent psychiatric disorders (emotional and conduct problems, management difficulties in the preschool period, anorexia nervosa, tics, psychosomatic disorders, etc.) *may* respond to family therapy. It should, however, be applied only when the presenting symptom can be seen on initial assessment to be maintained by mechanisms involving other members of the family. If, after assessment, the therapist strongly suspects that the presenting problem is not really centred in the child but with the family relationships more generally, then, if there are no contraindications (see below) family therapy is indicated.

## Contraindications

Family therapy should not be applied where:

1. The problem is mainly centred not in the family but in the child, for example in cases of childhood autism, severe personality problems, or where a child has clearly responded to a specific physical or psychological trauma unrelated to family life. Occasional family sessions may be helpful in such cases to help family members cope better together, but family therapy should not be the main mode of treatment.

2. One member of the family has a serious mental illness such as paranoid schizophrenia or a marked personality disorder which prevents a realistic perception of family relationships.

3. Effective resolution of the problem can be achieved by a cheaper form of treatment not involving the father or mother taking time off work or the other children time off school.

Relative but not absolute contraindications for using family therapy as an exclusive approach would include a family on the point of breakdown, the presence of a serious marital problem, the presence of 'legitimate' family secrets, and apparently unmodifiable scapegoating of one particular family member. When a family approach is used in these cases, family sessions may be combined with individual or marital therapy.

## Evidence for effectiveness

There is evidence that family therapy is helpful in a variety of cases in which the child is the presenting problem (Gurman and Kniskern 1979). Controlled studies have demonstrated slight but definite improvement in children with a range of behavioural and emotional disorders as well as psychosomatic conditions such as asthma. In a review of nineteen controlled studies of variable quality of the effectiveness of family therapy, fifteen of which involved children, about three-quarters of the patients treated did better than they would have done if the subjects had received no treatment or an alternative treatment (Markus *et al.* 1990).

## Case examples

### Informal family therapy—family practitioner

Jonathan was a 3-year-old brought to his GP for an opinion on a mole on his back, which turned out to be a pigmented naevus of no particular significance. However, examination of the mole turned out to be a traumatic experience, as Jonathan first would not allow his shirt to be taken off, then screamed, shouted, and wriggled while the doctor tried to examine him, and finally finished with a monumental temper tantrum when his mother tried to put his shirt back on again.

When calm was eventually restored, the family doctor enquired if this sort of thing happened often, and his young mother was soon in tears saying she could not manage the boy at all, her husband was little help, she felt she was going to lose control of herself with him, and she was on the verge of a 'breakdown'. The family doctor asked if she and her husband would bring Jonathan back in 10 days time for a further talk.

At this next appointment Jonathan was once again very difficult and started to play with the doctor's telephone and open his desk drawers while the parents were trying to talk to him. Neither parent seemed to make much attempt to stop Jonathan, though each looked at the other expectantly, obviously with this in mind. The family doctor asked who normally controlled him in this situation, and the father said he thought it was his wife's job. The mother said she had Jonathan all day and her view was that when father was home, he ought to take this responsibility. The doctor said he wondered if really the parents wanted to stay together because they were obviously making life impossible for each other. Both parents were very angry at this suggestion and jointly rounded on the doctor for making it. The doctor said in that case they looked as though they wanted the family to stay together but Jonathan to run it. Was he not a little young to take this responsibility?

The parents then both described their own childhoods of which the doctor knew something, as he had looked after them before they married. They had each had a very strict upbringing. They admitted they both hated the idea of bringing Jonathan up not to have what he wanted. The doctor did not comment on this, and the parents saw themselves that they would have to set firmer controls. No further appointments were made, and there was no systematic follow-up, but the doctor noted that when, 3 months later, he had to listen to Jonathan's chest, his shirt came off without demur.

### Informal family therapy—paediatric out-patients
An 8-year-old girl, Amanda, was referred by her general practitioner to paediatric out-patients with recurrent abdominal pain. The pain had begun a year previously—it was intermittent in frequency, occurring for a few hours every 8 or 9 days. It was not accompanied by vomiting or alteration of bowel habit. The pain was severe, and made Amanda cry. It never woke her at night. The pain stopped her going to school when it occurred, but she also had it at weekends and during school holidays.

She was an only child who lived with her parents in a residential area of the city. There did not seem to be any obvious stresses at home or at school. Physical examination was negative and, in particular, she had no abdominal tenderness.

The child was accompanied only by her mother. The paediatrician noted that the girl kept darting anxious glances at her mother. He also observed that Amanda was very neatly and smartly dressed and kept brushing imaginary flecks of dust from her skirt. The mother seemed rather obsessional in the way in which she seemed to have to get every detail of the occurrence of the pain precisely accurate.

In answer to questions it became clear that the mother thought the pain might be due to ulcerative colitis from which a maternal aunt had suffered. The paediatrician was reassuring, but ordered a blood count, urine examination, and straight X-ray of the abdomen. He asked to see the girl again in a fortnight and requested that her father should come as well on that occasion.

Investigations were negative. The paediatrician diagnosed non-organic recurrent abdominal pain and saw the family on three further occasions for about 20 minutes each time to monitor progress before referring back to the family practitioner. He discussed possible links between tension and pain with the family (of which they were already aware) and the family members talked about how Amanda could be helped to be less tense and anxious. When Amanda was asked to draw a picture, using felt pens, it was clear that she was unduly worried both about getting her hands dirty and about producing a perfect picture. The parents obviously disagreed about how relevant this behaviour might be to her pain, but could not express their disagreement to each other. Amanda looked more uncomfortable at this point. The paediatrician suggested that possibly parental disagreements might make Amanda less anxious if they were able to express them more clearly. Over the next few weeks Amanda's pains virtually ceased, although the family relationships appeared unchanged. The paediatrician was uncertain whether the interviews had resulted in pain alleviation or whether this was due to the natural history of the condition. A further possibility was that in fact the pains were persisting, but the parents denied them so that they need no longer attend interviews they themselves found painful.

*Family therapy—psychiatrist*
Helen, aged 12 years, was referred to a child psychiatrist with an established diagnosis of anorexia nervosa. At about the age of 10½ years, she had been a rather plump girl and a chance remark about her appearance by a physical education teacher had humiliated her. She began to diet and became preoccupied with her appearance. She exercised excessively and lost a stone in weight. Her parents eventually took her to the family practitioner who was alarmed at her appearance and referred her immediately. The paediatrician who saw her treated her on an out-patient basis over several months, but eventually referred her to a child psychiatrist. By now mealtimes were a nightmare, with the parents and Helen's 16-year-old sister Jasmine all shouting at her in order to get her to eat and then at each other when they failed to do so.

There was no significant family history. Helen was a very good girl who worked conscientiously at school. The parents said of their own relationship that they got on extremely well with each other until Helen's problems had begun. They said Jasmine presented no problems.

Helen was admitted to an in-patient unit and put on a behavioural regime to encourage her to gain weight. She was allowed increased activity for each 2 kg increase in weight. This programme was initially successful, but her weight began to plateau out before she reached the target at which she could go home for a weekend. Family interviews were arranged, although the father, in particular, set up obstacles to such sessions. The interviews were immediately illuminating. Father, who was a senior civil servant, sat slightly apart from his wife and two daughters. He was clearly excluded from much family life. It transpired that he worked very long hours and often did not know what had been going on at mealtimes. He undercut his wife's remarks at every opportunity. Mother was closely identified with her daughters and included herself when referring to their activities. For example she said to Jasmine, 'We are very good at maths, aren't we dear?' When the psychiatrist pointed out that she had repeated this phrasing in another connection, father, with heavy sarcasm, said 'Did you hear what the doctor said, dear, or are you *totally* deaf?' After three sessions of this nature a family meal was arranged in which it became clear that the parents could not agree on how to get Helen to eat and undercut each other's efforts constantly. The psychiatrist pointed this out, and was able to help them at the second meal to agree a strategy to encourage their daughter to eat. When her parents joined forces, Helen began to take notice of this and eat distinctly more than she had previously.

Marital sessions were arranged, and revealed that the parents had had an appalling relationship for years. No sexual relationships had occurred for over 2 years. The parents began to discuss separation without any apparent self-consciousness that their attitude now totally negated all they had said about their relationship when first seen. Further family interviews were held when the parents were encouraged to express some, though not all, of this problem to their children. Helen, who had previously said little in family interviews without looking at her mother for approval, began to speak more for herself. She reached a satisfactory weight and managed to maintain it, though she continued to have some personality problems. The parents refused further interviews once she was home from hospital. One year follow-up revealed that the girl had not relapsed and the parents were still together, with some improvement in the quality of their own relationship.

### Children of divorced or separating parents

The assessment and management of children with psychiatric disorders, developmental delay, or physical conditions, when the parents are divorced or separated, present special problems.

In assessing the child's difficulties, a decision has to be made whether to see the parents together or separately. The wishes of the parent who has control and custody should be respected and taken into account. Every effort, however, should be made to see the parent living apart, if he or she still has contact with the child, even if such contact is rather tenuous. Further, even though a divorced parent may not have a legal right to be involved in decision-making concerning a child's future, it is usually helpful to ascertain this parent's views, without in any way undermining the legal parent's position.

Children of parents living together face particular difficulties in their lives. They may be used as pawns in a continuing marital dispute by parents who, although physically separated, are still emotionally joined together by their dislike of each other. They may be used as message-bearers by parents who cannot bring themselves to communicate directly. They may be at risk of emotional upset arising because the separated parent does not keep to access times in a regular and predictable way. They may have to listen to their parents denigrating each other when they are fond of both of them. They often have to be flexible enough to make relationships with new partners their parents may find, and be able to cope with the ambivalent feelings these new parental figures elicit in them.

The rate of psychiatric disorders in the children of divorced parents is distinctly higher than that in the general population (Hetherington *et al.* 1982). Nevertheless, perhaps surprisingly, most children in this situation do not develop emotional disturbances, though of course their lives are always profoundly affected. It is clinically relevant that children whose parents have a disharmonious marriage are less likely to show disturbance if they have a good relationship with an adult outside the family, take part in an activity for which they receive positive recognition, and enjoy good relationships with their sibs (J. M. Jenkins and Smith 1990).

In the management of children whose parents have divorced or are in the process of separating or divorcing, it is likely to be particularly important to arrange separate interviews for the child, so that he can express his feelings and views about the situation privately. Counselling of parents can be undertaken separately or together, depending on the wishes of the parents, with priority being given to the legal parent. The role of the counsellor involved in providing advice concerning children during the process of divorce and separation is beyond the scope of this book, but is well discussed elsewhere (e.g. A. M. Levy 1985).

## 6.2.5 GROUP THERAPY

### Definition

The treatment of children and parents in groups can, like individual, family, and behavioural therapy, be both formal and informal. In formal therapy, groups are set up with a particular therapeutic purpose in mind. But group therapy can also be informal, and this occurs especially when, because individuals are already in group settings, such as schools, children's homes, or in-patient units, the opportunity can readily be taken to use group dynamics in a positive, therapeutic manner. Obviously, group processes also occur between staff working in different professional settings such as children's wards and special schools.

### Principles

Group therapists can use a number of different underlying theories to underpin their practices.

1. *Psychodynamic.* In groups run along psychoanalytic or psychodynamic lines, the emphasis is on the interpretation of unconscious aspects of group interaction. Thus a child who encouraged another in the group to get into trouble might be helped to see that it was he who had the wish to be disruptive himself.

2. *Behavioural.* Group settings can be used to apply behavioural therapies, taking advantage of the group in different ways. Thus an operant conditioning approach in a classroom might be more effective when applied to a particular child in that setting than when applied individually because reward or reinforcement could involve the whole class of children. A group setting may be particularly appropriate for the use of cognitive behavioural methods (see Section 6.2.2): group members assisting each other to check the validity of their self-cognitions.

In the application of social skills training (Section 6.2.2), a group provides a real-life opportunity to apply such skills with the therapist present and able to monitor the outcome. In a broad sense, social skills training forms part of all group treatments. The way in which limits are set and observed, the laying down of rules, the way in which aggressive behaviour and social isolation are dealt with—all provide possible opportunities for learning social relationships (Sands and Golub 1974).

3. *Nurturant.* Especially in young children, but also in older, deprived children, application of group treatments may allow children to receive care and attention, as well as experience in sharing, in ways they would otherwise have missed.

**Therapeutic groups** can also involve groups of parents and groups of families. Parent groups have been set up antenatally (to discuss and interact about issues relating to prospects of parenthood), and postnatally (to consider early experiences and problems in bringing up children). Parent groups have also been established for those whose children have similar behavioural or emotional problems. Thus behaviour modification principles and practice can be effectively and perhaps more economically imparted to groups of parents whose children show, for example, aggressive behaviour. Parent groups run on behavioural lines have also been effective for those with severely developmentally retarded young children. Finally, groups for parents whose children suffer from the same physical condition such as leukaemia or cystic fibrosis have also been found helpful (e.g. Bywater 1984). These provide a combination of support, practical self-help, and information.

**Informal group therapy** for children themselves has been described in nursery schools and other preschool educational intervention units, ordinary schools for older children, special schools for maladjusted children, children's homes, and in-patient and day psychiatric units for children and adolescents. More formal group therapy usually takes place in child and family psychiatry departments, but also in special schools and units.

**Voluntary self-help groups** have been set up for many purposes, either entirely by parents themselves or with initial professional stimulus. Paediatricians and paediatric nurses can play a helpful role here. Examples are groups for parents of very young children run by the National Childbirth Trust, and the very large numbers of parent organizations set up to provide support for parents with children suffering from particular physical conditions. Organizations now exist in the majority of economically developed countries for parents of children with most chronic physical conditions. Many such organizations have now achieved national or even international status, and also have active local branches. In the child mental health field in the UK, MIND, MENCAP, and the National Society for Autistic Children are particularly active.

### Practical issues

The setting up of formal groups in a clinical child psychiatric setting presents a number of practical problems, especially difficulties in finding a room of sufficient size and a suitable time when group participants can all attend. The age range of the children to be admitted to a group needs attention. It is helpful to think in terms of three main age groups whose needs will differ sufficiently to require age-separation: 4–7-year-olds, 8–12-year-olds, and adolescents (Lask 1989). Decisions have to be made about whether the group should be open (allowing newcomers to enter after the group has started) or closed. If inexperienced, group leaders or therapists need the opportunity for supervision. It is helpful to specify to participants the aims of the group, the nature of the limits to be set, the degree of confidentiality, and the likely number of sessions (usually 10–20) to be held. If group therapy is to be combined with individual sessions administered by the same therapist, the possibility of confusion needs to be considered.

### Indications and contraindications

Children most likely to benefit from formal group therapy are those with difficulties in their social relationships, especially undue shyness, social isolation, and other forms of anxiety in social situations. Antisocial children may benefit from behavioural techniques applied in a group setting. Children and adolescents who have suffered similar traumatic experiences (e.g. those who have suffered sexual abuse) may benefit from the opportunity to share their experiences and feelings in a group (see Section 1.3.3).

Those less likely or unlikely to benefit, at least from formal methods, include those who, like autistic children, do not have the necessary cognitive skills to form social relationships. For other children, such as those where, for example, faulty family communication or intrapsychic problems are predominant, group therapy will not be the preferred mode of treatment.

### Evidence for effectiveness

Most evaluative studies of group therapy suffer from serious methodological problems (Abramowitz 1976). However, Kolvin *et al.* (1981) in a careful controlled study were able to show that group therapy carried out in schools with groups of 7-year-olds and 12-year-olds resulted in improvement of neurotic problems (in the younger group) and antisocial as well as neurotic problems in the older age group. Follow-up confirmed that improvement was still detectable 3 years after the children were first assessed, and this form of therapy was more effective than others with which it was compared, such as parent counselling and teacher consultation regimes.

### 6.2.6 DRUG THERAPY

### Introduction

Medication has only a very limited place in the treatment of the psychiatric disorders of childhood and adolescence. Nevertheless, occasionally, appropriate use of medication can produce a highly beneficial effect, especially if used in combination with other forms of treatment. In general, drugs should be given in dosages appropriate to the child's body weight and, where facilities are available, it is desirable to check plasma levels.

Most parents are naturally anxious about the use of drugs to treat their children's disorders. Even if the clinician is convinced that a trial of medication would benefit a child, he will wish to take parental anxieties into account when recommending medication. A period of delay to try the effects of other forms of treatment is not usually harmful. Parents are most likely to be persuaded of the need for medication if the clinician explains that his use of drugs is very sparing, that a close eye will be kept for side-effects, that the child will be taken off the drugs as soon as possible, and that the drugs he is prescribing are not, as far as is known, addictive.

Because drugs have such a limited place in this age group, only a small number of them will be described here. Fuller information may be found elsewhere (M. Campbell *et al.* 1985; M. Campbell and Spencer 1988). Because recommended dosages change, all dosages provided here should be checked with a recent edition of an authoritative guide before use (e.g. British Medical Association and Pharmaceutical Society of Great Britain 1985).

### Hypnotics

Monosymptomatic sleep disorders should, in general, be treated with advice and behavioural management (Section 2.3.2). As a short-term measure it is reasonable to use:

Trimeprazine tartrate syrup (*Vallergan forte*).
*Dosage*: Approximate dose for 3–6 year old, 3 mg/kg in one dose. i.e. 30–60 mg (1–2 teaspoons).
*Side-effects*: Some children are made less sleepy and more irritable. Headaches, dry mouth.
*Comment*: Use only for periods up to 2 weeks because of habituation.

In older children use:

Nitrazepam.
*Dosage*: 5–10 mg (1–2 tablets), 30 minutes before bedtime.
*Side-effects*: Early morning drowsiness and irritability. Prolonged use may result in dependence.
*Comment*: Use only for periods of up to two weeks.

Other hypnotics include chloral (in younger children), dichloralphenazone and promethazine (in older children and adolescents). Antidepressant medication given at night (especially imipramine hydrochloride and amitriptyline hydrochloride—see below) will often improve insomnia in children suffering from depression.

### Minor tranquillizers (anxiolytic agents)

These are little used in child and adolescent psychiatric practice, partly because attacks of anxiety tend to be more episodic and less often 'free-floating' than in adulthood. However, short courses (1–3 weeks) of long-acting benzodiazepine medication (diazepam or chlordiazepoxide) may be indicated for subacute anxiety states in combination with counsel-ling, psychotherapy, behaviour modification, or other therapeutic methods. Longer courses should not be prescribed because of the risk of dependency. Diazepam may also be indicated in sleep disorders such as night terrors and sleep walking, and as a pre-anaesthetic for induction of sleep. Rectal diazepam may be indicated in status epilepticus, and has the advantage that parents can be instructed in its use.

Diazepam syrup or tablets.
*Mechanism*: Potentiates gamma aminobutyric acid (GABA) activity.

*Dosage*: 4–10 mg daily in two divided doses for children aged 6–12 years. 6–15 mg in three divided doses for 13–16-year-olds.

*Side-effects*: Drowsiness, irritability, dizziness, dry mouth.

*Comment*: Dependency occurs with prolonged usage. For short-term effects, such as to reduce morning anxiety in school-refusing children, it is advisable to use a benzodiazepine drug (e.g. Lorazepam) with more rapid effects, but not for more than a few days.

## Major tranquillizers (neuroleptic agents)

These are reserved for use in psychotic states, for multiple, complex tics, and for the treatment of very severe aggressive outbursts especially in mentally handicapped children. The relief of acute psychotic symptomatology is often dramatic. Maintenance treatment is also of value in the prevention of relapses of schizophrenic states. Drugs to be mentioned here are phenothiazines (chlorpromazine) and a butyrophenone (haloperidol).

Chlorpromazine syrup, tablets, or intramuscular injection.

*Mechanism*: Blocks activity of dopamine receptors in limbic system.

*Dosage*: 50 mg increasing to 150 mg in 2 or 3 divided doses— approximate maintenance dose for average-sized 13–16-year-old with a psychotic state. Higher doses may be necessary in acute states.

*Side-effects*: Dry mouth, postural hypotension, Parkinsonian symptoms (rigidity, tremor, salivation), restlessness, acute dystonia. Tardive dyskinesia (a syndrome characterized by stereotyped facial and tongue movements with choreoathetosis) may occur if usage continues over several years, and is sometimes irreversible.

*Comment*: In childhood and adolescence, for use only in schizophrenic and hypomanic psychotic states.

Haloperidol syrup, tablets or intramuscular injection.

*Mechanism*: Blocks dopamine receptors in limbic system.

*Dosage*: Initially 0.025 mg/kg body weight/day in divided doses, increasing to 0.05 mg/kg/day in divided doses. Higher doses may be prescribed in older adolescents with acute hypomanic or schizophrenic states.

*Side-effects*: Drowsiness. Extrapyramidal symptoms, especially acute dystonic reactions, sometimes occur with relatively low dosage. Apathy, depression. Dry mouth, nasal congestion. Orphenadrine 50 mg t.d.s. can be used to prevent recurrences of extrapyramidal symptoms if they occur, but should not be used routinely.

*Comment*: Main usage of this drug in child and adolescent psychiatry is in treatment of multiple, complex tics, especially Tourette syndrome, in which symptomatic improvement, though not usually complete relief, can often be achieved. It can be used in similar dosage to treat severe outbursts of aggressive behaviour, especially in the mentally handicapped.

It is also sometimes helpful in treating severe agitation and aggressive outbursts in adolescents with autism.

Other neuroleptic agents sometimes used in psychotic states include thioridazine, trifluoperazine, and fluphenazine. In maintenance treatment of schizophrenia, 'depot' injections of fluphenazine decanoate may be used at two-weekly intervals. Pimozide can be used with severe, multiple complex tics.

### Stimulant medication

Stimulant medication is of value in pervasive, severe forms of the hyperkinetic syndrome (attention-deficit disorder). It is also effective in narcolepsy. In hyperkinetic (attention-deficit hyperactivity) disorders it should be used in combination with counselling, behaviour modification, possibly dietary restriction, and, if necessary, special educational measures.

Methylphenidate.
*Mechanism*: Activates reticular arousal system. Children with the hyperkinetic syndrome may show underarousal, although findings are inconsistent. This explanation of mechanism is therefore inadequate, and the effects of stimulants remain largely unexplained.
*Dosage*: Initial dose 0.15 mg/kg in one dose before breakfast, increasing to 0.3–0.6 mg/kg in two divided doses, before breakfast and at lunchtime.
Approximate initial dose in 4–8-year-olds is 5 mg at breakfast. Increase stepwise by 5 mg increments every 3–4 days up to a maximum of 10 mg breakfast and lunchtime. It is of interest that in North America dosages used are generally higher (roughly double) those described here.
*Side-effects*: Reduction in appetite. Temporary growth retardation may occur. Excitement, muscular twitching. Insomnia.
*Comment*: This drug is now available in the UK by special arrangements with the manufacturers (Ciba). Dexamphetamine sulphate can be used as an alternative in approximately double the dosage recommended above for methylphenidate. Adverse reactions with stimulants may be reduced by using medication only on schooldays, i.e. not at weekends or in school holidays. Addiction has not been recorded with use of this medication in childhood. To avoid abuse, it is particularly important not to prescribe large quantities, and to monitor usage regularly. This medication should not be used for overactive children with tics or twitching may be worsened.

### Antidepressant medication

Antidepressant medication should only be used in children and adolescents in combination with counselling and psychotherapy. It should be reserved for those depressive disorders where there is severe and persistent depression of mood with either reduction in energy or agitation. In these circumstances, the tricyclic drugs (imipramine and amitriptyline hydrochloride) are indicated. In general, their use for this purpose in children under the age of 10 years is not advised.

Imipramine hydrochloride.
*Mechanism*: Affects action of neurotransmitters, especially serotonin and noradrenaline, probably by blockage of pre- and postsynaptic receptors. However, this is now regarded as only a partial explanation of their mechanism, which remains largely undetermined.
*Dosage*: 2–3.5 mg/kg/day in two or three divided doses. Approximate initial dose in 10–15-year-olds is 25 mg b.d., increasing by 25 mg increments to 50 mg t.d.s. in adolescents.
*Side effects*: Dry mouth and drowsiness are common. Blurred vision, constipation, nausea, postural hypotension, sweating, rashes. In high doses cardiotoxic effects occur.
*Comment*: Use in depressive disorders marked by apathy and anergia. ECG monitoring is required if doses of 3.5 mg/kg daily or higher are applied. Therapeutic effects occur only after 7–14 days' medication. Doses of 25–50 mg nocte may be used for symptomatic relief of *bed-wetting* over periods of 1–2 weeks in children aged 7 years or more.

Amitriptyline hydrochloride.
*Mechanism, dosage, side-effects* as with imipramine hydrochloride. Prescribing main dosage at night before sleep reduces the risk of drowsiness in the day.
*Comment*: Use in depressive disorders marked by agitation and restlessness. Therapeutic effects only occur after 7–14 days' medication. Dosage of 25–50 mg *nocte* may be used for symptomatic relief of bed-wetting over periods of 1–2 weeks in children aged 7 years or older. Relapse usually occurs when the drug is discontinued.

Clomipramine is another tricyclic drug used in similar dosage and found to be of especial value in obsessional disorders (Leonard *et al.* 1988) and in treatment of narcolepsy.

Other antidepressant drugs used in depressive disorders of childhood and adolescence include the monoamine oxidase inhibitors, especially phenelzine. Because of the risk of serious side-effects due to dietary non-compliance, use of these drugs is not advised in this age group.

In the rarely occurring bipolar affective disorders in older children and adolescents, lithium carbonate can be effective in the prevention of relapses (Delong and Aldersdorf 1987). Monitoring to avoid overdosage and side-effects is complex, and should be carried out by those experienced in its use in adult patients.

## Anticonvulsant medication

A description of the range of anticonvulsant medication is outside the scope of this book. The use of anticonvulsants in episodic psychiatric disorders in childhood without clinical evidence of seizure activity is not advised. The presence of EEG abnormalities alone is not thought to be sufficient ground for use of such medication for behavioural purposes.

All types of anticonvulsant medication may have adverse as well as occasionally beneficial effects on behaviour, emotional development, and learning in children with epilepsy. Their use should be closely monitored with this in mind (Section 5.7.2), and difficult behaviour or learning difficulties should indicate the possible need for a change in medication.

## Monitoring medication

All forms of medication require careful monitoring, especially in the initial stages of their use. Where appropriate, plasma levels should be measured. Observations by parents and children themselves will be a most valuable source of information. The clinician should, however, also make use of his own observations, as parents may, for one reason or another, be biased in favour of or against the use of medication, and this may affect their perception of its effects. Drugs are, however, unlikely to be effective if parents or children themselves are opposed to their use. When it is reported that children are refusing medication, this sometimes indicates that parents are ambivalent about its use.

If the child is at school, teachers should always be informed that a child is on medication. Their observation of the effect of drugs on the child's learning and school behaviour will always be valuable in monitoring dosage, and should be regularly obtained.

It is obviously not advisable to continue with medication unless it is producing benefit. This will usually be within 3 or 4 days with antipsychotic agents, and 2 or 3 weeks with stimulants and antidepressants. Even if a particular drug has had an initial beneficial effect, an attempt should be made to see how the child copes without the drug as soon as this is appropriate. The point in time at which discontinuation should be attempted will vary with the medication (e.g. after about 3 months with antidepressants, after about 6 months with stimulants, after 2–3 years fit-free with anticonvulsants).

Unfortunately, children sometimes stay on long-term medication for many years if periodic monitoring of this type does not occur.

## 6.2.7 MISCELLANEOUS THERAPIES

## Hypnotherapy

Hypnosis produces a special state of awareness in which an individual is particularly likely to accept from another person, or from himself, suggestions resulting in alterations of sensation or behaviour. Such alterations may then be generalized and experienced in ordinary awareness. There is some debate over whether hypnosis produces an altered state distinct from that produced by other relaxation therapies.

Although the popular view of hypnosis is that it involves one individual, the hypnotist, imposing his will on a subject, perhaps against the subject's will, the therapeutic use of hypnosis for children consists of a very different process. The therapist instead acts as a coach to teach the subject self-hypnosis in order to achieve for himself alterations in sensation or behaviour. The technique used usually involves relaxation—imagery and biofeedback as part of self-hypnosis (Olness 1989).

After an assessment procedure in which an indication for hypnosis is established, the procedure that is to be undertaken is explained to the child: 'I am going to help you to relax and then to imagine things that will (help your pain, improve your skin rash, make your headache less painful or less frequent, etc.)'. The child is then taught how to achieve muscle relaxation and encouraged to concentrate on images that will result in symptom relief. Imagery that might be effective is arrived at by a discussion between the child and the therapist. Thus the therapist might suggest the child imagines he has a lever inside his head that can control pain, and the child might indicate the lever could work better if, when he thought of it, he also imagined he was lying on a beach in the sun. The child is encouraged to practise self-hypnotic techniques by himself when symptoms occur in everyday life outside the consulting room.

Children of 5 years and even younger can be taught self-hypnosis and indeed, because imaginative skills peak at around the age of 9 years, children are particularly suitable subjects for this form of treatment. The techniques have been used either as the main form of treatment or as an adjunct to therapy in a wide range of common and less common disorders in childhood, including enuresis, headaches, fear of injections, and organically determined as well as psychogenic pain.

There is limited but clear-cut evidence for its effectiveness in a range of situations including pain control. More detailed information on the use of hypnosis in childhood (not to be undertaken without a period of training) is available in Olness and Gardner (1988). The establishment of a hypnotherapy service based in a psychiatric department in a children's hospital has been described by Sokel *et al.* (1990).

## Dietary treatment

Dietary measures are used in a wide variety of childhood disorders, and their use often has implications for psychosocial development. Thus, in metabolic disorders such as phenylketonuria and galactosaemia, diets low in phenylalanine and galactose are preventive against the development of mental retardation. In contrast, calorie restriction in obesity and calorie control in anorexia nervosa are examples of dietary treatment in conditions with a significant psychological component in aetiology (see Section 2.2). Additive-free and other more restrictive diets have also been used more recently in the treatment of the hyperkinetic syndrome (see Section 2.7.2). Such diets have also sometimes been found helpful in the treatment of other stress-related disorders, such as migraine and asthma, but their value in these conditions remains uncertain.

### Electroconvulsive therapy (ECT)

This form of treatment is very rarely indicated indeed in children and adolescents. However, it may be beneficial in youngsters with severe depressive illnesses, characterized by stupor, marked verbal or motor retardation, and nihilistic or paranoid delusions, where these have failed to respond to antidepressant medication. It is also effective in catatonic states occurring as manifestations of schizophrenia.

ECT should be administered to adolescents only by staff working in centres with experience of this form of treatment. Although the use of ECT in teenagers appropriately produces ethical concerns, there is no reason to withhold the treatment, merely because of age considerations, in those patients likely to benefit symptomatically from it. For a fuller discussion of the use of ECT in adults, see Gelder *et al.* (1989, pp. 679–89).

# Further reading

Baird, G. and Hall, D. M. B. (1985). Developmental paediatrics in primary care: what should we teach? *British Medical Journal* **219**, 583–6.

Goldsten, S. E., Yager, J., Heinicke, C. M., and Pynoos, R. S. (ed.) (1990). Preventing mental health disturbances in childhood. American Psychiatric, Washington, D.C.

Kolvin, I., Garside, R. F., Nicol, A. R., MacMillan, A., Wolstenholme, F., and Leitch, I. M. (1981). *Help starts here: the maladjusted child in the ordinary school.* Tavistock, London.

Offord, D. R. (1987). Prevention of behavioural and emotional disorders in children. *Journal of Child Psychology and Psychiatry* **28**, 9–20.

Rutter, M. (1982). Psychological therapies in childhood: issues and prospects. *Psychological Medicine* **12**, 723–40.

Shaffer, D., Ehrhardt, A., and Greenhill, L. (1985). *The clinical guide to child psychiatry.* Free Press, New York.

# 7

# Services

## 7.1 Primary and secondary health care

### Introduction

Children with psychosocial problems and their families are professionally served by health, education, and social welfare services. In this chapter, a description based on UK services is provided, but it is hoped that this will be of some relevance to those working in other countries. Within the developed world, the main variations in health care delivery lie in primary medical care. In some countries, such as the UK, some European (e.g. Scandinavian) countries, and Australasia, primary medical care is largely provided by a family practitioner service, whereas in others, such as the USA and other European countries such as France and Germany, primary care paediatricians mainly provide this service. The pattern of services throughout the developed world is otherwise rather similar, and the following account of UK services should be of relevance to those working in many other countries.

Primary health curative services in the UK, then, are provided by family practitioners and health visitors. They may refer to secondary health services, especially general paediatric and specialist paediatric services including child mental health facilities such as hospital departments of child psychiatry and child and family psychiatric clinics in the community. Preventive services, including the monitoring of children at risk, early detection of disorders such as hearing impairment before they present, and preventive procedures such as immunization are provided largely by community child health doctors. During the 1990s, this function will be gradually taken over by the primary care services, so that there will be closer integration of preventive and curative services. The detailed assessment and management of developmental delays and chronic disorders take place in hospital- or community-based child development centres, and the multidisciplinary service for such children is now usually directed by a consultant community paediatrician.

### Psychosocial work and primary care

Psychosocial problems and physical problems with an important psychosocial component form a large part of general practice consultations with children. One study (V. Bailey *et al.* 1978), found that 'pure' psychological problems accounted for 4 per cent of such consultations, physical symptoms with definite psychological problems for 7 per cent, and physical symptoms with possibly important psychological factors for a further 16 per cent. A further 9 per cent of consultations were for 'undifferentiated' disorders in which psychological factors were probably relevant. In all, including overlap, 32 per cent of child referrals to general practitioners were for one of these four reasons. As about 90 per cent of 0–4-year-olds and 60 per cent of 5–14-year-olds attend their general practitioners at least once a year, this means that family doctors regularly see a high proportion of their child patients for psychosocial reasons. Independent assessment of child attenders at family practitioners suggests that about a quarter show psychiatric disorder (M. Garralda and Bailey 1986) and, not surprisingly, disturbed children are more likely to present with complaints such as anxiety and bed-wetting. Parents of disturbed children are particularly likely to be socially stressed at the time of consultation.

There are a number of special characteristics of these primary care consultations. The child is usually accompanied only by the mother, there is usually a physical presenting symptom, and the consultation is usually relatively brief (in the UK lasting an average of about 10 minutes). The disorders that present are often difficult to classify using a formal psychiatric classification.

### Facilities
The frequency of primary care consultation for psychosocial reasons means that the facilities available to conduct these should be adequate. There should be a good supply of toys and books suitable to divert the child, and learn about his developmental level as well as something of his fantasy life (Section 1.2). Such toys and books need to be as carefully looked after as the sphygmomanometer and tendon hammer, and put away when not in use. There should be sufficient privacy for confidential material to be communicated without being overheard.

### The consultation
The principles of assessment for a psychosocial problem are described elsewhere (Section 1.2). In primary care, evaluation inevitably has to be rapid, though it is helpful if, when the need arises, the practitioner can put aside time for a longer consultation. Concentration on the reason for attendance at the present time, on the onset, nature, and severity of the presenting symptom, and on the presence or absence of other family problems is usually profitable. The nature of treatment suggested will depend on the type of problem presented. Encouragement to return if the problem persists will enable the practitioner to identify those chronic disorders that require referral. Many family practitioners and primary care paediatricians have difficulty in identifying appropriate children for referral (Dulcan *et al.* 1990), and there is an important need for the training of doctors in primary care for psychosocial aspects of their work.

Experienced health visitors and family practitioners will, for example, be alert to the fact that a mother who brings a child with an apparently trivial physical complaint may be using this symptom as an 'admission ticket' for a discussion of a problem that is more important to her (e.g. a behaviour problem in the child, or a marital difficulty) which she would be reluctant to raise without some prompting.

### Links with other agencies

Primary medical care involving psychosocial problems is likely to be much more effective if there are opportunities for close working with other professionals and agencies. The health visitor (district community nurse) working as part of a primary health care team will, in fact, often be able to take responsibility for a major component of this work. Attachment of social workers to general practice is becoming more frequent. Regular outreach visits from child psychiatric agency staff to primary care facilities have occasionally been described. Especially in small communities, the general practitioner will be able to make useful contacts with local schools and social workers.

## Community child health services

### Child health clinics

These are run both by the community health services and increasingly by family practitioners for surveillance and preventive activities. They are staffed professionally by doctors and health visitors. The preventive service offered includes vaccination and immunization, weighing and measuring, physical (especially sensory) examination, and developmental checks. Over 90 per cent of babies are brought in the first year of life to these clinics in the UK, but subsequently, in most areas, attendance drops off fairly sharply. Although most work in these clinics is directed towards preschool children, some authorities also run special clinics on the same premises for school-aged children. Special clinics for children with enuresis and obesity are those most commonly arranged.

While it is generally agreed that regular surveillance of a child population for checks on hearing, vision, etc. is desirable, there is less agreement about the value of identifying behaviour and emotional problems (Hall 1989). However, check-lists suitable for this purpose in young children have been developed (N. Richman *et al.* 1982), and found of value especially by health visitors. Bearing in mind their frequency, significance, and impact on family life, it is important for health visitors and doctors conducting surveillance examinations to enquire routinely about the presence of behaviour and emotional disorders.

There are significant psychosocial aspects to the delivery of a comprehensive, preventive programme. Children living in disorganized families, of low socio-economic status, and children from minority ethnic groups are most likely to be missed. An outreach system with careful record-keeping can, nevertheless, achieve good results even in deprived areas. There is evidence that child-centred counselling of mothers of infants and toddlers based in family practice can reduce rates of behaviour and emotional problems occurring subsequently (see Section 6.1).

There is in any event a large psychosocial component to work in child health clinics. Mild and moderate developmental delays are often produced by understimulation and poor social circumstances. Early feeding problems of psychosocial origin and failure to thrive are commonly identified as causes for concern in children seen in clinics. They offer an opportunity for counselling of mothers in child-rearing practices. In toddlers, behaviour problems such as sleep difficulties (Section 2.3) and other adjustment problems are often mentioned to doctors or health visitors by mothers in need of advice and counselling.

Such clinics should contain attractive and well-kept play material as well as adequate equipment for weighing and measuring, developmental assessment, sensory testing, and physical examination. There should be a room available that can be used for discussion of confidential matters. Good communication between community health doctors and nurses and general practitioners will ensure that if conditions requiring treatment are identified, therapy can be readily arranged.

### *The school medical service* (Whitmore 1985)

This is staffed largely by school nurses and doctors, speech therapists, and physiotherapists. In most authorities, all children are medically screened on school entry (Whitmore and Bax 1986) but, subsequently, screening examinations are usually selective. In some authorities even the school entry medical examination is selective, based on a screening procedure carried out by the school nurse (E. M. O'Callaghan and Colver 1987). Regular examinations are more likely to be held for children attending special schools, such as schools for the physically or mentally handicapped.

As well as the medical examination of children with learning problems and behavioural difficulties so as to identify underlying physical causes (e.g. sensory defects), staff of the school health service may also be asked to ensure that teachers understand the nature of physical disorders suffered by children they are teaching. Children may be referred to them because of suspicion of ill-treatment or neglect. Although the school psychologist is likely to be the main source of advice, school doctors and nurses may be asked about children with emotional and behaviour problems, especially if the child has an associated physical condition. They may also be asked to provide information to teachers to help them run classes on health education, sex education, birth control, and preparation for parenthood. Teachers may also require help in identifying and dealing with drug abusers. Members of the school health service are required to assess and produce medical reports on children thought to be in need of special education (Section 7.4.2).

For most of these tasks, school doctors and nurses require training in the psychosocial aspects of their work. Good communication with general practitioners, hospital paediatric services, social workers, educational psychologists, and other staff working in child psychiatric units is essential.

*The district handicap team*

In most districts in the UK, a district handicap team has been formed consisting of a community paediatrician, community nurse, educational psychologist, social worker, speech therapist, and physiotherapist. Other specialists, such as a hospital paediatrician, audiologist, orthopaedic surgeon, occupational therapist, and child psychiatrist may also attend as required, and sometimes do so regularly. Children suspected of chronic developmental disorders, physical or mental handicap, and sensory deficits are often referred to such teams for further assessment and advice, especially concerning future schooling. Such teams are often sited in a hospital- or community-based child development centre, where children are assessed and may attend subsequently on a daily, weekly, or less frequent basis, for further assessment and treatment. Mental handicap and learning disorders, epilepsy, and behaviour problems are the diagnoses most frequently made in children attending child development centres (Bax *et al.* 1988), so the contribution of mental health professionals is clearly essential.

As already indicated (Chapter 4), children suffering from physical and mental handicap have high rates of emotional and behavioural disorders. Parents of children with disabilities are also in need of counselling. There is therefore a strong case for those working in such units to have well-developed skills in the assessment of psychiatric disorders and in parent counselling. Close links with the child psychiatric service are desirable, and there are considerable advantages in a regularly occurring psychiatric liaison consultation service (Evered *et al.* 1989).

## Hospital out-patient care

In about 5–10 per cent of children attending general paediatric clinics, a mental health problem is the main reason for attendance (M. Garralda and Bailey 1989), though psychosocial factors are prominent in a much higher proportion. About a quarter of such children are showing concomitant psychiatric disorder at the time of referral, and the families of disturbed children are particularly likely to be under stress. Vomiting and feeding difficulties, failure to thrive, abdominal pain, constipation, and chronic headache all very frequently have strong psychosocial implications and are common reasons for referral.

There are many advantages to the provision of separate out-patient facilities for children and, in developed countries, these are increasingly available. The availability of trained paediatric nursing staff, and the provision of appropriate decoration, toys, and books are more likely to occur if there are separate out-patient facilities for children. As psycho-social problems are so common, it is important that privacy is available so that parents can feel comfortable when communicating confidential information.

The distress of anticipating painful procedures in child out-patients can be reduced in a number of ways. A satisfactory appointment system geared to the needs of families has been achieved in some out-patient departments. Obviously, medical staff may sometimes be required to attend to emer-gencies, but planning ahead can sometimes reduce this occurrence. The provision of play staff in the out-patient department can do much to reduce unnecessary anxiety, especially if there are unexpected delays. Preparation for procedures is as important in out-patient as in in-patient care. Children with chronic disorders should, if at all possible, be seen for review appoint-ments by the same member of staff whenever they attend. The continuing attendance of children merely because they have 'interesting' conditions should be discouraged. There should be clear aims for every out-patient attendance and, if junior staff are involved, these aims should be made explicit to them. If satisfactory care can be carried out by the general practitioner or health visitor, with a referral back to hospital as necessary, this should always be preferred to continuing hospital or out-patient appointments.

Ready availability of a paediatric medical social worker and member of a child psychiatric team for appropriate referrals will also improve the quality of care available for children seen in out-patients who have psycho-social components to their problems. Psychosocial aspects of the care of hospitalized children are discussed in Section 4.6.

The considerable psychosocial component in paediatric practice means that there is a real need for paediatricians to receive training in psycho-social aspects of their work (P. Graham and Jenkins 1985).

# 7.2 Mental health services for children

## Constituents of a comprehensive mental health service

A comprehensive mental health service for children and adolescents may be expected to include the following.

### A *child and family psychiatric clinic or out-patient department*
This may either be situated in the community, usually in a health centre or education authority premises, or form part of the local hospital service. The department is usually staffed by a child psychiatrist, psychologist (educational and/or clinical), and social worker. In some parts of the UK,

especially around London, child psychotherapists may also be members of clinic staff. The clinics generally take **referrals** from a range of other professionals—especially general practitioners, paediatricians, teachers, and social workers (including probation officers). Clinics situated in education authority premises usually have a high proportion of school referrals with behaviour problems and learning difficulties. Health-centre-based services see a high proportion of young children and children with physical complaints thought to be of psychological origin.

**Assessment** and **treatment** are usually provided by all the professionals working in the clinic, with a variable degree of specialization (see below). In some clinics, most families are seen by more than one professional (working conjointly); in others this is unusual. Most children and families are seen in the clinic, but home-visiting is a prominent feature of the work of some clinics. A relatively new development in the UK is the establishment in some districts of a community child psychiatric nursing service. In most clinics a good deal of emphasis is placed on outreach **consultation** with other agencies, so that often as much as one-third of a child mental health professional's time is spent discussing problems with other professionals, especially teachers and social workers (Black *et al.* 1974).

## In-patient units

The need for specialized psychiatric residential assessment and treatment facilities for children and adolescents will depend on the availability of other residential facilities. Thus, a high level of special residential educational provision for 'maladjusted' children and ready availability of substitute care by social service departments, for children whose parental care has broken down, will reduce the need for in-patient facilities. Nevertheless, a proportion of severely disturbed children will require such specialized psychiatric facilities. These include:

1. Children with complex paedo-psychiatric disorders (e.g. non-organic seizures, anorexia nervosa, encopresis) which have failed to respond to outpatient treatment or paediatric management.

2. Children where home care has broken down or threatened to break down and who are, for example, too aggressive or too anxious to be placed in social service or educational facilities.

3. Children with severe psychiatric disorders, for example psychotic depression and schizophrenia, who, it is thought, would benefit from the special skills available in an in-patient unit.

The age range of children admitted to psychiatric in-patient units usually ranges from 5–6 years up to puberty, and in adolescent units from puberty up to 17 or 18 years. There is, however, much variation. The main assessment and therapeutic input is usually provided by skilled nurses with psychiatric and/or paediatric training as well as the other mental health disciplines. Schooling is usually available on the premises of the unit, and other professionals, for example occupational therapists and speech therapists, may play a very active part.

The **assessment and treatment** philosophy of in-patient units varies considerably. Some admit only a highly selected group of disturbed children; others are much less selective. Some base their treatment mainly on psychoanalytic principles, others on behavioural principles. Most are eclectic, and put a good deal of emphasis on working with the family and trying to ensure that gains made in the unit are maintained at home. On discharge, a significant proportion of children are transferred to other residential units, especially residential schools for children with emotional and behaviour problems.

*Psychiatric day centres*
These are less frequently available, but such centres for the assessment and treatment of disturbed children have become more frequently established in the UK since the early 1970s (Bentovim and Lansdown 1973). Children are admitted for whom continuing home care is practicable and desirable, but who, it is thought, will benefit from a full daily programme of care. In the younger, preschool age group, these are often children with severe developmental disorders and management problems. In older children, school refusal and behavioural disorders in the school setting are common reasons for admission.

Most day centres cater for a fairly narrow age range—preschool (2–5 years), middle childhood (6–12), or teenage (13 years or older). Centres dealing with the preschool group usually involve the parents very actively in the treatment programme, and some will accept the child only if one or both parents attend at the same time. Treatment usually consists of both group and individual therapy, with the type of therapy depending on the orientation of the clinic. There is an increasing tendency to use focused behavioural methods.

The staffing pattern is very variable. Some are staffed mainly by social workers and others by nurses and nursing aides. There is also variable input from mental health professionals, but in many cases there is committed input from one or two psychiatrists or psychologists. There is a variable amount of contact with paediatric facilities, special schools, and day nurseries depending on where the centre is sited and the ages of the children admitted. Evidence is lacking for the effectiveness of such centres (N. Richman *et al.* 1983) in terms of symptom improvement, but they have an important and valuable assessment function, and delayed development may be accelerated (Cohen *et al.* 1987). Further, children and families are helped and supported over crises and escalations in chronic conditions by attendance at such centres.

*Walk-in centres*

There is a variety of other types of assessment and treatment facilities for disturbed children and adolescents. Walk-in centres for adolescents, who might not be prepared to accept conventional patterns of referral through general practitioners, etc., are becoming more common. These are most commonly staffed by social workers and volunteer counselling staff, sometimes with a mental health professional available for consultation. A small number of such facilities mainly staffed by psychologists and psychiatrists have been described.

## Mental health professionals

The following description is applicable in the UK, although a similar pattern of professional work and training exists in other developed countries.

**Child and adolescent psychiatrists.** Consultants have had specialized training in child psychiatry in this field for 3½–4½ years, following a medical education, general medical and surgical experience, and experience in general (adult) psychiatry. In the field of child mental health, although his work will cover the whole range of child psychiatric disorders, the psychiatrist has a special contribution to make where the presenting complaints are somatic, there is a possibility of psychosis, autism, or other serious mental disorder in the child, one or other parent is suffering from a mental illness, or there is a possibility that medication might be helpful to the child.

**Child psychologists.** There are two types of child psychologist in the UK—educational and clinical. Normally the educational psychologist will have initially obtained a teaching qualification and then taught in ordinary or special school for some time. He or she will then have taken a 1-year postgraduate course in educational psychology. He/she will spend a good deal of time working in schools, and will have special skills in the evaluation of learning and behavioural problems and in the development of remedial programmes. Following a first degree in psychology, and (usually) some practical experience in a health service or social service setting, clinical psychologists have taken a 2- or 3-year postgraduate course in clinical psychology. The special skills of the clinical psychologist lie in systematic assessment of behavioural deviance, and the development and evaluation of behavioural programmes, but many have also developed skills in counselling and psychotherapy. Many clinical psychologists work in the field of mental handicap, where their work particularly involves devising individualized and group programmes for the promotion of social and cognitive development.

**Social workers** have usually taken a first degree and had some practical experience before embarking on the certificate of qualification in social work (CQSW). Their special skills lie in the evaluation of family situations in which the child is at risk, and there is a need for family support, with concern that the child may not be receiving adequate care and protection. They have statutory responsibilities where a child is seriously at risk. Their counselling skills are likely to be highly developed.

**Child psychotherapists** are normally expected to have a first degree, followed by some practical experience in working with children in a field such as teaching, before embarking on a course lasting at least 4 years that includes a personal psychoanalysis. Their special skills are in individual psychotherapy with children, and assessment of the role psychodynamic factors are playing in psychological or psychosomatic problems, but increasingly they are using their skills in consultation and liaison services.

**Common skills.** All child mental health professionals (except for child psychotherapists, who are usually more specialized) have shared skills in the systematic assessment of behavioural and emotional disorders of children. They are all skilled in the counselling of both individual parents and children. Most are likely to be able to apply both psychodynamic and behavioural methods in the treatment of mental health problems in children and families. They are able to provide consultation with professionals working in other agencies. In many child and family psychiatric clinics and departments, a substantial proportion of the work is carried out interchangeably by members of all the disciplines represented.

## Organization of child mental health services

This varies considerably from country to country, and the following description applies only to the UK.

Health districts provide a child and adolescent psychiatric service. This is staffed by child psychiatrists (health authority employed), social workers (employed by the local authority's Social Services Department), and educational psychologists (employed by the local authority Education Department). Clinical psychologists and psychotherapists are employed by the health authority. These professionals all work together, usually in premises maintained by the health authority (hospital or health centre), or local education authority. Each health district and local authority determines its own pattern of service to ensure that, as far as possible, a full range of residential and community facilities is available, if not within the area at least reasonably close by. Disturbed children and adolescents may be placed outside the health district in a regional in-patient unit. A local education authority may arrange for disturbed pupils to attend a special school run by its own or another authority.

Each clinical department determines its own pattern of work. Thus the way of dealing with referrals, sharing of work, exercise of individual clinical responsibility, supervision, etc. is locally arranged. In most clinics, a substantial amount of work is 'extramural' and involves consultation with a number of different agencies outside the clinic, including ordinary and special schools, paediatric services, and the juvenile courts. In many areas, the demands made on the child mental health service mean that it is overstretched and priorities have to be established. This usually involves an emphasis on care for the most severely disturbed, on the use of short-term focused treatment procedures, and on consultation with other non-mental health professionals to enable them to work more successfully with disturbed children and their families.

## 7.3 Social welfare services and child protection

The majority of parents are able to provide conditions for their children that are adequate or more than adequate to enable them to thrive physically, emotionally, and intellectually. A significant minority are, however, unable to provide this level of care. Indications that children are not receiving adequate care include evidence of physical neglect, failure to thrive (Section 2.2.3), slow intellectual development (Section 2.6.2), as well as non-accidental injury and emotional and sexual abuse (see Section 1.3). Some of these problems may, of course, be due to factors other than inadequate parental care. Certain types of behaviour, such as primary attachment disorders (Section 2.1.3), may also indicate emotional deprivation.

As well as these signs in the child of inadequate parental care, there are also a number of risk factors in the family and environment that should alert the professional to the possibility that care might be inadequate. Professionals should, however, be wary of jumping to conclusions concerning inadequacy of care on the basis of the presence of one or more of these risk factors. Many children brought up in anomalous circumstances do in fact receive excellent care. There are, for example, many one-parent families in which the quality of child care is better than in some two-parent families. The following is a brief summary of family and environmental factors that suggest a child may be at risk of inadequate care.

1. *Parental marital disharmony.* About one in five children are brought up in families where the parental relationship is characterized by constant disagreements, quarrelling, and interpersonal tension.

2. *Marital breakdown and divorce.* About one in four children will have experienced the breakdown of the parental marriage by the age of 16 years. In many cases there will have been a prolonged period of marital disharmony, sometimes with parental separation, before the breakdown occurs.

3. *Single parent families.* About one child in ten is, in the early 1990s, brought up by one parent, and in about one case of single parenthood out of ten, the single parent is the father. Single parents vary greatly in the

amount of social support they receive from the other parent, from family members, friends, neighbours, and professional agencies.

4. *Parental mental illness*. The impact of parental mental health problems, such as depression, disturbed personality functioning, alcoholism, and psychosis is discussed in Section 1.3.

5. *Poverty*. It has been estimated that about one in eight children in the UK are living at or below the poverty line, calculated on the basis of their parents' entitlement to income support.

6. *Poor housing*. Many children live in unsatisfactory, cramped accommodation with inadequate indoor or outdoor space to play. About 1 per cent of children in the UK are living in accommodation made available to homeless families by local authorities.

7. *Racism*. Children of ethnic minorities readily identifiable by their colour are frequently subject to harassment by other children and neighbours. The widespread extent of racial harassment has been well documented.

It is a central task of the social worker, usually employed in the local authority's Social Services Department to monitor and improve the social welfare of children whose care is at risk or inadequate. The Children Act 1989, whose provisions are described in more detail below, describes parental inadequacy in terms of lack of 'reasonable parental care' resulting in the child suffering significant harm or being likely to suffer significant harm. However, all other professionals in contact with children and families have a duty to identify situations of unsatisfactory child care, to provide what help they can, and, in situations of serious risk and inadequacy, to refer the child to the local Social Services Department. Various means of informal and formal protection for the child are possible.

## 7.3.1 FAMILY SUPPORT

The family unit is the main source of protection for children. Consequently, the main aim of those involved in ensuring the care and protection of children is to support the cohesiveness of the family unit and prevent family breakdown. Various approaches to family support are possible. In all cases voluntary action is regarded as preferable to compulsory, legal measures, but in certain circumstances, the law has to be invoked for the protection of the child.

### Anticipatory action

The anticipation of crises in family life can diminish adverse effects on children. Thus, if a woman with schizophrenia becomes pregnant, anticipatory arrangements for the care of her baby, by the provision of extra support, fostering, or adoption will be necessary. Such action will be indicated in many other circumstances, for example on the release of a violent father from prison or when a single parent has to go into hospital for a planned operation.

## Support for families in crisis or under chronic stress

1. *Counselling and case work.* Case work has been defined by M. A. G. Brown (1966), as a 'helping activity which is made up of a very large number of constituent activities ranging from the giving of material assistance, through listening, expressing acceptance and reassurance, suggesting, advising and the setting of limits, to the making of comments that encourage the client to express his feelings, to examine his situation or to see connections between his present attitudes and behaviour and past experiences'. Clearly, case work has much in common with psychotherapy but, broadly speaking, has more supportive and fewer interpretive features. Since the 1960s, there has been a tendency for social workers to focus more on achievement of behavioural change in their clients, than on change of attitude or personality.

2. Professionals can ensure that families are in receipt of those *statutory benefits* to which they are entitled. Social workers are also able to assess financial need in families and, if appropriate, arrange for payment of discretionary benefits. Health professionals should be aware of benefits to which parents of handicapped children are entitled. For example, in the UK, attendance allowance is payable to parents of children aged 2–16 years who are severely disabled and require a great deal of extra attention.

3. *Other practical assistance.* Professionals can also assist children of families under stress by checking that they are receiving appropriate health care and education, or in other ways, for example by arranging placements for preschool children in day nurseries or school-age children in holiday play schemes. Children in problem families or with chronic disabilities may be considerably helped by the availability of family centres or the provision of a family aide who may visit the family daily and provide practical assistance as well as counselling.

4. *Brief respite care.* Families can also be supported in times of stress by temporary arrangements whereby a child spends a short period in a children's home, in day or short-term foster care, or, in the case of children with unusual nursing needs, in hospital. Such care may be particularly helpful for handicapped children who may place extra strain on a family's resources.

## Child-minding

Most child-minding is provided by other family members, but a substantial proportion is carried out for a daily charge by women who look after a small number of children while their mothers go out to work. The quality of such day care varies considerably, from the highly satisfactory to the very inadequate. Some children receive care that is a considerable improvement on the care they receive in their own home, while others may be looked after in a crowded, poorly illuminated room with no toys and little contact with adults. Child-minding is regulated in the UK by the Children Act 1989. Local authorities have a duty to keep a register of people who use their homes to look after children aged under 8 years for reward. Inspection of the premises and of the circumstances in which children are being looked after must be carried out at least once a year. Children whose mothers go out to work and arrange adequate substitute care with child minders do not appear to be significantly emotionally or intellectually advantaged or disadvantaged by the experience (P. Graham 1990), although clearly their opportunities for making a wider range of relationships with both adults and other children are likely to be increased.

### Day nurseries, nursery schools, playgroups

Day nurseries, run by Social Services Departments and staffed mainly by nursery nurses, provide care especially for children whose mothers are in special need, through social disadvantage or illness. They may take children of any age up to school entry. Nursery schools, run by local authority Education Departments, provide more formal learning experiences for children aged from 3 to 5 years. Playgroups, usually run by voluntary agencies, provide less formal experience often for a limited number of half days per week, for children usually aged between 2½ and 5 years.

The quality of care in all these types of establishment varies considerably, and consequently it has not been possible to draw general conclusions concerning the value of attendance.

### 7.3.2 SUBSTITUTE CARE FOR CHILDREN

Children may be placed in substitute care on residential basis either voluntarily or following court proceedings.

### Children's homes

Substitute care for children in children's homes is now generally seen as an unsatisfactory form of care, except on a short-term basis. Nevertheless, there remain a number of children for whom other forms of care are unavailable. In contrast to former times, children's homes now usually contain relatively small numbers of children (usually six to twelve) and provide much more personalized care. Unsatisfactory features of some children's homes in the UK include rapid turnover of staff and a high proportion of relatively inexperienced, untrained staff.

#### *Outcome*
Early descriptions of the cognitive and personality outcome of children reared in children's homes painted a dark picture. The mortality of infants

admitted into such homes was high. A high proportion of older children were seriously disturbed and had poor educational attainments. In adulthood, the rate of personality disorder was also raised. These findings related to children who had been brought up in large and often impersonal homes in the pre-war years.

Following the Second World War, the quality of care in children's homes has improved. Staffing ratios were improved, and staff showed increased awareness and sensitivity to psychological problems shown by the children themselves. Nevertheless, staff turnover usually remained high, so that children still often did not have the opportunity of a continuous relationship with one or two adults. Rates of behaviour and emotional problems in children in children's homes remained elevated (Wolkind and Rutter 1973). Early cognitive development, including language development (Tizard *et al.* 1972), was usually normal, though in later childhood educational retardation was common.

Follow-up studies of girls reared in children's homes in the 1960s into adulthood suggest that they have high rates of psychiatric problems (Wolkind and Kruk 1985) and show more inadequacy in caring for their own children than do controls (Quinton and Rutter 1985). Nevertheless, a high proportion have a reasonably favourable personality development; their capacity to look after their own children is much enhanced if they have a supportive husband.

These findings point to the need to provide family care and, if possible, adoptive care for children whose biological parents are unable to look after them. Inevitably, because of problems such as difficulty in obtaining parental consent, family care will not be feasible in some circumstances. The training and continuity of staff working in such homes remain a matter of concern.

### Permanency planning

The consequences of the provision of inadequate substitute care and delays in decision-making by social workers, courts, and indeed parents themselves, have led to a realization that there is always a need to make a long-term plan for children whose parents cannot take full responsibility for them. The concept of 'permanency planning' has been introduced with this in mind. Those concerned with the welfare of such children, including mental health professionals, need to ensure that children do not drift from one set of unsatisfactory temporary placements to another.

### Foster care

This involves the care and maintenance of a child by a person who is not a relative, guardian, or custodian. Long-term fostering is the usual arrangement when care is no longer possible for the child in his biological family, but arrangements for adoption cannot be made. Although, in some circumstances, the consent to adoption by biological parents who cannot themselves look after a child is no longer necessary, there are still a number of situations in which adoption is not feasible. Local authorities have statutory duties concerning the selection of foster parents and supervision of foster care. While most fostering arrangements are dictated by impending or actual family breakdown, fostering has also been employed positively as a means of rehabilitating adolescents with antisocial problems (Hazel 1977).

*Outcome*
Some children spend the whole of their childhood and adolescence in a single long-term fostering placement. Others have multiple foster placements over a long period of time, and, for yet others, a brief period of foster care may be a single interruption in a childhood otherwise spent with the biological family. It is not surprising therefore that follow-up studies of children who have been fostered are few and that outcome is very mixed. It seems likely that children who have multiple foster placements have a high likelihood of developing disturbance, but it is improbable that this can be attributed entirely to fostering. Such children usually experience, in addition, numerous separations, deprivation, and inconsistency of care (Berridge and Cleaver 1987).

Nevertheless, certain special problems concerned with fostering have emerged. Children undergoing this experience are particularly likely to be confused about their relationship to their original family as well as to the foster family. Their uncertain and temporary status often creates particular insecurity. They may be sensitive to rejection and respond to threats of further change with angry or anxious reactions.

There are various implications of these findings. The selection of foster parents needs careful appraisal, though in some communities, identification of sufficient, suitable families to foster children presents difficulties. The access of biological parents to their children while they are in foster care is potentially confusing, and children exposed to this experience need frequent explanations as well as the opportunity to express their own ideas and fantasies about the situation. As far as possible, where repeated foster placements are necessary for a child, they should be arranged with the same family. Foster parents who are temporarily fostering children with chronic physical disorders or mental handicap need to know as much about the management of the disorder as do the biological parents. If, as is often the case, the fostered children show signs of psychiatric disorder, their foster parents need to have easy access to appropriate mental health professional advice. Finally, the possibility of adoption should always be borne in mind as a preferable alternative to fostering if circumstances make it possible.

## Adoption

For children whose natural parents have no prospect of providing adequate care for them in the future, it is generally agreed that adoption into another family with permanent and legally binding transfer of parental rights is the most satisfactory procedure. Adoption proceedings can be taken by a local authority Social Services Department, a registered adoption agency, or a parent or guardian direct. Once parents have adopted a child, they have the same rights and duties as biological parents.

Applications to adopt a child are considered by a local authority Adoption Care Committee, which carries out social work and medical assessments of the child and of the prospective adoptive parents. Once a baby or older child suitable for adoption and acceptable to the parents has been identified, the child is placed for a probationary period of at least 3 months. The prospective parents apply for an adoption order, and a 'guardian ad litem', usually a social worker, is appointed by the court to ensure the child's interests are represented until a decision is made by the court. The consent of the biological parent to the adoption is required, though in certain special circumstances, for example mental illness, mental handicap, or prolonged parenting incapacity, consent may be dispensed with.

The easier availability of contraception and more permissive legislation for abortion has meant that there is a smaller number of babies available for adoption to infertile couples. Consequently, a much higher proportion of adoptions now involve more 'difficult to place' children. These include older, often disturbed children, those of mixed race or black parentage, and children with mental and/or physical handicaps. There are specialist adoption agencies to supervise arrangements for adoption of these children with special needs.

### Outcome

The educational and behavioural outcome of children adopted in the first few months of life and reared throughout childhood in the same adoptive family is very similar to that of a biological child brought up in the same circumstances. If the family is able to provide reasonably warm, consistent care and the child is not exposed to unusual stress, the chances of problematic behaviour are low. Nevertheless, there may be some differences. A great majority of adopted children have normal intelligence, but the correlation between the IQ of adopted children and their adoptive parents is lower than that between non-adopted children and their biological parents. Adopted children, especially adopted boys, have slightly higher rates of behavioural disturbance than non-adopted children, and these differences increase towards adolescence (Lambert and Streather 1980). However, rates in adoptive children are still distinctly lower than those in children brought up in homes broken by divorce or separation, or in single parent families. As adopted children reach adolescence, they do experience more uncertainty and confusion about their identity than children reared in biological families, but this does not usually lead to serious problems.

Children adopted later in childhood do not have such a favourable outcome. Nevertheless, such children are capable of making warm, positive attachments to adoptive parents, and late-adopted children who have previously been mainly reared in children's homes do better behaviourally than if they are returned to their biological families or stay in children's homes (Tizard 1977). At one point, it was not thought possible or desirable for physically and mentally handicapped children to be adopted, but this practice has now become much more common, and follow-up suggests that the great majority of such adoptions are highly successful, at least if they are carefully planned (Wolkind and Kozaruk 1983). There has been concern about the fate of black children adopted into white families. Behaviourally and intellectually, such children usually do very well. They are, however, more likely to show confusion over racial identity than other black children, though not necessarily other identity problems (Gill and Jackson 1984). Adoption does not, of course, affect the likelihood of children inheriting strongly genetically determined conditions such as schizophrenia and certain metabolic disorders from their biological parents.

The implications of these studies of adopted children are reasonably clear-cut. Adoption is, in general, a successful practice for children for whom it is obvious early on that care by biological parents is not feasible. Later adoption has a less good outcome, but is preferable to other alternatives. There is no reason why handicapped children should not be adopted, provided suitable parents can be identified, prepared to take the responsibility with knowledge of the problems they will face. Though this is a controversial area, it is better for black children to be adopted into suitable black families if they are available, but, if they are not, adoption into white families is much preferable to alternatives with less expectation of permanency.

Changes in legislation and social work practice in the 1980s mean that a number of biological parents are retaining contact with their children after they have been adopted. Adopted individuals now have the right, once they reach adulthood, to obtain information about their biological parents. These changes require careful monitoring to ensure that the generally excellent results of adoption are not prejudiced by them.

## 7.3.3 LEGISLATION

The system of legislation concerning the welfare of children in England and Wales (the Scottish system is somewhat different) has been simplified by the 1989 Children Act, which came into force in October 1991. The following legal procedures can be invoked for the protection and welfare of children.

1. *Emergency Protection Order (EPO).* This is a short-term order for a maximum of eight days that can be applied for if urgent action is required to protect a child. Anyone can make an application for an emergency protection order but, in fact, application will usually be made to a magistrate by a local authority or NSPCC social worker though it may also be made to a judge. The court must be satisfied that the child is likely to suffer significant harm unless he is removed from where he is to another place, or he is kept where he is (Department of Health 1989). Following the granting of an EPO, the applicant must allow reasonable contact with parents unless the court attaches extra provisions to the order relating to access.

2. *Child Assessment Order.* This can be made for up to seven days by a magistrate after receiving an application from a local authority or NSPCC social worker. The court must be satisfied that the applicant has reasonable cause to believe that the child is suffering, or is likely to suffer harm, that an assessment is required to enable the applicant to determine whether this is indeed the case, and that it is unlikely that a satisfactory assessment can be made unless an order is obtained.

3. *Care, interim care, and supervision orders.* These orders can be made for a longer period following application by a local authority or NSPCC social worker. The child must be suffering, or be likely to suffer, significant harm because of a lack of reasonable parental care, or because he is beyond parental control. The question what is 'reasonable' parental control is likely to give rise to considerable legal debate, but it is the intention that parents who are handicapped in their parenting by mental ill-health or low intelligence to such a significant degree that they require help with their children, will be expected to show that they are prepared to accept such help (Department of Health 1989). The philosophy of the Children Act is to expect parents to be responsible and, before parental rights can be removed to any degree, it has to be shown that significant harm is attributable to lack of parental care.

A 'Care Order' removes parental rights and puts the Social Services Department *in loco parentis*. However, some degree of parental responsibility is retained by the parents. They are expected to retain their duty to care for the child and raise him to moral, physical, and emotional health in all ways that are compatible with the care order. Even though a child may be 'in care' and removed from the family, social workers will make every attempt to work with the family, respect their wishes, etc. Children 'in care' may live at home under supervision of the Social Services Department.

A form of legal protection less extreme than a 'Care Order' is a 'Supervision Order' that a magistrate may also grant to a Social Services Department. This order, often made in relation to adolescents in trouble, provides a social worker such as a probation officer with the legal right to advise and assist the child and family while the child lives at home.

3. *Local authority accommodation.* Local authorities have a duty under Section 20 of the Children Act 1989 to provide accommodation for children whose parents are unable, for one reason or another, to look after them, or who have no one taking parental responsibility for them. The parents of children who are looked after voluntarily in local authority accommodation in this way do not lose their parental rights or responsibility and, except in certain unusual circumstances (for example if the child is over 16 and objects to removal), may remove their children from such accommodation.

4. *Wardship.* In cases where an individual has good reason to believe that a child's welfare is at risk, then an application can be made to the High Court for a child to be made a 'ward of court'. It is expected that wardship proceedings will become unusual when the provisions of the 1989 Children Act are fully operative, as most or perhaps all previous use of this procedure is now covered by the Act.

5. *Separation and divorce.* Children whose parents are separating or intending to divorce are entitled to the same legal protection as other children and indeed must be regarded as requiring special attention as far as their care is concerned. Parents engaged in divorce proceedings are encouraged to make a joint statement of the arrangements they propose for their children. These statements are considered by the district judge, who, perhaps assisted by a court welfare officer, decides whether it would be in the child's interests for a specific order under Section 8 of the 1989 Children Act to be made, or whether the parental arrangements proposed are satisfactory and can be agreed without further legal action.

Section 8 orders may specify various conditions on behalf of the child's welfare. A 'residence order' states with whom the child is to live. A 'contact' order requires the person with whom the child lives to allow the child to visit or stay with the person named in the order—usually, but not always, this will be the other parent.

Normally both parents will be expected to retain their parental responsibility to care for the child and raise him to physical, moral, and emotional health, in so far as this is compatible with any specific agreed arrangements or orders that are made.

Disputes regarding residence, contact, or other matters may be settled informally or by recourse to the courts. Professionals may be asked to provide expert advice in cases of conflict, and some principles on which advice can be given are provided below.

### Consent to and refusal of treatment

Normally, with children under the age of 16 (minors), parental consent is legally necessary to carry out any medical procedures such as prescribing medication, admitting the child to hospital, or performing an operation. It is also, in the great majority of cases, good practice for parents to be fully involved in such decisions concerning their children. In the UK, it was ruled in 1984 that, for certain procedures, i.e. the prescription of contraceptives to girls under the age of 16 years, practitioners could, in unusual circumstances, where this seemed in the best interests of the girl, act without the knowledge and consent of parents. Further, when a parent withholds consent to treatment in such a manner that a child's health or welfare are clearly endangered, then an application can be made for a Care Order as described above. If the delay involved in obtaining such an order would put the child's health at serious risk (e.g. when parental religious beliefs preclude a life-saving blood transfusion), then, in the UK, emergency treatment may be undertaken.

Normally, if doctors and parents are agreed on a course of action for a child involving the giving or withholding of treatment, it can be assumed that the agreed decision is in the child's best interests. In certain cases, however, such as those involving decisions to withhold life-saving treatment from babies with conditions that would result in later severe handicap, or very poor quality of life, it is legally open to Social Services Departments to take action, under the 1989 Children Act, such as application for an Emergency Protection Order. Most authorities believe that, at least in the great majority of cases, decisions concerning the treatment of young children should be left to parents and their professional advisers.

The position regarding consent to and refusal of treatment in older children is more complex. It is good practice to obtain both the permission of parents and the assent of children to all procedures. When a child thought to be capable of forming a sufficiently mature judgement refuses treatment, despite understanding all the implications of this decision, then parents cannot now override the wishes of the child. In these circumstances, the provisions of the 1983 Mental Health Act, which allows for compulsory admission and/or treatment, if the child's safety is at risk, may need to be invoked. This Act requires an application by the nearest relative or an approved social worker, as well as medical recommendations for compulsory admission or treatment.

### 7.3.4 ROLE OF PROFESSIONALS IN CHILD WELFARE

#### Social workers

Social work in relation to children and families has many aims. In the community, the social worker will see it as his duty to provide both practical and emotional support to families where the child is at risk of neglect or other forms of harm. He will invoke legal procedures for the protection of the child when this seems necessary, and will often act as the 'keyworker' when a number of professionals whose activities need to be coordinated, are concerned with caring for a family.

Specialism in social work was much reduced in the UK after the publication of the *Seebohm Report* (Department of Health and Social Security 1968), but in most local authority Social Services Departments, a significant amount of specialism continues. Hospital social workers attached to paediatric units from Social Service Departments, apply social work skills to families with children with physical and psychosocial problems. Some social service departments have specialist units concerned with family placement (adoption, fostering, etc.) and residential care. Child and family psychiatric departments (Section 7.2) usually have social workers attached full-time to their multidisciplinary teams.

Although the majority of social workers are employed by the local authority Social Services Department, voluntary agencies, such as the National Society for the Prevention of Cruelty to Children (NSPCC), also provide a significant number. Some are employed by voluntary bodies with a specific interest in a particular health problem, such as visual impairment or epilepsy.

### Health professionals

It is the responsibility of all health professionals to inform a social worker if they are seriously concerned for the future safety of a child, or have reason to believe that a child has been neglected, injured, or otherwise abused. Doctors and nurses are sometimes understandably reluctant to make referrals to social workers because they fear, often realistically, that, if they do, their positive relationship with the parents will be threatened. Clearly, there will be situations where there is only insubstantial suspicion of neglect etc. when a health professional may feel prepared to accept the heavy responsibility for not informing a social worker about his concerns.

**General practitioners and paediatricians** undertake responsibility for the medical examination of infants and children who are to be placed for adoption. It is important that prospective adoptive parents are aware of any special health risks of the child they are preparing to adopt and, if the child has a physical or mental condition, its likely course. In providing this information, the paediatrician will need to have access to all available data on the health of the parents, any other family history, and details of the pregnancy, birth, and any postbirth complications or illnesses. In past days, in questionable cases, the adoption of babies was often considerably delayed to see whether an infant's early development proceeded normally. It is now realized this is misguided, because of the benefits of early adoption (see above). Prospective parents should obviously be told if there are realistic doubts about a child's development, so that they can make as informed a decision as possible.

Paediatricians or general practitioners examine children admitted into the care of local authorities to determine whether the child needs medical treatment, has any transmissible disease or injuries, or shows evidence of neglect (Bamford 1979). Nude weight and height (or length) should also be carefully obtained as a baseline for future assessment.

**Psychiatrists** may be requested to advise concerning what type of placement for a child will provide the best conditions for personality development and least risk of behavioural or emotional disorder. They may also be asked to assess the relevance of a possible mental disorder, either in terms of the risk of such disorder in the child, or in relation to the fitness of a mentally ill adult to parent. Criteria that can be used in making these judgements are described below.

The psychiatrist may also be involved in consultation concerning procedures for assessing the suitability of substitute parents, especially prospective foster and adoptive parents. Finally, he may be asked to provide an opinion for children following separation and divorce arrangements, especially where there is dispute between parents concerning custody or access.

### Conflicts concerning child care

Although social workers have prime responsibility for decision-making concerning appropriate placement for children and for assisting the courts where legal procedures are taken, health professionals, such as paediatricians and psychiatrists, may also be asked for advice. In considering what advice to give, major considerations must be the identification of the environment which will be most favourable to personality development and reduce the risk of behavioural and emotional disorders (Wolkind 1984). The following considerations will be relevant:

1. The quality of the existing placement in terms of the health and personality of the main people caring for the child, the strength of attachment of the child to them, and the quality of health care and social stimulation they are providing. The assessment of the fitness of a parent suffering from mental illness to parent a child requires the special knowledge of a psychiatrist knowledgeable in the likely outcome of different disorders (Oates 1984).

2. The quality of the proposed alternative placement using the same criteria. This is often more difficult to appraise because of the hypothetical nature of the judgements involved. Nevertheless, it is usually possible to make some informed judgement of the quality of care likely to be available.

3. The age of the child and the length of time he has been in his existing placement. Children aged from 6 months to about 3 or 4 years are especially likely to show prolonged adverse reactions to a change in placement, unless this is very carefully arranged with the new parental figures only gradually taking over care.

4. The temperament of the child and the presence of already existing behaviour and emotional disorders. Children with adverse temperamental characteristics, especially those who are volatile, easily upset, and irregular in their eating and sleeping habits are more in need of continuous, warm care than other children, with perhaps more phlegmatic temperaments.

5. The presence of a physical disorder. Children with physical conditions, especially those involving brain dysfunction, are particularly liable to develop behavioural and emotional disorders when under stress.

6. The degree to which prospective 'parents' can accept their own limitations, and are motivated to turn to outside help if the child-care situation becomes intolerable is relevant. The availability of such help if it is required is also of relevance.

A fuller description of the factors that mental health professionals will wish to take into account in giving an opinion in cases of disputed adoption, residence, and contact orders, etc. is provided by Black *et al.* (1989).

## Preparation of reports to the court

Increasingly, when courts are dealing with cases of abuse or suspected abuse, or where there is conflict between parents over contact or the residence of a child, health professionals including psychiatrists and paediatricians are requested for their views. In these circumstances, as far as is possible, the professional should see his main duty as promoting the welfare of the child.

Useful guidelines on the writing of court reports are provided by Black *et al.* (1989). The report should open with the name and date of birth of the child, and with a statement of the professional position of the person writing the report and how he became involved in the case. The introduction should continue by stating the questions the person writing the report was asked to address, and the way the assessment was carried out. Usually this would involve interviews with the various members of the family and professionals concerned, but a report may be prepared entirely on the basis of an examination of documentation if this is agreed beforehand.

An account should be given of the events in the child's life before the assessment, the major problems that have occurred, and the measures previously taken to deal with them. The assessment interviews should be reported with particularly detailed accounts of the relevant sections of the interviews, making it clear whether remarks made were spontaneous or in response to specific questioning. An account of the results of any investigations, psychological or physical, that have been carried out should be added.

There should be a summary of the information, a conclusion and a list of recommendations. The feasibility of any recommendations concerning therapy should be checked out before they are made. The report should always be signed and a list of relevant documents that have been consulted appended.

# 7.4 School influences and special education

## 7.4.1 SCHOOL INFLUENCES

Attendance at preschool facilities, such as playgroups, nursery schools, etc. is often of social advantage to children while they are attending them, but evidence for long-lasting advantage in terms of better educational attainment later on is not clear-cut. There is also little evidence that, for most children, attendance makes a significant difference to emotional development. Probably a minority of young children find preschool experience more stressful than it is worth, while others, perhaps those with difficulties in achieving independence, may benefit considerably. Mothers of preschool children who go out to work (and whose children therefore require substitute care) are less likely to be depressed than those without employment. Maternal employment, however, makes little or no difference to rates of behavioural problems in preschool children (P. Graham 1990).

In older children, the characteristics of the schools they attend have been shown to have an important effect (Rutter *et al.* 1979*b*). Some schools obtain consistently better academic results for their pupils in examinations, as well as lower rates of delinquency and behaviour problems, and higher attendance rates than do other schools. The reasons lie partly in the type of pupil intake, but schools themselves also make a difference. The internal organization of schools is more important than physical features such as the size of the school and the age of the school buildings (Maughan 1988).

School characteristics that make a positive difference include (Rutter *et al.* 1979*b*).

(1) placing emphasis on rewards and praise for good work and behaviour;
(2) providing a comfortable environment for children in which they feel free to approach staff for help;
(3) ensuring that responsibility is shared out between different children and that all feel they are participating in an important aspect of school life;
(4) placing emphasis on the importance of academic work with teachers preparing lessons conscientiously, turning up punctually for lessons, marking homework promptly;
(5) ensuring teachers act as good models for behaviour;
(6) effective management of children in groups, rather than concentrating on them as individuals;
(7) a strong school organization with general agreement on goals and disciplinary rules to be applied.

In contrast, the level of physical resources, the age and structure of the buildings, size of class, and the degree of emphasis on punishment are of rather little importance.

The processes whereby certain school characteristics lead to better outcomes are unknown, but the impact they have on child self-esteem and sense of self-efficacy may well be relevant (Rutter 1989*b*).

## 7.4.2 SPECIAL EDUCATION

The first attempts to provide special schooling were made in the late nineteenth century when facilities were set up for children with physical and mental handicaps. From that time until the 1960s, in most developed countries an increasing amount of separate, special educational provision was created. Children were largely allocated to facilities on the basis of their medical diagnosis—deafness, blindness, etc. At this point it became clear that this approach was unsatisfactory (Department of Education and Science 1978) because:

(1) medical diagnosis was often a poor guide to educational need;
(2) many children had multiple handicaps, and their needs were poorly met in schools set up for one disability alone;
(3) many children in ordinary schools were discovered to have special needs that were not being met at all;
(4) separate education seemed to have many disadvantages for children; the main dangers were seen to be stigmatization, lack of opportunity to mix with non-handicapped children, and unnecessarily low expectations for educational performance.

As a result, there has been a move towards integrating handicapped children into ordinary schools, though it is recognized that some children with very special needs will still need to be educated separately. Placement is now based on educational need rather than medical diagnosis. This trend is exemplified by the passing of Public Law 94/142 in the USA, and the Education Act (1981) in the UK.

### Types of special educational need

As many as one in six children will have some sort of special educational need at some stage in their school careers (Department of Education and Science 1978). Types of help needed may include:

(1) a more favourable staff ratio to provide extra time to give to the child;
(2) teachers with special training to deal with children with particular handicaps;
(3) the ready availability of other professionals, for example speech therapists, psychologists;

(4) special equipment, for example to assist children with sensory dis-
    orders;
(5) a special classroom regime; for example, some children may benefit
    from a highly structured, and others from a particularly permissive
    atmosphere;
(6) special building features, for example the building may need to have
    ramps to be suitable for a child in a wheelchair.

## Special educational provision

It is agreed that ideally there should be a 'continuum of special educational
provision' to match the continuum of special educational needs that children
show (E. Anderson 1976). A useful review of types of educational interventions for children with learning disabilities in the USA is provided by Lerner
(1989). Currently, in the UK such provision consists, or should consist of:

1. Extra help provided in the ordinary classroom. This might involve,
for example, the employment of a teaching aide to work part-time, or the
application of a behaviour modification programme in the classroom, by
the teacher in consultation with a psychologist.

2. Part-time withdrawal of the child to a small remedial group held
within the school. Extra help for children behindhand with reading is often
provided in this way.

3. Part-time withdrawal to a class held in a separate unit or school.
Units of this type exist for children with more serious learning or behavioural
problems as well as for those with hearing problems.

4. Full-time special education in a separate day school or unit. Special
schools are available in most education authorities for the following types
of problem:

(a) Mild mental retardation. Schools for children with moderate learn-
    ing difficulties.
(b) Moderate and severe mental retardation. Schools for children with
    severe learning difficulties.
(c) Psychiatric disorders. Schools for children with emotional and be-
    haviour difficulties.
(d) Physical disability. Schools for the physically handicapped (P. H.
    schools), and for specific sensory disabilities.
(e) Autism and communication disorders. Schools for autistic and aphasic
    children.

In addition, so-called 'schools for the delicate' also exist in some areas for
children with milder physical, learning, and behavioural disorders. Some
ordinary schools and schools for children with learning difficulties have
'assessment units' where young preschool children with handicaps can be
placed, assessed, and educated before moving on to ordinary or special
schools.

5. Full-time special education in a separate residential school. These exist for the same categories as day schools. Residence may be on a weekly or termly boarding basis.

## Assessment for special education

As part of their work, teachers assess children's progress and identify those with learning and behavioural problems. Occasionally they feel the need for outside help, and in these circumstances will first call on help that is available from within the school, for example from a specialist teacher, a year head, head teacher, or the school nurse. If it seems as though a child requires further assessment, then outside professionals—the school doctor, an outside advisory teacher, psychologist, or psychiatrist—may be invoked. These professionals are, however, likely to be involved with only a minority of children with problems, though in the USA in particular there is increasing demand for psychologists and psychiatrists to work in a consultation role in schools (Jellinek 1990). The type of involvement will obviously depend on the nature of the problem, but it may include a personal assessment of the child, or merely a discussion between the outside professional and members of the school staff. Consultation with teachers can be carried out along similar lines to those followed with other professionals.

Most children with educational problems will be educated with relatively minor special arrangements in ordinary schools, though in a number of such children, behavioural problems of some significance are not identified and an opportunity for mental health consultation is lost (Mattison *et al.* 1986). In a small minority (perhaps about 2 per cent), it becomes clear that this is insufficient. If children have a need for special education or provision, then a full assessment must be carried out leading to a 'statement' of the child's special educational needs (Education Act 1981). The aim of the assessment should be to make clear what the child needs and how this can be provided, while retaining as far as possible components of normal education and contact with ordinary children without special needs.

There are two main pathways to a full assessment. In some children, mainly those with severe learning difficulties and physical disabilities, it is clear well before they go to school that they will need special education. In others, especially those with mild and moderate learning difficulties, the need becomes apparent only when the child has been at school for some time. Health professionals have obligations especially involving the first group (A. MacFarlane 1985). If a doctor thinks a child he is seeing is likely to require special education, then he should:

(1) discuss the matter with the parents;
(2) in the case of children over the age of 2 years, inform the education authority of his view of the child's needs;

(3) if appropriate, suggest the part a voluntary organization might play in helping.

In children under 2 years, the education authority should only be contacted with parental agreement. With a child over the age of 2 years, parental agreement is highly desirable, but not mandatory.

The outcome of the full assessment should consist of five main components. There should be reports from the child's teacher, an educational psychologist, and health professionals, a statement from the parents about what they consider to be the child's special needs, and finally a statement from the education authority summarizing the child's needs and what special provision is thought necessary to be made available. Parents are expected to see the final statement as well as the reports on which it is based. If they do not agree with the statement they have the right to appeal.

Health professionals should provide relevant information that will both assist in the preparation of the statement of the child's educational needs and be useful to the teachers responsible for the child's further education. Their main task is to summarize the child's physical and psychological needs, particularly for any special professional help (e.g. speech therapy, psychotherapy), medication, diet, and aids to daily living. If the child needs any special classroom regime such as a highly structured or permissive setting, this should be stated.

It may be important for those receiving the statement of educational need to be aware of family problems that might be affecting the child's learning. For example, continuous social work support may need to be arranged. In these circumstances it is often difficult to phrase an account of the situation in a way that is acceptable to parents who, of course, will read the report. However, it is usually possible to frame statements in a way that makes it clear that, though the parents are finding it difficult to be, for example, as tolerant and appropriately protective to their children as they should be, they are nevertheless doing the best they can for their children. Communication of potentially shameful or embarrassing information may pose problems. Such information can only be disclosed with parental consent, and indeed it is always helpful to tell parents roughly what one is intending to put into a report before one writes it. It is usually possible to phrase statements in such a way that teachers receiving the information can know that there are serious domestic problems in a child's background, without going into details of desertion, alcoholism, marital infidelity, or other sensitive matters.

Health professionals should not attempt to make *educational* recommendations themselves—that is the responsibility of the education authority. They may, however, sometimes wish to engage in behind-the-scenes advocacy on behalf of a child to ensure that delay in placement is not too long, or that an apparently very appropriate placement is brought to the attention of the education authority.

Contributions to assessment are, of course, only a small part of the role played by health professionals, paediatricians, nurses, psychiatrists, etc. in educational medicine. In particular, such professionals are important in helping teachers and others in ordinary and special schools to understand and cope with the physical and psychiatric problems of children in their care. With parental permission, information relevant to teacher's concerns needs to be transmitted to school doctors via the school health service following in-patient admissions and out-patient attendances. Visits to schools by health professionals working in hospitals can be especially helpful when a child's physical or psychiatric problems are causing unusual concern.

### The value of special education for disturbed children

A discussion of the teaching methods used with children with emotional and behavioural problems is outside the scope of this book (but see Laslett 1977). A wide diversity of methods is used, often strongly influenced by psychodynamic or behavioural theories. The personal warmth and concern of the teacher are major influences. Other important factors are thought to include firm discipline not based on punishment, a well-thought-out academic programme, the opportunity of continuity of relationships provided by staff stability, and the availability of personal counselling for the student.

There are few evaluative studies of special education. However, Wrate *et al.* (1985) studied the effects of different types of education on a sizeable group of children with neurotic and conduct disorders. They found that children treated in ordinary schools did better than those placed in special ESN(M) schools and schools for the maladjusted, though children with conduct disorders did not fare well in any setting. This study needs replication, for obviously, if confirmed, the findings have major implications for the organization of special education. Rutter and Bartak (1973) were able to show that autistic children in schools with a structured, goal-setting approach to education made more cognitive gains than those in schools with a more permissive relationship-fostering approach. The capacity of the children to make relationships did not differ between the two types of school. There is a need for further evaluative studies—perhaps a fitting expression to end this book because it applies to so many aspects of child psychiatry.

## Further reading

Department of Education and Science (1978). *Special educational needs*. HMSO, London.

Graham, P., Pearson, L., and Sheldon, B. (1982). National case study (England and Wales). Child mental health and psychosocial development. *Newsletter, Association for Child Psychology and Psychiatry, London* 7, 2–18.

Modell, M. and Boyd, R. (1982). *Paediatric problems in general practice*. Oxford University Press, Oxford.

National Children's Bureau (1987). *Investing in the future*, Report of the Children's Policy and Practice Review Group. National Children's Bureau, London.

# References

Abramowitz, C. V. (1976). The effectiveness of group therapy with children. *Archives of General Psychiatry* **33**, 320–6.

Agnarsson, U. and Clayden, G. (1990). Constipation in childhood. *Maternal and Child Health* **15**, 252–6.

Ainsworth, M., Blehar, M. C., Waters, E., and Wall, S. (1978). *Patterns of attachment: a psychological study of the strange situation.* Lawrence Erlbaum, Hillsdale, New Jersey.

American Academy of Pediatrics, Section of Urology (1975). The timing of elective surgery on the genitalia of male children with particular reference to undescended testes and hypospadias. *Pediatrics* **56**, 479–83.

American Psychiatric Association (1987). *Diagnostic and statistical manual of mental disorders (3rd edn—revised).* American Psychiatric Association, Washington, DC.

Anderson, E. (1976). Special schools or special schooling for the handicapped child. The debate in perspective. *Journal of Child Psychology and Psychiatry* **17**, 151–5.

Anderson, E., Clarke, L., and Spain, B. (1982). *Disability in adolescence.* Methuen, London.

Anderson, H. R., Dick, B., MacNair, R. S., Palmer, J. C., and Ramsey, J. C. (1982). An investigation of one hundred and forty deaths associated with volatile substance abuse in the United Kingdom (1971–1981). *Human Toxicology* **1**, 207–21.

Anderson, J., Williams, S., McGee, R., and Silva, P. A. (1986). The prevalence of DSM III disorder in a large sample of pre-adolescent children from the general population. *Archives of General Psychiatry* **44**, 69–74.

Apley, J. and Hale, B. (1973). Children with recurrent abdominal pain: how do they grow up? *British Medical Journal* **3**, 7–9.

Apley, J. and MacKeith, R. (1968). *The child and his symptoms.* Blackwell Scientific Publications, Oxford.

ASBAH (1983). *Sex for young people with spina bifida and cerebral palsy.* Association for Spina Bifida and Hydrocephalus with the co-operation of the Spastics Society. Scottish Spina Bifida Association and the Committee for Sexual and Personal Relations of the Disabled, London.

Assessment of Performance Unit (1978). Assessing the performance of children. *DES Report on Education No. 93.* Department of Education and Science, London.

Atkins, D. M. and Patenaude, A. F. (1987). Psychosocial preparation and follow-up for pediatric bone marrow transplant patients. *American Journal of Orthopsychiatry* **57**, 246–52.

Bailey, G. W. (1989). Current perspectives on substance abuse in youth. *Journal of the American Academy of Child and Adolescent Psychiatry* **28**, 151–62.

Bailey, V., Graham, P., and Boniface, D. (1978). How much child psychiatry does a general practitioner do? *Journal of the Royal College of General Practitioners* **28**, 621–6.

Baird, G. and Hall, D. M. B. (1985). Developmental paediatrics in primary care. What should we teach? *British Medical Journal* **291**, 583–6.

Baker, A. W. and Duncan, S. P. (1985). Child sexual abuse: a study of prevalence in Great Britain. *Child Abuse and Neglect* **9**, 457–68.

Baldwin, J. A. and Oliver, J. E. (1975). Epidemiology and family characteristics of severely abused children. *Journal of Preventive and Social Medicine* **29**, 205–21.

Bamford, F. N. (1979). Medical examinations of children admitted to care. In *Medical aspects of adoption and foster care* (ed. S. N. Wolkind), pp. 16–21. Heinemann, London.

Bamford, F. N., Jones, P. M., Pearson, D., Ribeiro, G. G., Shalet, M., and Beardwell, C. G. (1976). Residual disabilities in children treated for intracranial space-occupying lesions. *Cancer* **37**, 1149–51.

Bancroft, J., Axworthy, D., and Ratcliffe, S. (1982). The personality and psychological development of boys with 47 XXY chromosome constitutions. *Journal of Child Psychology and Psychiatry* **23**, 169–80.

Banks, M. and Jackson, P. (1982). Unemployment and risk of minor psychiatric disorder in young people: cross-sectional and longitudinal evidence. *Psychological Medicine* **12**, 786–98.

Barker, P. (1981). *Basic family therapy*. Granada, London.

Barkley, R. A. (1989). Hyperactive girls and boys: stimulant drug effects on mother–child interactions. *Journal of Child Psychology and Psychiatry* **30**, 379–90.

Bassuk, E. K. and Rosenberg, L. (1990). Psychosocial characteristics of homeless children and children with homes. *Pediatrics* **85**, 257–61.

Baumer, J. H., David, T. J., Valentine, S. J., Roberts, S. J., and Hughes, R. (1981). Many parents think their child is dying when having a first febrile convulsion. *Developmental Medicine and Child Neurology* **23**, 462–4.

Bax, M., Hart, H., and Jenkins, S. M. (1990). *Child development and child health: the pre-school years*. Blackwell Scientific Publications, Oxford.

Bax, M. C. O., Robinson, R. J., and Gath, A. (1988). The reality of handicap. In *Child health in a changing society* (ed. J. A. Forfar), p. 116. Oxford University Press, Oxford.

Bayley, N. (1969). *Bayley scales of infant development: birth to two years*. Psychological Corporation, New York.

Beales, J. G., Keen, J. H., and Holt, P. J. (1983). The child's perception of the disease and the experience of pain in juvenile chronic arthritis. *Journal of Rheumatology* **10**, 61–5.

Beck, A. T. (1967). *Depression: clinical, theoretical and experimental aspects*. Harper and Row, New York.

Beck, A. T., Rush, A. J., Shaw, B. F., and Emery, G. (1979). *Cognitive therapy of depression*. John Wiley, New York.

Bell, R. Q. (1974). Contributions of human infants to caregiving and social interaction. In *The effect of the infant on its caregiver* (ed. M. Lewis and L. A. Rosenblum). Wiley, New York.

Bellman, M. (1966). Studies in encopresis. *Acta Paediatrica Scandinavica*, Suppl. 170.

Belson, W. (1975). *Juvenile theft: the causal factors*. Harper and Row, London.

Belson, W. (1978). *Television violence and the adolescent boy*. Saxon House, Farnborough.

Bender, B. G., Belleau, L., Fukuhara, J. T., Mrazek, D., and Strunk, R. C. (1987). Psychomotor adaptation in children with chronic severe asthma. *Pediatrics* **79**, 723–7.

Bentovim, A. (1988). Understanding the phenomenon of sexual abuse: a family systems view of causation. In *Child sexual abuse within the family* (ed. A. Bentovim, A. Elton, J. Hildebrand, M. Tranter, and E. Vizard), pp. 40–58. Wright, London.

Bentovim, A. and Lansdown, R. (1973). Day hospitals and centres for children in the London area. *British Medical Journal* **4**, 536–8

Berg, I. (1970). A follow-up study of school phobic adolescents admitted to an in-patient unit. *Journal of Child Psychology and Psychiatry* **11**, 37–47.

Berg, I., Marks, I., McGuire, R., and Lipsedge, M. (1974). School phobia and agoraphobia. *Psychological Medicine* **4**, 428–34.

Berg, I., Butler, D., and Hall, G. (1976). The outcome of adolescent school phobia. *British Journal of Psychiatry* **128**, 80–5.

Berg, I., Consterdine, M., Hullin, R., McGuire, R., and Tyrer, S. (1978). A randomly controlled trial of two court procedures in truancy. *British Journal of Criminology* **18**, 232–44.

Berg, R., Svensson, J., and Astrom, G. (1981). Social and sexual adjustment of men operated for hypospadias during childhood: a controlled study. *Journal of Urology* **125**, 313–17.

Berger, M., Yule, W., and Rutter, M. (1975). Attainment and adjustment in two geographic areas. II. The prevalence of specific reading retardation. *British Journal of Psychiatry* **126**, 510–19.

Berridge, D. and Cleaver, H. (1987). *Fostering breakdown*. Basil Blackwell, London.

Berrueta-Clement, J. R., Schweinhart, L. J., Barnett, W. S., Epstein, A. S., and Weikart, D. P. (1984). *Changed lives: the effects of the Perry pre-school program on youth through age 19*. The High Scope Press, Ypsilanti, Michigan.

Bertagnoli, M. W. and Borchardt, C. M. (1990). A review of ECT for children and adolescents. *Journal of the American Academy of Child and Adolescent Psychiatry* **29**, 302–7.

Berwick, D. M., Levy, J. C., and Kleinerman, R. (1982). Failure to thrive: diagnostic yield of hospitalisation. *Archives of Disease in Childhood* **57**, 347–51.

Bewley, B. R. (1979). Teachers smoking. *British Journal of Preventive and Social Medicine* **33**, 219–22.

Bewley, B. R., Bland, J. M., and Harris, R. (1974). Factors associated with the start of cigarette smoking by primary schoolchildren. *British Journal of Preventive and Social Medicine* **28**, 37–44.

Bijur, P. E., Golding, J., and Haslum, M. (1988). Persistence of occurrence of injury: can injuries of pre-school children predict injuries of school-aged children? *Pediatrics* **82**, 707–12.

Bille, B. (1962). Migraine in schoolchildren. *Acta Paediatrica Scandinavica*, Suppl. 136.

Billings, A. G., Moos, R. H., Miller, J. J., and Gottleib, J. E. (1987). Psychosocial adaptation in juvenile rheumatic disease: a controlled evaluation. *Health Psychology* **6**, 343–59.

Bingley, L., Leonard, J., Hensman, S., and Lask, B. (1980). Comprehensive management of children on a paediatric ward. *Archives of Disease in Childhood* **55**, 555–61.

Bird, H. R. and Kestenbaum, C. J. (1988). A semi-structured approach to clinical assessment. In *Handbook of clinical assessment of children and adolescents*. (ed. C. J. Kestenbaum and D. T. Williams), pp. 19–30. New York University Press, New York.

Bird, H. R., Gould, S. M., Yager, T., Staghezza, B., and Canino, G. (1989). Risk factors for maladjustment in Puerto Rican children. *Journal of the American Academy of Child and Adolescent Psychiatry* **28**, 846–50.

Birley, J. and Brown, G. (1970). Crises and life change preceding the onset or relapse of acute schizophrenia: clinical aspects. *British Journal of Psychiatry* **116**, 327–33.

Bishop, D. V. M. (1983). *Test for reception of grammar*. National Foundation for Educational Research, Slough.

Bishop, D. V. M. (1985). Age of onset and outcome in acquired aphasia with convulsive disorder. *Developmental Medicine and Child Neurology* **27**, 705–12.

Bishop, D. V. M. (1987). The causes of specific developmental aphasia ('developmental dysphasia'). *Journal of Child Psychology and Psychiatry* **28**, 1–8.

Black, D. and Urbanowicz, M. (1987). Family intervention with bereaved children. *Journal of Child Psychology and Psychiatry* **28**, 467–76.

Black, D., Black, M., and Martin, F. (1974). A pilot study of the use of consultant time in psychiatry. *British Journal of Psychiatry, Supplement. News and Notes*, September, pp. 3–5.

Black, D., Wolkind, S., and Harris Hendriks, J. (1991). *Child psychiatry and the law*. Gaskell, Royal College of Psychiatrists, London.

Blagg, N. R. and Yule, W. (1984). The behavioural treatment of school refusal: a comparative study. *Behaviour, Research and Therapy* **22**, 119–27.

Blair, H. (1977). Natural history of childhood asthma. *Archives of Disease in Childhood* **52**, 613–19.

Bluma, S., Shearer, M., Frohman, A., and Hilliard, J. (1976). *Portage guide to early education: checklist*. NFER-Nelson, Windsor.

Blurton-Jones, N., Rossetti Ferreira, N. C., Farquhar Brown, M., and Macdonald, L. (1978). The association between perinatal factors and later night waking. *Developmental Medicine and Child Neurology* **20**, 427–34.

Bohman, M. (1978). Some genetic aspects of alcoholism and criminality. *Archives of General Psychiatry* **35**, 269–76.

Bolton, D., Collins, S., and Steinberg, D. (1983). The treatment of obsessive-compulsive disorder in adolescence. *British Journal of Psychiatry* **142**, 456–64.

Botvin, G. J., Erg, A., and Williams, C. L. (1980). Preventing the onset of cigarette smoking through life-skills training. *Preventive Medicine* **9**, 135–43.

Bowlby, J. (1971). *Attachment and loss. Volume 1. Attachment*. Penguin Press, Harmondsworth.

Bowlby, J. (1975). *Attachment and loss. Volume 2. Separation, anxiety and anger*. Penguin Press, Harmondsworth.

Bowlby, J. (1980). *Attachment and loss. Volume 3. Loss.* Penguin Press, Harmondsworth.

Boyle, M. H. and Offord, D. R. (1990). Primary prevention of conduct disorder. *Journal of the American Academy of Child and Adolescent Psychiatry* 29, 227–33.

Brazelton, T. B. (1962). A child-oriented approach to toilet training. *Pediatrics* 29, 121–8.

Brazelton, T. B., Koslowski, O., and Main, M. (1974). The origin of reciprocity: the early mother–infant interaction. In *The effect of the infant on its caregiver* (ed. M. Lewis and L. A. Rosenblum), pp. 49–76. John Wiley, New York.

Brent, D. A. *et al.* (1988). Risk factors for adolescent suicide. *Archives of General Psychiatry* 45, 581–8.

Breslau, N. (1985). Psychiatric disorder in children with physical disabilities. *Journal of the American Academy of Child Psychiatry* 24, 87–94.

Breslau, N. (1990). Does brain dysfunction increase children's vulnerability to environmental stress? *Archives of General Psychiatry* 47, 15–20.

Breslau, N. and Prabucki, K. (1987). Siblings of disabled children. *Archives of General Psychiatry* 44, 1040–6.

Breslau, N., Weitzman, M., and Messenger, K. (1981). Psychosocial functioning of siblings of disabled children. *Pediatrics* 67, 344–53.

Brett, E. (1983). *Paediatric neurology.* Churchill Livingstone, Edinburgh.

Brewster, A. B. (1982). Chronically ill children's concepts of their illness. *Pediatrics* 69, 355–62.

British Medical Association and Pharmaceutical Society of Great Britain (1985). *British national formulary.* British Medical Association, London.

Brooke, O. G., Anderson, H. R., Bland, J. M., Peacock, J. L., and Stewart, C. M. (1985). Effects on birthweight of smoking, alcohol, caffeine, socio-economic factors and psychosocial stress. *British Medical Journal* 298, 795–801.

Brown, B. (1979). Beyond separation: some new evidence on the impact of brief hospitalization in young children. In *Medicine, illness and society. Beyond separation: further studies of children in hospital* (ed. D. Hall and M. Stacey). Routledge and Kegan Paul, London.

Brown, D. G. (1972). Stress as a precipitant factor of eczema. *Journal of Psychosomatic Research* 16, 321–7.

Brown, G. W. and Davidson, S. (1978). Social class, psychiatric disorder of mother and accidents to children. *Lancet* i, 378–80.

Brown, G. W. and Harris, T. (1978). *Social origins of depression.* Tavistock Publications, London.

Brown, J. B. and Lloyd, H. (1975). A controlled study of children not speaking at school. *Journal of the Association of Workers with Maladjusted Children* 10, 49–63.

Brown, M. A. G. (1966). A review of case-work methods. In *New developments in casework* (ed. E. Younghusband), p. 15. George Allen and Unwin, London.

Bruininks, R. H. (1978). *Bruininks–Oseretsky test of motor proficiency.* American Guidance Service, Circle Pines, Minnesota.

Bryant-Waugh, R., Knibbs, J., Fosson, A., Kaminski, Z., and Lask, B. (1988). Long-term follow-up of patients with early onset anorexia nervosa. *Archives of Disease in Childhood* 63, 5–9.

Burke, P., Meyer, V., Kocoshis, S., Orenstein, D., Chandra, R., and Sauer, J. (1989). Obsessive–compulsive symptoms in childhood inflammatory disease and cystic fibrosis. *Journal of the American Academy of Child and Adolescent Psychiatry* **28**, 525–7.

Burton, L. (1974). *Care of the child facing death*. Routledge and Kegan Paul, London.

Burton, L. (1975). *The family life of sick children*. Routledge and Kegan Paul, London.

Buss, A. and Plomin, R. (1984). *Temperament. Early developing personality traits*. Erlbaum, Hillsdale, New Jersey.

Butler, N. and Alberman, E. (1969). The effects of smoking in pregnancy. In *Perinatal problems* (ed. N. Butler and E. Alberman), pp. 72–84. Livingstone, Edinburgh.

Byrne, E. A., and Cunningham, C. C. (1985). The effect of mentally handicapped children on families: a conceptual review. *Journal of Child Psychology and Psychiatry* **26**, 847–64.

Bywater, M. (1981). Adolescents with cystic fibrosis: psychosocial adjustment. *Archives of Disease in Childhood* **56**, 538–43.

Bywater, M. (1984). Coping with a life-threatening illness: an experiment in parents' groups. *British Journal of Social Work* **14**, 117–27.

Cade, B. (1984). Paradoxical techniques in therapy. *Journal of Child Psychology and Psychiatry* **25**, 509–16.

Cadman, D., Boyle, M., Szatmari, P., and Offord, D. R. (1987). Chronic illness, disability, and mental and social well-being: findings of the Ontario Child Health Study. *Pediatrics* **79**, 805–13.

Cadoret, R. J., Cain, C., and Crowe, R. R. (1983). Evidence for gene–environment interaction in the development of adolescent antisocial behaviour. *Behavioral Genetics* **13**, 301–10.

Calnan, M. W., Dale, J. W., and Fonseka, C. P. de (1976). Suspected poisoning in children. Study of the incidence of true poisoning and poisoning scare in a defined population. *Archives of Disease in Childhood* **51**, 180–5.

Cameron, R. J. (1990). Curriculum-related assessment. In *Educational assessment in the primary school* (ed. J. Beech and L. Harding). NFER-Nelson, Windsor.

Campbell, A. G. M. (1984). Failure to thrive. In *Textbook of paediatrics* (ed. A. O. Forfar and G. C. Arneil), pp. 475–9. Churchill Livingstone, Edinburgh.

Campbell, M. and Spencer, E. K. (1988). Psychopharmacology in child and adolescent psychiatry: a review of the past five years. *Journal of the American Academy of Child Psychiatry* **27**, 269–79.

Campbell, M., Green, W. H., and Deutsch, S. L. (1985). *Child and adolescent psychopharmacology*. Sage Publications, Beverley Hills, London.

Cantwell, D. (1975). Genetics of hyperactivity. *Journal of Child Psychology and Psychiatry* **16**, 261–4.

Caplan, H. (1970). *Hysterical 'conversion' symptoms in childhood*. M.Phil Dissertation, University of London.

Caplan, M. G. and Douglas, V. I. (1969). Incidence of parental loss in children with depressed mood. *Journal of Child Psychology and Psychiatry* **10**, 225–36.

Capute, A. J., Shapiro, B. K., Palmer, F. B., Ross, A., and Wachtel, R. C. (1985). Normal gross motor development: the influence of race, sex and socioeconomic status. *Developmental Medicine and Child Neurology* **27**, 635–43.

Carlson, G. A. and Cantwell, D. P. (1980). Unmasking masked depression in children and adolescents. *American Journal of Psychiatry* 137, 445–9.

Carr, J. (1970). Mental and motor development in young mongol children. *Journal of Mental Deficiency Research* 14, 205–20.

Carr, J. (1988). Six weeks to twenty-one years old: a longitudinal study of children with Down's syndrome and their families. *Journal of Child Psychology and Psychiatry* 29, 407–32.

Cassell, S. and Paul, M. (1967). The role of puppet therapy on the emotional responses of children hospitalised for cardiac catheterisation. *Journal of Pediatrics* 71, 233–9.

Cauce, A. M., Comer, J. P., and Schwartz, D. (1987). Long-term effects of a systems-oriented school prevention program. *American Journal of Orthopsychiatry* 57, 127–31.

Caveness, W. F., Merritt, H. F., and Gallup, G. H. (1974). A survey of public attitudes towards epilepsy in 1974 with an indication of trends over the past twenty-five years. *Epilepsia* 15, 523–6.

Cederblad, M. (1968). A child psychiatric study on Sudanese Arab children. *Acta Psychiatrica Scandinavica*, Suppl. 200.

Cerreto, M. C. and Travis, L. B. (1984). Implications of psychological and family factors in the treatment of diabetes. *Pediatric Clinics of North America* 31, 689–710.

Chamberlain, R. N., Christie, P. N., Holt, K. J., Huntley, R. M., Pollard, R., and Roche, M. C. (1983). A study of schoolchildren who had identified virus infection during infancy. *Child: Care, Health and Development* 9, 29–47.

Chaplais, J. du Z. and Macfarlane, J. A. (1984). A review of 404 'late walkers'. *Archives of Disease in Childhood* 59, 512–16.

Chazan, M. (1962). School phobia. *British Journal of Educational Psychology* 32, 209–17.

Chess, S. and Fernandez, P. (1980). Neurologic damage and behavior disorders in rubella children. *American Annals of the Deaf* 125, 998–1001.

Chess, S. and Thomas, A. (1990). Continuities and discontinuities in temperament. In *Straight and devious pathways from childhood to adulthood* (ed. L. Robins and M. Rutter), pp. 205–20. Cambridge University Press, Cambridge.

Child Accident Prevention Trust (1989). *Basic principles of child accident prevention: a guide to action.* Child Accident Prevention Trust, London.

Christensen, A. P. and Sanders, M. R. (1987). Habit reversal and differential reinforcement of other behaviour in the treatment of thumb-sucking: an analysis of generalisation and side-effects. *Journal of Child Psychology and Psychiatry* 28, 281–95.

Christensen, M. F. and Mortensen, O. (1975). Long-term prognosis in children with recurrent abdominal pain. *Archives of Disease in Childhood* 50, 110–14.

Christodoulou, G. N., Gargoulas, A., Papalouvkasa, A., Marinopoulou, A., and Sideris, E. (1977). Primary peptic ulcer in childhood. Psychosocial, psychological and psychiatric aspects. *Acta Psychiatrica Scandinavica* 56, 215–22.

Clark, D. A. and Bolton, D. (1985). Obsessive-compulsive adolescents and their parents: a psychometric study. *Journal of Child Psychology and Psychiatry* 26, 267–76.

Clausen, J. A. (1975). The social meaning of differential physical and sexual maturation. In *Adolescence in the life cycle: psychological change and social content* (ed. S. E. Dragastin and G. H. Elder), pp. 25–48. Halsted Press, London.

Clayden, G. S. (1988). Reflex anal dilatation associated with severe chronic constipation in children. *Archives of Disease in Childhood* 63, 832–6.

Clements, S. (1966). Minimal brain dysfunction in children. *NINDS Monograph No. 3*, US Public Health Services, Washington, DC.

Coates, S. and Zucker, K. J. (1988). Gender identity disorders in children. In *Handbook of clinical assessment of children and adolescents, Volume 2* (ed. C. J. Kestenbaum and D. T. Williams), pp. 893–914. New York University Press, New York.

Cohen, N. J., Bradley, S., and Kolers, N. (1987). Outcome evaluation of a therapeutic day treatment program for delayed and disturbed pre-schoolers. *Journal of the American Academy of Child Psychiatry* 26, 687–93.

Cohn, A. H. (1983). *An approach to preventing child abuse.* National Committee for the Prevention of Child Abuse, Chicago.

Cohn, S. (1988). Assessing the gifted child and adolescent. In *Handbook of clinical assessment of children and adolescents. Volume 1* (ed. C. J. Kestenbaum and D. T. Williams), pp. 355–76. New York University Press, New York.

Cohn, J. F. and Tronick, E. (1989). Specificity of infants' response to mothers affective behaviour. *Journal of the American Academy of Child and Adolescent Psychiatry* 28, 242–8.

Colley, J. R. T. and Reid, D. D. (1970). Urban and social class origins of childhood bronchitis in England and Wales. *British Medical Journal* 2, 213.

Comer, J. P. (1985). The Yale–New Haven Primary Prevention Project: a follow-up study. *Journal of the American Academy of Child Psychiatry* 24, 154–60.

Condon, J. T. (1987). Psychological and physical symptoms during pregnancy: a comparison of male and female expectant parents. *Journal of Reproductive and Infant Psychology* 5, 207–19.

Connell, H. M. and McConnel, T. S. (1981). Hydrocephalus in infancy: psychiatric sequelae. *Developmental Medicine and Child Neurology* 23, 505–17.

Cooper, C. E. (1974). Child abuse and neglect—medical aspects. In *The maltreatment of children* (ed. S. M. Smith). M.T.P., Lancaster.

Cooper, P. J., Campbell, A. E., Day, A., Kennerly, H., and Bond, A. (1988). Non-psychotic psychiatric disorder after childbirth. *British Journal of Psychiatry* 152, 799–806.

Cooper, S. (1987). The fetal alcohol syndrome. *Journal of Child Psychology and Psychiatry* 28, 223–8.

Corbett, J. A. (1979). Psychiatric morbidity and mental retardation. In *Psychiatric illness and mental handicap* (ed. F. E. James and R. P. Snaith). Gaskill Press, London.

Corbett, J. A., Harris, R., and Robinson, R. (1975). Epilepsy. In *Mental retardation and developmental disabilities* (ed. J. Wortis), pp. 79–111. Brunner-Mazel, New York.

Corbett, J. A., Harris, R., Taylor, E., and Trimble, M. (1977). Progressive disintegrative psychosis of childhood. *Journal of Child Psychology and Psychiatry* 18, 211–19.

Corbett, J. A., Trimble, M. R., and Nicol, T. C. (1985). Behavioral and cognitive impairment in children with epilepsy: the long-term effects of anti-convulsant therapy. *Journal of the American Academy of Child Psychiatry* **24**, 17–23.

Costello, A. J. (1989). Reliability in diagnostic interviewing. In *Handbook of child psychiatric diagnosis* (ed. C. G. Last and M. Hersen), pp. 28–40. John Wiley, Chichester.

Costello, A. J., Edelbrock, C., Dulcan, M. K., Kalas, R., and Klaric, S. A. (1984). *Final report to NIMH on the Diagnostic Interview Schedule for Children.* NIMH, Washington, DC.

Costello, A. M. de L. *et al.* (1989). Prediction of neurodevelopmental impairment at four years from brain ultrasound appearances of very pre-term infants. *Developmental Medicine and Child Neurology* **30**, 711–22.

Cousens, P., Waters, B., Said, J., and Stevens, M. (1988). Cognitive effects of cranial irradiation in leukaemia: a survey and meta-analysis. *Journal of Child Psychology and Psychiatry* **29**, 839–52.

Cowen, L., Mok, J., Corey, M., MacMillan, H., Simmons, R., and Levison, H. (1986). Psychologic adjustment of the family with a member who has cystic fibrosis. *Pediatrics* **77**, 745–53.

Cox, A., Hopkinson, K., and Rutter, M. (1981). Psychiatric interviewing techniques. II. Naturalistic study: eliciting factual information. *British Journal of Psychiatry* **138**, 283–91.

Cox, J. L., Connor, Y., and Kendell, R. E. (1982). Prospective study of the psychiatric disorders of childbirth by personal interview. *British Journal of Psychiatry* **140**, 111–19.

Crisp, A. H., Palmer, R. L., and Kalucy, R. S. (1976). How common is anorexia nervosa? A prevalence study. *British Journal of Psychiatry* **128**, 549–54.

Crowell, J., Keener, M., Ginsburg, N., and Anders, T. (1987). Sleep habits in toddlers 18 to 36 months old. *Journal of the American Academy of Child and Adolescent Psychiatry* **26**, 510–15.

Cullen, K. J. (1976). A six year controlled trial of prevention of children with behavioural disorders. *Journal of Pediatrics* **88**, 662–6.

Cunningham, C. C., Morgan, P. A., and McGucken, R. B. (1984). Down's syndrome: is dissatisfaction with disclosure of diagnosis inevitable? *Developmental Medicine and Child Neurology* **26**, 33–9.

Cytryn, L., Cytryn, E., and Rieger, E. (1967). Psychological implications of cryptorchidism. *Journal of the American Academy of Child Psychiatry* **6**, 131–42.

Dahl, J. A., Melin, L., and Leissner, P. (1988). Effects of a behavioural intervention on epileptic seizure behaviour and paroxysmal activity: a systematic replication of three cases of children with intractable epilepsy. *Epilepsia* **29**, 172–83.

Davie, R., Butler, N., and Goldstein, H. (1972). *From birth to seven: a report of the National Child Development Study.* Longmans, London.

Davies, K. E. (1989). *The Fragile X syndrome.* Oxford University Press, Oxford.

Davison, I. S., Faull, C., and Nicol, A. R. (1986). Research note: temperament and behaviour in six year olds with recurrent abdominal pain: a follow-up. *Journal of Child Psychology and Psychiatry* **27**, 539–44.

Deaton, A. V. (1985). Adaptive non-compliance in pediatric asthma: the parent as expert. *Journal of Pediatric Psychology* **10**, 1–14.

De Jonge, G. A. (1973). Epidemiology of enuresis: a survey of the literature. In *Bladder Control and Enuresis. Clinics in Developmental Medicine Nos. 48/49* (ed. I. Kolvin, R. MacKeith, and S. R. Meadow), pp. 109–17. Heinemann, Spastics International Medical Publications, London.

Delong, G. R. and Aldersdorf, A. L. (1987). Long-term experience with lithium treatment in childhood: correlation with clinical diagnosis. *Journal of the American Academy of Child and Adolescent Psychiatry* **26**, 389–94.

Department of Education and Science (1978). *Special educational needs*. HMSO, London.

Department of Health (1989). *An introduction to the Children Act, 1989*. HMSO, London.

Department of Health and Social Security (1968). *Report of the commission on local authority and allied personal social services*. HMSO, London.

Department of Health and Social Security (1976). *Fit for the future. Report of the Child Health Services Committee*. HMSO, London.

Department of Health and Social Security (1980). *Inequalities in health*. HMSO, London.

Department of Health and Social Security (1988). *Working together*. HMSO, London.

Diamond, L. J. and Jaudes, P. K. (1983). Child abuse in a cerebral-palsied population. *Developmental Medicine and Child Neurology* **25**, 169–74.

Dietz, W. H. and Gortmaker, S. L. (1985). Do we fatten our children at the television set? *Pediatrics* **75**, 807–12.

Dische, S., Yule, W., Corbett, J., and Hand, D. (1983). Childhood noctural enuresis: factors associated with outcome of treatment with an enuretic alarm. *Developmental Medicine and Child Neurology* **25**, 67–80.

Dodge, J. A. (1972). Psychosomatic aspects of infantile pyloric stenosis. *Journal of Psychosomatic Research* **16**, 1–5.

Dorman, C., Hurley, A. D., and D'Avignon, J. (1988). Language and learning disorders of older boys with Duchenne Muscular Dystrophy. *Developmental Medicine and Child Neurology* **28**, 316–27.

Dorner, S. (1976). Adolescents with spina bifida: how they see their situation. *Archives of Diseases in Childhood* **51**, 439–44.

Douglas, J. W. B., Ross, J. M., and Simpson, H. R. (1968). *All our future: a longitudinal study of secondary education*. Peter Davies, London.

Downey, J., Ehrhardt, A. A., Morishima, A., Bell, J. J., and Gruen, R. (1987). Gender role development in two clinical syndromes. Turner Syndrome versus Constitutional Short Stature. *Journal of the American Academy of Child and Adolescent Psychiatry* **26**, 566–73.

Drotar, D., Baskiewicz, A., Irvin, N., Kennell, J., and Klaus, M. (1975). The birth of an infant with a congenital malformation: a hypothetical model. *Pediatrics* **56**, 710–17.

Drotar, D., Doershuk, C., Stern, R., Boat, T., Boyer, W., and Matthews, L. (1981). Psychological functioning of children with cystic fibrosis. *Pediatrics* **67**, 338–43.

Dubowitz, V. and Hersov, L. (1976). Management of children with non-organic (hysterical) disorders of motor function. *Developmental Medicine and Child Neurology* **18**, 358–68.

Dulcan, M. K., Costello, E. J., Costello, A. J., Edelbrock, C., Brent, D., and Janiszewski, S. (1990). The pediatrician as gatekeeper to mental health care for children: do parents' concerns open the gate? *Journal of the American Academy of Child and Adolescent Psychiatry* **29**, 453–8.

Dunn, J. (1988). Sibling influence and child development. *Journal of Child Psychology and Psychiatry* **29**, 119–28.

Dunnell, K. (1990). Monitoring children's health. *Population Trends* No. 60, 16–22.

Dutton, P. V., Furnell, J. R. G., and Speirs, A. L. (1985). Environmental stress factors associated with toddler diarrhoea. *Journal of Psychosomatic Research* **29**, 85–8.

Earls, F. and Richman, N. (1980). The prevalence of behavior problems in three year old children of West Indian born parents. *Journal of Child Psychology and Psychiatry* **21**, 99–106.

Eaton-Evans, J. and Dugdale, A. E. (1988). Sleep patterns of infants in the first year of life. *Archives of Dieases in Childhood* **63**, 647–9.

Edwards, G. (1982). *The treatment of drinking problems: a guide for the helping professions*. Grant McIntyre, London.

Egger, J., Carter, C., Graham, P., Gumley, D., and Soothill, J. (1985). A controlled trial of oligoantigenic treatment in the hyperkinetic syndrome. *Lancet* i, 540–5.

Eisenberg, L. (1958). School phobia: a study in the communication of anxiety. *American Journal of Psychiatry* **114**, 712–18.

Eisenberg, L., Conners, K., and Sharpe, L. (1965). A controlled study of the differential application of out-patient psychiatric treatment for children. *Japanese Journal of Child Psychiatry* **6**, 125–32.

Eiser, C. and Town, C. (1987). Teacher's concerns about chronically sick children: implications for paediatricians. *Developmental Medicine and Child Neurology* **29**, 56–63.

Elliott, C., Murray, D. J., and Pearson, L. (1983). *The British Ability Scales*, New end. National Foundation for Educational Research/Nelson, Windsor.

Emery, A. E. H. (1985). Infanticide, filicide and cot death. *Archives of Disease in Childhood* **66**, 505–7.

Eriksen, E. (1965). *Childhood and society*. Penguin Press, Harmondsworth.

Eth, S. R. and Pynoos, R. S. (1985). *Post-traumatic stress disorder in children*. American Psychiatric Association, Washington, DC.

Evan-Jones, L. G. and Rosenbloom, L. (1978). Disintegrative psychosis in childhood. *Developmental Medicine and Child Neurology* **20**, 462–70.

Evered, C. J., Hill, P. D., Hall, D. M., and Hollins, S. C. (1989). Liaison psychiatry in a child development clinic. *Archives of Disease in Childhood* **64**, 754–8.

Farber, J. M. (1987). Psychopharmacology of self-injurious behaviour in the mentally retarded. *Journal of the American Academy of Child and Adolescent Psychiatry* **26**, 296–302.

Farrington, D. P. (1977). The effects of public labelling. *British Journal of Criminology* **17**, 112–25.

Farrington, D. P. (1981). The prevalence of convictions. *British Journal of Criminology* **21**, 173–5.

Feingold, B. F. (1975). Hyperkinesis and learning difficulties linked to artificial food and colors. *American Journal of Nursing* **75**, 797–803.

Feldman, F., Cantor, D., Soll, S., and Bachrach, W. (1967). Psychiatric study of a consecutive series of thirty-four patients with ulcerative colitis. *British Medical Journal* **3**, 14–17.

Ferrari, M., Matthews, S., and Barabas, G. (1983). The family and child with epilepsy. *Family Process* **22**, 53–60.

Ferri, E. (1976). *Growing up in a one-parent family*. NFER Publishing Co., Slough.

Fine, S. (1979). Incidence of visual handicap in childhood. In *Visual handicap in children* (ed. V. Smith and J. Keen). Spastics International Medical Publications, Heinemann, London.

Finkelhor, D. (1979). *Sexually victimised children*. Free Press, New York.

Firth, M., Gardner-Medwin, D., Hosking, G., and Wilkinson, E. (1983). Interviews with parents of boys suffering from muscular dystrophy. *Developmental Medicine and Child Neurology* **25**, 466–71.

Fishler, K., Koch, R., Donnell, G., and Graliker, B. V. (1966). Psychological correlates in galactosemia. *American Journal of Mental Deficiency* **71**, 116–25.

Fishman, M. and Palkes, H. (1974). The validity of psychometric testing in children with congenital malformations of the central nervous system. *Developmental Medicine and Child Neurology* **16**, 180–5.

Fitzpatrick, C., Barry, C., and Garvey, C. (1986). Psychiatric disorder among boys with Duchenne Muscular Dystrophy. *Developmental Medicine and Child Neurology* **28**, 589–95.

Flament, M. F. *et al.* (1988). Obsessive compulsive disorder in adolescence: an epidemiological study. *Journal of the American Academy of Child and Adolescent Psychiatry* **27**, 764–71.

Flament, M. F. *et al.* (1990). Childhood obsessive–compulsive disorder. *Journal of Child Psychology and Psychiatry* **31**, 363–80.

Fogelman, D., Tibbenham, A., and Lambert, L. (1980). Absence from school: findings from the National Child Development Study. In *Out of school* (ed. L. Hersov and I. Berg), pp. 25–48. John Wiley, Chichester.

Fonagy, P., Moran, G. S., Lindsay, M. K. M., Kurtz, A. B., and Brown, R. (1987). Psychological adjustment and diabetic control. *Archives of Disease in Childhood* **62**, 1009–13.

Fordham, K. E. and Meadow, S. R. (1988). Controlled trial of standard pad and bell alarm against mini-alarm for nocturnal enuresis. *Archives of Disease in Childhood* **64**, 651–6.

Foreman, D. M. and Goodyer, I. M. (1988). Salivary cortisol hypersecretion in juvenile depression. *Journal of Child Psychology and Psychiatry* **29**, 311–21.

Forrest, G. C., Claridge, R. S., and Baum, J. D. (1981). Practical management of perinatal death. *British Medical Journal* **282**, 31–2.

Forrest, G. C., Standish, E., and Baum, D. (1982). Support after perinatal death: a study of support and counselling after perinatal bereavement. *British Medical Journal* **285**, 1475–8.

Forssman, H. and Thuwe, I. (1966). One hundred and twenty children born after application for therapeutic abortion refused. *Acta Psychiatrica Scandinavica* **42**, 71–88.

Fosson, A., Knibbs, J., Bryant-Waugh, R., and Lask, B. (1987). Early onset anorexia nervosa. *Archives of Disease in Childhood* **62**, 114–18.

Foster, R. and Nihira, K. (1969). Adaptive behavior as a measure of psychiatric impairment. *American Journal of Mental Deficiency* **74**, 401–4.

Fournier, J-P., Garfinkel, B. D., Bond, A., Beauchesne, H., and Shapiro, S. K. (1987). Pharmacological and behavioral management of enuresis. *Journal of the American Academy of Child and Adolescent Psychiatry* **26**, 845–53.

Fraiberg, S. (1977). *Insights from the blind*. Basic Books, New York.

Frank, D. A. and Zeisel, S. H. (1988). Failure to thrive. *Pediatric Clinics of North America* **35**, 1187–206.

Frankenburg, W. K., Dodds, J. B., Fandal, A. W., Kazuk, E., and Cohrs, M. (1975). *Denver Developmental Screening Test*. Ladoca Project and Publishing Foundation, Denver.

Freeman, J. (1983). Emotional problems of the gifted child. *Journal of Child Psychology and Psychiatry* **24**, 481–5.

Freeman, R. D. (1970). Psychiatric problems in adolescents with cerebral palsy. *Developmental Medicine and Child Neurology* **12**, 64–70.

Freeman, R. D. (1977). Psychiatric aspects of sensory disorders and intervention. In *Epidemiological approaches in child psychiatry* (ed. P. J. Graham), pp. 275–304. Academic Press, London.

Freeman, R. D., Malkin, S. F., and Hastings, J. O. (1975). Psychosocial problems of deaf children and their families: a comparative study. *American Annals of the Deaf* **120**, 391–405.

Freeman, R. D. *et al.* (1989). Blind children's early emotional development: do we know enough to help? *Child: Care, Health and Development* **15**, 3–28.

Freud, A. (1966). *Normality and pathology in childhood*. Hogarth Press, London.

Fundudis, T., Kolvin, I., and Garside, R. (eds.) (1979). *Speech retarded and deaf children: their psychological development*. Academic Press, London.

Furniss, T., Bingley-Miller, L., and Bentovim, A. (1984). Therapeutic approach to sexual abuse. *Archives of Disease in Childhood* **59**, 865–70.

Galante, R. and Foa, D. (1986). An epidemiological study of psychic trauma and treatment effectiveness for children after a natural disaster. *Journal of the American Academy of Child Psychiatry* **25**, 357–63.

Galatzer, A., Nofar, E., Beit Halachmi, N., Aran, O., Shalit, M., Roitman, A., and Laron, Z. (1981). Intellectual and psychosocial functions of children, adolescents and young adults before and after operations for craniopharyngioma. *Child: Care, Health and Development* **7**, 307–16.

Gambrill, E. (1983). Behavioral intervention with child abuse and neglect. *Progress in Behaviour Modification* **15**, 1–56.

Garland, A., Shaffer, D., and Whittle, B. (1989). A national survey of school based, adolescent suicide prevention programs. *Journal of the American Academy of Child and Adolescent Psychiatry* **28**, 931–4.

Garralda, M. and Bailey, D. (1989). Psychiatric disorders in general practice referrals. *Archives of Disease in Childhood* **64**, 1727–33.

Garralda, M. E. and Bailey, D. (1986). Children with psychiatric disorders in primary care. *Journal of Child Psychology and Psychiatry* **27**, 611–24.

Garralda, M. E. (1984). Hallucinations in children with conduct and emotional disorders. *Psychological Medicine* **14**, 589–96.

Garralda, M. E., Jameson, R. A., Reynolds, J. M., and Postlethwaite, R. J. (1988). Psychiatric adjustment in children with chronic renal failure. *Journal of Child Psychology and Psychiatry* **29**, 79–91.

Garrison, C. Z., Schluchter, M. D., Schoenbach, V. J., and Kaplan, B. H. (1989).

Epidemiology of depressive symptoms in young adolescents. *Journal of the American Academy of Child and Adolescent Psychiatry* **28**, 343–51.

Garson, A., Benson, R. S., Ivler, L., and Patton, C. (1978). Parental reactions to children with congenital heart disease. *Child Psychiatry and Human Development* **9**, 86–94.

Gath, A. (1989). Living with a mentally handicapped brother or sister. *Archives of Disease in Childhood* **64**, 513–16.

Gath, A. and Gumley, D. (1984). Down's syndrome and the family: follow-up of children first seen in infancy. *Developmental Medicine and Child Neurology* **26**, 500–8.

Gath, A. and Gumley, D. (1987). Retarded children and their siblings. *Journal of Child Psychology and Psychiatry* **28**, 715–30.

Gatherer, A., Parfitt, J., Porter, E., and Vessey, M. (1979). *Is health education effective?* Health Education Council, 78 New Oxford Street, London.

Gayton, W. F., Friedman, S. B., Tavormina, J. F., and Tucker, F. (1977). Children with cystic fibrosis: psychological test findings of patients, siblings and parents. *Pediatrics* **59**, 888–94.

Gelder, M., Gath, D., and Mayou, R. (1989). *Oxford textbook of psychiatry*, (2nd edn. Oxford University Press, Oxford.

Gergen, P. J., Mullaly, P. I., and Evans, R. (1988). National survey of prevalence of asthma among children in the United States, 1976–1980. *Pediatrics* **81**, 1–7.

Gibbens, T. C. N. (1974). Preparing psychiatric case reports. *British Journal of Hospital Medicine* **11**, 278–84.

Gill, O. and Jackson, B. (1984). *Adoption and race*. Batsford, London.

Gillberg, C., Persson, U., Grufman, M., and Temner, U. (1986). Psychiatric disorders in mildly and severely mentally retarded urban children and adolescents. Epidemiological aspects. *British Journal of Psychiatry* **149**, 68–74.

Gillies, D. R. N. and Forsythe, W. I. (1984). Treatment of multiple tics and the Tourette Syndrome. *Developmental Medicine and Child Neurology* **26**, 830–3.

Giovannoni, J. M. and Becerra, R. M. (1979). *Defining child abuse*. Free Press, New York.

Gittelman-Klein, R. and Mannuzza, S. (1989). The long-term outcome of the attention deficit disorder/hyperkinetic syndrome. In *Attention deficit disorder: clinical and basic research* (ed. T. Sagvolden and T. Archer), pp. 71–91. Lawrence Erlbaum, Hillsdale, New Jersey.

Glaser, D. and Bentovim, A. (1987). Psychological aspects of congenital heart disease. In *Paediatric cardiology* (ed. R. H. Anderson, F. J. Macartney, E. A. Shinebourne, and M. Tynan). Churchill Livingstone, Edinburgh. (In press).

Golombok, S., Spencer, A., and Rutter, M. (1983). Children in lesbian and single-parent households: psychosexual and psychiatric appraisal. *Journal of Child Psychology and Psychiatry* **24**, 551–72.

Goodman, J. D. and Sours, J. A. (1967). *The child mental status examination*. Basic Books, New York.

Goodman, R. and Stevenson, J. (1989). A twin study of hyperactivity. II. The aetiological role of genes, family relationships and perinatal adversity. *Journal of Child Psychology and Psychiatry* **30**, 691–709.

Goodyer, I. and Taylor, D. C. (1985). Hysteria. *Archives of Disease in Childhood* **60**, 680–1.

Goodyer, I. M. (1990). *Life experiences, development and childhood psycho-pathology.* John Wiley, Chichester.

Goodyer, I. M., Wright, C., and Altham, P. M. E. (1987). The impact of recent life events in psychiatric disorders of childhood and adolescence. *British Journal of Psychiatry* **151**, 179–85.

Gordon, N. and McKinlay, I. (1980). *Helping clumsy children.* Churchill Living-stone, Edinburgh.

Goyette, C. H., Conners, C. K., Petti, T. A., and Curtis, L. E. (1978). Effects of artificial colours on hyperactive children: a double-blind challenge study. *Psychopharmacology Bulletin* **14**, 39–40.

Graham, J. M., Bashir, J. S., Stark, R. E., Silbert, A., and Waltzer, S. (1988). Oral and written language of XXY boys: implications for anticipatory guidance. *Pediatrics* **81**, 795–806.

Graham, P. (1979). Epidemiological approaches to child mental health in develop-ing countries. In *Psychopathology and youth: a cross-cultural perspective* (ed. E. F. Purcell), pp. 28–45. Josiah Macey Jr. Foundation, New York.

Graham, P. (1982). Child psychiatry in relation to primary health care. *Social Psychiatry* **17**, 109–16.

Graham, P. (1984). Paediatric referral to a child psychiatrist. *Archives of Disease in Childhood* **59**, 1103–5.

Graham, P. (1986). Behavioural and intellectual development. *British Medical Bulletin* **42**, 155–62.

Graham, P. (1990). Maternal employment. *Archives of Disease in Childhood* **65**, 565–6.

Graham, P. and Jenkins, S. (1985). Training of paediatricians for psychosocial aspects of their work. *Archives of Disease in Childhood* **60**, 777–80.

Graham, P. and Rutter, M. (1973). Psychiatric disorder in the young adolescent: a follow-up study. *Proceedings of the Royal Society of Medicine* **66**, 1226–9.

Graham, P. and Stevenson, J. (1985). A twin study of genetic influences of be-havioral deviance. *Journal of the American Academy of Child Psychiatry* **24**, 33–41.

Graham, P. and Stevenson, J. (1987). Temperament and psychiatric disorder: the genetic contribution to behaviour in childhood. *Australian and New Zealand Journal of Psychiatry* **21**, 267–74.

Graham, P., Rutter, M., and George, S. (1973). Temperamental characteristics as predictors of behavioural disorders in children. *American Journal of Ortho-psychiatry* **43**, 328–39.

Green, M. and Solnit, A. J. (1964). Reactions to the threatened loss of a child: a vulnerable child syndrome. *Pediatrics* **34**, 58–66.

Green, R. (1974). *Sexual identity and conflict,* pp. 212–13. Duckworth, London.

Green, R. (1976). One hundred and ten feminine and masculine boys: behavioural contrasts and demographic similarities. *Archives of Sexual Behaviour* **5**, 435–9.

Green, R., Williams, K., and Goodman, M. (1982). Ninety nine 'tomboys' and 'non tomboys': behavioural contrasts and demographic similarities. *Archives of Sexual Behaviour* **11**, 247–66.

Greenwood, R., Bhalla, A., Gordon, A., and Roberts, J. (1983). Behavior disturb-ances during recovery from herpes simplex encephalitis. *Journal of Neurology, Neurosurgery and Psychiatry* **46**, 809–17.

Grey, M. J., Genel, M., and Tamborlane, W. V. (1980). Psychosocial adjustment of latency aged diabetics: determinants and relationship to control. *Pediatrics* **65**, 69–73.

Griffiths, R. (1954). *The abilities of babies*. McGraw-Hill, New York.

Gross, I., Wheeler, M., and Hess, K. (1976). The treatment of obesity in adolescents using behavioral self-control. *Clinical Pediatrics* **15**, 920–4.

Gross, M. (1979). Pseudo-epilepsy: a study in adolescent hysteria. *American Journal of Psychiatry* **136**, 210–13.

Guilleminault, C., Carskadon, M., and Dement, W. C. (1974). On the treatment of rapid eye movement narcolepsy. *Archives of Neurology* **30**, 90–3.

Gurman, A. and Kniskern, D. P. (1979). Research on marital and family therapy. In *Handbook of psychotherapy and behavior change* (ed. S. Garfield and A. Berger). Brunner Mazel, New York.

Gustafsson, P. A., Kjellman, N.-I. M., Ludvigsson, J., and Cederblad, M. (1987). Asthma and family interaction. *Archives of Disease in Childhood* **62**, 258–63.

Gustavson, K. H., Hagberg, B., Hagberg, G., and Sars, K. (1977). Severe mental retardation in a Swedish county. II. Etiological and pathogenetic aspects of children born 1959–1970. *Neuropadiatrie* **8**, 293–304.

Gutelius, M. F., Kirsch, A. D., MacDonald, S., Brooks, M. R., and McErlean, T. (1977). Controlled study of child health supervision: behavioral results. *Pediatrics* **60**, 294–304.

Gyulay, J. E. (1978). *The dying child*. McGraw-Hill, New York.

Hagberg, B. (1978). The epidemiological panorama of major neuropaediatric handicap in Sweden. In *Care of the handicapped child. Clinics in Developmental Medicine 67* (ed. J. Apley), pp. 111–24. Heinemann Medical Books, London.

Hagberg, B., Hagberg, G., Lewerth, A., and Lindberg, V. (1981). Mild mental retardation in Swedish school children. II. Etiological and pathogenetic aspects. *Acta Paediatrica Scandinavica* **70**, 445–52.

Hagberg, B. and Hagberg, G. (1984). Prenatal and perinatal risk factors in a survey of 681 Swedish cases. In *The Epidemiology of the Cerebral Palsies* (ed. F. Stanley and E. Alberman). Blackwell Scientific Publications, Oxford.

Hagerman, R. (1989). Behaviour and treatment of the fragile X syndrome. In *The fragile X syndrome* (ed. K. E. Davies). Oxford University Press, Oxford.

Haggerty, R. J. (1980). Life stress, illness and social support. *Developmental Medicine and Child Neurology* **22**, 391–400.

Hall, D. M. B. (ed.) (1989). *Health for all children*. Oxford Medical Publications, Oxford.

Hamblin, T. J., Hussain, J., Akbar, A. N., Tang, Y. C., Smith, J. L., and Jones, D. B. (1983). Immunological reason for chronic ill health after infectious mononucleosis. *British Medical Journal* **287**, 85–9.

Hamburg, D. and Hamburg, B. (1980). A life-span perspective on adaptation and health. In *Family and health: epidemiological approaches* (ed. B. Kaplan and M. Ibrahim). University of South Carolina Press, Chapel Hill.

Hannaway, P. J. (1970). Failure to thrive. A study of one hundred infants and children. *Clinical Pediatrics* **9**, 96–9.

Harbord, M. G. and Manson, J. I. (1987). Temporal lobe epilepsy in childhood: re-appraisal of etiology and outcome. *Pediatric Neurology* **3**, 263–8.

Haring, N. G., Lovitt, T. C., Eaton, M. D., and Hanson, C. L. (1978). *The 4th R: research in the classroom*. Charles Merrill, Columbus, Ohio.

Harris, B. H., Schwaitzberg, S. D., Seman, T. M., and Herrmann, C. (1989). The hidden morbidity of pediatric trauma. *Journal of Pediatric Surgery* 24, 103–5.

Harris, D. B. (1963). *Children's drawings as measures of intellectual maturity: a revision and extension of the Goodenough Draw-A-Man Test*. Harcourt, Brace and World, New York.

Hawton, K. (1982). Attempted suicide in children and adolescents. *Journal of Child Psychology and Psychiatry* 23, 497–503.

Hayden, T. L. (1980). The classification of elective mutism. *Journal of the American Academy of Child Psychiatry* 19, 118–33.

Hazel, N. (1977). How family placements can combat delinquency. *Social Work Today* 8, 6–7.

Hemsley, R., Howlin, P., Berger, M., Hersov, L., Holbrook, D., Rutter, M., and Yule, W. (1978). Teaching autistic children in a family context. In *Autism: a reappraisal of concepts and treatment* (ed. M. Rutter and E. Schopler), pp. 379–411. Plenum, New York.

Henderson, S. E. and Hall, D. (1982). Concomitants of clumsiness in young schoolchildren. *Developmental Medicine and Child Neurology* 24, 448–60.

Hensey, O. J., Williams, J. K., and Rosenbloom, L. (1983). Intervention in child abuse: experience in Liverpool. *Developmental Medicine and Child Neurology* 25, 606–11.

Herbert, M. (1981). *Behavioural treatment of problem children: a practice manual*. Academic Press, London.

Herbert, M., Sluckin, W., and Sluckin, A. (1982). Mother to infant 'bonding'. *Journal of Child Psychology and Psychiatry* 23, 205–21.

Hersov, L. A. (1960). Persistent non-attendance at school. *Journal of Child Psychology and Psychiatry* 1, 130–6.

Herzog, D. B. and Copeland, P. M. (1985). Eating disorders. *New England Journal of Medicine* 313, 294–303.

Heston, L. (1966). Psychiatric disorder in foster home-reared children of schizophrenic mothers. *British Journal of Psychiatry* 113, 819–25.

Hetherington, E. M., Cox, M., and Cox, R. (1982). Effects of divorce on parents and children. In *Non-traditional families* (ed. M. E. Lamb), pp. 233–88. Lawrence Erlbaum, Hillsdale, New Jersey.

Hewison, J. and Tizard, J. (1980). Parental involvement and reading attainment. *British Journal of Educational Psychology* 50, 209–15.

Hill, R. A., Standen, P. J., and Tattersfield, A. E. (1989). Asthma, wheezing and school absence in primary schools. *Archives of Disease in Childhood* 64, 246–51.

Hoare, P. (1984*a*). The development of psychiatric disorder among schoolchildren with epilepsy. *Developmental Medicine and Child Neurology* 26, 3–13.

Hoare, P. (1984*b*). Psychiatric disturbance in the families of epileptic children. *Developmental Medicine and Child Neurology* 26, 14–19.

Hobson, R. F. (1986). The autistic child's appraisal of expressions of emotion. *Journal of Child Psychology and Psychiatry* 27, 321–42.

Hochberg, Z., Gardos, M., and Benderly, A. (1987). Psychosexual outcome of assigned females and males with 46XX virilising congenital adrenal hyperplasia. *European Journal of Paediatrics* 146, 497–9.

Hodges, J. and Tizard, B. (1989). Social and family relationships of ex-institutional adolescents. *Journal of Child Psychology and Psychiatry* 30, 77–97.

Hollingworth, C. E., Tanguay, P. E., Grossman, G., and Pabst, P. (1980). Long-term outcome of obsessive-compulsive disorders in childhood. *Journal of the American Academy of Child Psychiatry* 19, 134–44.

Holtzman, N. A., Kronmal, R. A., Doorninck, W., Azen, C., and Koch, R. (1986). Effect of age at loss of dietary control on intellectual performance and behavior of children with phenylketonuria. *New England Journal of Medicine* 314, 593–8.

Honzik, M. P., Macfarlane, J. W., and Allen, L. (1948). Stability of mental test performances between two and eighteen years. *Journal of Experimental Education* 17, 309.

Howlin, P. and Rutter, M. (1987). *Treatment of autistic children*. John Wiley, New York.

Hsu, L. K. G., Crisp, A. H., and Harding, B. (1979). Outcome of anorexia nervosa. *Lancet* i, 61–5.

Hunt, A. and Dennis, J. (1987). Psychiatric disorder among children with tuberous sclerosis. *Developmental Medicine and Child Neurology* 29, 190–8.

Hunt, G. M. and Holmes, A. E. (1976). Factors related to intelligence in treated cases of spina bifida. *American Journal of Diseases in Childhood* 130, 823–7.

Hunt, H. and Wills, D. M. (1983). The family and the young visually handicapped child. In *Pediatric ophthalmology: current aspects* (ed. K. Wybar and D. Taylor), pp. 95–105. Marcel Dekker, New York.

Hutchings, B. and Mednick, S. A. (1974). Registered criminology in the adoptive and biological parents of registered male adoptees. In *Genetics, environment and psychopathology* (ed. S. A. Mednick, F. Schulsinger, F. Higgins, and B. Bell), pp. 215–27. North-Holland, Amsterdam.

Iwaniec, D., Herbert, M., and McNeish, A. J. (1985). Social work intervention with failure to thrive children and their families. II. Behavioural social work intervention. *British Journal of Social Work* 15, 357–89.

Jannoun, L. (1983). Are cognitive and educational development influenced by age at which prophylactic therapy is given in acute lymphoblastic leukaemia? *Archives of Disease in Childhood* 58, 953–8.

Jarvelin, M. R., Moilanen, I., Vikevainen-Tervonen, L., and Huttunen, N.-P. (1990). Life changes and protective capacities in enuretic and non-enuretic children. *Journal of Child Psychology and Psychiatry* 31, 763–74.

Jastak, J. F. and Jastak, S. (1978). *The Wide Range Achievement Test,* Revised edn. Jastak Associates, Wilmington, Delaware.

Jay, B. (1979). Genetic causes of visual handicap: prevalence and prevention. In *Visual handicap in children* (ed. V. Smith and J. Keen), pp. 94–101. Spastics International Medical Publications, Heinemann, London.

Jellinek, M. S. (1990). School consultation: evolving issues. *Journal of the American Academy of Child and Adolescent Pyschiatry* 29, 311–14.

Jenkins, J. and Gray, O. P. (1983). Changing clinical pictures of non-accidental injury to children. *British Medical Journal* 287, 1767–9.

Jenkins, J. M. and Smith, M. A. (1990). Factors protecting children living in disharmonious homes: maternal reports. *Journal of the American Academy of Child and Adolescent Psychiatry* 29, 60–9.

Jenkins, R. L. (1973). *Behavior disorders of childhood and adolescence.* C. Thomas, Springfield, Illinois.

Jenkins, S., Bax, M., and Hart, H. (1980). Behaviour problems in preschool children. *Journal of Child Psychology and Psychiatry* **21**, 5–17.

Johnson, S. B. (1988). Psychosocial aspects of juvenile diabetes. *Journal of Child Psychology and Psychiatry* **29**, 729–38.

Johnstone, F. D. and Forfar, J. O. (1984). Pre-natal paediatrics. In *Textbook of Paediatrics*, 3rd edn. (ed. J. O. Forfar and G. C. Arneil), p. 104. Churchill Livingstone, London.

Kahn, A., Sottiaux, B. P., Appelboom-Fondu, J., Blum, D., Rebuffat, E., and Levitt, J. (1989). Long-term development of children monitored as infants for an apparent life-threatening event during sleep: a ten year follow-up study. *Pediatrics* **83**, 668–73.

Kales, A., Soldatos, C. R., Bixler, E. O., Ladda, R. L., Charney, D. S., Webber, G., and Schweitzer, P. K. (1980). Hereditary factors in sleepwalking and night terrors. *British Journal of Psychiatry* **137**, 111–18.

Kallarackal, A. M. and Herbert, M. (1976). The happiness of Indian immigrant children. *New Society* **34**, 422–4.

Kandel, D. B., Davies, M., Karus, D., and Yamaguchi, K. (1986). The consequences in young adulthood of adolescent drug involvement. *Archives of General Psychiatry* **43**, 746–54.

Kannoy, K. W. and Schroeder, C. S. (1985). Suggestions to parents about common behavior problems in a paediatric primary care office: 5 years of follow-up. *Journal of Pediatric Psychology* **10**, 15–30.

Kaplan, B. J., McNichol, J., Conte, R. A., and Moghadam, H. K. (1989). Dietary replacement in pre-school aged hyperactive boys. *Pediatrics* **83**, 7–17.

Kashani, J. and Simmonds, J. F. (1979). The incidence of depression in children. *American Journal of Psychiatry* **136**, 1203–5.

Kashani, J. H., Konig, P., Shepperd, J. A., Wilfley, D., and Morris, D. A. (1988). Psychopathology and self-concept in asthmatic children. *Journal of Pediatric Pschology* **13**, 509–20.

Kazak, A. E. and Clarke, M. W. (1986). Stress in families of children with myelomeningocoele. *Developmental Medicine and Child Neurology* **28**, 220–8.

Kazak, A. E., Reber, M., and Snitzer, L. (1988). Childhood chronic disease and family functioning: a study of phenylketonuria. *Pediatrics* **81**, 224–30.

Kazdin, A. E. (1987). *Conduct disorders in childhood and adolescence*. Sage Publications, Newbury Park.

Kemmer, F. W., Bisping, R., Steingruber, H. J. *et al.* (1986). Psychological stress and metabolic control in patients with type 1 diabetes mellitus. *New England Journal of Medicine* **314**, 1078–84.

Kempe, C. H. and Helfer, R. E. (1980). *The battered child*. University of Chicago Press, Chicago.

Kendell, R. E., Rennie, D., Clark, J. A., and Dean, C. (1981). The social and obstetric correlates of psychiatric admission in the puerperium. *Psychological Medicine* **11**, 341–50.

Kennedy, W. A. (1965). School phobia: rapid treatment of fifty cases. *Journal of Abnormal Psychology* **70**, 285–9.

King, M. D., Day, R. E., Oliver, J. S., Lush, M., and Watson, J. M. (1981). Solvent encephalopathy. *British Medical Journal* **283**, 663–5.

Kirk, S. A., McCarthy, J. J., and Kirk, W. D. (1968). *The Illinois Test of Psycholinguistic Abilities (Revised)*. University of Illinois Press, Urbana.

Klaus, M. and Kennell, J. (1976). *Maternal–infant bonding. The impact of early separation or loss on family development.* Mosby, St. Louis.

Klaus, M. H., Kennell, J. H., Robertson, S. S., and Sosa, R. (1986). Effects of social support during parturition on maternal and infant morbidity. *British Medical Journal* 293, 585–7.

Klein, N. C., Alexander, J. F., and Parsons, B. V. (1977). Impact of family systems intervention on recidivism and sibling delinquency: a model of primary prevention and program evaluation. *Journal of Consulting and Clinical Psychology* 45, 469–74.

Klein, S. D., Simmons, R. G., and Anderson, C. R. (1984). Chronic kidney disease and transplantation in childhood and adolescents. In *Chronic illness and disabilities in childhood and adolescence* (ed. R. Blum), pp. 1429–1557. Grune and Stratton, London.

Klonoff, H., Low, M. D., and Clark, C. (1977). Head injuries in children: a prospective five year follow-up. *Journal of Neurology, Neurosurgery and Psychiatry* 40, 1211–19.

Koblenzer, C. S. and Koblenzer, P. J. (1988). Chronic intractable atopic eczema. Its occurrence as a sign of impaired parent–child relationship and psychologic developmental arrest: improvement through parent insight and education. *Archives of Dermatology* 124, 1673–7.

Koegel, R., Schreibman, L., O'Neil, R. E., and Burke, J. C. (1983). The personality and family-interaction characteristics of parents of autistic children. *Journal of Consulting and Clinical Psychology* 51, 683–92.

Kolvin, I. (1971). Psychoses in childhood—a comparative study. In *Infantile autism: concepts, characteristics and treatment* (ed. M. Rutter), pp. 7–26. Churchill Livingstone, London.

Kolvin, I. and Fundudis, T. (1981). Elective mute children: psychological development and background factors. *Journal of Child Psychology and Psychiatry* 22, 219–32.

Kolvin, I., Garside, R. F., Nicol, A. R., MacMillan, A., Wolstenholme, F., and Leitch, I. M. (1981). *Help starts here: the maladjusted child in the ordinary school.* Tavistock, London.

Komrower, G. M. and Lee, D. H. (1970). Long-term follow-up of galactosaemia. *Archives of Disease in Childhood* 45, 367–73.

Kovacs, M. and Gatsonis, C. (1989). Stability and change in childhood-onset depressive disorders: longitudinal course as a diagnostic validator. In *The validity of psychiatric diagnosis* (ed. L. N. Robins and J. E. Barrett), pp. 57–75. Raven Press, New York.

Kramer, H. H., Awiszus, D., Sterzel, U., Van Halteren, A., and Classen, R. (1989). Development of personality and intelligence in children with congenital heart disease. *Journal of Child Psychology and Psychiatry* 30, 299–308.

Kreitman, N. (1977). *Parasuicide.* Wiley, Chichester.

Krener, P. and Miller, F. B. (1989). Psychiatric response to HIV spectrum disease in children and adolescents. *Journal of the American Academy of Child and Adolescent Psychiatry* 28, 596–605.

Kumar, R. and Robson, K. (1978). Previous induced abortion and ante-natal depression in primiparas: preliminary report of a survey of mental health in pregnancy. *Psychological Medicine* 8, 711–15.

Kun, L. E., Mulhern, R. K., and Crisco, J. J. (1983). Quality of life in children treated for brain tumours: intellectual, emotional and academic function. *Journal of Neurosurgery* **58**, 1–6.

Kupst, M. J. and Schulman, J. L. (1988). Long-term coping with pediatric leukemia: a 6 year follow-up study. *Journal of Pediatric Psychology* **13**, 7–22.

Kupst, M. J., Schulman, J. L., Davis, A. T., and Richardson, C. C. (1983). The psychological impact of pediatric bacterial meningitis on the family. *Pediatric Infectious Diseases* **2**, 12–17.

Kurtz, Z., Pilling, D., Blau, J. N., and Peckham, C. S. (1983). Migraine in children. Findings from the National Child Development Study. In *Progress in migraine research II* (ed. F. C. Rose). Pitman Medical, Tunbridge Wells.

Lambert, L. and Streather, J. (1980). *Children in changing families: a study of adoption and illegitimacy.* Macmillan, London.

Lamont, M. A. and Dennis, N. R. (1988). Aetiology of mild mental retardation. *Archives of Disease in Childhood* **63**, 1032–8.

Lannering, B., Markey, I., Lundberg, A., and Olsson, E. (1990). Long-term sequelae after pediatric brain tumors: their effect on disability and quality of life. *Medical Pediatric Oncology* **18**, 304–10.

Lansdown, R. (1981). Cleft lip and palate: a prediction of psychological disfigurement? *British Journal of Orthodontics* **8**, 83–8.

Lansdown, R. and Benjamin, G. (1985). The development of the concept of death in children aged five to nine years. *Child: Care, Health and Development* **11**, 13–20.

Lansdown, R. and Polak, L. (1975). A study of the psychosocial effects of facial deformity in children. *Child: Care, Health and Development* **1**, 885–91.

Lapouse, R. and Monk, M. A. (1968). An epidemiological study of behaviour characteristics in children. In *Children's behavior disorders* (ed. H. Quay). Van Nostrand, Princetown, New Jersey.

Lask, B. (1987). Family therapy. *British Medical Journal* **294**, 203–4.

Lask, B. (1988). Novel and non-toxic treatment for night terrors. *British Medical Journal* **297**, 592.

Lask, B. (1988). Psychological aspects of gastro-intestinal disorders. In *Essentials of paediatric gastroenterology* (ed. P. Milla and D. Muller). Churchill Livingstone, Edinburgh.

Lask, B. (1989). Family therapy and group therapy. In *Studies on child psychiatry* (ed. B. J. Tonge, G. D. Burrows, and J. S. Werry), pp. 431–42. Elsevier, Amsterdam.

Lask, B. and Matthew, D. (1979). Childhood asthma—a controlled trial of family psychotherapy. *Archives of Diseases in Childhood* **54**, 116–19.

Laslett, R. (1977). *Educating maladjusted children.* Crosby Lockwood Staples, London.

Lassman, L. and Arjona, V. E. (1967). Pontine gliomas of childhood. *Lancet* **i**, 913–15.

Laurance, B. (1985). Prader-Willi syndrome. *Journal of Maternal and Child Health* **10**, 106.

Lawson, D., Metcalfe, M., and Pampiglione, G. (1965). Meningitis in childhood. *British Medical Journal* **1**, 557–62.

Laxova, R., Ridler, M., and Bowen-Bravery, M. (1977). An aetiological survey of the severely retarded Hertfordshire children who were born between January 1st, 1965 and December 31st, 1967. *American Journal of Medical Genetics* **1**, 75–86.

Leach, D. (1980). Assessing children with learning difficulties: an alternative model for psychologists and teachers. *Journal of the Association of Educational Psychologists* 5, 16–23.

Le Couteur, A., Bailey, A. J., Rutter, M., and Gottesman, I. (1989). An epidemiologically based twin study of autism. Paper given at the First World Congress on Psychiatric Genetics, Churchill College, Cambridge, 3–5 August, 1989.

Lee, C. M. and Mash, E. J. (1989). Behaviour therapy. In *Studies on child psychiatry* (eds. B. J. Tonge, G. D. Burrows, and J. Werry), p. 424. Elsevier, Amsterdam.

Leff, J. (1978). Social and psychological causes of the acute attack. In *Schizophrenia: towards a new synthesis* (ed. J. K. Wing). Academic Press, London.

Leicester, J. (1982). Temper-tantrums, epilepsy and episodic dyscontrol. *British Journal of Psychiatry* 141, 262–6.

Lenke, R. R. and Levy, H. L. (1980). Maternal phenylketonuria and hyperphenylalanaemia. An international survey of the outcome of treated and untreated pregnancies. *New England Journal of Medicine* 303, 1202–8.

Leonard, H., Swedo, S., Rapoport, J., Coffey, M., and Cheslow, D. (1988). Treatment of obsessive–compulsive disorder with clomipramine and desmethylimipramine: a double-blind cross-over comparison. *Psychopharmacology Bulletin* 24, 252–66.

Lerner, J. W. (1989). Educational interventions in learning disabilities. *Journal of the American Academy of Child and Adolescent Psychiatry* 28, 326–33.

Leslie, S. A. (1974). Psychiatric disorder in the young adolescents of an industrial town. *British Journal of Psychiatry* 125, 113–24.

Leventhal, J. M. (1988). Can child maltreatment be predicted during the peri-natal period: evidence from longitudinal cohort studies. *Journal of Reproductive Infant Psychology* 6, 139–61.

Levine, M. D. (1982). Encopresis: its potentiation, evaluation and alleviation. *Pediatric Clinics of North America* 29, 315–29.

Levine, M. D. and Bakow, H. (1976). Children with encopresis: a study of treatment outcome. *Pediatrics* 58, 845–52.

Levy, A. M. (1985). The divorcing family: its evaluation and treatment. In *The clinical guide to child psychiatry* (ed. D. Shaffer, A. E. Ehrhardt, and L. L. Greenhill), pp. 353–70. Free Press, New York.

Levy, D. (1943). *Maternal overprotection*. Columbia University Press, New York.

Liepmann, M. C. (1979). Mentally handicapped children in Mannheim. In *Estimating needs for mental health care* (ed. H. Hafner), pp. 113–24. Springer-Verlag, Berlin.

Linde, L., Rasof, B., and Dunn, O. J. (1967). Mental development in congenital heart disease. *Pediatrics* 38, 92–101.

Lindsay, J., Glaser, G., Richards, P., and Ounsted, C. (1984). Developmental aspects of focal epilepsies of childhood treated by neurosurgery. *Developmental Medicine and Child Neurology* 26, 574–87.

Lishman, W. A. (1978). *Organic psychiatry*, p. 244. Blackwell Scientific Publications, Oxford.

Lloyd, J. K. and Wolff, O. H. (1976). Obesity. In *Recent advances in paediatrics* (ed. D. Hull), pp. 306–31. Churchill Livingstone, Edinburgh.

Lorber, J. (1971). Results of treatment of myelomeningocoele. An analysis of 524 unselected cases. *Developmental Medicine and Child Neurology* 13, 279–303.

Lotter, V. (1978). Follow-up studies. In *Autism: a reappraisal of concepts and treatments* (ed. M. Rutter and E. Schopler), pp. 475–95. Plenum Press, New York.

Lowe, M. and Costello, A. J. (1976). *Symbolic play test.* NFER Publishing Company, Windsor.

Luk, S. L., Leung, P. W. L., and Lee, P. L. M. (1988). Conners Teacher Rating Scale in Chinese children in Hong Kong. *Journal of Child Psychology and Psychiatry* 29, 165–74.

Lynch, M. A. and Roberts, J. (1982). *Consequences of child abuse.* Academic Press, London.

McAnarney, E. R., Pless, I. B., Satterwhite, B., and Friedman, S. B. (1974). Psychological problems of children with chronic juvenile arthritis. *Pediatrics* 53, 523–8.

McBurney, A. K. and Eaves, L. C. (1986). Evolution of developmental and psychological test scores. In *Sequelae of low birthweight: the Vancouver Study, Clinics in Developmental Medicine* Nos. 95/96 (ed. H. G. Dunn), pp. 54–67. MacKeith Press, London.

McCauley, E., Ito, J., and Kay, T. (1986). Psychosocial functioning in girls with Turner's Syndrome and short stature: social skills, behavior problems and self-concept. *Journal of the American Academy of Child Psychiatry* 25, 105–12.

McClure, G. M. G. (1988). Suicide in children in England and Wales. *Journal of Child Psychology and Psychiatry* 29, 345–9.

Maccoby, E. E. and Jacklin, L. N. (1980). Psychological sex differences. In *Scientific foundations of developmental psychiatry* (ed. M. Rutter). Heinemann Medical, London.

McDermott, J. and Finch, S. (1967). Ulcerative colitis in children: reassessment of a dilemma. *Journal of the American Academy of Child Psychiatry* 6, 512–25.

McFarlane, A. C. (1988). Recent life events and psychiatric disorder in children: the interaction with preceding extreme adversity. *Journal of Child Psychology and Psychiatry* 29, 677–90.

Macfarlane, A. (1985). The Education Act 1981. *British Medical Journal* 290, 1848–9.

Macfarlane, J. W., Allen, L., and Honzik, M. R. (1954). *A developmental study of the behavior problems of normal children between 21 months and 14 years.* University of California Press, Berkeley and Los Angeles.

McGrath, P. J., Goodman, J. T., Firestone, P., Shipman, R., and Peters, S. (1983). Recurrent abdominal pain. A psychogenic disorder? *Archives of Disease in Childhood* 58, 888–90.

McGrath, P. J. *et al.* (1988). Relaxation prophylaxis for childhood migraine: a randomised placebo controlled trial. *Developmental Medicine and Child Neurology* 30, 626–31.

McGreevy, P. and Arthur, M. (1987). Effective behavioural treatment of self-biting by a child with Lesch–Nyhan syndrome. *Developmental Medicine and Child Neurology* 29, 536–40.

McNichol, K. N., Williams, H. C., Allan, J., and McAndrews, I. (1973). Spectrum of asthma in children. III: Psychological and social components. *British Medical Journal* 4, 16–20.

McQueen, P. C., Spence, M. W., Winsor, E., Garner, J. B., and Pereira, L. (1986).

Causal origins of major mental handicap in the Canadian maritime provinces. *Developmental Medicine and Child Neurology* **28**, 697–707.

Macy, C. (1986). Psychological factors in nausea and vomiting in pregnancy: a review. *Journal of Reproductive and Infant Psychology* **4**, 23–55.

Madge, N. and Tizard, J. (1980). Intelligence. In *Scientific foundation of developmental psychiatry* (ed. M. Rutter), pp. 245–65. Heinemann Medical, London.

Maguire, P. (1983). The psychological sequelae of childhood leukaemia. *Recent Results in Cancer Research* **88**, 47–56.

Mannino, F. V. and Shore, M. F. (1975). The effects of consultation: a review of empirical studies. *American Journal of Community Psychology* **3**, 1–21.

Marchi, M. and Cohen, P. (1990). Early childhood eating behaviors and adolescent eating disorders. *Journal of the American Academy of Child and Adolescent Psychiatry* **29**, 112–17.

Markova, I., Macdonald, K., and Forbes, C. (1980*a*). Impact of haemophilia on child-rearing practices and parental cooperation. *Journal of Child Psychology and Psychiatry* **21**, 153–62.

Markova, L., Macdonald, K., and Forbes, C. (1980*b*). Integration of haemophiliac boys into normal schools. *Child: Care, Health and Development* **6**, 101–9.

Markova, L., Phillips, J. S., and Forbes, C. D. (1984). The use of tools by children with haemophilia. *Journal of Child Psychology and Psychiatry* **25**, 261–71.

Marks, I. (1987). The development of normal fear: a review. *Journal of Child Psychology and Psychiatry* **28**, 667–98.

Markus, E., Lange, A., and Pettigrew, T. F. (1990). Effectiveness of family therapy—a meta analysis. *Journal of Family Therapy* **12**, 205–22.

Marlow, N., Roberts, B. L., and Cooke, R. W. I. (1988). Motor skills in extremely low birthweight children at the age of six years. *Archives of Disease in Childhood* **64**, 839–47.

Marteau, T. M., Bloch, S., and Baum, D. (1987). Family life and diabetic control. *Journal of Child Psychology and Psychiatry* **28**, 823–34.

Martin, P. (1975). Marital breakdown in families of patients with spina bifida cystica. *Developmental Medicine and Child Neurology* **17**, 757–64.

Matheny, A. P. (1986). Injuries among toddlers: contributions from child, mother and family. *Journal of Pediatric Psychology* **11**, 163–76.

Mattison, R. E., Humphrey, F. J., Kales, S. N., Handford, H. A., Finkenbinder, R. L., and Hernit, R. C. (1986). Psychiatric background and diagnosis of children. Evaluation for special class placement. *Journal of the American Academy of Child Psychiatry* **25**, 514–20.

Mattson, A. and Gross, S. (1966). Social and behavioral studies on haemophiliac children and their families. *Journal of Pediatrics* **68**, 952–64.

Matus, I. (1981). Assessing the nature and clinical significance of psychological contributions to childhood asthma. *American Journal of Orthopsychiatry* **51**, 327–41.

Maughan, B. (1988). School experiences as risk/protective factors. In *Studies of psychosocial risk* (ed. M. Rutter). Cambridge University Press, Cambridge.

Meadow, R. (1982). Munchausen syndrome by proxy. *Archives of Disease in Childhood* **57**, 92–8.

Meadow, R. (1985). Management of Munchausen syndrome by proxy. *Archives of Disease in Childhood* **60**, 385–93.

Mechanic, D. (1978). *Medical sociology*, 2nd edn. Free Press, Glencoe, Illinois.

Meyer, R. J. and Haggerty, R. J. (1962). Streptococcal infection in families. *Pediatrics* **29**, 539–49.

Meyers, A. M. and Craighead, W. E. (1984). *Cognitive behaviour therapy with children*. Plenum Press, New York.

Mills, M., Puckering, C., Pound, A., and Cox, A. (1984). What is it about depressed mothers that influences their children's functioning? In *Recent research in developmental psychopathology* (ed. J. Stevenson), pp. 11–17. Pergamon, Oxford.

Minde, K. K. (1978). Coping styles of twenty-four adolescents with cerebral palsy. *American Journal of Psychiatry* **135**, 1344–9.

Minde, K. K., Whitelaw, A., Brown, J., and Fitzhardinge, P. (1983). Effect of neonatal complications in premature infants on early parent–infant interactions. *Developmental Medicine and Child Neurology* **25**, 763–77.

Minuchin, S. (1974). *Families and family therapy*. Tavistock, London.

Mitchell, W. G., Gorrell, R. W., and Greenberg, R. A. (1980). Failure to thrive in primary care. *Pediatrics* **65**, 971–7.

Monck, E. and Graham, P. (1988). Suicide ideation in a total population of 15–19 year old girls. In *Tentatives de Suicide à l'Adolescence*, pp. 167–76. Centre Internationale de l'Enfance, Paris.

Money, J. (1975). Psychologic counselling: hermaphroditism. In *Endocrine and genetic diseases of childhood and adolescence*, 2nd edn. W. B. Saunders, Philadelphia.

Money, J. and Ehrhardt, A. (1972). *Man and woman: boy and girl. The differentiation and dimorphism of gender identity from conception to maturity*. Johns Hopkins University Press, Baltimore, Maryland.

Money, J., Schwartz, J., and Lewis, V. G. (1984). Adult erotosexual status and fetal hormonal masculinization and demasculinization: 46 XX congenital virilizing adrenal hyperplasia and 46 XY androgen insensitivity syndrome compared. *Psychoneuroendrocrinology* **9**, 405–14.

Mrazek, D. A., Casey, B., and Anderson, I. (1987). Insecure attachment in severely asthmatic pre-school children: is it a risk factor? *Journal of the American Academy of Child and Adolescent Psychiatry* **26**, 516–20.

Mulhern, R. K., Crisio, J. J., and Camitta, B. (1981). Patterns of communication among paediatric patients with leukaemia, parents and physicians: prognostic disagreements and misunderstandings. *Journal of Pediatrics* **99**, 480–3.

Murphy, G. (1985). Update—Self-injuring behaviour in the mentally handicapped. *Newsletter, Association for Child Psychology and Psychiatry* **7**, 2–11.

Murphy, G., Hulse, J. A., Jackson, D., Tyrer, P., Glossop, J., Smith, I., and Grant, D. (1986). Early treated hypothyroidism: development at 3 years. *Archives of Disease in Childhood* **61**, 761–5.

Murray, D. M. and Lewis, S. W. (1987). Is schizophrenia a developmental disorder? *British Medical Journal* **295**, 681–2.

Naditch, M. P. (1974). Acute adverse reactions to psychoactive drugs. *Journal of Abnormal Psychology* **83**, 394–403.

Neale, M. D. (1988). *Neale analysis of reading ability manual*, Revised British edn. Macmillan, London.

National Children's Bureau (1985). *Help starts here*. Voluntary Council for Handicapped Children, London.

Needleman, H., Gunnoe, C., Leviton, A., Reed, R., Peresie, H., Maher, C., and Barrett, P. (1979). Deficits in psychologic and classroom performances of children with elevated dentine lead levels. *New England Journal of Medicine* **300**, 689–95.

Neligan, G. A., Kolvin, I., Scott, D. McI., and Garside, R. F. (1976). Born too soon or too small. *Clinics in Developmental Medicine, No. 61.* Heinemann, London.

Neuchterlein, K. H. (1986). Childhood precursors of adult schizophrenia. *Journal of Child Psychology and Psychiatry* **27**, 133–44.

Newacheck, P. W., Budetti, P. P., and Halfon, N. (1986). Trends in activity-limiting chronic conditions among children. *American Journal of Public Health* **76**, 178–84.

Newberger, C. S. and White, K. M. (1989). Cognitive foundations for parental care. In *Child maltreatment* (ed. D. Cicchetti and V. Carlson), pp. 302–16. Cambridge University Press, Cambridge.

Newson, J. and Newson, E. (1963). *Patterns of infant care in an urban community.* Penguin Press, Harmondsworth.

Newton, R. W. (1988). Psychosocial aspects of pregnancy: the scope for intervention. *Journal of Reproductive and Infant Psychology* **6**, 23–39.

Nicol, A. R. and Eccles, M. (1985). Psychotherapy for Munchausen syndrome by proxy. *Archives of Disease in Childhood* **60**, 344–8.

Nielsen, J. (1969). Klinefelter's syndrome and the XYY syndrome. A genetic, endocrinological and psychiatric/psychological study of thirty-three severely hypogonadal male patients and two patients with karyotype 47 XYY. *Acta Psychiatrica Scandinavica*, Suppl. 209.

Nunn, K. (1986). The episodic dyscontrol syndrome in childhood. *Journal of Child Psychology and Psychiatry* **27**, 439–46.

Nurcombe, B., Seifer, R., Scioli, A., Tramontana, M. G., Grapentine, W. L., and Beauchesne, H. C. (1989). Is major depressive disorder in adolescence a distinct diagnostic entity? *Journal of the American Academy of Child and Adolescent Psychiatry* **28**, 333–42.

Nyhan, W. L. (1976). Behaviour in the Lesch–Nyhan syndrome. *Journal of Autism and Childhood Schizophrenia* **6**, 235–42.

Oakley, A. (1988). Is social support good for the health of mothers and babies. *Journal of Reproductive and Infant Psychology* **6**, 3–21.

Oates, M. (1984). Assessing fitness to parent. In *Taking a stand*, pp. 29–41. British Agencies for Adoption and Fostering, London.

Oates, M. (1988). The development of an integrated community orientated service for severe post-natal mental illness. In *Motherhood and mental illness. 2. Causes and consequences* (ed. I. F. Brockington). Wright, London.

O'Brien, S., Ross, L. V., and Christopherson, E. R. (1986). Primary encopresis: evaluation and treatment. *Journal of Applied Behavioral Analysis* **19**, 137–45.

O'Callaghan, E. M. and Colver, A. F. (1987). Selective medical examinations on starting school. *Archives of Disease in Childhood* **62**, 1041–3.

O'Callaghan, M. and Hull, D. (1978). Failure to thrive or failure to rear. *Archives of Disease in Childhood* **53**, 788–93.

O'Connor, N. and Hermelin, B. (1984). Idiot savant calendrical calculations. *Psychological Medicine* **14**, 801–6.

Oddy, M. (1984). Head injury during childhood: the psychological implications. In *Closed head injury: psychological, social, and family consequences* (ed. N. Brooks), pp. 179–94. Oxford University Press, Oxford.

Office of Population Censuses and Surveys (1983). *Social trends*. HMSO, London.

Office of Population Censuses and Surveys (1986). *Birth statistics 1985, England and Wales*. HMSO, London.

Offord, D., Boyle, M. H., and Racine, Y. (1989). Ontario Child Health Study: correlates of disorder. *Journal of the American Academy of Child and Adolescent Psychiatry* **28**, 856–60.

Offord, D. R. (1987). Prevention of behavioural and emotional disorders in children. *Journal of Child Psychology and Psychiatry* **28**, 9–20.

Offord, D. R. *et al.* (1987). Ontario Child Health Study. II. Six months prevalence of disorder and rates of serious utilisation. *Archives of General Psychiatry* **44**, 832–6.

Olness, K. (1989). Hypnotherapy: a cyberphysiologic strategy in pain management. *Pediatric Clinics of North America* **36**, 873–84.

Olness, K. and Gardner, G. G. (1988). *Hypnosis and hypnotherapy with children*. Grune and Stratton, Philadelphia.

Olsson, B. and Rett, A. (1987). Autism and Rett Syndrome: behavioural investigation and differential diagnosis. *Developmental Medicine and Child Neurology* **29**, 429–41.

Olweus, D. (1989). Bully/victim problems among schoolchildren: basic facts and effects of a school based intervention program. In *The development and treatment of childhood aggression*. Erlbaum, Hillsdale, New Jersey.

Orlansky, H. (1949). Infant care and personality. *Psychological Bulletin* **46**, 1–48.

Orvaschel, H. (1988). Structured and semi-structured psychiatric interviews for children. In *Handbook of clinical assessment of children and adolescents* (ed. C. J. Kestenbaum and D. T. Williams), pp. 31–42. New York University Press, New York.

Osofsky, H. J., Osofsky, J. D., Kendall, N., and Rajan, R. (1973). Adolescents as mothers: an inter-disciplinary approach to a complex problem. *Journal of Youth and Adolescence* **2**, 233–49.

Ottinger, D. and Simmons, J. (1964). Behaviour of human neonates and prenatal maternal anxiety. *Psychological Reports* **14**, 391–4.

Palmai, G., Storey, P. B., and Briscoe, O. (1967). Social class and the young offender. *British Journal of Psychiatry* **113**, 1073–82.

Panitch, H. S. and Berg, B. O. (1970). Brain stem tumors of childhood and adolescence. *American Journal of Diseases in Childhood* **119**, 465–72.

Parmelee, A. H. (1986). Children's illnesses: their beneficial effects on behavioral development. *Child Development* **57**, 1–10.

Parsons, T. (1951). *The social system*. Free Press, Glencoe.

Patterson, G. R. (1982). *Coercive family process*. Castalia Publishing Company, Eugene, Oregon.

Payton, J. B., Steele, M. W., Wenger, S. L., and Minshew, N. J. (1989). The fragile X marker and autism in perspective. *Journal of the American Academy of Child and Adolescent Psychiatry* **28**, 417–21.

Pelligrini, D. S. and Urbain, E. S. (1985). An evaluation of inter-personal cognitive problem-solving training with children. *Journal of Child Psychology and Psychiatry* **26**, 17–41.

Pennington, B. F. (1990). The genetics of dyslexia. *Journal of Child Psychology and Psychiatry* **31**, 192–202.

Peterson, L. (1989). Coping by children undergoing stressful medical procedures: some conceptual, methodological and therapeutic issues. *Journal of Consulting and Clinical Psychology* **57**, 380–7.

Petty, L. K., Ornitz, E. M., Michelman, J. D., and Zimmerman, E. G. (1984). Autistic children who become schizophrenic. *Archives of General Psychiatry* **41**, 129–35.

Plant, M. A., Peck, D. F., and Stuart, R. (1982). Self-reported drinking habits and alcohol-related consequences amongst a cohort of Scottish teenagers. *British Journal of Addiction* **77**, 75–90.

Pless, I. B. and Douglas, J. W. B. (1971). Chronic illness in childhood. I. Epidemiological and clinical characteristics. *Pediatrics* **47**, 405–14.

Pless, I. B., Cripps, H. A., Davies, J. M. C., and Wadsworth, M. (1989). Chronic physical illness in childhood: psychological and social effects in adolescence and adult life. *Developmental Medicine and Child Neurology* **31**, 747–55.

Porter, R. (ed.) (1984). *Child sexual abuse within the family*. Tavistock, London.

Pot-Mees, C. (1989). *The psychosocial effects of bone marrow transplantation in children*. Eburon, Delft.

Prendergast, M. *et al.* (1988). The diagnosis of childhood hyperactivity: a US UK cross-national study of DSM III and ICD-9. *Journal of Child Psychology and Psychiatry* **29**, 289–300.

Price, R. A., Pauls, D. L., Kruger, S. D., and Caine, E. D. (1988). Family data support a dominant major gene for Tourette syndrome. *Psychiatry Research* **24**, 251–61.

Purcell, K. and Weiss, J. (1970). Asthma. In *Symptoms of psychopathology* (ed. C. Costello), pp. 597–623. Wiley, New York.

Query, J. M., Reichelt, C., and Christoferson, L. A. (1990). Living with chronic illness: a retrospective study of patients shunted for hydrocephalus and their families. *Developmental Medicine and Child Neurology* **32**, 119–28.

Quinton, D. and Rutter, M. (1976). Early hospital admission and later disturbances of behaviour: an attempted replication of Douglas' findings. *Developmental Medicine and Child Neurology* **18**, 447–59.

Quinton, D. and Rutter, M. (1985). Parenting behaviour of mothers raised 'in care'. In *Longitudinal studies in child psychology and psychiatry* (ed. A. R. Nicol), pp. 157–201. John Wiley, Chichester.

Quinton, D., Rutter, M., and Gulliver, L. (1990). Continuities in psychiatric disorders from childhood to adulthood in the children of psychiatric patients. In *Straight and deviant pathways from childhood to adulthood* (ed. L. Robins and M. Rutter), pp. 259–78. Cambridge University Press, Cambridge.

Rachelefsky, G. S. *et al.* (1986). Behavior abnormalities and poor school performance due to oral theophylline use. *Pediatrics* **78**, 1133–8.

Raphael, B. (1975). The management of pathological grief. *Australian and New Zealand Journal of Psychiatry* **9**, 173–80.

Raphael, B. (1984). *The anatomy of bereavement*. Hutchinson, London.

Rapoport, J. L. (1986). Childhood obsessive compulsive disorders. *Journal of Child Psychology and Psychiatry* **27**, 289–95.

Ratcliffe, S. G. and Field, M. A. (1982). Emotional disorder in XYY children: four case reports. *Journal of Child Psychology and Psychiatry* **23**, 401–6.

Ratcliffe, S. G., Butler, G. E., and Jones, M. (1990). Edinburgh study of growth and development of children with sex chromosome abnormalities. In *Birth defects original article series* (ed. J. Hamerton and A. Robinson). Alan R. Liss Inc., New York.

Rauch, P. and Jellinek, M. S. (1988). Psychosocial development in children with cutaneous disease. In *Pediatric dermatology* (ed. L. A. Schachner and R. C. Hanson), pp. 139–58. Churchill Livingstone, New York.

Raymer, D., Weininger, O., and Hamilton, J. R. (1984). Psychological problems in children with abdominal pain. *Lancet* i, 439–40.

Reed, G. F. (1963). Elective mutism in children: a re-appraisal. *Journal of Child Psychology and Psychiatry* 4, 99–107.

Reid, W. J. and Shyne, A. W. (1969). *Brief and extended casework*. Columbia University Press, New York.

Reisman, J. M. (1973). *Principles of psychotherapy with children*. John Wiley, New York.

Rekers, G. A. (1977). Assessment and treatment of childhood gender problems. In *Advances in clinical child psychology*, Vol. 1 (ed. B. B. Lahey and A. E. Kazdin), pp. 267–306. Plenum Press, New York.

Revill, S. and Blunden, R. (1979). A home training service for pre-school, developmentally handicapped children. *Behaviour Research and Therapy* 17, 207–14.

Reynell, J. (1969). *Reynell Developmental Language Scales*. NFER Publishing Company, Windsor.

Reynell, J. and Zinkin, P. (1975). New procedures for the developmental assessment of young children with severe visual handicaps. *Child: Care, Health and Development* 1, 61–9.

Richardson, S. A. and Royce, J. (1968). Race and physical handicap in children's preference for other children. *Child Development* 39, 467–80.

Richman, L. C. (1978). The effects of facial disfigurement on teachers' perceptions of ability in cleft palate children. *Cleft Palate Journal* 15, 115–60.

Richman, N. (1977a). Behaviour problems in pre-school children. Family and social factors. *British Journal of Psychiatry* 131, 523–7.

Richman, N. (1977b). Is a behaviour check-list for pre-school children useful? In *Epidemiological approaches in child psychiatry* (ed. P. Graham), pp. 125–38. Academic Press, London.

Richman, N. (1981). A community survey of one to two year olds with sleep disturbances. *Journal of the American Academy of Child Psychiatry* 20, 281–91.

Richman, N., Stevenson, J., and Graham, P. (1975). Prevalence of behaviour problems in three year old children: an epidemiological study in a London borough. *Journal of Child Psychology and Psychiatry* 16, 277–87.

Richman, N., Stevenson, J., and Graham, P. (1982). *Pre-school to school: a behavioural study*. Academic Press, London.

Richman, N., Graham, P., and Stevenson, J. (1983). Long term effects of treatment in a pre-school day centre. *British Journal of Psychiatry* 142, 71–7.

Richman, N., Douglas, J., Hunt, H., Lansdown, R., and Levere, R. (1985). Behavioural methods in the treatment of sleep disorders—a pilot study. *Journal of Child Psychology and Psychiatry* 26, 581–90.

Riikonen, R. and Amnell, G. (1981). Psychiatric disorders in children with earlier infantile spasms. *Developmental Medicine and Child Neurology* 23, 747–60.

Rivinus, T. M., Jamison, D. L., and Graham, P. J. (1975). Childhood organic neurological disease presenting as psychiatric disorder. *Archives of Disease in Childhood* 50, 115–19.

Robins, L. and Rutter, M. (eds.) (1990). *Straight and deviant pathways from childhood to adulthood.* Cambridge University Press, Cambridge.

Robins, L. N. (1966). *Deviant children grown up.* Williams and Wilkins, Baltimore.

Robins, L. N. (1978). Sturdy childhood predictors of adult antisocial behaviour: replications from longitudinal studies. *Psychological Medicine* 8, 611–22.

Roche, A., Lipman, R., Overall, J., and Hung, H. (1979). The effects of stimulant medication on the growth of hyperkinetic children. *Pediatrics* 63, 647–50.

Rodeck, C. H. (ed.) (1987). Fetal diagnosis of congenital defects. *Clinical Obstetrics and Gynaecology*, Volume 1, No. 3. Ballière & Tindall, London.

Rojahn, J. (1986). Self-injurious and stereotypic behavior of non-institutionalised mentally retarded people. *American Journal of Mental Deficiency* 91, 268–76.

Rosenfeld, A. A., Siegel-Gorlick, B., and Haavik, D. (1984). Parental perceptions of children's modesty: a cross-sectional survey of ages 2–10 years. *Psychiatry* 47, 351–65.

Rosenthal, A. and Levine, S. V. (1971). Brief psychotherapy with children: process and therapy. *American Journal of Psychiatry* 128, 141–5.

Rosett, H. L., Weiner, L., Lee, A., Zuckerman, B., Dooling, E., and Oppenheimer, E. (1983). Patterns of alcohol consumption and fetal development. *Obstetrics and Gynaecology* 61, 539–46.

Runciman, W. G. (1972). *Relative deprivation and social justice.* Penguin Press, Harmondsworth.

Russell, A. T., Bott, L., and Sammons, C. (1989). The phenomenology of schizophrenia occurring in childhood. *Journal of the American Academy of Child and Adolescent Psychiatry* 28, 399–407.

Russell, G. F. M. (1979). Bulimia nervosa: an ominous variant of anorexia nervosa. *Psychological Medicine* 9, 429–48.

Russell, G. F. M., Szmukler, G. I., Dare, C., and Eisler, I. (1987). An evaluation of family therapy in anorexia nervosa and bulimia nervosa. *Archives of General Psychiatry* 44, 1047–56.

Russell, J. K. (1974). Sexual activity and its consequences in teenagers. *Clinics in Obstetrics and Gynaecology* 3, 683.

Rutter, M. (1982). Psychological therapies in childhood: issues and prospects. *Psychological Medicine* 12, 723–40.

Rutter, M. (1983). Cognitive deficits in the pathogenesis of autism. *Journal of Child Psychology and Psychiatry* 24, 513–31.

Rutter, M. (1985). The treatment of autistic children. *Journal of Child Psychology and Psychiatry* 26, 193–214.

Rutter, M. (1989a). Child psychiatric disorders in ICD-10. *Journal of Child Psychology and Psychiatry* 30, 499–513.

Rutter, M. (1989b). Pathways from childhood to adult life. *Journal of Child Psychology and Psychiatry* 30, 23–51.

Rutter, M. (1989c). Isle of Wight revisited. Twenty-five years of child psychiatric epidemiology. *Journal of the American Academy of Child and Adolescent Psychiatry* 28, 633–53.

Rutter, M. and Bartak, L. (1973). Special educational treatment of autistic children: a comparative study. II. Follow-up findings and implications for services. *Journal of Child Psychology and Psychiatry* **14**, 241–70.

Rutter, M. and Giller, H. (1983). *Juvenile delinquency: trends and perspectives.* Penguin Press, Harmondsworth.

Rutter, M. and Madge, N. (1976). *Cycles of disadvantage*, p. 180. Heinemann, London.

Rutter, M. and Quinton, D. (1984). Parental psychiatric disorder: effects on children. *Psychological Medicine* **14**, 835–80.

Rutter, M., Greenfield, D., and Lockyer, L. (1967). A five to fifteen year follow-up study of infantile psychosis. II. Social and behavioural outcome. *British Journal of Psychiatry* **113**, 1183–99.

Rutter, M., Tizard, J., and Whitmore, K. (1970a). *Education, health and behaviour.* Longmans, London.

Rutter, M., Graham, P., and Yule, W. (1970b). *A neuropsychiatric study in childhood.* Heinemann, London.

Rutter, M., Yule, W., Berger, M., Yule, B., Morton, J., and Bagley, C. (1974). Children of West Indian immigrants. I. Rates of behavioural deviance and of psychiatric disorder. *Journal of Child Psychology and Psychiatry* **15**, 241–62.

Rutter, M., Cox, A., Tupling, C., Berger, M., and Yule, W. (1975). Attainment and adjustment in two geographical areas. I. Prevalence of psychiatric disorders. *British Journal of Psychiatry* **126**, 493–509.

Rutter, M., Tizard, J., Yule, W., Graham, P., and Whitmore, K. (1976a). Isle of Wight studies 1964–1974. *Psychological Medicine* **6**, 313–32.

Rutter, M., Graham, P., Chadwick, O., and Yule, W. (1976b). Adolescent turmoil: fact or fiction? *Journal of Child Psychology and Psychiatry* **17**, 35–56.

Rutter, M., Shaffer, D., and Sturge, C. (1979a). *A guide to a multi-axial classification scheme for psychiatric disorders in childhood and adolescence.* Institute of Psychiatry, London.

Rutter, M., Maughan, B., Mortimore, P., and Ouston, J. (1979b). *Fifteen thousand hours.* Open Books, London.

Rutter, M., Chadwick, O., and Shaffer, D. (1983). Head injury. In *Developmental neuropsychiatry* (ed. M. Rutter), pp. 83–111. Guilford Press, New York.

Sabatino, D. and Cramblett, H. (1968). Behavioral sequelae of California encephalitis virus infection in children. *Developmental Medicine and Child Neurology* **10**, 331–7.

Sabbeth, B. and Leventhal, J. M. (1984). Marital adjustment to chronic childhood illness. *Pediatrics* **73**, 762–8.

Sands, A. M. and Golub, S. (1974). Breaking the bonds of tradition: a reassessment of group treatment of latency age children. *American Journal of Psychiatry* **131**, 662–5.

Sattler, J. M. (1982). *Assessment of children's intelligence and special abilities*, 2nd edn. Allyn and Bacon, New York.

Sawyer, M. G., Minde, K., and Zuker, R. (1982). The burned child—scarred for life? A study of the psychosocial impact of a burn injury at different developmental stages. *Burns* **9**, 205–14.

Saylor, C. F., Pallmeyer, T. P., Finch, A. J., Eason, L., Treiber, F., and Folger, C.

(1987). Predictors of psychological distress in hospitalized pediatric patients. *Journal of the American Academy of Child and Adolescent Psychiatry* **26**, 232–6.

Sbriglio, R., Hartman, N., Millman, R. B., and Khuri, E. T. (1988). Drug and alcohol abuse in children and adolescents. In *Handbook of clinical assessment of children and adolescents, Volume II* (ed. C. J. Kestenbaum and D. T. Williams), pp. 915–37. New York University Press, New York.

Schechter, A. (1978). *Treatment aspects of drug dependence.* CRC Press, Florida.

Schechter, M. D. and Roberge, L. (1976). Sexual exploitation. In *Child abuse and neglect: the family and the community* (ed. R. E. Helfer and C. H. Kempe). Ballinger, Cambridge, Massachusetts.

Schlieper, A. (1985). Chronic illness and school achievement. *Developmental Medicine and Child Neurology* **27**, 69–79.

Schonell, F. J. and Schonell, F. E. (1950). *Diagnostic and attainment testing.* Oliver and Boyd, Edinburgh.

Schulman, J. L. (1988). Use of a coping approach in the management of children with conversion reactions. *Journal of the American Academy of Child and Adolescent Psychiatry* **27**, 785–788.

Schultz, J. A. (1983). Timing of elective hypospadias repair in children. *Pediatrics* **71**, 342–51.

Seligman, M. (1975). *Helplessness: on depression, development and death.* Freeman, San Francisco.

Sell, S. H. (1987). Long-term sequelae of bacterial meningitis in children. *Pediatric Infectious Disease* **6**, 775–8.

Sewell, W. H. and Mussen, P. H. (1952). The effect of feeding, weaning and scheduling procedures on childhood adjustment on the formation of oral symptoms. *Child Development* **23**, 185–91.

Shaffer, D. and Fisher, P. (1981). The epidemiology of suicide and attempted suicide. *Journal of the American Academy of Child Psychiatry* **20**, 545–65.

Shaffer, D., Gardner, A., and Hedge, B. (1984). Behaviour and bladder disturbance of enuretic children: a rational classification of a common disorder. *Developmental Medicine and Child Neurology* **26**, 781–92.

Shaffer, D., Garland, A., Gould, M., Fisher, P., and Trautman, P. (1988). Preventing teenage suicide: a critical review. *Journal of the American Academy of Child and Adolescent Psychiatry* **27**, 675–87.

Shannon, F. T., Fergusson, D. M., and Dimond, M. E. (1984). Early hospital admissions and subsequent behaviour problems in six year olds. *Archives of Diseases in Childhood* **59**, 815–19.

Shapiro, A. K., Shapiro, E., and Wayne, H. L. (1973). The symptomatology and diagnosis of Gilles de la Tourette's syndrome. *Journal of the American Academy of Child Psychiatry* **12**, 703–23.

Shaw, M. T. (1977). Accidental poisoning in children: a psychosocial study. *New Zealand Medical Journal* **85**, 269–72.

Shneidman, E. S. (1947). *Make-a-picture story test.* Psychological Corporation, New York.

Siebelink, B. M., Bakker, D. J., Binnie, C. D., and Kasteleijn-Nolst-Trenite, D. G. (1988). Psychological effects of sub-clinical epileptiform EEG discharges in children. II. General intelligence tests. *Epilepsy Research* **2**, 117–21.

Silbert, A., Wolff, P. H., and Lilienthal, J. (1977). Spatial and temporal processing in patients with Turner's syndrome. *Behavioral Genetics* **7**, 11–21.

Sillanpaa, M. (1983). Changes in the prevalence of migraine and other headaches during the first seven school years. *Headache* **23**, 15–19.

Silva, P., Williams, S., and McGee, R. (1987). A longitudinal study of children with developmental language delay at age 3. Later intelligence, reading and behaviour problems. *Developmental Medicine and Child Neurology* **29**, 630–40.

Silverman, C. L., Palkes, H., Talent, B., Kovnar, E., Clouse, J. W., and Thomas, P. R. (1984). Late effects of radiotherapy on patients with cerebellar medulloblastoma. *Cancer* **54**, 825–9.

Simonds, J. F. and Parraga, O. (1982). Prevalence of sleep disorders and sleep behaviors in children and adolescents. *Journal of the American Academy of Child Psychiatry* **21**, 383–8.

Singh, N. N. and Millichamp, C. J. (1985). Pharmacological treatment of self-injurious behaviour in mentally retarded persons. *Journal of Autism and Developmental Disorders* **15**, 257–67.

Skuse, D. (1989). Psychosocial adversity and impaired growth: in search of causal mechanisms. In *The scope of epidemiological psychiatry: essays in honour of Michael Shepherd*, pp. 240–63. Routledge, London.

Skuse, D. H. and Burrell, S. (1982). A review of solvent abusers and their management by a child psychiatry out-patient service. *Human Toxicology* **1**, 321–9.

Skynner, R. (1974). School phobia: a reappraisal. *British Journal of Medical Psychology* **47**, 1–16.

Slater, E., Beard, A. W., and Clithero, E. (1963). The schizophrenia-like psychoses of epilepsy. *British Journal of Psychiatry* **109**, 95–150.

Sluckin, W., Herbert, M., and Sluckin, A. (1983). *Maternal bonding*. Blackwell, London.

Smith, I., Beasley, M. G., Wolff, O. H., and Ades, A. A. (1988). Behavior disturbance in 8 year old children with early treated phenylketonuria. *Journal of Pediatrics* **112**, 403–8.

Smith, M. and Grocke, M. (In Press). Normal family sexuality and sexual knowledge in children. *British Journal of Psychiatry Monograph*. Gaskell Press, London.

Smith, M., Delves, T., Lansdown, R., Clayton, B., and Graham, P. (1983). The effects of lead exposure on urban children. *Developmental Medicine and Child Neurology*, Supplement No. 47. Spastics International Medical Publications, Heinemann, London.

Smithells, R. W. and Smith, I. J. (1984). Alcohol and the fetus. *Archives of Disease in Childhood* **59**, 1113–14.

Sobel, R. (1970). Psychiatric implications of accidental poisoning in childhood. *Pediatric Clinics of North America* **17**, 653–85.

Sokel, B., Lansdown, R., and Kent, A. (1990). The development of a hypnotherapy service for children. *Child: Care, Health and Development* **16**, 227–33.

Sollee, N. D. and Kindlon, D. J. (1987). Lateralised brain injury and behavior problems in children. *Journal of Abnormal Child Psychology* **15**, 479–91.

Sonksen, P. M. (1985). A developmental approach to sensory disabilities in early childhood. *International Rehabilitation Medicine* **7**, 27–32.

Sorensen, R. C. (1973). *Adolescent sexuality in contemporary America: personal values and sexual behavior, ages 13–19*. World Press, New York.

Sourindrhin, I. (1985). Solvent misuse. *British Medical Journal* 290, 94–5.

Spain, B. (1974). Verbal and performance ability in pre-school spina bifida children. *Developmental Medicine and Child Neurology* 16, 773–80.

Sparrow, S., Balla, D., and Cicchetti, D. (1984). *The Vineland Adaptive Behavior Scales*. American Guidance Service, Circle Pines, Minnesota.

Spaulding, B. R. and Morgan, S. B. (1986). Spina bifida children. Their parents: a population prone to family dysfunction? *Journal of Pediatric Psychology* 11, 359–74.

Spence, S. H. (1983). Teaching social skills to children. *Journal of Child Psychology and Psychiatry* 24, 621–7.

Spivack, G., Platt, J. J., and Shure, M. B. (1976). *The problem-solving approach to adjustment*. Jossey-Bass, San Francisco.

Stark, O., Atkins, E., Wolff, O. H., and Douglas, J. W. B. (1981). Longitudinal study of obesity in the National Survey of Health and Development. *British Medical Journal* 283, 13–17.

Steffenburg, S. and Gillberg, C. (1986). Autism and autistic-like conditions in Swedish rural and urban areas: a population study. *British Journal of Psychiatry* 149, 81–7.

Stein, Z. and Susser, M. (1985). Effects of early nutrition on neurological and mental competence in human beings. *Psychological Medicine* 15, 717–26.

Stevens, J. R. (1959). The emotional activation of the electroencephalogram in patients with convulsive disorders. *Journal of Nervous and Mental Disorders* 128, 339.

Stevenson, J. and Richman, N. (1976). The prevalence of language delay in a population of three year old children and its association with general retardation. *Developmental Medicine and Child Neurology* 18, 431–41.

Stevenson, J., Graham, P., Fredman, G., and McLoughlin, V. (1987). A twin study of genetic influences in reading ability and disability. *Journal of Child Psychology and Psychiatry* 28, 229–48.

Stewart, M. A. and Culver, K. W. (1982). Children who set fires: the clinical picture and a follow-up. *British Journal of Psychiatry* 140, 357–63.

Stewart, M. A., Pitts, F., Craig, A., and Dieruf, W. (1966). The hyperactive child syndrome. *American Journal of Orthopsychiatry* 36, 861–7.

Stewart, M. A., DeBlois, C. J., and Cummings, C. (1980). Psychiatric disorder in the parents of hyperactive boys and those with conduct disorder. *Journal of Child Psychology and Psychiatry* 21, 283–92.

Stoddard, F. J., Norman, D. K., Murphy, J. M., and Beardslee, W. R. (1989). Psychiatric outcome of burned children and adolescents. *Journal of the American Academy of Child and Adolescent Psychiatry* 28, 589–95.

Stolberg, A. L. and Garrison, K. M. (1985). Evaluating a primary prevention programme for children of divorce. *American Journal of Community Psychology* 13, 111–24.

Stores, G. (1978). Schoolchildren with epilepsy at risk for learning and behaviour problems. *Developmental Medicine and Child Neurology* 20, 502–8.

Stores, G. (1985). Clinical and EEG evaluation of seizures and seizure-like disorders. *Journal of the American Academy of Child Psychiatry* **24**, 10–16.

Straus, M. A., Gelles, R. J., and Steinmetz, S. K. (1980). *Behind closed doors: violence in the American family.* Anchor Doubleday, Garden City, New York.

Strober, M. and Carlson, G. (1982). Bipolar illness in adolescents with major depression: clinical, genetic and psychopharmacologic predictors in a three to four year prospective follow-up investigation. *Archives of General Psychiatry* **39**, 549–55.

Stuart, R. (1967). Behavioural control of overeating. *Behaviour Research and Therapy* **5**, 357–65.

Stunkard, A. J., Craighead, L. W., and O'Brien, R. (1980). Controlled trial of behaviour therapy, pharmacotherapy and their combination in the treatment of obesity. *Lancet* **ii**, 1045–7.

Stunkard, A. J. *et al.* (1986). An adoption study of human obesity. *New England Journal of Medicine* **314**, 193–8.

Sullivan, B. J. (1979). Adjustment in diabetic adolescent girls. II. Adjustment, self-esteem and depression in diabetic adolescent girls. *Psychosomatic Medicine* **41**, 127–38.

Swisher, L. P. and Pinsker, E. J. (1971). The language characteristics of hyper-verbal, hydrocephalic children. *Developmental Medicine and Child Neurology* **13**, 746–55.

Szajnberg, N. M., Skrinjaric, J., and Moore, A. (1989). Affect attunement, attachment, temperament and zygosity: a twin study. *Journal of the American Academy of Child and Adolescent Psychiatry* **28**, 249–53.

Szatmari, P., Offord, D. R., and Boyle, M. H. (1989). Ontario Child Health Study: prevalence of attention deficit disorder with hyperactivity. *Journal of Child Psychology and Psychiatry* **30**, 219–30.

Tanner, J. M. and Whitehouse, R. H. (1975). Revised standards for triceps and subcapsular skinfolds in British children. *Archives of Disease in Childhood* **50**, 142–5.

Tanner, J. M., Whitehouse, R. H., and Takaishi, M. (1966). Standards from birth to maturity for height, weight, height velocity and weight velocity: British children in 1965. *Archives of Disease in Childhood* **41**, 454–71.

Taylor, D. C. (1982). Counselling the parents of handicapped children. *British Medical Journal* **284**, 1027–8.

Taylor, E. (1984). Diet and behaviour. *Archives of Disease in Childhood* **59**, 97–8.

Taylor, E. (1985). *The hyperactive child: a parents' guide.* Dunitz, London.

Taylor, E. (1986). Overactivity, hyperactivity and hyperkinesis: problems and prevalence. In *The overactive child* (ed. E. Taylor), pp. 1–18. McKeith Press, London.

Taylor, E. M. and Emery, J. L. (1988). Trends in unexpected infant death in Sheffield. *Lancet* **2**, 1121–3.

Taylor, H. G., Michaels, R. H., Mazur, P. M., Bauer, R. E., and Liden, C. B. (1984). Intellectual, neuropsychological and achievement outcomes in children six to eight years after recovery from haemophilus influenzal meningitis. *Pediatrics* **74**, 198–205.

Taylor, I. G. (1984). Deafness. In *Textbook of paediatrics* (ed. G. C. Arneil), pp. 1714–25. Churchill Livingstone, Edinburgh.

Tejani, A., Dobias, B., and Sambursky, J. (1982). Long-term prognosis after H. influenzae meningitis: prospective evaluation. *Developmental Medicine and Child Neurology* **24**, 338–43.

Telch, M. J., Killen, J. D., McAlister, A. L., Perry, C. L., and Maccoby, N. (1982). Long-term follow-up of a pilot project on smoking prevention with adolescents. *Journal of Behavioural Medicine* **5**, 1–8.

Teplin, S. W., Howard, J. A., and O'Connor, M. J. (1981). Self-concept of young children with cerebral palsy. *Developmental Medicine and Child Neurology* **23**, 730–8.

Thake, A., Todd, J., Webb, T., and Bundey, S. (1987). Children with the fragile X chromosome at schools for the mildly mentally retarded. *Developmental Medicine and Child Neurology* **29**, 711–19.

Theut, S. K., Pedersen, F. A., Zaslow, M. J., and Rabinovich, B. A. (1988). Pregnancy subsequent to parental loss: parental anxiety and depression. *Journal of the American Academy of Child and Adolescent Psychiatry* **27**, 289–92.

Thomas, A. and Chess, S. (1977). *Temperament and development*. Brunner Mazel, New York.

Thomas, A., Chess, S., and Birch, H. G. (1968). *Temperament and behavior disorders in childhood*. New York University Press, New York.

Thomas, A., Bax, M., Coombes, K., Goldson, E., Smyth, D., and Whitmore, K. (1985). The health and social needs of physically handicapped young adults. Are they being met by the statutory services? Supplement No. 50 27, No. 4. SIMP. Heinemann, London.

Thomson, G. O. B., Raab, G. M., Hepburn, W. S., Hunter, R., Fulton, M., and Laxen, D. P. H. (1989). Blood lead levels and children's behaviour. *Journal of Child Psychology and Psychiatry* **30**, 515–28.

Thorndike, R. L. (1986). *Stanford–Binet intelligence scale*. Houghton Mifflin, Boston.

Till, K. (1975). *Paediatric neurosurgery*. Blackwell Scientific Publications, Oxford.

Tizard, B. (1977). *Adoption: a second chance*. Open Books, London.

Tizard, B. and Hodges, J. (1978). The effect of early institutional rearing on the development of eight year old children. *Journal of Child Psychology and Psychiatry* **19**, 99–118.

Tizard, B. and Hughes, M. (1984). *Young children learning*. Fontana, London.

Tizard, B., Cooperman, O., Joseph, A., and Tizard, J. (1972). Environmental effects on language development: a study of young children in long-stay residential nurseries. *Child Development* **43**, 337–58.

Tobias, A. L. and Gordon, J. B. (1980). Social consequences of obesity. *Journal of the American Dietetic Association* **76**, 338–42.

Tobiasen, J. M. (1984). Psychosocial correlates of congenital facial cleft: a conceptualisation and model. *Cleft Palate* **21**, 131–9.

Tobin-Richards, M. H., Boxer, A. M., and Petersen, A. (1982). The psychological significance of pubertal change: sex differences in perception of self during early adolescence. In *Girls at puberty: biological and psychological perspectives* (ed. J. Brooks-Gunn and A. Petersen), pp. 127–54. Plenum Press, New York.

Toedter, L. J., Lasker, J. N., and Alhadeff, J. M. (1988). The perinatal grief scale: development and initial validation. *American Journal of Orthopsychiatry* **58**, 435–49.

Torup, E. (1962). A follow-up study of children with tics. *Acta Paediatrica* **51**, 261–8.

Trimble, M. R. (1987). Anticonvulsant drugs and cognitive function: a review of the literature. *Epilepsia* **28**, 37–45.

Truax, C. B. and Carkhuff, R. R. (1967). *Towards effective counselling and psychotherapy: training and practice*. Aldin, Chicago.

Tsai, M., Feldman-Summers, S., and Edgar, M. (1979). Child molestation: variables related to differential impacts on psychosexual functioning in adult women. *Journal of Abnormal Psychology* **88**, 407–17.

Tsiantis, J., Xypolita-Tsantili, D., and Papadakou-Lagoyianni, S. (1982). Family reactions and their management in a parents' group with beta-thalassaemia. *Archives of Diseases in Childhood* **57**, 860–3.

Ultmann, M. H., Belman, A. L., Ruff, H. A., Novick, B. E., Cohen, H. J., and Rubinstein, A. (1985). Developmental abnormalities in infants and children with acquired immune deficiency syndrome (AIDS) and Aids-related complex. *Developmental Medicine and Child Neurology* **27**, 563–71.

Ungerer, J. A., Horgan, B., Chaiton, J., and Champion, G. D. (1988). Psychosocial functioning in children and young adults with juvenile arthritis. *Pediatrics* **81**, 195–202.

Van Eerdewegh, M. M., Clayton, P. J., and Van Eerdewegh, P. (1985). The bereaved child: variables influencing early psychopathology. *British Journal of Psychiatry* **147**, 188–94.

Verhulst, F. C., Berden, G. F. M., and Sanders-Woudstra, J. (1986). Mental health in Dutch children. II. The prevalence of psychiatric disorder and relationship between measures. *Acta Psychiatrica Scandinavica* **72**, Suppl. No. 324.

Vernon, D. T. A., Foley, J. M., Sipowicz, R. R., and Schulman, J. L. (1965). *The psychological responses of children to hospitalization and illness*. Charles C. Thomas, Springfield, Illinois.

Voeller, K. S. and Rothenburg, M. B. (1973). Psychosocial aspects of the management of seizures in children. *Journal of Pediatrics* **51**, 1072–82.

Volkmar, F. R. and Cohen, D. J. (1989). Disintegrative disorder or 'late onset' autism. *Journal of Child Psychology and Psychiatry* **30**, 717–24.

Wahlstrom, J., Steffenburg, S., Hallgren, L., and Gillberg, C. (1989). Chromosome findings in twins with early onset autistic disorder. *American Journal of Medical Genetics* **32**, 19–21.

Walk, R. D. (1980). Perception. In *Developmental psychiatry* (ed. M. Rutter), pp. 177–84. Heinemann Medical Books, London.

Walker, L. S., Ford, M. B., and Donald, W. D. (1987). Cystic fibrosis and family stress: effects of age and severity of illness. *Pediatrics* **79**, 239–46.

Wallace, S. J. (1984). Febrile convulsions: their significance for later intellectual development and behaviour. *Journal of Child Psychology and Psychiatry* **25**, 15–21.

Wallander, J. L., Varni, J. W., Babani, L., Banis, H., and Wilcox, K. T. (1988). Children with chronic physical disorders. Maternal reports of their psychological adjustment. *Journal of Pediatric Psychology* **13**, 197–212.

Waller, D. and Eisenberg, L. (1980). School refusal in childhood—a psychiatric–pediatric perspective. In *Out of school—modern perspectives in school refusal and truancy* (ed. L. Hersov and I. Berg), pp. 207–29. Wiley, Chichester.

Wasserman, A. L. (1984). A prospective study of the impact of home monitoring on the family. *Journal of Pediatrics* 74, 323–9.

Wasserman, G. A., Lennon, M. C., Allen, R., and Shilansky, M. (1987). Contributions to attachment in normal and physically handicapped infants. *Journal of the American Academy of Child and Adolescent Psychiatry* 26, 9–15.

Waters, E., Matas, L., and Sroufe, L. A. (1975). Infants' reactions to an approaching stranger: description, validation and functional significance of wariness. *Child Development* 44, 348–56.

Watkins, J. M., Asarnow, R. F., and Tanguay, P. E. (1988). Symptom development in childhood onset schizophrenia. *Journal of Child Psychology and Psychiatry* 29, 865–78.

Webster, A., Bamford, J. M., Thyer, N. J., and Ayles, R. (1989). The psychological, educational and auditory sequelae of early, persistent, secretory otitis media. *Journal of Child Psychology and Psychiatry* 30, 529–46.

Wechsler, D. (1974). *Manual for the Wechsler Intelligence Scale for Children— revised*. Psychological Corporation, New York.

Wechsler, D. (1989). *Manual for the Wechsler Pre-School and Primary Scale of Intelligence*. Psychological Corporation, New York.

Weissman, M. and Paykel, E. (1974). *The depressed woman: a study of social relationships*. University of Chicago Press, Chicago.

Werner, E., Simonian, K., Bierman, J. E., and French, F. E. (1967). Cumulative effect of perinatal complications and deprived environment on physical, intellectual and social development of pre-school children. *Pediatrics* 39, 480–505.

Werry, J. S. and Wollersheim, J. P. (1989). Behavior therapy with children and adolescents: a twenty-year overview. *Journal of the American Academy of Child and Adolescent Psychiatry* 28, 1–18.

West, D. J. and Farrington, D. P. (1973). *Who becomes delinquent?* Heinemann Educational Books, London.

Whitam, F. (1977). Childhood indications of male homosexuality. *Archives of Sexual Behaviour* 6, 89–96.

White, K., Kolman, M. L., Wexler, P., Polin, G., and Winter, R. J. (1984). Unstable diabetes and unstable families: a psychosocial evaluation of diabetic children with recurrent ketoacidosis. *Pediatrics* 73, 749–55.

Whitman, B. Y. and Accardo, P. J. (1987). Behavioral symptomatology in Prader-Willi syndrome adolescents. *American Journal of Medical Genetics* 28, 897–905.

Whitmore, K. (1985). *Health services in schools—a new look*. Spastics International Medical Publications, London.

Whitmore, K. and Bax, M. C. O. (1986). The school entry medical examination. *Archives of Disease in Childhood* 61, 807–17.

Wilkinson, P. W. (1975). Obesity in childhood. A community study in Newcastle-upon-Tyne. *Archives of Disease in Childhood* 50, 826.

Wilkinson, V. A. (1981). Juvenile chronic arthritis in adolescence: facing the reality. *International Rehabilitation Medicine* 3, 11–17.

Williamson, W. D., Desmond, M. M., LaFevers, N., Taber, L. H., Catlin, F. I., and Weaver, T. G. (1982). Symptomatic congenital cytomegalovirus. Disorders of language, learning and hearing. *American Journal of Diseases in Childhood* 136, 902–5.

Wing, L. (1981). Asperger's syndrome: a clinical account. *Psychological Medicine* **11**, 115–29.

Winnicott, D. W. (1953). Transitional objects and transitional phenomena. *International Journal of Psychoanalysis* 87–9.

Wiseman, H. M., Guest, K., Murray, G., and Volans, G. N. (1987). Accidental poisoning in childhood: a multi-centre survey. I. General epidemiology. *Human Toxicology* **6**, 293–301.

Wittkower, E. D. and Hunt, B. A. (1958). Psychological aspects of atopic dermatitis in children. *Canadian Medical Association Journal* **79**, 810–17.

Wolfer, J. A. and Visintainer, M. A. (1979). Pre-hospital psychological preparation for tonsillectomy patients: effects on children's and parents' adjustment. *Pediatrics* 646–55.

Wolff, S. (1984). The concept of personality disorder in childhood. *Journal of Child Psychology and Psychiatry* **25**, 5–13.

Wolff, S. and Chick, J. (1980). Schizoid personality in childhood: a controlled follow-up study. *Psychological Medicine* **10**, 85–100.

Wolff, S., Narayan, S., and Moyes, B. (1988). Personality characteristics of parents of autistic children: a controlled study. *Journal of Child Psychology and Psychiatry* **29**, 143–53.

Wolke, D. (1987). Environmental neonatology. *Archives of Disease in Childhood* **62**, 987–8.

Wolkind, S. (1984). A child psychiatrist in court: using the contributions of developmental psychology. In *Taking a stand*, pp. 7–17. British Agencies for Adoption and Fostering, London.

Wolkind, S. and Kozaruk, A. (1983). *Children with special needs: a review of children with medical problems placed by the Adoption Resources Exchange from 1974–1977*. Report to the Department of Health and Social Security, London.

Wolkind, S. and Kruk, S. (1985). From child to parent: early separation and the adaptation to motherhood. In *Longitudinal studies in child psychology and psychiatry* (ed. A. R. Nicol). John Wiley, Chichester.

Wolkind, S. and Rutter, M. (1973). Children who have been 'in care'—an epidemiological study. *Journal of Child Psychology and Psychiatry* **14**, 97–105.

Wood, B., Watkins, J. B., Boyle, J. T., Nogueira, J., Zimand, E., and Carroll, L. (1989). Psychological functioning in children with Crohn's Disease and ulcerative colitis: implications for models of psychobiological interaction. *Journal of the American Academy of Child and Adolescent Psychiatry* **26**, 774–81.

Wood, C. B. S. and Walker-Smith, J. A. (1981). *MacKeith's infant feeding and feeding difficulties*. Churchill Livingstone, Edinburgh.

Woodmansey, A. C. (1967). Emotion and the motions: an inquiry into the causes and perceptions of functional disorders of defaecation. *British Journal of Medical Psychology* **40**, 207–23.

Woodward, S., Pope, A., Robson, W. J., and Hagan, O. (1985). Bereavement counselling after sudden infant death. *British Medical Journal* **290**, 363–5.

Woolston, J. L. (1987). Obesity in infancy and early childhood. *Journal of the American Academy of Child and Adolescent Psychiatry* **26**, 123–6.

World Health Organization (1990). *ICD-10: 1990 draft of chapter 5*. World Health Organization, Geneva.

Wrate, R. M., Kolvin, I., Garside, R. F., Wolstenholme, F., Hulbert, C. M., and Leitch, I. M. (1985). Helping seriously disturbed children. In *Longitudinal studies in child psychology and psychiatry* (ed. A. R. Nicol), pp. 265–318. John Wiley, Chichester.

Yule, W. (1985). Behavioural approaches. In *Child and adolescent psychiatry* (ed. M. Rutter and L. Hersov), pp. 794–808. Blackwell Scientific Publications, Oxford.

Zeanah, C. H. (1989). Adaptation following perinatal loss: a critical review. *Journal of the American Academy of Child and Adolescent Psychiatry* 28, 467–80.

Zeitlin, H. (1986). *The natural history of psychiatric disorder in childhood.* Maudsley Monographs, Oxford University Press, London.

Zucker, K. J. (1985). Cross-gender identified children. In *Gender dysphoria: development, research, management* (ed. B. H. Steiner). Plenum Press, New York.

Zucker, K. J., Bradley, S. J., Doering, R. W., and Lozinski, J. A. (1985). Sex-typed behavior in cross-gender-identified children: stability and change at a one year follow-up. *Journal of the American Academy of Child Psychiatry* 24, 710–19.

# Index